Operator Theory: Advances and Applications
Vol. 153

Editor:
I. Gohberg

Recent Advances in Operator Theory, Operator Algebras, and their Applications

XIXth International Conference on Operator Theory,
Timişoara (Romania), 2002

D. Gaşpar
I. Gohberg
D. Timotin
F.H. Vasilescu
L. Zsidó
Editors

Birkhäuser Verlag
Basel · Boston · Berlin

Editors:

Dumitru Gaşpar
Department of Mathematics
University of the West Timişoara
Bul. V. Parvan Nr. 4
1900 Timişoara
Romania
e-mail: gaspar@math.uvt.ro

Israel Gohberg
School of Mathematical Sciences
Raymond and Beverly Sackler Faculty of Exact Sciences
Tel Aviv University
Ramat Aviv 69978
Israel
e-mail: gohberg@post.tau.ac.il

Dan Timotin
Institute of Mathematics of the Romanian Academy
P.O. Box 1-764
Bucharest 70700
Romania
Dan.Timotin@imar.ro

Florian-Horia Vasilescu
Laboratoire AGAT-UMR 8524
U.F.R. de mathématiques
Université des Sciences et Technologie de Lille
59655 Villeneuve d'Ascq Cedex
France
fhvasil@agat.univ-lille1.fr

László Zsidó
Dipartimento di Matematica
Università di Roma „Tor Vergata"
Via della Ricerca Scientifica 1
00133 Roma
Italy
zsido@mat.uniroma2.it

2000 Mathematics Subject Classification 46-xx, 47-xx, 34-xx, 93-xx

A CIP catalogue record for this book is available from the
Library of Congress, Washington D.C., USA

Bibliographic information published by Die Deutsche Bibliothek
Die Deutsche Bibliothek lists this publication in the Deutsche Nationalbibliografie; detailed bibliographic data is available in the Internet at <http://dnb.ddb.de>.

ISBN 3-7643-7127-7 Birkhäuser Verlag, Basel – Boston – Berlin

Part of Springer Science+Business Media
Printed on acid-free paper produced from chlorine-free pulp. TCF ∞
Cover design: Heinz Hiltbrunner, Basel
Printed in Germany
ISBN 3-7643-7127-7

9 8 7 6 5 4 3 2 1

www.birkhauser.ch

Contents

Foreword

The Romanian conferences in operator theory, as they are now commonly called, have started in the year 1976 as an annual workshop on operator theory held at the University of Timişoara, originally only with Romanian attendance. The meeting soon evolved into an international conference, with an increasingly larger participation. It has been organized jointly, initially by the Department of Mathematics of INCREST and by the Faculty of Sciences of the University of Timişoara, then (since 1990) by the Institute of Mathematics of the Romanian Academy and the Faculty of Mathematics of the West University of Timişoara. The venue was usually Timişoara (and, occasionally, Herculane, Bucharest or Predeal). Since 1986 the conference has been regularly held biannually at the beginning of the summer.

The *19th Conference on Operator Theory* (OT 19) took place between June 27th and July 2nd 2002, at the West University of Timişoara. It is a pleasure to acknowledge the considerable financial support received through the programme EURROMMAT of the European Community, under contract ICA1-CT-2000-70022. Partial support has also been provided by the Romanian Ministry of Education, Research and Youth, grants CERES 152/2001 and 153/2001.

The full programme of the conference is included in the sequel. It is worth mentioning also a special event that has taken place during the conference: professor Israel Gohberg has been awarded the title of Doctor Honoris Causa of the West University of Timişoara.

This volume is a careful selection of papers authored by participants at the 19th Conference on Operator Theory. Traditionally, these conferences are open to a broad range of contributions from operator theory, operator algebras and their applications. This feature is also shared by the proceedings volume, covering a large variety of topics, such as single operator theory, C^*-algebras, differential operators, integral transforms, stochastic processes and operators, quantum systems, special classes of operators, holomorphic operator functions, interpolation problems, and system theory.

Last but not least, special thanks are due to Barbara Ionescu, of the Theta Foundation, for her excellent work in preparing the final version of the manuscripts.

The Editors

Programme

THURSDAY, JUNE 27

		Morning Session
9:00– 9:30	Opening	
Plenary Section	Chairman:	D. Gaşpar
9:50–10:30	I. Gohberg	*Infinite systems of linear equations*
10:30–11:30	F.-H. Vasilescu	*Existence of unitary dilations as a moment problem*
11:40–12:30	G. Weiss	*Traces, ideals and arithmetic means*
		Afternoon Session
Plenary Section	Chairman:	A. Gheondea
15:00–15:40	J. Partington	*Semigroups, functional models and Hankel operators*
Section A	Chairman:	J. Partington
16:00–16:30	T. Constantinescu	*Szegö kernels and polynomials in several commuting variables*
16:35–17:05	E. Fricain	*Functional models and asymptotically orthonormal sequences*
17:20–17:50	I. Chalendar	*Overcompleteness of sequences of reproducing kernels in the model space* K_θ
17:55–18:25	A. Halanay	*Controlled factorization for some commuting pairs of contractions with thin spectrum*
Section B	Chairman:	G. Weiss
16:00–16:30	Z. Jablonski	*Completely hyperexpansive operators*
16:35–17:05	M. Kaltenbaeck	*On Hermite-Biehler functions of finite order*
17:20–17:50	H. Woracek	*De Branges space subject to growth conditions*
17:55–18:25	P. Găvruţă	*Atomic decomposition of linear operators*

Section C	Chairman:	V. Müller
16:00–16:30	H. Akça	*On existence of solutions of semilinear impulsive functional differential equations with nonlocal conditions*
16:35–17:05	M. Möller	*Some operator models for a linearized equation describing oscillations of plasma*
17:20–17:50	A. Pechentsov	*Regularized traces of differential operators*
17:55–18:25	L. Zielinski	*Asymptotic distribution of eigenvalues for Schrödinger type operators*

Friday, June 28

Morning Session

Section C	Chairman:	F.-H. Vasilescu
9:00– 9:40	L. Kérchy	*Reflexive subspaces of Toeplitz-type operators*
9:50–10:30	V. Müller	*Power bounded operators and supercyclic vectors*
10:50–11:30	G.F. Popescu	*Multivariable Nehari problem and interpolation*

Aula Magna

12:00 Presentation of the title of Doctor Honoris Causa of the West University of Timişoara to Professor I. Gohberg

Afternoon Session

Plenary Section	Chairman:	F. Rădulescu
15:00–15:40	L. Zsidó	*Group and quantum group actions having particular fixed point algebras*
Section A	Chairman:	L. Kérchy
16:00–16:30	M. Bakonyi	*Page's theorem for ordered groups*
16:35–17:05	D. Timotin	*The intertwining lifting theorem for ordered groups*
17:20–17:50	F. Turcu	*On the dual algebras generated by spherical contractions*
17:55–18:25	M. Kosiek	*Invariant subspaces for commuting contractions*

Section B	Chairman:	T. Schlumprecht
16:00–16:30	A. Stroh	*The weak mixing property for C^*-dynamical systems*
16:35–17:05	R. Duvenhage	*Recurrence and ergodicity in unital $*$-algebras*
17:20–17:50	F. Fidaleo	*The investigation of ergodic properties of quantum systems by a perturbative analysis systems by a perturbative analysis*
17:55–18:25	A. Gheondea	*Sequential quantum measurements*
Section C	Chairman:	L.G. Brown
16:00–16:30	A. Dahlner	*Norm controlled inversion in quasi-Banach algebras*
16:35–17:05	F. Kittaneh	*Bounds for the zeros of polynomials from matrix inequalities*
17:20–17:50	H. Winkler	*Canonical systems with selfadjoint interface conditions*
17:55–18:25	T. Bînzar	*Commuting triples of subnormal operators and related moments*

Saturday, June 29

Morning Session

Plenary Section	Chairman:	G.K. Pedersen
9:00– 9:40	R. Nest	*Connes-Kasparov conjecture*
9:50–10:30	F. Rădulescu	*On Connes' embedding conjecture*
10:50–11:30	L.G. Brown	*Murray-von Neumann equivalence of projections C^*-algebras*
11:40–12:30	P. Goldstein	*Stable isomorphism of certain continuous fields of Cuntz-Krieger algebras*

Afternoon Session

Plenary Section	Chairman:	D.R. Larson
15:00–15:40	T. Schlumprecht	*How many operators do there exist on a Banach space?*

Section A	Chairman:	A. Atzmon
16:00–16:30	D. Pik	*The Kalman-Yakubovich-Popov inequality and infinite-dimensional discrete time dissipative systems*
16:35–17:05	M. Dritschel	*A completely positive approach to Ando's theorem*
17:20–17:50	C. Badea	*Hankel operators and similarity problems*
17:55–18:25	A. Siskakis	*Classical matrices and composition operators*
Section B	Chairman:	J. Janas
16:00–16:30	D. Cichon	*Weighted approximation of entire functions and Toeplitz operators in Segal-Bargmann spaces*
16:35–17:05	N. Tiţa	*On the distance between an operator and an operator ideal*
17:20–17:50	I. Suciu	*Hyperbolic structures on the Harnack parts of contractions*
17:55–18:25	I. Valuşescu	*An operatorial view on periodic correlated processes*
Section C	Chairman:	M. Şabac
16:00–16:30	L. Carrot	*Computation of the p-numerical radius for truncated shifts*
16:35–17:05	T. Yamamoto	*Finite-dimensional Q-algebras and von Neumann inequality*
17:20–17:50	S. Czerwik	*Nonlinear set-valued contraction mappings in b-metric spaces*
17:55–18:25	M.B. Ghaemi	*The sums and products of commuting AC-operators*

Monday, July 1

Morning Session

Plenary Section	Chairman:	E. Christensen
9:00– 9:40	G.K. Pedersen	*Trace inequalities for functions of several variables*
9:50–10:30	J. Esterle	*Asymptotic behavior at the origin of the distance between elements of a strongly continuous semigroup*

10:50–11:30	A. Atzmon	*Reducible representations of abelian groups*
11:40–12:20	D.R. Larson	*Wavelets, frames and operator theory*

Afternoon Session

Plenary Section Chairman: B. Chevreau

15:00–15:40	J. Janas	*Spectral theory of Jacobi matrices*

Section A Chairman: Y. Kawahigashi

16:00–16:30	C. Pop	*Topological entropy and crossed products*
16:35–17:05	G. Popescu	*Non-commutative inequalities in operator algebras*
17:20–17:50	M. Buneci	*The equality of the reduced and the full C^*-algebras and the amenability of a topological groupoid*
17:55–18:25	B. Balogun	*The three test problems of Kaplansky for Hilbert C^*-modules*

Section B Chairman: R. Nest

16:00–16:30	M. Măntoiu	*Spectral analysis by algebraic and topological methods*
16:35–17:05	J.-L. Tu	*The gamma element for discrete groups which admit a uniform embedding into Hilbert space*
17:20–17:50	C. Antonescu	*Approaches for the study of some classes generated by symmetric norming functions*
17:55–18:25	A. Tikhonov	*Functional model for operators with spectrum on a curve*

Section C Chairman: M. Bakonyi

16:00–16:30	J. Stochel	*Domination and normality*
16:35–17:05	P. Niemiec	*Separate and joint similarity to families of (bounded) normal operators on Hilbert space*
17:20–17:50	D. Popovici	*Moment problems and unitary dilations*
17:55–18:25	L. Sasu	*Dichotomy concepts for evolution operators and cocycles*

Tuesday, July 2

Morning Session

Plenary Section	Chairman:	J. Esterle
9:00– 9:40	G. Cassier	*Power boundedness, invariant subspaces and similarity to contractions*
9:50–10:30	B. Chevreau	*A multicontraction version of a theorem of Apostol*
10:50–11:30	E. Christensen	*Property gamma, cohomology, complemented subspaces, similarities and length*
11:40–12:30	Y. Kawahigashi	*Classification of local conformal nets. Case* $c < 1$

Afternoon Session

Section A	Chairman:	L. Zsidó
16:00–16:30	M. Martin	*Integral operators with general measurable kernels dominated by maximal operators*
16:35–17:05	B. Prunaru	*Approximately reflexive algebras*
17:20–17:50	D. Beltiţă	*Several variables spectral theory and complex structures*
17:55–18:25	M. Şabac	*Commutators and Dunford spectral projectors*
Section B	Chairman:	P. Goldstein
16:00–16:30	A. Terescenco	*Some remarks on quotient Hilbert spaces*
16:35–17:05	P. Gaşpar	*On finite variate periodically correlated processes*
17:20–17:50	A. Crăciunescu	*Multicontractions avec le spectre de Harte dominant*
17:55–18:25	R. Negrea	*On a class of McShane's stochastic integral equations*
Section C	Chairman:	G. Cassier
16:00–16:30	L.D. Lemle	*The Lie-Trotter formula for semigroups*
16:35–17:05	V. Ungureanu	*Uniform exponential stability and the uniform observability of time-varying linear stochastic systems in Hilbert space*
17:20–17:50	I. Şerban	*Compact perturbations of isometries*
17:55–18:25	C.G. Ambrozie	*Remarks on Nevanlinna-Pick interpolation*

List of Participants

H. Akça, *Dhahran*
R. Archer, *Newcastle*
C.-G. Ambrozie, *Bucharest*
C. Antonescu, *Braşov*
C. D'Antoni, *Rome*
A. Atzmon, *Tel Aviv*
C. Badea, *Lille*
M. Bakonyi, *Atlanta*
B. Balogun, *Lesotho*
S. T. Belinschi, *Bucharest*
D. Beltiţă, *Bucharest*
T. Bînzar, *Timişoara*
L. de Branges, *West Lafayette*
L. G. Brown, *West Lafayette*
M. Buneci, *Târgu-Jiu*
L. Carrot, *Lyon*
G. Cassier, *Lyon*
I. Chalendar, *Lyon*
B. Chevreau, *Bordeaux*
E. Christensen, *Copenhague*
D. Cichon, *Krakow*
T. Constantinescu, *Dallas*
S. Czerwik, *Gliwice*
A. Dahlner, *Lund*
M. Dritschell, *Newcastle*
R. Duvenhage, *Pretoria*
J. Esterle, *Bordeaux*
F. Fidaleo, *Rome*
E. Fricain, *Lyon*
P. Găvruţă, *Timişoara*
M. B. Ghaemi, *Birjand*
A. Ghanmi, *Rabat*
Th. Giordano, *Ottawa*
I. Gohberg, *Tel Aviv*
P. Goldstein, *Zagreb*
Z. Jablonski, *Krakow*
J. Janas, *Krakow*
M. Kaltenbaeck, *Wien*
Y. Kawahigashi, *Tokyo*
L. Kérchy, *Szeged*
E. Kirchberg, *Muenster*
F. Kittaneh, *Amman*
M. Kosiek, *Krakow*
J. Kustermans, *Leuven*
S. H. Kye, *Seoul*
D. Larson, *College Station*
L. D. Lemle, *Hunedoara*
M. Măntoiu, *Bucharest*
L. Marcoci, *Bucharest*
M. Martin, *Baldwin*
M. Mbekhta, *Lille*
M. Möller, *Johannesburg*
V. Müller, *Prague*
R. Munteanu, *Bucharest*
R. Nest, *Copenhague*
R. Nicoară, *Bucharest*
P. Niemiec, *Krakow*
J. R. Partington, *Leeds*
D. Păunescu, *Timişoara*
A. Pechentsov, *Moscow*
G. K. Pedersen, *Copenhagen*
M. Petcu, *Bucharest*
D. Pik, *Leiden*
C. Pop, *Bucharest*
M. Popa, *Bucharest*
N. Popa, *Bucharest*

G. F. Popescu, *San Antonio*
G. Popescu, *Craiova*
D. Popovici, *Timişoara*
B. Prunaru, *Bucharest*
B. Ramazan, *Tempe*
Z.-J. Ruan, *Urbana-Champaign*
A.-M. Sâmbotin, *Lund*
T. Schlumprecht, *College Station*
I. Şerban, *Timişoara*
M. Shchukin, *Minsk*
A. Siskakis, *Thessaloniki*
J. Stochel, *Krakow*
A. Stroh, *Pretoria*
I. Suciu, *Bucharest*
A. Tikhonov, *Simferopol*
N. Ţiţa, *Braşov*
J.-L. Tu, *Paris*
V. Ungureanu, *Târgu-Jiu*
I. Valuşescu, *Bucharest*
A. Voinea, *Bucharest*
H. Winkler, *Groningen*
H. Woracek, *Wien*
G. Weiss, *Cincinnati*
T. Yamamoto, *Sapporo*
L. Zielinski, *Calais*
L. Zsidó, *Rome*

Operator Theory:
Advances and Applications, Vol. 153, 1–11

On Existence of Solutions of Semilinear Impulsive Functional Differential Equations with Nonlocal Conditions

Haydar Akça, Valéry Covachev and Eada Al-Zahrani

Abstract. The existence, uniqueness and continuous dependence of a mild solution of a semilinear impulsive functional-differential evolution nonlocal Cauchy problem in general Banach spaces are studied. Methods of fixed point theorems, of a C_0 semigroup of operators and the Banach contraction theorem are applied.

Mathematics Subject Classification (2000). 34A37, 34G20, 34K30, 34K99.

Keywords. Semilinear, Impulsive, Functional-differential equations, Nonlocal conditions, Mild solution.

1. Introduction

In this paper we study the existence, uniqueness and continuous dependence of a mild solution of a nonlocal Cauchy problem for a semilinear impulsive functional-differential evolution equation. Such problems arise in some physical applications as a natural generalization of the classical initial value problems. The results for a semilinear functional-differential evolution nonlocal problem ([2], [3], [4]) are extended for the case of impulse effect.

We consider a nonlocal Cauchy problem in the form:

$$\begin{cases} \dot{u}(t) + Au(t) = f(t, u(t), u(b_1(t)), \dots, u(b_m(t))), & t \in (t_0, t_0 + a],\ t \neq \tau_k, \\ u(\tau_k + 0) = Q_k u(\tau_k) \equiv u(\tau_k) + I_k u(\tau_k), & k = 1, 2, \dots, \kappa, \\ u(t_0) = u_0 - g(u), \end{cases} \tag{1.1}$$

where $t_0 \geq 0$, $a > 0$ and $-A$ is the infinitesimal generator of a compact C_0 semigroup of operators on a Banach space E. I_k $(k = 1, 2, \dots, \kappa)$ are linear operators acting in the Banach space E. The functions f, g, b_i $(i = 1, 2, \dots, m)$ are given functions satisfying some assumptions and u_0 is an element of the Banach space E. $I_k u(\tau_k) = u(\tau_k + 0) - u(\tau_k - 0)$ and the impulsive moments τ_k are such that $t_0 < \tau_1 < \tau_2 < \cdots < \tau_k < \cdots < \tau_\kappa < t_0 + a$, $\kappa \in \mathbb{N}$.

Theorems about the existence, uniqueness and stability of solutions of differential and functional-differential abstract evolution Cauchy problems were studied in [2], [3], and [4]. The results presented in this paper are a generalization and a continuation of some results reported in publication [1]. We consider a classical semilinear impulsive functional-differential equation in the case of a nonlocal condition, reduced to the classical impulsive initial functional value problem. The nonlinearity f in problem (1.1) is of a more general type (involves more than one delay which may be variable) than the respective function in [1]. Also, in the present paper the *existence* of a mild solution (Theorem 2.1) is proved under less restrictive conditions using Schauder's fixed point theorem. The Lipschitz conditions are introduced later to prove the existence and *uniqueness* of the classical solution of problem (1.1), and the continuous dependence of the mild solution of problem (1.1) on the initial condition.

As usual in the theory of impulsive differential equations, at the points of discontinuity τ_i of the solution $t \mapsto u(t)$ we assume that $u(\tau_i) \equiv u(\tau_i - 0)$. It is clear that, in general, the derivatives $\dot{u}(\tau_i)$ do not exist. On the other hand, according to the first equality of (1.1) there exist the limits $\dot{u}(\tau_i \mp 0)$. According to the above convention, we assume $\dot{u}(\tau_i) \equiv \dot{u}(\tau_i - 0)$.

Throughout the paper we assume that E is a Banach space with norm $\|\cdot\|$, $-A$ is the infinitesimal generator of a C_0 semigroup $\{T(t)\}_{t\geq 0}$ on E, $D(A)$ is the domain of A. A C_0 semigroup $\{T(t)\}_{t\geq 0}$ is said to be a compact C_0 semigroup of operators on E if $T(t)$ is a compact operator for every $t > 0$. We denote $I := [t_0, t_0+a]$, $M := \sup_{t\in[0,a]} \{\|T(t)\|_{BL(E,E)}\}$ and X is the space of piecewise continuous functions $I \to E$ with discontinuities of the first kind at $\tau_1, \tau_2, \ldots, \tau_\kappa$. Let $f : I \times E^{m+1} \to E$, $g : X \to E$ (for instance, we can have $g(u) = \tilde{g}(u(t_1), u(t_2), \ldots, u(t_p))$, where $\tilde{g} : E^p \to E$, $t_0 < t_1 < t_2 < \cdots < t_p < t_0 + a$, $p \in \mathbb{N}$), $b_i : I \to I$ $(i = 1, 2, \ldots, m)$ and $u_0 \in E$. In the sequel, the operator norm $\|\cdot\|_{BL(E,E)}$ will be denoted by $\|\cdot\|$. We need the following sets:

$$E_\rho := \{z \in E, \|z\| \leq \rho\} \quad \text{and} \quad X_\rho := \{w \in X, \|w\|_X \leq \rho\}, \quad \rho > 0.$$

Introduce the following assumptions:

A1: $f \in C(I \times E^{m+1}, E)$, $g \in C(X, E)$ and $b_i \in C(I, I)$, $i = 1, 2, \ldots, m$ and there are constants $C_i > 0$, $i = 1, 2, 3$, such that

$$\begin{cases} \|f(s, z_0, z_1, \ldots, z_m)\| \leq C_1 \text{ for } s \in I,\ z_i \in E_r,\ i = 0, 1, \ldots, m, \\ \|g(w)\| \leq C_2 \text{ and } \max\limits_{k=1,2,\ldots,\kappa} \|I_k w\| \leq C_3 \text{ for } w \in X_r, \end{cases} \tag{1.2}$$

where $r := M(aC_1 + \|u_0\| + C_2 + \kappa C_3)$.

A2: $g(\lambda w_1 + (1-\lambda)w_2) = \lambda g(w_1) + (1-\lambda)g(w_2)$ for $w_i \in X_r$, $i = 1, 2$, and $\lambda \in (0, 1)$ and r is given by (1.2).

A3: The set

$$\{w(t_0) = u_0 - g(w) : w \in X_r\},$$

where r is given by (1.2), is precompact in E.

Example 1.1. Consider the scalar problem

$$\begin{cases} \dot{u}(t) + Au(t) = \frac{1}{4a}\left(u^2(t) + \sum_{i=1}^{m} 2^{-i}u^2(b_i(t))\right), & t \in (t_0, t_0 + a],\, t \neq \tau_k, \\ u(\tau_k + 0) = (1 + c_k)u(\tau_k), & k = 1, 2, \dots, \kappa, \\ u(t_0) = \frac{1}{6} - \frac{1}{2}\sum_{j=1}^{p} 2^{-j}u(t_j), \end{cases}$$

where $E = \mathbb{R}$, $A > 0$, the constants c_k satisfy $|c_k| \leq \frac{1}{4\kappa}$, $k = 1, 2, \dots, \kappa$.

In this case $T(t) = e^{-At}$, $M = 1$ and it is easy to see that condition **A1** is satisfied with $r = 1$. Assumptions **A2** and **A3** are obviously satisfied, thus Theorem 2.1 can be applied to this problem.

Consider the initial value problem (see [3])

$$\begin{cases} \dot{u}(t) + Au(t) = f(t), & t \in (t_0, t_0 + a], \\ u(t_0) = x, \end{cases} \tag{1.3}$$

where $f : I \to E$, $-A$ is the infinitesimal generator of a C_0 semigroup $T(t)$, $t \geq 0$, and $x \in E$.

Definition 1.2. A function u is said to be a **strong solution** of problem (1.3) on I if u is differentiable almost everywhere on I, so that $(du/dt) \in L^1((t_0, t_0 + a); E)$, $u(t_0) = x$ and $\dot{u}(t) + Au(t) = f(t)$ a.e. on I.

The unique strong solution u on I is given by the formula

$$u(t) = T(t - t_0)x + \int_{t_0}^{t} T(t - s)f(s)\, ds, \quad t \in I. \tag{1.4}$$

Definition 1.3. A function $u : I \to E$ is said to be a **classical solution** of the problem (1.3) on I if u is continuous on I and continuously differentiable on $(t_0, t_0 + a]$, such that $u(t) \in D(A)$ for $t_0 < t \leq t_0 + a$ and the problem (1.3) is satisfied on I.

If E is a Banach space and $-A$ is the infinitesimal generator of a C_0 semigroup $T(t)$, $t \geq 0$, $f : I \to E$ is continuous on I and $x \in D(A)$, then the problem (1.3) has a classical solution u on I given by (1.4).

Next consider the initial value problem for the impulsive linear system

$$\begin{cases} \dot{u}(t) + Au(t) = f(t), & t \in (t_0, t_0 + a],\, t \neq \tau_k, \\ u(\tau_k + 0) = u(\tau_k) + I_k u(\tau_k), & k = 1, 2, \dots, \kappa, \\ u(t_0) = x, \end{cases} \tag{1.5}$$

where A, f and x are as in problem (1.3), and τ_k and I_k are as in problem (1.1).

Definition 1.4. A function $u : I \to E$ is said to be a **classical solution** of the problem (1.5) on I if u is piecewise continuous on I with discontinuities of the first kind at $\tau_1, \tau_2, \ldots, \tau_\kappa$ and continuously differentiable on $(t_0, t_0 + a] \setminus \{\tau_k\}_{k=1}^{\kappa}$, such that $u(t) \in D(A)$ for $t_0 < t \leq t_0 + a$ and the problem (1.5) is satisfied on I.

If A, f and x are as above and $I_k : D(A) \to D(A)$, then the problem (1.5) has a classical solution u on I given by the formula

$$u(t) = T(t - t_0)x + \int_{t_0}^{t} T(t - s)f(s)\, ds + \sum_{t_0 \leq \tau_k < t} T(t - \tau_k)I_k u(\tau_k). \tag{1.6}$$

Formula (1.6) motivates us to give the following definition.

Definition 1.5. A function $u \in X$ satisfying the following integro-summary equation

$$\begin{aligned} u(t) &= T(t - t_0)u_0 - T(t - t_0)g(u) \\ &+ \int_{t_0}^{t} T(t - s)f(s, u(s), u(b_1(s)), \ldots, u(b_m(s)))\, ds \\ &+ \sum_{t_0 \leq \tau_k < t} T(t - \tau_k)I_k u(\tau_k), \qquad t \in [t_0, t_0 + a] \end{aligned}$$

is said to be a **mild solution** of the nonlocal Cauchy problem (1.1).

2. Existence and uniqueness theorems

Theorem 2.1. *Suppose that assumptions* **A1**–**A3** *are satisfied, then the impulsive nonlocal Cauchy problem* (1.1) *has a mild solution.*

Proof. The mild solution of the impulsive system (1.1) with nonlocal condition satisfies the operator equation

$$u(t) = (Fu)(t),$$

where

$$\begin{aligned} (Fw)(t) &:= T(t - t_0)u_0 - T(t - t_0)g(w) \\ &+ \int_{t_0}^{t} T(t - s)f(s, w(s), w(b_1(s)), \ldots, w(b_m(s)))\, ds \\ &+ \sum_{t_0 \leq \tau_k < t} T(t - \tau_k)I_k w(\tau_k), \qquad t \in [t_0, t_0 + a], \end{aligned} \tag{2.1}$$

so that

$$\|(Fw)(t)\| \leq M\|u_0\| + MC_2 + aMC_1 + \kappa MC_3 = r, \tag{2.2}$$

where the impulsive moments τ_k are such that $t_0 < \tau_1 < \tau_2 < \cdots < \tau_k < \cdots < \tau_\kappa < t_0 + a$, $\kappa \in \mathbb{N}$.

Let $Y := \{w \in X_r : u_0 = w(t_0) + g(w)\}$. According to assumption **A2**, Y is a convex subset of X_r. We have $F : Y \to Y$. Moreover, **A1** implies that $F \in C(Y, Y)$. Now we will show that $F(Y) = \{F(w)(t) : w \in Y,\, t \in I\}$ is precompact in E.

Observe that

$$Y(t_0) = \{(Fw)(t_0) : w \in Y,\, t \in I\} = \{u_0 - g(w) : w \in X_r\} = \{w(t_0) : w \in X_r\}.$$

Thus according to the assumption **A3**, $Y(t_0)$ is precompact in E. Let $t > t_0$ be fixed. For an arbitrary $\varepsilon \in (t_0, t)$, define a mapping F_ε on Y by the expression

$$\begin{aligned}(F_\varepsilon w)(t) &:= T(t-t_0)u_0 - T(t-t_0)g(w) \\ &\quad + \int_{t_0}^{t-\varepsilon} T(t-s)f(s, w(s), w(b_1(s)), \ldots, w(b_m(s)))\,ds \\ &\quad + \sum_{t_0 \le \tau_k < t-\varepsilon} T(t-\tau_k)I_k w(\tau_k) = T(t-t_0)u_0 - T(t-t_0)g(w) \\ &\quad + T(\varepsilon)\int_{t_0}^{t-\varepsilon} T(t-s-\varepsilon)f(s, w(s)w(b_1(s)), \ldots, w(b_m(s)))\,ds \\ &\quad + T(\varepsilon)\sum_{t_0 \le \tau_k < t-\varepsilon} T(t-\varepsilon-\tau_k)I_k w(\tau_k).\end{aligned} \tag{2.3}$$

Since $T(t)$ is compact for every $t > t_0$, then the set

$$Y_\varepsilon := \{(F_\varepsilon w)(t) : w \in Y\}$$

is precompact in E for every $\varepsilon \in (t_0, t)$. Moreover, from the formulae (1.2), (2.1) and (2.3) we have

$$\begin{aligned}&\|(Fw)(t) - (F_\varepsilon w)(t)\| \\ &\le \left\| \int_{t-\varepsilon}^{t} T(t-s)f(s, w(s), w(b_1(s)), \ldots, w(b_m(s)))\,ds \right\| \\ &\quad + \left\| \sum_{t-\varepsilon \le \tau_k < t} T(t-\tau_k)I_k w(\tau_k) \right\| \le \varepsilon M C_1 + M C_2 i(t-\varepsilon, t),\end{aligned} \tag{2.4}$$

where $i(t-\varepsilon, t)$ is the number of impulses on the interval $(t-\varepsilon, t)$ and $i(t-\varepsilon, t) \to 0$ as $\varepsilon \to 0$, so $\|(Fw)(t) - (F_\varepsilon w)(t)\| \to 0$ as $\varepsilon \to 0$, and consequently the set $F(Y)$ is a uniformly bounded on each interval of continuity family of functions. From formulae (1.2) and (2.1) we observe that

$$\begin{aligned}
&\|(Fw)(t_1) - (Fw)(t_2)\| \\
&\quad \le \|(T(t_1 - t_0) - T(t_2 - t_0))u_0\| + \|(T(t_1 - t_0) - T(t_2 - t_0))g(w)\| \\
&\qquad + \left\| \int_{t_0}^{t_1} (T(t_1 - s) - T(t_2 - s)) f(s, w(s), w(b_1(s)), \dots, w(b_m(s)))\, ds \right\| \\
&\qquad + \left\| \int_{t_1}^{t_2} T(t_2 - s) f(s, w(s), w(b_1(s)), \dots, w(b_m(s)))\, ds \right\| \\
&\qquad + \left\| \sum_{t_0 \le \tau_k < t_1} T(t_1 - \tau_k) I_k w(\tau_k) - \sum_{t_0 \le \tau_k < t_2} T(t_2 - \tau_k) I_k w(\tau_k) \right\| \\
&\quad \le \|T(t_1 - t_0) - T(t_2 - t_0)\| (\|u_0\| + C_2) \\
&\qquad + C_1 \int_{t_0}^{t_1} \|T(t_1 - s) - T(t_2 - s)\|\, ds + MC_1(t_2 - t_1) \\
&\qquad + C_3 \sum_{t_0 \le \tau_k < t_1} \|T(t_1 - \tau_k) - T(t_2 - \tau_k)\| + MC_3 i(t_1, t_2).
\end{aligned} \tag{2.5}$$

The coefficients of $\|u_0\|$, C_1, C_2 and C_3 in (2.5) are independent of $w \in Y$ and those terms tend to zero when $t_2 \to t_1$, except for the case $t_1 = \tau_k$ for some $k = 1, 2, \dots, \kappa$ and $t_2 \to \tau_k + 0$. As a consequence of the continuity of $T(t)$ in the uniform operator topology for $t > 0$, which follows from the compactness of $T(t)$ for $t > 0$, $F(Y)$ is an equicontinuous on each interval of continuity family of functions. Since all the assumptions of Arzela–Ascoli's theorem are satisfied on each interval of continuity, then $F(Y)$ is a precompact subset of Y. Finally, applying Schauder's fixed point theorem to X, Y and F, it follows that F has a fixed point in Y and any fixed point of F is a mild solution of the nonlocal Cauchy problem (1.1). This completes the proof of the theorem. □

Theorem 2.2. *Suppose that the functions f, g and b_i $(i = 1, 2, \dots, m)$ satisfy assumptions* **A1** *and* **A2**, *where $u_0 \in E$. Then in the class of all the functions w, for which assumption* **A3** *holds, the nonlocal Cauchy problem* (1.1) *has a mild solution u. If in addition:*

(i) *E is a reflexive Banach space,*

(ii) *there exists a constant $L > 0$ such that*

$$\begin{cases} \|f(s, u_0, u_1, \dots, u_m) - f(\tilde{s}, \tilde{u_0}, \tilde{u_1}, \dots, \tilde{u_m})\| \le L_1\Big(|s - \tilde{s}| + \sum_{k=0}^{m} \|u_k - \tilde{u_k}\|\Big), \\ \|I_k \nu\|_E \le L_2 \|\nu\|_E \quad \text{for } \nu \in E,\ k = 1, 2, \dots, \kappa, \end{cases}$$

where $s, \tilde{s} \in I$, $u_i, \tilde{u_i} \in E_r$ $(i = 0, 1, 2, \dots, m)$ and $L = \max\{L_1, L_2\}$,

(iii) *u is the unique mild solution of the problem* (1.1) *and there is a constant* $K > 0$ *such that*

$$\|u(b_i(s)) - u(b_i(\tilde{s}))\| \le K\|u(s) - u(\tilde{s})\| \quad \text{for } s, \tilde{s} \in I,$$

(iv) *the element* $u_0 \in D(A)$ *and* $g(u) \in D(A)$,

then u is the unique classical solution of the impulsive nonlocal Cauchy problem (1.1).

Proof. Since all the assumptions of Theorem 2.1 are satisfied, then the nonlocal impulsive Cauchy problem (1.1) possesses a mild solution u which, according to assumption (iii), is the unique mild solution of the problem (1.1). Now, we will show that u is the unique classical solution of semilinear, nonlocal and impulsive, Cauchy problem (1.1). Therefore observe that

$$\begin{aligned}
u(t+h) - u(t) &= [T(t+h-t_0)u_0 - T(t-t_0)u_0] \\
&\quad - [T(t+h-t_0)g(u) - T(t-t_0)g(u)] \\
&\quad + \int_{t_0}^{t_0+h} T(t+h-s)f(s,u(s),u(b_1(s)),u(b_2(s)),\ldots,u(b_m(s)))\,ds \\
&\quad + \int_{t_0+h}^{t+h} T(t+h-s)f(s,u(s),u(b_1(s)),u(b_2(s)),\ldots,u(b_m(s)))\,ds \\
&\quad - \int_{t_0}^{t} T(t-s)f(s,u(s),u(b_1(s)),u(b_2(s)),\ldots,u(b_m(s)))\,ds \\
&\quad + \sum_{t_0 \le \tau_k < t+h} T(t+h-\tau_k)I_k(u(\tau_k)) - \sum_{t_0 \le \tau_k < t} T(t-\tau_k)I_k(u(\tau_k)) \\
&= T(t-t_0)[T(h)-I]u_0 \;-\; T(t-t_0)[T(h)-I]g(u) \\
&\quad + \int_{t_0}^{t_0+h} T(t+h-s)f(s,u(s),u(b_1(s)),u(b_2(s)),\ldots,u(b_m(s)))\,ds \\
&\quad + \int_{t_0}^{t} T(t-s)[f(s+h,u(s+h),u(b_1(s+h)),\ldots,u(b_m(s+h))) \\
&\quad - f(s,u(s),u(b_1(s)),u(b_2(s)),\ldots,u(b_m(s)))]\,ds \\
&\quad + \sum_{t \le \tau_k < t+h} T(t+h-\tau_k)I_k(u(\tau_k)) \\
&\quad + \sum_{t_0 \le \tau_k < t} \big(T(t+h-\tau_k) - T(t-\tau_k)\big)I_k(u(\tau_k))
\end{aligned}$$

for $t \in [t_0, t_0+a)$, $h > 0$ and $t+h \in (t_0, t_0+a]$.

Consequently we have

$$\begin{aligned}
&\|u(t+h)-u(t)\| \\
&\quad \le hM\|Au_0\| + hM\|Ag(u)\| + hMC_1 + MLah \\
&\qquad + ML\int_{t_0}^{t}\Big(\|u(s+h)-u(s)\| + \sum_{k=1}^{m}\|u(b_k(s+h)) - u(b_k(s))\|\Big)ds \\
&\qquad + \sum_{t\le\tau_k<t+h}\|T(t+h-\tau_k)I_k(u(\tau_k))\| \qquad\qquad (2.6)\\
&\qquad + \sum_{t_0\le\tau_k<t}\|\big(T(t+h-\tau_k)-T(t-\tau_k)\big)I_k(u(\tau_k))\| \\
&\quad \le C_*h + ML(1+mK)\int_{t_0}^{t}\|u(s+h)-u(s)\|\,ds + MC_3 i(t,t+h),
\end{aligned}$$

where

$$C_* := M\big(\|Au_0\| + \|Ag(u)\| + C_1 + aL + \|AI_k(u(\tau_k))\| i(t_0,t)\big).$$

By applying Gronwall's inequality and from (2.6) we have

$$\|u(t+h)-u(t)\| \le \big(C_*h + MC_3 i(t,t+h)\big)\exp(aML(1+mK)).$$

Thus $\|u(t+h)-u(t)\| \to 0$ as $h \to 0$, and u is Lipschitz continuous on each interval of continuity in I. The Lipschitz continuity of u on each interval of continuityin I combined with the Lipschitz continuity of f on $I \times E^{m+1}$ imply that $t \mapsto f(t,u(t),u(b_1(t)),\dots,u(b_m(t)))$ is Lipschitz continuous on each interval of continuity in I. This property of $t \mapsto f(t,u(t),u(b_1(t)),\dots,u(b_m(t)))$, together with the assumptions of Theorem 2.2, implies that the linear Cauchy problem

$$\begin{aligned}
\dot v(t) + Av(t) &= f(t,u(t),u(b_1(t)),\dots,u(b_m(t))), && t\in I,\ t\ne\tau_k,\\
v(\tau_k+0) &\equiv u(\tau_k) + I_k(u(\tau_k)), && k=1,2,\dots,\kappa,\\
v(t_0) &= u_0 - g(u),
\end{aligned}$$

has a unique classical solution v such that

$$\begin{aligned}
v(t) &= T(t-t_0)u_0 - T(t-t_0)g(u)\\
&\quad + \int_{t_0}^{t} T(t-s)f(s,u(s),u(b_1(s)),\dots,u(b_m(s)))\,ds\\
&\quad + \sum_{t_0\le\tau_k<t} T(t-\tau_k)I_k(u(\tau_k)), \quad t\in I.
\end{aligned}$$

Consequently, u is the unique classical solution of the nonlocal impulsive Cauchy problem (1.1), and the proof is complete. □

3. Continuous dependence of a mild solution on the initial condition

Theorem 3.1. *Suppose that the functions f, g and $I(u)$ satisfy the assumptions* **A1–A3** *and there exist constants μ_1, μ_2, μ_3 such that*

(i) $\|g(u) - g(\tilde{u})\| \leq \mu_1 \|u - \tilde{u}\|$,

(ii) $\|f(s, u(s), \dots, u(b_m(s))) - f(s, \tilde{u}(s), \dots, \widetilde{u(b_m(s))})\| \leq \mu_2 \|u - \tilde{u}\|$,

(iii) $\|I_k(u(\tau_k)) - I_k(\tilde{u}(\tau_k))\| \leq \mu_3 \|u(\tau_k) - \tilde{u}(\tau_k)\|$,

where $u, \tilde{u} \in C(I, E)$. If u and $\tilde{u}$ are mild solutions of the problem (1.1) *with the respective initial values $u_0, \tilde{u}_0$ and the constants μ_1 and $\mu = \max\{\mu_2, \mu_3\}$ satisfy the inequality*

$$\mu_1 < \frac{\exp(-(t_0 + a)M\mu)(1 + M\mu)^{-\kappa}}{M}, \tag{3.1}$$

then the following inequality holds:

$$\|u(t) - \tilde{u}(t)\| \leq \frac{M \exp((t_0 + a)M\mu)(1 + M\mu)^{\kappa}}{1 - M\mu_1 \exp((t_0 + a)M\mu)(1 + M\mu)^{\kappa}} \|u_0 - \tilde{u}_0\|. \tag{3.2}$$

Proof. Assume that $u, \tilde{u}$ are the mild solutions of problem (1.1). Then

$$\begin{aligned} u(t) - \tilde{u}(t) = {} & [T(t - t_0)(u_0 - \tilde{u}_0) - T(t - t_0)(g(u) - g(\tilde{u}))] \\ & + \int_{t_0}^{t} T(t - s)\big(f(s, u(s), u(b_1(s)), u(b_2(s)), \dots, u(b_m(s))) \\ & - f(s, \tilde{u}(s), \tilde{u}(b_1(s)), \tilde{u}(b_2(s)), \dots, \tilde{u}(b_m(s)))\big)\, ds \\ & + \sum_{t_0 \leq \tau_k < t+h} T(t - \tau_k)[I_k(u(\tau_k)) - I_k(\tilde{u}(\tau_k))], \end{aligned}$$

where $t \in [t_0, t_0 + a]$. From **A1–A3** and the hypotheses of the theorem, we have

$$\begin{aligned} \|u(\zeta) - \tilde{u}(\zeta)\| \leq {} & M\|u_0 - \tilde{u}_0\| + M\mu_1 \|u - \tilde{u}\| \\ & + M\mu_2 \int_{t_0}^{\zeta} \|u - \tilde{u}\|\, ds + M\mu_3 \sum_{t_0 \leq \tau_k < \zeta} \|u(\tau_k) - \tilde{u}(\tau_k)\| \end{aligned}$$

for $t_0 \leq \zeta \leq t_0 + a$. Using this result, it follows that

$$\begin{aligned} \sup_{\zeta \in I} \|u(\zeta) - \tilde{u}(\zeta)\| \leq {} & M\|u_0 - \tilde{u}_0\| + M\mu_1 \|u - \tilde{u}\| \\ & + M\mu_2 \int_{t_0}^{t} \|u - \tilde{u}\|\, ds + M\mu_3 \sum_{t_0 \leq \tau_k < t} \|u(\tau_k) - \tilde{u}(\tau_k)\|. \end{aligned}$$

Thus

$$\|u(t) - \tilde{u}(t)\| \le M\|u_0 - \tilde{u}_0\| + M\mu_1\|u - \tilde{u}\| + M\mu\Big\{\int_{t_0}^{t} \|u - \tilde{u}\|\, ds + \sum_{t_0 \le \tau_k < t} \|u(\tau_k) - \tilde{u}(\tau_k)\|\Big\}.$$

Applying Gronwall's inequality for discontinuous functions (see [5]), it follows that

$$\|u(t) - \tilde{u}(t)\| \le \{\|u_0 - \tilde{u}_0\| + \mu_1\|u - \tilde{u}\|\}\, M \exp((t_0 + a)M\mu)(1 + M\mu)^{\kappa}.$$

We can also write this inequality in the form

$$\begin{aligned} &[1 - M\mu_1 \exp\big((t_0 + a)M\mu\big)(1 + M\mu)^{\kappa}]\|u(t) - \tilde{u}(t)\| \\ &\quad \le M \exp\big((t_0 + a)M\mu\big)(1 + M\mu)^{\kappa}\|u_0 - \tilde{u}_0\|. \end{aligned} \tag{3.3}$$

Additionally, if equality (3.1) is valid, then inequality (3.3) is equivalent to inequality (3.2). This completes the proof of Theorem 3.1. □

Remark 3.2. If $\mu_1 = \kappa = 0$, then inequality (3.2) is reduced to the classical inequality

$$\|u(t) - \tilde{u}(t)\| \le M \exp((t_0 + a)M\mu)\|u_0 - \tilde{u}_0\|,$$

which is characteristic for the continuous dependence of the semilinear functional-differential evolution Cauchy problem with the classical initial condition.

Acknowledgment

The first author would like to thank King Fahd University of Petroleum and Minerals, Department of Mathematical Sciences for providing excellent research facilities.

References

[1] H. Akça, A. Boucherif, and V. Covachev, *Impulsive Functional-differential Equations with Nonlocal Conditions.* Int. J. of Math. and Math. Sci. **29**:5 (2002), 251–256.

[2] L. Byszewski, and H. Akça, *On a Mild Solution of a Semilinear Functional-Differential Evolution Nonlocal Problem.* J. Appl. Math. Stoch. Anal. **10**:3 (1997), 265–271.

[3] L. Byszewski, and H. Akça, *Existence of Solutions of a Semilinear Functional-Differential Evolution Nonlocal Problem.* Nonlinear Analysis **34**:1 (1998), 65–72.

[4] V. Lakshmikantham, D.D. Bainov, and P.S. Simeonov, *Theory of Impulsive Differential Equations.* Series in Modern Applied Mathematics, vol. 6, World Scientific, Singapore, 1989.

[5] A.M. Samoilenko, and N.A. Perestyuk, *Impulsive Differential Equations.* World Scientific Series on Nonlinear Science. Ser. A: Monographs and Treatises **14**, World Scientific, Singapore, 1995.

Haydar Akça
King Fahd University of Petroleum and Minerals
Department of Mathematical Science
Dhahran 31261
Saudi Arabia
e-mail: akca@kfupm.edu.sa

Valéry Covachev
Institute of Mathematics
Bulgarian Academy of Sciences
Sofia
Bulgaria
Current address:
Department of Mathematics
and Statistics
College of Science
Sultan Qaboos University
Sultanate of Oman
e-mail: vcovachev@hotmail.com

Eada Al-Zahrani
Science College for Girls
Department of Mathematics
Dammam
Saudi Arabia

Operator Theory:
Advances and Applications, Vol. 153, 13–38

On Banach-Lie Algebras, Spectral Decompositions and Complex Polarizations

Daniel Beltiţă

Abstract. Complex Kähler polarizations are constructed for a class of real Banach-Lie algebras that are not necessarily L^*-algebras but include all the real compact L^*-algebras. The approach is based on the theory of spectral decompositions of Banach space operators, and particularly on Dunford scalar operators. The main results are illustrated by means of a family of examples that are constructed starting from the Schatten-von Neumann classes of Hilbert space operators $\mathcal{C}_p$ with $p \geq 2$.

Mathematics Subject Classification (2000). Primary 58B12; Secondary 17B65, 46H70, 47B40.

Keywords. Banach-Lie algebra, Spectral decomposition, Complex polarization.

Introduction

Our aim in the present paper is to extend to a Banach setting certain ideas that arose in [20] in the study of a special kind of infinite-dimensional Lie groups modeled over Hilbert spaces. That local study was motivated by the orbit method in infinite dimensions, and what we are doing here is essentially to explore some spectral theoretic aspects of that method. To explain this in more detail, let us begin with the case of finite-dimensional groups.

In the representation theory of Lie groups, an important problem related to the orbit method is to construct complex structures compatible with the natural symplectic structures of the coadjoint orbits of a finite-dimensional Lie group G, and to this end one has to make use of complex polarizations (see, e.g., [15] and [19]). As it is well known, these are complex subalgebras of the complexified Lie algebra of G, and are closely related to root-space decompositions (see, e.g., Proposition IX.2.9 in [19]), thus essentially leaning on *spectral theory of matrices.*

On the other hand, in connection with the investigation of highest weight representations of certain Hilbert-Lie groups (see [7], [8] and [20]), complex polarizations of the corresponding Hilbert-Lie algebras are required again, and a central role in their construction is played by the *spectral theory of self-adjoint operators.* And the latter theory is just a Hilbert space extension of the diagonalization theory of Hermitian matrices.

In view of some recent successful studies in representation theory of Banach-Lie groups modeled on certain ideals of compact operators (see [16] and [18]), the problem to relate that representation theory to complex polarizations of the corresponding Lie algebras thus arises naturally. It is just the *aim* of the present paper to do a first step in that direction, by providing a method to construct complex polarizations for a class of Banach-Lie algebras which are more general than the L^*-algebras dealt with in [20] (see Theorem 2.10 below). To this end, we make use of the *theory of spectral decompositions* – also called local spectral theory – for bounded linear operators on Banach spaces (see, e.g., [9], [10], [26], [12]).

Before proceeding to a more detailed description of the contents of the present paper, let us briefly explain why we are working here only on the level of Lie algebras, without making any attempt to draw consequences which our results can have for the corresponding Lie groups. The motivation for this fact is two-fold. Firstly, the spectral theoretic flavor of the methods we use fits the best to the world of "linear objects", i.e., of Lie algebras. Secondly, it is one of the basic principles in Lie theory that things should be understood on the Lie algebraic level first, and the corresponding Lie group problems should be approached afterwards, starting from the already available Lie algebraic information. Accordingly, it seems more appropriate to postpone to another paper ([3]) the consequences of the Lie algebraic results in the present note. The paper [3] (see also [4]) includes in particular a method to provide invariant complex structures for homogeneous spaces of certain Banach-Lie groups, heavily leaning on the complex polarizations constructed in the present note.

In Section 1 we are concerned with spectral decompositions for bounded derivations of (not necessarily associative) Banach algebras. The motivation for considering spectral decompositions in this special instance stems from the essential role played by the spectral theory of derivations in order to understand certain most interesting objects arising in the infinite-dimensional Lie theory (see, e.g., [20] and [21]). More specifically, the main result of Section 1 is Theorem 1.4, which essentially asserts that each normal derivation with closed range of a complex Banach-Lie algebra leads to a "triangular decomposition" of that algebra. A version of this result without the closed-range hypothesis is contained in Remark 1.6.

The core of the present paper is Section 2, where we prove our main result concerning the construction of weak Kähler polarizations by means of the spectral decompositions (Theorem 2.10). The class of real Banach-Lie algebras to which this result applies is described by the set of hypotheses $1^\circ - 5^\circ$ that are stated before Lemma 2.8. An algebra $\mathfrak{g}(D)$ of this class is constructed by means of a bounded derivation D of a real Banach-Lie algebra $\mathfrak{g}$ for which certain conditions are satisfied. In particular, we require that the coadjoint representation of $\mathfrak{g}$ should be embedded into the adjoint one (see hypothesis 2°). The idea of construction of $\mathfrak{g}(D)$ comes from the construction of restricted Lie algebras (cf. Definition III.1 in [22]). The last result of Section 2 (namely Proposition 2.12) shows when the Lie algebra $\mathfrak{g}$ we are working with is a compact L^*-algebra. In this special case, our construction of polarizations agrees with the one described in Lemma VII.4 in [20]. Though, we note that, in our more general situation, we can obtain only *weak* Kähler polarizations. (See the precise definitions in the following.)

In Section 3 we take into consideration as illustrating examples the family of Banach-Lie algebras $\{\mathcal{C}_p(\mathcal{H})\}_{1\le p\le\infty}$ (the Schatten-von Neumann classes of operators on the complex Hilbert space $\mathcal{H}$). As we previously mentioned, the present note is a preliminary step of the attempt to extend the representation theory of the Hilbert-Lie groups in [7] and [20] to the more general Banach-Lie groups investigated in [16] and [18]. The latter groups are modeled on the Schatten-von Neumann classes, and that is why we apply the results of Section 1 and Section 2 only to Lie algebras derived from these classical Banach-Lie algebras (in the terminology of [13]). In particular, we conclude Section 3 by working out the details of a family of examples which fall under the hypotheses 1°–5° of Section 2 but are not L^*-algebras (see Example 3.5).

Preliminaries

Throughout the paper we denote by $\mathrm{Der}(\mathfrak{A})$ the Banach-Lie algebra of all bounded derivations of a Banach (not necessarily associative) algebra $\mathfrak{A}$. (We say that $\mathfrak{A}$ is a Banach algebra if $\mathfrak{A}$ is a Banach space endowed with a bounded bilinear map $\mathfrak{A}\times\mathfrak{A}\to\mathfrak{A}$, $(a,b)\mapsto a\cdot b$.) For a real or complex Banach space $\mathfrak{X}$ we denote by $\mathrm{id}_{\mathfrak{X}}$ the identical map on $\mathfrak{X}$, by $\mathfrak{X}^*$ the topological dual of $\mathfrak{X}$, by $\mathcal{B}(\mathfrak{X})$ the algebra of all bounded linear operators on $\mathfrak{X}$ and, when $\mathfrak{X}$ is complex, for $D\in\mathcal{B}(\mathfrak{X})$ we denote by $\sigma(D)$ the spectrum of D. In this case, for every $x\in\mathfrak{X}$ we denote by $\sigma_D(x)$ the *local spectrum* of x with respect to D. We recall that $\sigma_D(x)$ is a closed subset of $\sigma(D)$ and by definition, a complex number w belongs to $\mathbb{C}\setminus\sigma_D(x)$ if and only if there exists an open neighborhood W of w and a holomorphic function $\xi\colon W\to\mathfrak{X}$ such that

$$(z\,\mathrm{id}_{\mathfrak{X}}-D)\xi(z)=x\qquad\text{for every } z\in W.$$

If F is a subset of $\mathbb{C}$ we further denote

$$\mathfrak{X}_D(F)=\{x\in\mathfrak{X}\mid\sigma_D(x)\subseteq F\}.$$

We note that, in the case when $\mathfrak{X}$ has *finite dimension* m, we have

$$\mathfrak{X}_D(F)=\bigoplus_{\lambda\in F\cap\sigma(D)}\mathrm{Ker}\,((D-\lambda\,\mathrm{id}_{\mathfrak{X}})^m)\qquad\text{for every } F\subseteq\mathbb{C},$$

while in the case when $\mathfrak{X}$ is a *Hilbert space* and D is a *normal operator with the spectral measure* $E_D(\cdot)$ we have

$$\mathfrak{X}_D(F)=\mathrm{Ran}\,E_D(F)\qquad\text{whenever } F \text{ is a closed subset of } \mathbb{C}.$$

We refer to Section 12 in [5] for a review of the few elements of local spectral theory we need. (For more details see the Notes of Chapter I in [5].)

For the sake of completeness, we now recall some elementary facts on complex polarizations (see, e.g., Section VI in [20]). Let $\mathfrak{g}$ be a real Banach-Lie algebra and ω a *continuous* 2*-cocycle* of $\mathfrak{g}$, that is $\omega : \mathfrak{g} \times \mathfrak{g} \to \mathbb{R}$ is a continuous skew-symmetric bilinear form such that

$$\omega([X,Y],Z) + \omega([Y,Z],X) + \omega([Z,X],Y) = 0$$

for every $X, Y, Z \in \mathfrak{g}$. In this case,

$$\mathfrak{h} := \{X \in \mathfrak{g} \mid \omega(X, \mathfrak{g}) = \{0\}\}$$

is a closed subalgebra of $\mathfrak{g}$. A *complex polarization* of $\mathfrak{g}$ in ω is a closed subspace $\mathfrak{p}$ of the complexification $\mathfrak{g}_{\mathbb{C}}$ of $\mathfrak{g}$ such that there exists a closed subspace of $\mathfrak{g}_{\mathbb{C}}$ complementary to $\mathfrak{p}$ and such that the following properties hold:

(C1) $\mathfrak{p}$ is a (complex closed) subalgebra of the complex Banach-Lie algebra $\mathfrak{g}_{\mathbb{C}}$ such that $[\mathfrak{h}, \mathfrak{p}] \subseteq \mathfrak{p}$,
(C2) $\mathfrak{p} \cap \overline{\mathfrak{p}} = \mathfrak{h}_{\mathbb{C}}$,
(C3) $\mathfrak{p} + \overline{\mathfrak{p}} = \mathfrak{g}_{\mathbb{C}}$,
(C4) $\omega(\mathfrak{p}, \mathfrak{p}) = \{0\}$,

where $Z \mapsto \overline{Z}$ denotes the complex conjugation on $\mathfrak{g}_{\mathbb{C}}$ whose set of fixed points is just $\mathfrak{g}$. In this case, the natural inclusion map $\mathfrak{g} \to \mathfrak{g}_{\mathbb{C}}$ induces an isomorphism $\mathfrak{g}/\mathfrak{h} \cong \mathfrak{g}_{\mathbb{C}}/\mathfrak{p}$ of real Banach spaces (see Section VI in [20]), thus $\mathfrak{g}/\mathfrak{h}$ is actually a complex Banach space. On the other hand, note that ω induces a symplectic form on $\mathfrak{g}/\mathfrak{h}$ by

$$\mathfrak{g}/\mathfrak{h} \times \mathfrak{g}/\mathfrak{h} \to \mathbb{R}, \quad (X + \mathfrak{h}, Y + \mathfrak{h}) \mapsto \omega(X, Y).$$

This symplectic form is the imaginary part of the continuous Hermitian sesquilinear form $(\cdot \mid \cdot)$ which is defined on the complex Banach space $\mathfrak{g}/\mathfrak{h} \cong \mathfrak{g}_{\mathbb{C}}/\mathfrak{p} = (\overline{\mathfrak{p}} + \mathfrak{p})/\mathfrak{p}$ by

$$(Z_1 + \mathfrak{p} \mid Z_2 + \mathfrak{p}) = \mathrm{i}\omega(Z_1, \overline{Z_2}) \quad \text{for } Z_1, Z_2 \in \overline{\mathfrak{p}}.$$

If the following condition holds,

(C5) $\mathrm{i}\omega(Z, \overline{Z}) > 0$ for every $Z \in \mathfrak{p} \setminus \mathfrak{h}_{\mathbb{C}}$,

then $(\cdot \mid \cdot)$ is even a scalar product (i.e., it is positively definite) and $\mathfrak{p}$ is called a *weak Kähler polarization*. In the case when the map

$$\mathfrak{g}/\mathfrak{h} \to (\mathfrak{g}/\mathfrak{h})^*, \quad X + \mathfrak{h} \mapsto \omega(X, \cdot),$$

is moreover invertible, one says that $\mathfrak{p}$ is a *strong Kähler polarization*. (An equivalent condition is that $(\cdot \mid \cdot)$ defines the topology of $\mathfrak{g}/\mathfrak{h}$, i.e., $\mathfrak{g}/\mathfrak{h}$ is actually a complex Hilbert space; see stage I in the proof of Proposition 2.12 below.)

1. Spectral decompositions and derivations

The main aim of the present section is to get a sort of triangular decomposition theorems for Banach-Lie algebras (see Theorem 1.4 and Remark 1.6). To this end, we need the following result, whose proof uses the method of proof of Proposition 6.3 from Chapter IV of [26] (or Proposition 2 in Section 13 of [5]).

Proposition 1.1. *If $\mathfrak{A}$ is a complex Banach (not necessarily associative) algebra, $D \in \mathrm{Der}(\mathfrak{A})$, F_1 and F_2 are subsets of $\mathbb{C}$, and either*

(i) *each of the sets $F_1 \cap \sigma(D)$ and $F_2 \cap \sigma(D)$ is at the same time open and closed with respect to the topology of $\sigma(D)$, or*

(ii) *D has the single-valued extension property and each of $F_1 \cap \sigma(D)$ and $F_2 \cap \sigma(D)$ is a closed subset of $\sigma(D)$,*

then we have

$$\sigma_D(a_1 a_2) \subseteq F_1 + F_2$$

for every $a_1 \in \mathfrak{A}_D(F_1)$ and $a_2 \in \mathfrak{A}_D(F_2)$.

Proof. We can suppose without loss of generality that both F_1 and F_2 are contained in $\sigma(D)$. Pick $z_0 \notin F_1 + F_2$. Then there exists a compact neighborhood Q_0 of z_0 such that $Q_0 \cap (F_1 + F_2) = \emptyset$. Next choose open subsets V_0 and V of $\mathbb{C}$ that are relatively compact and moreover have the properties $F_2 \subseteq V_0 \subseteq V$, $Q_0 \cap (F_1 + \overline{V}) = \emptyset$ and the boundary Γ_0 of V_0 is contained in V and consists of a finite number of piecewise smooth Jordan curves.

On the other hand, in both the situations (i) and (ii) there exist the holomorphic functions $\alpha_1 : \mathbb{C} \setminus F_1 \to \mathfrak{A}$ and $\alpha_2 : \mathbb{C} \setminus F_2 \to \mathfrak{A}$ such that

$$(t - D)\alpha_1(t) = a_1 \quad \text{for } t \in \mathbb{C} \setminus F_1$$

and

$$(s - D)\alpha_2(s) = a_2 \quad \text{for } s \in \mathbb{C} \setminus F_2.$$

Note that, in view of the way Q_0, V_0 and V have been chosen, we have $t - s \notin F_1$ whenever $t \in Q_0$ and $s \in \Gamma_0$. In particular, we can define a holomorphic function $\gamma : Q_0 \to \mathfrak{A}$ by

$$\gamma(t) = \frac{1}{2\pi \mathrm{i}} \int_{\Gamma_0} \alpha_1(t - s)\alpha_2(s)\, \mathrm{d}s \qquad \text{for } t \in Q_0.$$

By making use of the fact that D is a bounded derivation of $\mathfrak{A}$, for every $t \in Q_0$ we get

$$\begin{aligned}
(t - D)\gamma(t) &= \frac{1}{2\pi \mathrm{i}} \int_{\Gamma_0} (t - D)(\alpha_1(t - s)\alpha_2(s))\, \mathrm{d}s \\
&= \frac{1}{2\pi \mathrm{i}} \int_{\Gamma_0} ((t - D)\alpha_1(t - s))\alpha_2(s)\, \mathrm{d}s - \frac{1}{2\pi \mathrm{i}} \int_{\Gamma_0} \alpha_1(t - s)(D\alpha_2(s))\, \mathrm{d}s \\
&= \frac{1}{2\pi \mathrm{i}} \int_{\Gamma_0} ((t - s - D)\alpha_1(t - s))\alpha_2(s)\, \mathrm{d}s + \frac{1}{2\pi \mathrm{i}} \int_{\Gamma_0} \alpha_1(t - s)((s - D)\alpha_2(s))\, \mathrm{d}s.
\end{aligned}$$

Since $(t-s-D)\alpha_1(t-s)=a_1$ and $(s-D)\alpha_2(s)=a_2$, we further deduce

$$\begin{aligned}(t-D)\gamma(t) &= a_1\cdot\left(\frac{1}{2\pi\mathrm{i}}\int_{\Gamma_0}\alpha_2(s)\,\mathrm{d}s\right)+\left(\frac{1}{2\pi\mathrm{i}}\int_{\Gamma_0}\alpha_1(t-s)\,\mathrm{d}s\right)\cdot a_2\\ &= a_1\cdot a_2-0\cdot a_2=a_1a_2,\end{aligned}$$

where we have used the fact that Γ_0 surrounds F_2, while the $\mathfrak{A}$-valued function $s\mapsto\alpha_1(t-s)$ is holomorphic on a neighborhood of the closure of V_0 and Γ_0 is the boundary of V_0. □

The next result is a version of Example 4.9 in [25]. See also Lemma 1 in Section 17 of [5].

Corollary 1.2. *Under the hypothesis of Proposition* 1.1 *we have*

$$\mathfrak{A}_D(F_1)\cdot\mathfrak{A}_D(F_1)\subseteq\mathfrak{A}_D(F_1+F_2).$$

In particular, if the derivation D has the single-valued extension property and S is an arbitrary subsemigroup of $(\mathbb{C},+)$, then $\mathfrak{A}_D(S)$ is a subalgebra of $\mathfrak{A}$. If moreover $S\cap(-S)=\{0\}$, then $\mathfrak{A}_D(S\setminus\{0\})$ is also a subalgebra of $\mathfrak{A}$.

Proof. Use Proposition 1.1. In order to prove the last assertion note that, if S is a subsemigroup of $(\mathbb{C},+)$ such that $S\cap(-S)=\{0\}$, then for every subsets F_1 and F_2 of $S\setminus\{0\}$ we have $F_1+F_2\subseteq S\setminus\{0\}$. □

Remark 1.3. In the important special situation when S is a closed half-plane, one can prove that $\mathfrak{A}_D(S)$ is a subalgebra by making use of the remark that D generates a 1-parameter group of automorphisms of $\mathfrak{A}$. To this end, recall that for every $s\in\mathbb{R}$

$$\mathfrak{A}_D([s,\infty)+\mathrm{i}\mathbb{R})=\Big\{a\in\mathfrak{A}\mid\limsup_{t\to\infty}\frac{\log\|\mathrm{e}^{-tD}a\|}{t}\le -s\Big\},$$

$$\mathfrak{A}_D((-\infty,s]+\mathrm{i}\mathbb{R})=\Big\{a\in\mathfrak{A}\mid\limsup_{t\to\infty}\frac{\log\|\mathrm{e}^{tD}a\|}{t}\le s\Big\}$$

(see Proposition 4.1 in Chapter V of [26]). Since e^{tD} are automorphisms of $\mathfrak{A}$, it then easily follows that

$\mathfrak{A}_D(\pm[s_1,\infty)+\mathrm{i}\mathbb{R})\cdot\mathfrak{A}_D(\pm[s_2,\infty)+\mathrm{i}\mathbb{R})\subseteq\mathfrak{A}_D(\pm[s_1+s_2,\infty)+\mathrm{i}\mathbb{R})$ for $s_1,s_2\in\mathbb{R}$.

In particular, $\mathfrak{A}_D(\pm[s,\infty)+\mathrm{i}\mathbb{R})$ are subalgebras of $\mathfrak{A}$ whenever $s\ge 0$. (See also [21].) □

Now we can prove the main result of the present section. It might be thought of as a version of the "triangular decomposition" which is obtained from a root-space decomposition of a complex finite-dimensional semisimple Lie algebra by fixing an order on the corresponding set of roots.

Theorem 1.4. *Let $\mathfrak{A}$ be a complex Banach-Lie algebra and assume that $D\in\mathrm{Der}(\mathfrak{A})$ is a normal derivation with closed range. If for some closed subsemigroup S of $(\mathbb{C},+)$ we have $S\cap(-S)=\{0\}$, S is contained in a closed half-plane and $\sigma(D)\subseteq$*

$S \cup (-S)$, then $\mathfrak{A}$ has the following decomposition into a direct sum of closed Lie subalgebras,

$$\mathfrak{A} = \mathfrak{A}_- \oplus \mathfrak{A}_0 \oplus \mathfrak{A}_+,$$

where $\mathfrak{A}_0 = \operatorname{Ker} D$, $\mathfrak{A}_- = \mathfrak{A}_D((-S) \setminus \{0\})$ and $\mathfrak{A}_+ = \mathfrak{A}_D(S \setminus \{0\})$. Moreover we have

$$[\mathfrak{A}_0, \mathfrak{A}_\pm] \subseteq \mathfrak{A}_\pm,$$

each of the Lie algebras $\mathfrak{A}_\pm$ is nilpotent and the operator $\operatorname{ad} a : \mathfrak{A} \to \mathfrak{A}$ is nilpotent for every $a \in \mathfrak{A}_- \cup \mathfrak{A}_+$.

Proof. Since D is a normal operator with closed range, Theorem 4.12 in [17] shows that 0 is an isolated point of the spectrum of D. Thus we get the following decomposition of $\mathfrak{A}$ into a direct sum of closed linear subspaces of it,

$$\mathfrak{A} = \mathfrak{A}_D((-S) \setminus \{0\}) \oplus \mathfrak{A}_D(\{0\}) \oplus \mathfrak{A}_D(S \setminus \{0\}). \tag{1.1}$$

Next note that D has the single-valued extension property by Remark 6(a) and Theorem 2 in Section 14 of [5], hence Corollary 1.2 above implies that each term in the decomposition (1.1) is a Lie subalgebra of $\mathfrak{A}$. Since D is normal, we have $\mathfrak{A}_D(\{0\}) = \operatorname{Ker} D$ (see Corollary 4 in Section 14 from [5]).

To see that the Lie algebra $\mathfrak{A}_+$ is nilpotent, assume without loss of generality that $\mathfrak{A}_+ \neq \{0\}$. By the hypothesis concerning S, we may suppose that $S \subseteq \{0\} \cup \{z \in \mathbb{C} \mid \operatorname{Re} z > 0\}$ (after a rotation of the complex plane, that is working with ζD instead of D for some suitable $\zeta \in \mathbb{C}$ with $|\zeta| = 1$). Denote

$$\lambda_+ = \inf\{\operatorname{Re} z \mid 0 \neq z \in \sigma(D) \cap S\}.$$

Since 0 is an isolated point in the spectrum of D it then follows $\lambda_+ > 0$. Now

$$([n\lambda_+, \infty) + \mathrm{i}\mathbb{R}) \cap \sigma(D) = \emptyset \quad \text{for } n > \frac{1}{\lambda_+} \sup_{z \in \sigma(D)} \operatorname{Re} z,$$

hence Corollary 1.2 immediately implies that the nth term of the central series of $\mathfrak{A}_+$ vanishes for $n > \sup\limits_{z \in \sigma(D)} \operatorname{Re} z / \lambda_+$.

A similar reasoning shows that the Lie algebra $\mathfrak{A}_-$ is nilpotent as well. Next take $a \in \mathfrak{A}_+$. With λ_+ as above we then have $a \in \mathfrak{A}_D([\lambda_+, \infty) + \mathrm{i}\mathbb{R})$. Since $\mathfrak{A} = \mathfrak{A}_D(\sigma(D))$, for every positive integer n we have

$$(\operatorname{ad} a)^n \mathfrak{A} \subseteq \underbrace{[\mathfrak{A}_D([\lambda_+, \infty) + \mathrm{i}\mathbb{R}), [\ldots, [\mathfrak{A}_D([\lambda_+, \infty) + \mathrm{i}\mathbb{R})}_{n \text{ times}}, \mathfrak{A}_D(\sigma(D))] \ldots]],$$

hence by Corollary 1.2 it follows that

$$(\operatorname{ad} a)^n \mathfrak{A} \subseteq \mathfrak{A}_D([n\lambda_+, \infty) + \mathrm{i}\mathbb{R} + \sigma(D)).$$

On the other hand, recall that $\lambda_+ > 0$. This implies that for $n > \sup\limits_{z, w \in \sigma(D)} (\operatorname{Re} z - \operatorname{Re} w)/\lambda_+$ we have

$$([n\lambda_+, \infty) + \mathrm{i}\mathbb{R} + \sigma(D)) \cap \sigma(D) = \emptyset,$$

and thus for such n we get $(\operatorname{ad} a)^n \mathfrak{A} = \{0\}$. One can similarly treat the case $a \in \mathfrak{A}_-$. □

Remark 1.5. The above Theorem 1.4 can also be proved by making use of Remark 1.3 instead of Corollary 1.2. (See also the remarks preceding Theorem IV.5 in [21].) □

As we shall see in Section 3, the closed-range condition imposed to the derivation in Theorem 1.4 turns out to be a rather restrictive one in many concrete situations. In view of this fact, we discuss below a version of Theorem 1.4, where the closed-range condition is replaced by the requirement that the considered derivation is Dunford scalar (see either [10] or Definition 3 in Section 14 of [5] for the definition of the latter notion). This is the case, for example, when one deals with a normal derivation of a complex Banach-Lie algebra whose underlying Banach space is actually a Hilbert space, in particular when one deals with a normal derivation of a complex L^*-algebra (see, e.g., [23] and [24]). We point out however that, according to the results of [14], if $N \in \mathcal{B}(\mathcal{H})$ is a normal operator and $p \neq 2$, then the normal derivation $\operatorname{ad} N|_{\mathcal{C}_p(\mathcal{H})}$ of $\mathcal{C}_p(\mathcal{H})$ is Dunford scalar if and only if the spectrum of N is finite (see also Corollary 3.4(iii) below).

Remark 1.6. Consider a Dunford scalar derivation $D \in \operatorname{Der}(\mathfrak{A})$ of the complex Banach-Lie algebra $\mathfrak{A}$ and denote by $E_D(\cdot)$ the spectral measure of D. As in Theorem 1.4, assume that S is a closed subsemigroup of $(\mathbb{C}, +)$ such that $S \cap (-S) = \{0\}$, S is contained in a closed half-plane and $\sigma(D) \subseteq S \cup (-S)$. Then

$$\operatorname{id}_{\mathfrak{A}} = E_D((-S) \setminus \{0\}) + E_D(\{0\}) + E_D(S \setminus \{0\})$$

and all of the pairwise products of the idempotent operators $E_D((-S) \setminus \{0\})$, $E_D(\{0\})$ and $E_D(S \setminus \{0\})$ vanish. Then

$$\mathfrak{A} = \widetilde{\mathfrak{A}}_- \oplus \mathfrak{A}_0 \oplus \widetilde{\mathfrak{A}}_+, \tag{1.2}$$

where

$$\mathfrak{A}_0 = \operatorname{Ran} E_D(\{0\}) = \operatorname{Ker} D,\ \widetilde{\mathfrak{A}}_- = \operatorname{Ran} E_D((-S) \setminus \{0\}),\ \widetilde{\mathfrak{A}}_+ = \operatorname{Ran} E_D(S \setminus \{0\}).$$

To see that $\widetilde{\mathfrak{A}}_+$ is a Lie subalgebra of $\mathfrak{A}$, assume just as in the proof of Theorem 1.4 that $S \subseteq \{0\} \cup ((0, \infty) + \mathrm{i}\mathbb{R})$. Next note that, since the spectral measure $E_D(\cdot)$ is countably additive (see [10]), the operator $E_D(S \setminus \{0\})$ equals the pointwise limit of the sequence $\{E_D(S_n)\}_{n \geq 1}$, where $S_n = S \cap ([\frac{1}{n}, \infty) + \mathrm{i}\mathbb{R})$ whenever $n \geq 1$.

Then for every $x, y \in \widetilde{\mathfrak{A}}_+$ we have

$$[x, y] = \left[\lim_{n \to \infty} E_D(S_n)x, \lim_{n \to \infty} E_D(S_n)y\right] = \lim_{n \to \infty} \left[E_D(S_n)x, E_D(S_n)y\right].$$

Since $\operatorname{Ran} E_D(F) = \mathfrak{A}_D(F)$ for each closed subset F of $\mathbb{C}$, by either Corollary 1.2 or Remark 1.3 we get for each positive integer n

$$\begin{aligned}
[E_D(S_n)x, E_D(S_n)y] &\in [\mathfrak{A}_D(S_n), \mathfrak{A}_D(S_n)] \\
&\subseteq \mathfrak{A}_D(S_n + S_n) \subseteq \mathfrak{A}_D\left(S \cap \left(\left[\frac{2}{n}, \infty\right) + \mathrm{i}\mathbb{R}\right)\right) \\
&= \operatorname{Ran} E_D\left(S \cap \left(\left[\frac{2}{n}, \infty\right) + \mathrm{i}\mathbb{R}\right)\right) \\
&\subseteq \widetilde{\mathfrak{A}}_+.
\end{aligned}$$

Since $\widetilde{\mathfrak{A}}_+$ is closed, we thus get $[x, y] \in \widetilde{\mathfrak{A}}_+$ and it follows that $\widetilde{\mathfrak{A}}_+$ is indeed a Lie subalgebra of $\mathfrak{A}$. One can similarly prove that also $\widetilde{\mathfrak{A}}_-$ is a Lie subalgebra of $\mathfrak{A}$.

In order to compare the decompositions (1.1) and (1.2), one should note that $\mathfrak{A}_- \subseteq \widetilde{\mathfrak{A}}_-$ and $\mathfrak{A}_+ \subseteq \widetilde{\mathfrak{A}}_+$. □

2. Complex polarizations

In this section we are going to make use of Theorem 1.4 and Remark 1.6 for constructing complex polarizations of certain Banach-Lie algebras (see Theorem 2.10 below). To begin with, we prove a result which will help us to show that our candidates for complex polarizations indeed fulfill the condition (C4) from the Preliminaries.

Proposition 2.1. *Let $\mathfrak{X}$ be a complex Banach space, $\omega : \mathfrak{X} \times \mathfrak{X} \to \mathbb{C}$ a bounded bilinear form, and $D \in \mathcal{B}(\mathfrak{X})$ such that $\omega(\operatorname{Ker} D, \operatorname{Ker} D) = \{0\}$ and*

$$\omega(D(X), Y) = -\omega(X, D(Y)) \quad \textit{for every } X, Y \in \mathfrak{X}.$$

Let S be a closed subset of $\mathbb{C}$ such that $S \cap (-S) = \{0\}$ and $\sigma(\mathrm{i}D) \subseteq S \cup (-S)$. Suppose furthermore that the operator $\mathrm{i}D$ is either

(i) *normal with closed range, or*
(ii) *Dunford scalar, having the spectral measure denoted by $E_{\mathrm{i}D}(\cdot)$,*

and denote accordingly

$$\mathfrak{X}_\pm = \begin{cases} \mathfrak{X}_{\mathrm{i}D}(\pm S \setminus \{0\}) & \textit{in the case (i)}, \\ E_{\mathrm{i}D}(\pm S \setminus \{0\}) & \textit{in the case (ii)}, \end{cases}$$

and $\mathfrak{X}_0 = \operatorname{Ker} D$ in both cases. Then $\mathfrak{X}$ has the following decomposition into a direct sum of closed subspaces

$$\mathfrak{X} = \mathfrak{X}_- \oplus \mathfrak{X}_0 \oplus \mathfrak{X}_+,$$

and moreover

$$\omega(\mathfrak{p}, \mathfrak{p}) = \{0\},$$

where $\mathfrak{p} = \mathfrak{X}_- \oplus \mathfrak{X}_0$.

Proof. In the case (i), Theorem 4.12 in [17] implies that 0 is an isolated point in the spectrum of iD and, in view of the hypothesis concerning S, one immediately gets $\mathfrak{X} = \mathfrak{X}_- \oplus \mathfrak{X}_0 \oplus \mathfrak{X}_+$. In the case (ii), just use the reasoning by which the decomposition (2) in Remark 1.6 has been constructed.

In order to prove that $\omega(\mathfrak{p},\mathfrak{p}) = \{0\}$, first consider the dual operator $D^* \in \mathcal{B}(\mathfrak{X}^*)$. We have $\sigma(\mathrm{i}D^*) = \sigma(\mathrm{i}D) \subseteq S \cup (-S)$ and, in the case (i), iD^* is a normal operator with closed range, while in the case (ii) iD^* is a Dunford scalar operator whose spectral measure $E_{\mathrm{i}D^*}(\cdot)$ is given by $E_{\mathrm{i}D^*}(\Delta) = (E_{\mathrm{i}D}(\Delta))^*$ for each Borel subset Δ of $\mathbb{C}$. By the above reasoning, $\mathfrak{X}^*$ then decomposes into a direct sum of closed subspaces

$$\mathfrak{X}^* = \mathfrak{X}^*_- \oplus \mathfrak{X}^*_0 \oplus \mathfrak{X}^*_+,$$

where $\mathfrak{X}^*_\varepsilon$ is naturally isometrically isomorphic to the dual of the Banach space $\mathfrak{X}_\varepsilon$ for every $\varepsilon \in \{-,0,+\}$, while

$$\langle (\mathfrak{X}_\varepsilon)^*, \mathfrak{X}_\delta \rangle = \{0\} \quad \text{whenever } \varepsilon, \delta \in \{-,0,+\} \text{ with } \varepsilon \neq \delta, \tag{2.1}$$

where $\langle\,\cdot\,,\cdot\,\rangle : \mathfrak{X}^* \times \mathfrak{X} \to \mathbb{C}$ is the natural pairing. (See the relations between duality and spectral decompositions in Proposition 5.6 in Chapter IV of [26].)

Next define

$$\theta : \mathfrak{X} \to \mathfrak{X}^*, \quad X \mapsto \omega(X, \cdot).$$

We then get

$$\omega(X,Y) = \langle \theta(X), Y \rangle \quad \text{for every } X, Y \in \mathfrak{X}.$$

By making use of the assumed relation between ω and D, we then have

$$\begin{aligned} \langle \theta(D(X)), Y \rangle &= \omega(D(X), Y) = -\omega(X, D(Y)) \\ &= -\langle \theta(X), D(Y) \rangle = -\langle D^*(\theta(X)), Y \rangle \end{aligned}$$

for arbitrary $X, Y \in \mathfrak{X}$. Consequently

$$\theta \circ D = (-D^*) \circ \theta,$$

and this in turn implies

$$\theta(\mathfrak{X}_-) \subseteq \mathfrak{X}^*_+ \quad \text{and} \quad \theta(\mathfrak{X}_0) \subseteq \mathfrak{X}^*_0.$$

(In the case (i), this is an easy consequence of the definition of the spectral subspaces, see also Corollary 1 in Section 13 of [5]. In the case (ii), one moreover makes use of a reasoning similar to the one used, e.g., in the proof of Lemma 7.23 in Chapter IV of [26].) Then by (3) we get

$$\begin{aligned} \omega(\mathfrak{p},\mathfrak{p}) &= \langle \theta(\mathfrak{p}), \mathfrak{p} \rangle = \langle \theta(\mathfrak{X}_-) + \theta(\mathfrak{X}_0), \mathfrak{X}_- + \mathfrak{X}_0 \rangle \subseteq \langle \mathfrak{X}^*_+ + \theta(\mathfrak{X}_0), \mathfrak{X}_- + \mathfrak{X}_0 \rangle \\ &= \langle \theta(\mathfrak{X}_0), \mathfrak{X}_0 \rangle = \omega(\mathrm{Ker}\, D, \mathrm{Ker}\, D) = \{0\}. \end{aligned}$$ □

The following corollary will help us to show that the space $\mathfrak{p}$ constructed in Proposition 2.1 has the properties (C2) and (C3) needed for a complex polarization (see the Preliminaries).

Corollary 2.2. *In the setting of Proposition* 2.1, *if moreover* $S = [0,\infty)$ *and* $^- : \mathfrak{X} \to \mathfrak{X}$ *is a bounded conjugation commuting with* D, *then* $\overline{\mathfrak{X}_0} = \mathfrak{X}_0$ *and* $\overline{\mathfrak{X}_\pm} = \mathfrak{X}_\mp$.

Proof. One has just to remark that $\overline{\mathrm{i}D(X)} = -\mathrm{i}D(\overline{X})$. □

The construction in the following proposition is suggested by the construction of restricted Lie algebras (see, e.g., Exercises VII.3–4 in [20] and Definition III.1 in [22]).

Proposition 2.3. *Let $(\mathfrak{Y}, \|\cdot\|_{\mathfrak{Y}})$ and $(\mathfrak{V}, \|\cdot\|_{\mathfrak{V}})$ be two Banach spaces over the field $\mathbb{K} \in \{\mathbb{R}, \mathbb{C}\}$, and $\iota : \mathfrak{V} \to \mathfrak{Y}$ be a bounded linear one-to-one operator. For $D \in \mathcal{B}(\mathfrak{Y})$ define*

$$\mathfrak{Y}(D) := \{Y \in \mathfrak{Y} \mid D(Y) \in \operatorname{Ran} \iota\} = D^{-1}(\operatorname{Ran} \iota),$$
$$\|Y\|_{\mathfrak{Y}(D)} := \|Y\|_{\mathfrak{Y}} + \|\iota^{-1}(D(Y))\|_{\mathfrak{V}} \quad \text{for all } Y \in \mathfrak{Y}(D).$$

Then $(\mathfrak{Y}(D), \|\cdot\|_{\mathfrak{Y}(D)})$ is a Banach space and the following assertions are equivalent:

(i) *there exists $\tilde{D} \in \mathcal{B}(\mathfrak{V})$ such that $\iota \circ \tilde{D} = D \circ \iota$;*
(ii) $\operatorname{Ran} \iota \subseteq \mathfrak{Y}(D)$;
(iii) $D(\operatorname{Ran} \iota) \subseteq \operatorname{Ran} \iota$.

If (i) *holds, then we further have*

(iv) *$\tilde{D}$ is uniquely determined, in fact $\tilde{D} = \iota^{-1} \circ (D|_{\operatorname{Ran} \iota}) \circ \iota$;*
(v) *$\mathfrak{Y}(D)$ is invariant to D;*
(vi) *there exists a linear isometry $\psi : \mathfrak{Y}(D) \to \mathfrak{V} \times \mathfrak{Y}$ such that $(\tilde{D} \times D) \circ \psi = \psi \circ (D|_{\mathfrak{Y}(D)})$, where the Banach space $\mathfrak{V} \times \mathfrak{Y}$ is endowed with the sum-norm, that is $\|(V, Y)\|_{\mathfrak{V} \times \mathfrak{Y}} = \|V\|_{\mathfrak{V}} + \|Y\|_{\mathfrak{Y}}$ for every $V \in \mathfrak{V}$ and $Y \in \mathfrak{Y}$.*

Proof. The proof has several steps.

1° The proof of the fact that $(\mathfrak{Y}(D), \|\cdot\|_{\mathfrak{Y}(D)})$ is a Banach space is straightforward and we omit it.

2° In order to prove that the assertions (i)–(iii) are equivalent, first note that the implication (i) ⇒ (ii) is obvious, and (ii) ⇔ (iii) by the very definition of $\mathfrak{Y}(D)$. Hence it only remains to prove the implication (ii) ⇒ (i).

If (ii) holds, we can define

$$\tilde{D} = \iota^{-1} \circ (D|_{\operatorname{Ran} \iota}) \circ \iota = \iota^{-1} \circ D \circ \iota,$$

which is a linear map of $\mathfrak{V}$ into itself, satisfying $\iota \circ \tilde{D} = D \circ \iota$. We have to prove that the linear map $\tilde{D}$ is bounded on the Banach space $\mathfrak{V}$, and this follows by the Closed Graph Theorem.

3° Next assume that the equivalent assertions (i)–(iii) hold. To prove (iv), just recall that ι is one-to-one, hence $\iota \circ \tilde{D} = D \circ \iota$ together with $D(\operatorname{Ran} \iota) \subseteq \operatorname{Ran} \iota$ imply $\tilde{D} = \iota^{-1} \circ (D|_{\operatorname{Ran} \iota}) \circ \iota$. For (v) note that, by the very definition, the image of $\mathfrak{Y}(D)$ under D is contained in $\operatorname{Ran} \iota$ and then use (ii).

4° Finally, for proving (vi), define

$$\psi : \mathfrak{Y}(D) \to \mathfrak{V} \times \mathfrak{Y}, \quad Y \mapsto \big(\iota^{-1}(D(Y)), Y\big).$$

Then ψ is obviously linear and for every $Y \in \mathfrak{Y}(D)$ we have

$$\|\psi(Y)\|_{\mathfrak{V} \times \mathfrak{Y}} = \|\iota^{-1}(D(Y))\|_{\mathfrak{V}} + \|Y\|_{\mathfrak{Y}} = \|Y\|_{\mathfrak{Y}(D)}$$

as well as

$$
\begin{aligned}
(\tilde{D} \times D)(\psi(Y)) &= \big(\tilde{D}(\iota^{-1}(D(Y))), D(Y)\big) \\
&\overset{\text{(i)}}{=} \big(\iota^{-1}(D(D(Y))), D(Y)\big) = \psi(D(Y)).
\end{aligned}
$$

□

Corollary 2.4. *In the setting of Proposition* 2.3, *assume* $\mathbb{K} = \mathbb{C}$. *If* (i) *holds and* D *and* $\tilde{D}$ *are either both of them Hermitian operators, or both of them Hermitian operators with closed ranges, or both of them Dunford scalar operators with* $\sigma(D) \cup \sigma(\tilde{D}) \subseteq \mathrm{i}\mathbb{R}$, *then* $D|_{\mathfrak{Y}(D)}$ *has the same property as a bounded linear operator on the Banach space* $\mathfrak{Y}(D)$.

Proof. Use Proposition 2.3(vi). In the case of Dunford scalar operators see also Section 4 in Chapter IV of [9]. □

We now note a topological property which $\mathfrak{Y}(D)$ can inherit from $\mathfrak{Y}$ and $\mathfrak{V}$.

Corollary 2.5. *In the setting of Proposition* 2.3, *if both* $\mathfrak{Y}$ *and* $\mathfrak{V}$ *are reflexive Banach spaces, then the Banach space* $\mathfrak{Y}(D)$ *is reflexive as well.*

Proof. Use Proposition 2.3(vi). □

Remark 2.6. In the setting of Proposition 2.3, we always have $\operatorname{Ker} D \subseteq \mathfrak{Y}(D)$ and $\|Y\|_{\mathfrak{Y}(D)} = \|Y\|_{\mathfrak{Y}}$ for every $Y \in \operatorname{Ker} D$. In particular, if $D = 0$, then $(\mathfrak{Y}(D), \|\cdot\|_{\mathfrak{Y}(D)}) = (\mathfrak{Y}, \|\cdot\|_{\mathfrak{Y}})$.

We also note that Proposition 2.3 can be extended to the case when the hypothesis $\operatorname{Ker} \iota = \{0\}$ is replaced by the more general hypothesis that there exists a closed subspace $\mathfrak{V}_0$ of $\mathfrak{V}$ such that $\mathfrak{V}$ is the direct sum of $\operatorname{Ker} \iota$ and $\mathfrak{V}_0$. To this end, in the statement and the proof of Proposition 2.3 one just has to replace throughout ι^{-1} by $\big(\iota|_{\mathfrak{V}_0}\big)^{-1}$. □

We now describe a natural framework where the Banach space $\mathfrak{Y}(D)$ from Proposition 2.3 has a structure of Banach-Lie algebra.

Proposition 2.7. *In the setting of Proposition* 2.3, *assume that* $\mathfrak{Y}$ *is a Banach-Lie algebra, that* D *is a derivation of* $\mathfrak{Y}$, *and that we have a bounded representation* $\rho\colon \mathfrak{Y} \to \mathcal{B}(\mathfrak{V})$ *such that* $\iota \circ \rho(Y) = (\operatorname{ad} Y) \circ \iota$ *for every* $Y \in \mathfrak{Y}$. *Then*

(j) $\mathfrak{Y}(D)$ *is a Lie subalgebra of* $\mathfrak{Y}$ *which is a Banach-Lie algebra with the norm* $\|\cdot\|_{\mathfrak{Y}(D)}$;
(jj) $\operatorname{Ran} \iota$ *is an ideal of* $\mathfrak{Y}$;
(jjj) *if* (i) *in Proposition* 2.3 *holds, then* $D|_{\mathfrak{Y}(D)}$ *is a bounded derivation of the Banach-Lie algebra* $\mathfrak{Y}(D)$; *more precisely,* $\|D|_{\mathfrak{Y}(D)}\| \le \max\{\|\tilde{D}\|, \|D\|\}$.

Proof. (j) For arbitrary $Y_1, Y_2 \in \mathfrak{Y}(D)$ there exist $V_1, V_2 \in \mathfrak{V}$ such that $D(Y_j) = \iota(V_j)$ for $j \in \{1,2\}$. Then

$$\begin{aligned} D([Y_1, Y_2]) &= [D(Y_1), Y_2] + [Y_1, D(Y_2)] \\ &= [\iota(V_1), Y_2] + [Y_1, \iota(V_2)] \\ &= -((\operatorname{ad} Y_2) \circ \iota)(V_1) + ((\operatorname{ad} Y_1) \circ \iota)(V_2) \\ &= -(\iota \circ (\rho(Y_2)))(V_1) + (\iota \circ (\rho(Y_1)))(V_2) \\ &= \iota(-(\rho(Y_2))(V_1) + (\rho(Y_1))(V_2)). \end{aligned}$$

In particular $D([Y_1, Y_2]) \in \operatorname{Ran} \iota$, that is $[Y_1, Y_2] \in \mathfrak{Y}(D)$. Thus $\mathfrak{Y}(D)$ is indeed a Lie subalgebra of $\mathfrak{Y}$.

On the other hand by the above computation we have

$$\iota^{-1}(D([Y_1, Y_2])) = -(\rho(Y_2))(V_1) + (\rho(Y_1))(V_2), \tag{2.2}$$

so that

$$\begin{aligned} \|[Y_1, Y_2]\|_{\mathfrak{Y}(D)} &= \|[Y_1, Y_2]\|_{\mathfrak{Y}} + \|\iota^{-1}(D([Y_1, Y_2]))\|_{\mathfrak{V}} \\ &= \|[Y_1, Y_2]\|_{\mathfrak{Y}} + \| - (\rho(Y_2))(V_1) + (\rho(Y_1))(V_2)\|_{\mathfrak{V}} \\ &\le M\|Y_1\|_{\mathfrak{Y}} \cdot \|Y_2\|_{\mathfrak{Y}} + \|\rho\| \cdot \|Y_2\|_{\mathfrak{Y}} \cdot \|V_1\|_{\mathfrak{V}} + \|\rho\| \cdot \|Y_1\|_{\mathfrak{Y}} \cdot \|V_2\|_{\mathfrak{V}} \\ &\le C(\|Y_1\|_{\mathfrak{Y}} + \|V_1\|_{\mathfrak{V}})(\|Y_2\|_{\mathfrak{Y}} + \|V_2\|_{\mathfrak{V}}) \\ &= C\|Y_1\|_{\mathfrak{Y}(D)} \cdot \|Y_2\|_{\mathfrak{Y}(D)}, \end{aligned}$$

where $C = \max\{M, \|\rho\|\}$, M being the norm of the adjoint representation $\operatorname{ad} : \mathfrak{Y} \to \mathcal{B}(\mathfrak{Y})$.

(jj) For every $Y \in \mathfrak{Y}$ and $V \in \mathfrak{V}$ we have

$$[Y, \iota(V)] = ((\operatorname{ad} Y) \circ \iota)(V) = (\iota \circ (\rho(Y)))(V) \in \operatorname{Ran} \iota.$$

(jjj) One applies (vi) in Proposition 2.3 together with the observation that, since $\mathfrak{V} \times \mathfrak{Y}$ is endowed with the sum-norm, we have $\|\tilde{D} \times D\| = \max\{\|\tilde{D}\|, \|D\|\}$. □

We are now going to show how Theorem 1.4 and Remark 1.6 can be used to construct complex polarizations in certain 2-cocycles of a special class of real Banach-Lie algebras of the type of $\mathfrak{Y}(D)$ from Proposition 2.7 above. The appropriate framework in order to do so will be provided by Banach-Lie algebras for which the coadjoint representation is "embedded" into the adjoint one (see hypothesis 2° below). The examples we should have in mind are the following.

(A) The real L^*-algebras. In this case the coadjoint representation is equivalent to the adjoint one by means of the topological isomorphism

$$\iota : \mathfrak{g}^* \to \mathfrak{g}$$

defined by $\iota^{-1}(X) = \langle X, \cdot \rangle$ for every $X \in \mathfrak{g}$, where $\langle \cdot, \cdot \rangle$ stands for the scalar product of the real L^*-algebra $\mathfrak{g}$. (See also Proposition 2.12 below.)

(B) The unitary real form $\mathfrak{u}_p$ of the complex Banach-Lie algebra $\mathcal{C}_p(\mathcal{H})$ for $2 \leq p < \infty$, that is

$$\mathfrak{u}_p = \{X \in \mathcal{C}_p(\mathcal{H}) \mid \mathrm{i}X \text{ is self-adjoint}\}.$$

It is well known that the topological dual of $\mathcal{C}_p(\mathcal{H})$ is naturally isomorphic to $\mathcal{C}_q(\mathcal{H})$, where $q = p/(p-1)$. Since $2 \leq p < \infty$, we have $q \leq p$, and thus $\mathcal{C}_q(\mathcal{H})$ is naturally embedded into $\mathcal{C}_p(\mathcal{H})$. On the level of the real forms, we get that the topological dual of $\mathfrak{u}_p$ is isomorphic to $\mathfrak{u}_q$, which is embedded into $\mathfrak{u}_p$. In particular we have an injective operator

$$\iota \colon (\mathfrak{u}_p)^* \to \mathfrak{u}_p$$

which obviously intertwines the coadjoint representation of $\mathfrak{u}_p$ and the adjoint one.

In order to simplify the statements we are going to prove, let us now make some hypotheses which will be assumed until the end of the present section.

1° Let $\mathfrak{g}$ be a real Banach-Lie algebra whose underlying Banach space is reflexive. We denote by

$$\langle \cdot, \cdot \rangle \colon \mathfrak{g}^* \times \mathfrak{g} \to \mathbb{R}$$

the natural pairing and perform the natural identification $\mathfrak{g} = (\mathfrak{g}^*)^*$.

2° Consider a bounded linear operator $\iota \colon \mathfrak{g}^* \to \mathfrak{g}$ which *intertwines the coadjoint representation of* $\mathfrak{g}$ *and the adjoint one*, that is, the diagram

$$\begin{array}{ccc} \mathfrak{g}^* & \xrightarrow{\ \iota\ } & \mathfrak{g} \\ {\scriptstyle \mathrm{ad}^* X}\downarrow & & \downarrow{\scriptstyle \mathrm{ad}\, X} \\ \mathfrak{g}^* & \xrightarrow{\ \iota\ } & \mathfrak{g} \end{array}$$

is commutative for every $X \in \mathfrak{g}$. Furthermore assume that $\operatorname{Ker} \iota = \{0\}$ and $\iota^* = \iota$, where $\iota^* \colon \mathfrak{g}^* \to (\mathfrak{g}^*)^* = \mathfrak{g}$ is the operator dual to ι.

3° Pick $D \in \operatorname{Der}(\mathfrak{g})$ such that

$$\iota \circ (-D^*) = D \circ \iota,$$

where $D^* \colon \mathfrak{g}^* \to \mathfrak{g}^*$ stands of course for the operator dual to D. Next denote (see Proposition 2.7)

$$\mathfrak{g}(D) = D^{-1}(\operatorname{Ran} \iota)$$

and define the bilinear form

$$\omega \colon \mathfrak{g}(D) \times \mathfrak{g}(D) \to \mathbb{R}, \quad \omega(Y_1, Y_2) = \langle \iota^{-1}(D(Y_1)), Y_2 \rangle.$$

4° Consider the complexification $\mathfrak{A} = \mathfrak{g}_{\mathbb{C}}$ of $\mathfrak{g}$ and denote by $Z \mapsto \overline{Z}$ the conjugation of $\mathfrak{A}$ corresponding to $\mathfrak{g}$, so that $\mathfrak{g} = \{Z \in \mathfrak{A} \mid Z = \overline{Z}\}$. We will always use the same notation both for a linear operator or (multi)linear functional on a real vector space and for its extension to the complexified space. In particular, we have the complex Banach-Lie algebra $\mathfrak{A}(D) = D^{-1}(\operatorname{Ran} \iota)$ (see Proposition 2.7), which is the complexification of $\mathfrak{g}(D)$.

5° Assume that

$$\langle V, \iota(V)\rangle \leq 0 \text{ for every } V \in \mathfrak{g}^*$$

and $\langle V, \iota(V)\rangle = 0$ if and only if $V = 0$.

We will see that, under favorable circumstances, the bilinear form ω in 3° is a continuous 2-cocycle of the Banach-Lie algebra $\mathfrak{g}(D)$. To begin with, let us prove the continuity of ω.

Lemma 2.8. *There exists a positive constant M such that*

$$|\omega(Y_1, Y_2)| \leq M\|Y_1\|_{\mathfrak{g}(D)}\|Y_2\|_{\mathfrak{g}(D)}$$

for every $Y_1, Y_2 \in \mathfrak{g}(D)$.

Proof. In view of the definition of ω, it obviously suffices to prove that the linear operator

$$\iota^{-1} \circ D : \left(\mathfrak{g}(D), \|\cdot\|_{\mathfrak{g}(D)}\right) \to \left(\mathfrak{g}^*, \|\cdot\|_{\mathfrak{g}^*}\right)$$

is bounded. And this follows by the very definition of $\|\cdot\|_{\mathfrak{g}(D)}$ (see Proposition 2.3). □

We now prove other basic properties of the bilinear form ω. The first of these properties will allow us to make use of the Proposition 2.1, while the other two properties are cocycle features of ω.

Proposition 2.9. *For every $Y_1, Y_2 \in \mathfrak{g}(D)$ and $Z_1, Z_2 \in \operatorname{Ker} D + \operatorname{Ran}\left(D|_{\mathfrak{g}(D)}\right)$ we have*

(i) $\omega(D(Y_1), Y_2) = -\omega(Y_1, D(Y_2))$, *and*
(ii) $\omega(Z_1, Z_2) = -\omega(Z_2, Z_1)$.

If we assume moreover that the operator $\mathrm{i}D$ is Hermitian on $\mathfrak{A}$, then for every $Z_1, Z_2, Z_3 \in \mathfrak{g}(D)$ we have the above relation (ii) *and moreover*

(iii) $\omega([Z_1, Z_2], Z_3) + \omega([Z_2, Z_3], Z_1) + \omega([Z_3, Z_1], Z_2) = 0$.

Proof. (i) Denote $V_j = \iota^{-1}(D(Y_j)) \in \mathfrak{g}^*$ for $j \in \{1, 2\}$. Then $D(Y_j) = \iota(V_j)$, so $D^2(Y_j) = D(\iota(V_j)) = -\iota(D^*(V_j))$ (see the above hypothesis 3°). Consequently

$$\omega(D(Y_1), Y_2) = \langle \iota^{-1}(D^2(Y_1)), Y_2\rangle = -\langle D^*(V_1), Y_2\rangle = -\langle V_1, D(Y_2)\rangle.$$

On the other hand,

$$\omega(Y_1, D(Y_2)) = \langle \iota^{-1}(D(Y_1)), D(Y_2)\rangle = \langle V_1, D(Y_2)\rangle,$$

and the desired equality follows.

(ii) Since ω is bilinear, it obviously suffices to prove that $\omega(Z, Z) = 0$ for an arbitrary $Z \in \operatorname{Ker} D + \operatorname{Ran}\left(D|_{\mathfrak{g}(D)}\right)$.

Given such a Z, there exist $Z_0, \tilde{Z} \in \mathfrak{g}(D)$ such that

$$Z = D(\tilde{Z}) + Z_0$$

and $D(Z_0) = 0$. Taking into account first the bilinearity of ω and then the already proved property (i), we get

$$\begin{aligned}\omega(Z,Z) &= \omega(D(\tilde{Z}), D(\tilde{Z})) + \omega(D(\tilde{Z}), Z_0) + \omega(Z_0, D(\tilde{Z})) + \omega(Z_0, Z_0)\\ &= \omega(D(\tilde{Z}), D(\tilde{Z})) - \omega(\tilde{Z}, D(Z_0)) - \omega(D(Z_0), \tilde{Z}) - \langle \iota^{-1}(D(Z_0)), Z_0\rangle\\ &= \omega(D(\tilde{Z}), D(\tilde{Z})).\end{aligned}$$

Furthermore, since $\tilde{Z} \in \mathfrak{g}(D)$, there exists $\tilde{V} \in \mathfrak{g}^*$ such that $D(\tilde{Z}) = \iota(\tilde{V})$. Thus

$$D^2(\tilde{Z}) = D(\iota(\tilde{Z})) = -\iota(D^*(\tilde{V}))$$

and we get

$$\begin{aligned}\omega(D(\tilde{Z}), D(\tilde{Z})) &= \langle \iota^{-1}(D^2(\tilde{Z})), D(\tilde{Z})\rangle = -\langle D^*(\tilde{V}), D(\tilde{Z})\rangle = -\langle \tilde{V}, D^2(\tilde{Z})\rangle\\ &= \langle \tilde{V}, \iota(D^*(\tilde{V}))\rangle = \langle D^*(\tilde{V}), \iota(\tilde{V})\rangle\\ &= \langle D^*(\tilde{V}), D(\tilde{Z})\rangle = -\langle \iota^{-1}(D^2(\tilde{Z})), D(\tilde{Z})\rangle\\ &= -\omega(D(\tilde{Z}), D(\tilde{Z})).\end{aligned}$$

where the fifth equality follows since $\iota^* = \iota$ by the hypothesis 2°. Consequently $\omega(D(\tilde{Z}), D(\tilde{Z})) = 0$ and thus

$$\omega(Z,Z) = \omega(D(\tilde{Z}), D(\tilde{Z})) = 0.$$

(iii) Since iD is Hermitian, its dual iD^* is also Hermitian, hence Corollary 2.4 above shows that $\mathrm{i}D|_{\mathfrak{A}(D)}$ is Hermitian as well. Then Corollary 4.5 in [17] implies in particular that the subspace

$$\mathrm{Ker}\, D + \mathrm{Ran}\, (D|_{\mathfrak{A}(D)})$$

is dense in $\mathfrak{A}(D)$, where D is viewed as an operator on $\mathfrak{A}(D)$ (see hypothesis 4°). Since D commutes with the conjugation $^-$ (see 4° again), it then easily follows that $\mathrm{Ker}\, D + \mathrm{Ran}\, \big(D|_{\mathfrak{g}(D)}\big)$ is dense in $\big(\mathfrak{g}(D), \|\cdot\|_{\mathfrak{g}(D)}\big)$, hence the already proved property (ii) extends to all $Z_1, Z_2 \in \mathfrak{g}(D)$. On the other hand, by the same reason, in view of Lemma 2.8 it suffices to prove the property (jjj) only for $Z_1, Z_2, Z_3 \in \mathrm{Ker}\, D + \mathrm{Ran}\, \big(D|_{\mathfrak{g}(D)}\big)$.

Since $Z_j \in \mathfrak{g}(D)$, there exists $W_j \in \mathfrak{g}^*$ such that $D(Z_j) = \iota(W_j)$ for each $j \in \{1,2,3\}$. Now recall the relation (2.2) from the proof of Proposition 2.7. With the present notations, it says that

$$\iota^{-1}(D([Z_i, Z_j])) = -(\mathrm{ad}^* Z_j)(W_i) + (\mathrm{ad}^* Z_i)(W_j)$$

for every $i, j \in \{1, 2, 3\}$. It then follows that

$$\begin{aligned}
&\omega([Z_1, Z_2], Z_3) + \omega([Z_2, Z_3], Z_1) + \omega([Z_3, Z_1], Z_2) \\
&\quad = \langle (\mathrm{ad}^* Z_1)(W_2) - (\mathrm{ad}^* Z_2)(W_1), Z_3 \rangle \\
&\qquad + \langle (\mathrm{ad}^* Z_2)(W_3) - (\mathrm{ad}^* Z_3)(W_2), Z_1 \rangle \\
&\qquad + \langle (\mathrm{ad}^* Z_3)(W_1) - (\mathrm{ad}^* Z_1)(W_3), Z_2 \rangle \\
&\quad = \langle W_2, [Z_3, Z_1] \rangle - \langle W_1, [Z_3, Z_2] \rangle \\
&\qquad + \langle W_3, [Z_1, Z_2] \rangle - \langle W_2, [Z_1, Z_3] \rangle \\
&\qquad + \langle W_1, [Z_2, Z_3] \rangle - \langle W_3, [Z_2, Z_1] \rangle \\
&\quad = 2\big(\langle W_1, [Z_2, Z_3] \rangle + \langle W_2, [Z_3, Z_1] \rangle + \langle W_3, [Z_1, Z_2] \rangle \\
&\quad = 2\big(\omega(Z_1, [Z_2, Z_3]) + \omega(Z_2, [Z_3, Z_1]) + \omega(Z_3, [Z_1, Z_2])\big) \\
&\quad \overset{\text{(jj)}}{=} -2\big(\omega([Z_2, Z_3], Z_1) + \omega([Z_3, Z_1], Z_2) + \omega([Z_1, Z_2], Z_3)\big).
\end{aligned}$$

Thus $3\big(\omega([Z_1, Z_2], Z_3) + \omega([Z_2, Z_3], Z_1) + \omega([Z_3, Z_1], Z_2)\big) = 0$ and we are done. □

We can now prove the main result of the present paper. It extends Theorem VII.6 in [20].

Theorem 2.10. *Under the above hypotheses* 1°–5°, *if* $\mathrm{i}D$ *is a Dunford scalar operator on* $\mathfrak{A}$ *with the spectral measure* $E_{\mathrm{i}D}(\cdot)$ *and with*

$$\sigma(\mathrm{i}D) \subseteq (-\infty, -\alpha] \cup \{0\} \cup [\alpha, \infty)$$

for some $\alpha > 0$, *then* $\mathrm{i}D|_{\mathfrak{A}(D)}$ *is a Dunford scalar operator on* $\mathfrak{A}(D)$ *whose spectral measure is denoted by* $E_{\mathrm{i}D|_{\mathfrak{A}(D)}}(\cdot)$, *and*

$$\mathfrak{p} = \mathrm{Ran}\, E_{\mathrm{i}D|_{\mathfrak{A}(D)}}\big((-\infty, 0]\big)$$

is a weak Kähler polarization of the real Banach-Lie algebra $\mathfrak{g}(D)$ *in the cocycle* ω.

Proof. First note that $\mathrm{i}D$ is in particular a Hermitian operator on $\mathfrak{A}$ with respect to some equivalent norm of $\mathfrak{A}$, hence ω is a continuous 2-cocycle of $\mathfrak{g}(D)$ by Proposition 2.9.

Next note that $\mathrm{i}D|_{\mathfrak{A}(D)}$ is a Dunford scalar operator by Corollary 2.4. Moreover we have $\sigma(\mathrm{i}D|_{\mathfrak{A}(D)}) \subseteq \mathbb{R}$ since $\mathrm{i}D|_{\mathfrak{A}(D)}$ is similar to the restriction of $\mathrm{i}D^* \times \mathrm{i}D : \mathfrak{A}^* \times \mathfrak{A} \to \mathfrak{A}^* \times \mathfrak{A}$ and $\sigma(\mathrm{i}D^* \times \mathrm{i}D) = \sigma(\mathrm{i}D^*) \cup \sigma(\mathrm{i}D) \subseteq \mathbb{R}$. (See, e.g., Lemma 4.8 in Chapter IV of [26] for the relation between the spectrum of an operator and the spectrum of its restriction to a closed invariant subspace.)

In view of Proposition 2.9, we can now apply Proposition 2.1(ii) with $\mathfrak{X} = \mathfrak{A}(D)$ and $S = \{0\} \cup [\alpha, \infty)$ to get a decomposition

$$\mathfrak{A}(D)\ (= \mathfrak{X}) \ = \mathfrak{X}_- \oplus \mathfrak{X}_0 \oplus \mathfrak{X}_+$$

where $\mathfrak{p} = \mathfrak{X}_- \oplus \mathfrak{X}_0$, $\omega(\mathfrak{p}, \mathfrak{p}) = \{0\}$, and $\mathfrak{p}$ is a closed subalgebra of the Banach-Lie algebra $\mathfrak{A}(D)$ (see Remark 1.6). By Corollary 2.2 it further follows that

$$\mathfrak{p} \cap \overline{\mathfrak{p}} = \mathfrak{X}_0 \quad \text{and} \quad \mathfrak{p} + \overline{\mathfrak{p}} = \mathfrak{A}(D).$$

Thus, for $\mathfrak{p}$ to be a Kähler polarization of $\mathfrak{g}(D)$ in ω, we have to prove that

$$\mathrm{i}\omega(Z,\overline{Z}) > 0 \quad \text{whenever } 0 \neq Z \in \mathfrak{X}_-. \tag{2.3}$$

To this end, first note the commutative diagrams

$$\begin{array}{ccc} \mathfrak{A}^* & \xrightarrow{\iota} & \mathfrak{A}(D) \\ {\scriptstyle -\mathrm{i}D^*}\downarrow & & \downarrow{\scriptstyle \mathrm{i}D|_{\mathfrak{A}(D)}} \\ \mathfrak{A}^* & \xrightarrow{\iota} & \mathfrak{A}(D) \end{array} \qquad \text{and} \qquad \begin{array}{ccc} \mathfrak{A}^* & \xrightarrow{\iota} & \mathfrak{A} \\ {\scriptstyle -\mathrm{i}D^*}\downarrow & & \downarrow{\scriptstyle \mathrm{i}D} \\ \mathfrak{A}^* & \xrightarrow{\iota} & \mathfrak{A} \end{array}$$

(see Proposition 2.3(i)–(ii), with $\mathfrak{V} = \mathfrak{A}^*$, $\mathfrak{Y} = \mathfrak{A}$ and $\tilde{D} = -D^*$). If we denote the spectral measures of $\mathrm{i}D$, $\mathrm{i}D^*$, $-\mathrm{i}D^*$ and $\mathrm{i}D|_{\mathfrak{A}(D)}$ by $E_{\mathrm{i}D}$, $E_{\mathrm{i}D^*}$, $E_{-\mathrm{i}D^*}$ and $E_{\mathrm{i}D|_{\mathfrak{A}(D)}}$, respectively, then we have the following relations

(j) $\quad E_{\mathrm{i}D^*}(\Delta) = (E_{\mathrm{i}D}(\Delta))^*$,

(jj) $\quad E_{\mathrm{i}D|_{\mathfrak{A}(D)}}(\Delta) \circ \iota = \iota \circ E_{-\mathrm{i}D^*}(\Delta) = \iota \circ E_{\mathrm{i}D^*}(-\Delta)$,

(jjj) $\quad E_{\mathrm{i}D}(\Delta) \circ \iota = \iota \circ E_{-\mathrm{i}D^*}(\Delta) = \iota \circ E_{\mathrm{i}D^*}(-\Delta)$

for every Borel subset Δ of $\mathbb{R}$ (see Section 4 in Chapter IV of [9]). In particular, if we denote (as in the proof of Proposition 2.1)

$$\mathfrak{A}^*_{\pm} = \operatorname{Ran} E_{\mathrm{i}D^*}\big(\pm [\alpha,\infty)\big),\ \mathfrak{A}^*_0 = \operatorname{Ran} E_{\mathrm{i}D^*}\big(\{0\}\big) = \operatorname{Ker} D^*,$$

then (jj) implies

$$\iota(\mathfrak{A}^*_\varepsilon) \subseteq \mathfrak{X}_{-\varepsilon}, \quad \text{in fact } \iota^{-1}(\mathfrak{X}_{-\varepsilon}) = \mathfrak{A}^*_\varepsilon \text{ for } \varepsilon \in \{-,0,+\}, \tag{2.4}$$

because $\mathfrak{A}^* = \mathfrak{A}^*_- \oplus \mathfrak{A}^*_0 \oplus \mathfrak{A}^*_+$ and $\mathfrak{X} = \mathfrak{X}_- \oplus \mathfrak{X}_0 \oplus \mathfrak{X}_+$.

Since $\mathfrak{X}_- = \operatorname{Ran} E_{\mathrm{i}D|_{\mathfrak{A}(D)}}\big((-\infty,-\alpha]\big)$, it easily follows that $D(\mathfrak{X}_-) = \mathfrak{X}_-$. But $\mathfrak{X}_- \subseteq \mathfrak{A}(D)$, hence

$$\mathfrak{X}_- = D(\mathfrak{X}_-) \subseteq D\big(\mathfrak{A}(D)\big) \subseteq \operatorname{Ran} \iota.$$

On the other hand, $\iota^{-1}(\mathfrak{X}_-) = \mathfrak{A}^*_+$ (see (2.4)), so that $\mathfrak{X}_- \subseteq \iota(\mathfrak{A}^*_+)$. This implies that $\iota(\mathfrak{A}^*_+) = \mathfrak{X}_-$, hence (2.3) will follow from

$$\mathrm{i}\omega\big(\iota(V),\overline{\iota(V)}\big) > 0 \quad \text{whenever } 0 \neq V \in \mathfrak{A}^*_+. \tag{2.5}$$

To prove this fact, fix $V \in \mathfrak{A}^*_+ \setminus \{0\}$ and note that

$$\mathrm{i}D\big(\iota(V)\big) = \int_{-\infty}^{-\alpha} tE_{\mathrm{i}D}(\mathrm{d}t)\big(\iota(V)\big),$$

because $\iota(V) \in \operatorname{Ran} E_{\mathrm{i}D}\big((-\infty,-\alpha]\big)$ by (2.4). On the other hand,

$$\begin{aligned} \omega\big(\iota(V),\overline{\iota(V)}\big) &= \langle \iota^{-1}D\iota(V),\overline{\iota(V)}\rangle = -\langle D^*(V),\iota(\overline{V})\rangle \\ &= \langle \overline{V},\big(\iota\circ(-D^*)\big)(V)\rangle = \langle \overline{V}, D\big(\iota(V)\big)\rangle, \end{aligned}$$

where the last but one equality follows since $\iota^* = \iota$. Consequently

$$\begin{aligned}
\mathrm{i}\omega\big(\iota(V),\overline{\iota(V)}\big) &= \langle \overline{V}, \mathrm{i}D\big(\iota(V)\big)\rangle \\
&= \int_{-\infty} -\alpha t\langle \overline{V}, E_{\mathrm{i}D}(\,\mathrm{d}t)\big(\iota(V)\big)\rangle \\
&= \int_{-\infty}^{-\alpha} t\langle \overline{V}, \big(E_{\mathrm{i}D}(\,\mathrm{d}t)\big)^2\big(\iota(V)\big)\rangle \\
&= \int_{-\infty}^{-\alpha} t\langle E_{\mathrm{i}D^*}(\,\mathrm{d}t)\overline{V}, \iota\big(E_{-\mathrm{i}D^*}(\,\mathrm{d}t)V\big)\rangle \qquad \big(\text{by (j) and (jj)}\big) \\
&= \int_{-\infty}^{-\alpha} t\langle \overline{E_{-\mathrm{i}D^*}(\,\mathrm{d}t)V}, \iota\big(E_{-\mathrm{i}D^*}(\,\mathrm{d}t)V\big)\rangle \\
&\geq -\alpha\langle \overline{E_{-\mathrm{i}D^*}\big((-\infty,-\alpha]\big)V}, \iota\big(E_{-\mathrm{i}D^*}\big((-\infty,-\alpha]\big)V\big)\rangle \\
&= -\alpha\langle \overline{V}, \iota(V)\rangle \\
&> 0.
\end{aligned}$$

(The first inequality follows since $\langle \overline{E_{-\mathrm{i}D^*}(\cdot)V}, \iota(E_{-\mathrm{i}D^*}(\cdot)V)\rangle$ is a non-negative measure by hypothesis 5°, and the last inequality is an easy consequence of the same hypothesis 5° preceding Lemma 2.8.) Thus (2.5) is proved and we are done. □

Remark 2.11. Under the hypothesis of Theorem 2.10, $\operatorname{Ran} D$ is closed in $\mathfrak{A}$ and we have the following description of $\mathfrak{A}(D)$:

$$\mathfrak{A}(D) = \operatorname{Ker} D \oplus \iota(\operatorname{Ran} D^*).$$

To prove this fact we use the notation of the proof of Theorem 2.10. Recall that $\iota(\mathfrak{A}^*_\pm) = \mathfrak{X}_\pm$ and

$$\sigma(\mathrm{i}D^*) = \sigma(\mathrm{i}D) \subseteq (-\infty, -\alpha] \cup \{0\} \cup [\alpha, \infty).$$

On the other hand, $\mathfrak{X}_0 = \operatorname{Ker} D$, thus

$$\begin{aligned}
\mathfrak{A}(D) &= \mathfrak{X}_- \oplus \mathfrak{X}_0 \oplus \mathfrak{X}_+ = \iota(\mathfrak{A}^*_+) \oplus \operatorname{Ker} D \oplus \iota(\mathfrak{A}^*_-) \\
&= \operatorname{Ker} D \oplus \iota(\mathfrak{A}^*_- \oplus \mathfrak{A}^*_+) = \operatorname{Ker} D \oplus \iota(\operatorname{Ran} D^*).
\end{aligned}$$ □

We conclude the present section by a result connected with Example (A) preceding the above hypotheses 1°–5°. It shows that the framework described by the hypotheses 1°–5° reduces to the special situation when $\mathfrak{g}$ is a compact real L^*-algebra (see, e.g., Section 2 in [2]) precisely when ι in hypothesis 2° is onto.

Proposition 2.12. *Under the hypotheses* 1°–5°, *if* $\iota : \mathfrak{g}^* \to \mathfrak{g}$ *is onto, then the Banach-Lie algebra* $\mathfrak{g}$ *is a compact real* L^**-algebra.*

Proof. The proof has two stages.

I) At this stage we consider a real Banach space $\mathfrak{Z}$ and denote as usual by $\langle\cdot,\cdot\rangle$ the natural duality pairing $\mathfrak{Z}^* \times \mathfrak{Z} \to \mathbb{R}$. We prove that, if there exists a bounded invertible map $\varphi : \mathfrak{Z} \to \mathfrak{Z}^*$ such that the bilinear form $(\cdot \mid \cdot)$,

$$\mathfrak{Z} \times \mathfrak{Z} \to \mathbb{R}, \qquad (X, Y) \mapsto (X \mid Y) := \langle \varphi(X), Y \rangle,$$

is a real scalar product, then this scalar product defines the topology of $\mathfrak{Z}$, thus turning it into a real Hilbert space.

What we have to prove is that, for some positive constants m and M, we have

$$m\|X\| \le (X \mid X)^{1/2} \le M\|X\| \quad \text{whenever } X \in \mathfrak{Z}. \tag{2.6}$$

The right-hand side of this inequality is a consequence of the continuity of φ; more precisely,

$$(Y \mid Y)^{1/2} \le \|\varphi\|^{1/2}\|Y\| \quad \text{whenever } Y \in \mathfrak{Z}.$$

As for the left-hand side, note that the Schwartz inequality for $(\cdot \mid \cdot)$ and then the above inequality imply

$$\begin{aligned}\|\varphi(X)\| &= \sup_{0\ne Y\in\mathfrak{Z}} \frac{|\langle\varphi(X), Y\rangle|}{\|Y\|} = \sup_{0\ne Y\in\mathfrak{Z}} \frac{|(X \mid Y)|}{\|Y\|} \le \sup_{0\ne Y\in\mathfrak{Z}} \frac{(X \mid X)^{1/2}(Y \mid Y)^{1/2}}{\|Y\|} \\ &= (X \mid X)^{1/2} \sup_{0\ne Y\in\mathfrak{Z}} \frac{(Y \mid Y)^{1/2}}{\|Y\|} \le \|\varphi\|^{1/2}(X \mid X)^{1/2}\end{aligned}$$

for every $X \in \mathfrak{Z}$. Since $\|X\| \le \|\varphi^{-1}\| \cdot \|\varphi(X)\|$, we finally get the desired inequality (2.6) with $m = \|\varphi^{-1}\|^{-1} \cdot \|\varphi\|^{-1/2}$ and $M = \|\varphi\|^{1/2}$.

II) We now come back to the proof. Note that the hypotheses 2° and 5° allow us to apply the result of stage I) with $\mathfrak{Z} = \mathfrak{g}$ and $\varphi = -\iota^{-1}$ to deduce that the scalar product $(\cdot \mid \cdot)$ defined on $\mathfrak{g}$ by

$$(X \mid Y) = -\langle \iota^{-1}(X), Y \rangle \quad \text{whenever } X, Y \in \mathfrak{g}$$

defines the topology of $\mathfrak{g}$. (The symmetry of $(\cdot \mid \cdot)$ follows by $\iota^* = \iota$ from hypothesis 2°, while its positive definiteness follows by the hypothesis 5°.)

Since for every $X, Y, Z \in \mathfrak{g}$ we have by the intertwining property of ι from the hypothesis 2°

$$\begin{aligned}([X, Y] \mid Z) &= -\langle(\iota^{-1} \circ \operatorname{ad} X)(Y) \mid Z\rangle = -\langle\big((\operatorname{ad}^* X) \circ \iota^{-1}\big)(Y), Z\rangle \\ &= \langle \iota^{-1}(Y), (\operatorname{ad} X)Z\rangle = -(Y \mid [X, Z]),\end{aligned}$$

it follows that $\mathfrak{g}$ is indeed a compact real L^*-algebra when endowed with this scalar product. □

3. Classical Banach-Lie algebras

This last section is essentially devoted to applying Theorem 2.10 in the special situation of the example (B) preceding the hypotheses 1°–5° in Section 2 (see Example 3.5 below). We denote by $\mathcal{H}$ a complex Hilbert space and by $\{\mathcal{C}_p(\mathcal{H})\}_{1\le p\le\infty}$ the family of Schatten-von Neumann classes of operators on $\mathcal{H}$, with the convention that $\mathcal{C}_\infty(\mathcal{H})$ stands for the set of all compact operators on $\mathcal{H}$.

Proposition 3.1. *Let $1 \le p \le \infty$, denote by $\mathfrak{A}$ the Banach-Lie algebra $\mathcal{C}_p(\mathcal{H})$ and consider an arbitrary Hermitian derivation $D \in \mathrm{Der}(\mathfrak{A})$. Then there exists a self-adjoint operator $A \in \mathcal{B}(\mathcal{H})$ such that*

$$D(C) = [A, C] \quad \text{for every } C \in \mathfrak{A}$$

and we have

$$\sigma(D) = \{t - s \mid t, s \in \sigma(A)\}.$$

Proof. Since D is a derivation of $\mathfrak{A}$, Proposition 9 in Chapter II of [13] shows that there exists $A_1 \in \mathcal{B}(\mathcal{H})$ such that $D(C) = [A_1, C]$ for every $C \in \mathfrak{A}$.

Now, if $p = 2$, then $\mathfrak{A} = \mathcal{C}_2(\mathcal{H})$ is a Hilbert space and the Hilbert space adjoint of D is $C \mapsto [A_1^*, C]$ (A_1^* standing here for the Hilbert space adjoint of A_1). Since D is Hermitian, hence self-adjoint, we get $[A_1 - A_1^*, C] = 0$ for every $C \in \mathcal{C}_2(\mathcal{H})$, hence there exists $w \in \mathbb{C}$ such that $A_1 - A_1^* = w\,\mathrm{id}_{\mathcal{H}}$. Then w has to be purely imaginary and $A = A_1 - (w/2)\,\mathrm{id}_{\mathcal{H}}$ is the self-adjoint operator we are looking for.

In the case $p \ne 2$, since D is Hermitian, the main result of [6] implies that there exist self-adjoint operators $B_1, B_2 \in \mathcal{B}(\mathcal{H})$ with

$$D(C) = B_1C - CB_2 \quad \text{for every } C \in \mathcal{C}_p(\mathcal{H}).$$

Then

$$(A_1 - B_1)C = C(A_1 - B_2) \quad \text{for every } C \in \mathcal{C}_p(\mathcal{H})$$

and it is straightforward to deduce that there exists $z \in \mathbb{C}$ with $A_1 - B_1 = A_1 - B_2 = z\,\mathrm{id}_{\mathcal{H}}$. In particular $B_1 = B_2$ and we can take B_1 for the desired self-adjoint operator A.

The last assertion in the statement is well known: the spectrum of D can be computed by means of Corollary 4.3 in [11]. □

Before proceeding to develop Example 3.5, let us briefly mention some facts which put in a proper perspective the closed-range Hermitian derivations of the classical Banach-Lie algebras $\mathcal{C}_p(\mathcal{H})$. The following result extends Lemma VII.3 in [20] (see also Theorem IV.4 in [21]).

Corollary 3.2. *Let $D \in \mathrm{Der}(\mathfrak{A})$ be a Hermitian derivation of the Banach-Lie algebra $\mathfrak{A} = \mathcal{C}_p(\mathcal{H})$, where $1 \le p \le \infty$ is arbitrary. Then D has closed range if and only if its spectrum is finite, and in this case D is diagonalizable.*

Proof. We use the notation of Proposition 3.1. Since D is Hermitian, by Theorem 4.12 in [17] we get (as in the proof of Theorem 1.3 above) that D has closed range if and only if 0 is an isolated point of its spectrum. Since Proposition 3.1 shows that $\sigma(D) = \{t - s \mid t, s \in \sigma(A)\}$, it easily follows that 0 is an isolated point of $\sigma(D)$ if and only if $\sigma(A)$ is finite. The last assertion of the corollary is a consequence of the fact that every Hermitian operator with finite spectrum is diagonalizable (see, e.g., Corollary 4 in Section 14 of [5]). □

Remark 3.3. It is a well-known phenomenon that not every derivation of a classical complex Banach-Lie algebra of compact operators on $\mathcal{H}$ is inner when $\dim \mathcal{H} = \infty$, as it is the case for the complex finite-dimensional semisimple Lie algebras (see [13]). See, e.g., the characterization of Hermitian derivations of $\mathcal{C}_p(\mathcal{H})$ for $1 \leq p \leq \infty$ in Proposition 3.1 above, where the operator $A \in \mathcal{B}(\mathcal{H})$ need not belong to $\mathcal{C}_p(\mathcal{H})$. Although, every derivation D of $\mathcal{C}_p(\mathcal{H})$ extends to a (inner) derivation $\operatorname{ad} A$ of $\mathcal{B}(\mathcal{H})$. □

From the perspective of Remark 3.3, it is interesting to note that there exist some rather strong connections between an arbitrary derivation of $\mathcal{C}_p(\mathcal{H})$ and its extension to a derivation of $\mathcal{B}(\mathcal{H})$. The following corollary is intended to justify this observation.

Corollary 3.4. *Let $1 \leq p \leq \infty$ and $A \in \mathcal{B}(\mathcal{H})$. Denote by $\mathcal{D}$ the bounded inner derivation* $\operatorname{ad} A$ *of $\mathcal{B}(\mathcal{H})$ and by D the derivation of $\mathcal{C}_p(\mathcal{H})$ which is the restriction of $\mathcal{D}$. Then the following assertions hold.*

(i) $\sigma(D) = \sigma(\mathcal{D}) = \{t - s \mid t, s \in \sigma(A)\}$.

(ii) *D is Hermitian if and only if $\mathcal{D}$ is Hermitian if and only if there exists $z \in \mathbb{C}$ such that $A - z \operatorname{id}_{\mathcal{H}}$ is self-adjoint.*

(iii) *If D is Hermitian, then it has closed range if and only if $\mathcal{D}$ has closed range if and only if the spectrum of D is finite if and only if the spectrum of $\mathcal{D}$ is finite.*

Proof. (i) Use Corollary 4.3 in [11], as in the proof of Proposition 3.1 above.

(ii) If D is Hermitian, use Proposition 3.1 and its proof to deduce the asserted properties of $\mathcal{D}$ and A.

On the other hand, recall that $\mathcal{B}(\mathcal{H})$ is naturally isomorphic to the topological dual of the Banach space $\mathcal{C}_1(\mathcal{H})$ and note that the dual of the operator

$$\mathcal{D}_* : \mathcal{C}_1(\mathcal{H}) \to \mathcal{C}_1(\mathcal{H}), \quad K \mapsto [K, A]$$

is just $\mathcal{D}$. Now, if $\mathcal{D}$ is Hermitian it follows at once that also $\mathcal{D}_*$ is Hermitian, and the above argument used in the proof of Proposition 3.1 shows that $A = A_0 + z \operatorname{id}_{\mathcal{H}}$ for some self-adjoint operator $A_0 \in \mathcal{B}(\mathcal{H})$ and some $z \in \mathbb{C}$.

The other implications from (ii) are obvious.

(iii) Since D is Hermitian, in view of (ii) we may assume without loss of generality that $A = A^*$. First recall the main result of [1] which asserts that $\mathcal{D}$ has closed range if and only if A is similar to an operator generating a finite-dimensional C^*-algebra. Then it follows that $\mathcal{D}$ has closed range if and only if the spectrum of the self-adjoint operator A is finite (which also follows by (i) and (ii) along with Corollary 3.2). This is further equivalent by (i) to the fact that either the spectrum of D or the spectrum of $\mathcal{D}$ is finite. Now it only remains to make use of Corollary 3.2 and we are done. □

We now finally turn to the example (B), which suggested us to introduce the hypotheses 1°–5° in Section 2.

Example 3.5. Let $2 \leq p < \infty$ and consider the real Banach-Lie algebra

$$\mathfrak{g} = \mathfrak{u}_p$$

of all skew-adjoint operators belonging to the Schatten-von Neumann class

$$\mathfrak{A} = \mathcal{C}_p(\mathcal{H})$$

on a complex Hilbert space $\mathcal{H}$. Then $\mathfrak{A} = \mathfrak{g}_{\mathbb{C}}$ and $\mathfrak{g}$ is the set of fixed points of the conjugation

$$Z \mapsto \overline{Z} = -Z^*$$

of the complex Banach-Lie algebra $\mathfrak{A}$ (where Z^* denotes the Hilbert space adjoint of the operator Z). For the Banach space $\mathfrak{A}$ (resp. $\mathfrak{g}$), the dual Banach space is $\mathfrak{A}^* = \mathcal{C}_q(\mathcal{H})$ (respectively $\mathfrak{g}^* = \mathfrak{u}_q$) with $q = p/(p-1)$, the duality pairing $\langle \cdot, \cdot \rangle$ being defined by

$$\mathfrak{A}^* \times \mathfrak{A} \to \mathbb{C}, \quad (V, X) \mapsto \langle V, X \rangle = \mathrm{Tr}(VX).$$

Since $2 \leq p < \infty$, we have $1 < q \leq 2 \leq p < \infty$, so that $\mathcal{C}_q(\mathcal{H}) \subseteq \mathcal{C}_p(\mathcal{H})$. We denote the corresponding inclusion map of $\mathfrak{u}_q$ into $\mathfrak{u}_p$ by

$$\iota \colon \mathfrak{g}^* \to \mathfrak{g}.$$

(We use the same notation ι for the inclusion map of $\mathfrak{A}^*$ into $\mathfrak{A}$.) We have $\iota^* = \iota$ because $\mathrm{Tr}(VX) = \mathrm{Tr}(XV)$ for every $V \in \mathcal{C}_q(\mathcal{H})$ and $X \in \mathcal{C}_p(\mathcal{H})$. The intertwining property of ι (see hypothesis 2° in Section 2) is obvious.

To see that the hypothesis 5° from Section 2 is satisfied, note that, for $X \in \mathfrak{g} = \mathfrak{u}_p$, $-X^2$ is a nonnegative trace-class operator, hence

$$\langle X, \iota(X) \rangle = \mathrm{Tr}(X^2) \leq 0$$

and $\mathrm{Tr}(X^2) = 0$ if and only if $X = 0$.

We now turn to the way hypothesis 3° from Section 2 can be realized in the present setting. To this end, take a bounded derivation D of the complex Banach-Lie algebra $\mathfrak{A}$ such that $\mathrm{i}D$ is a Hermitian operator on $\mathfrak{A}$. As Proposition 3.1 shows, then there exists a bounded skew-adjoint operator A on $\mathcal{H}$ such that $D = \mathrm{ad}\, A|_{\mathfrak{A}}$ and

$$\sigma(\mathrm{i}D) = \sigma(\mathrm{i}A) - \sigma(\mathrm{i}A).$$

For every $V \in \mathcal{C}_q(\mathcal{H})$ and $X \in \mathcal{C}_p(\mathcal{H})$ one has

$$\langle V, D(X) \rangle = \mathrm{Tr}(V[A, X]) = \mathrm{Tr}([V, A]X) = \langle (\mathrm{ad}\,(-A))(V), X \rangle,$$

so that the operator dual to D is $D^* = \mathrm{ad}\,(-A)|_{\mathfrak{A}^*}$. (Recall that $\mathfrak{A}^* = \mathcal{C}_q(\mathcal{H})$.) This formula for D^* shows in particular that $\iota \circ (-D^*) = D \circ \iota$. Next note that

$$\mathfrak{A}(D) = \{X \in \mathcal{C}_p(\mathcal{H}) \mid [A, X] \in \mathcal{C}_q(\mathcal{H})\}, \quad \|\cdot\|_{\mathfrak{A}(D)} = \|\cdot\|_{\mathcal{C}_p(\mathcal{H})} + \|[A,\cdot]\|_{\mathcal{C}_q(\mathcal{H})},$$
$$\mathfrak{g}(D) = \{X \in \mathfrak{A}(D) \mid X \text{ is skew-adjoint}\},$$

and

$$\omega(Y_1, Y_2) = \mathrm{Tr}([A, Y_1]Y_2) \quad \text{whenever } Y_1, Y_2 \in \mathfrak{g}(D).$$

(Note that here we have $\mathrm{Tr}([A, Y_1]Y_2) \in \mathbb{R}$ because all of the operators A, Y_1, Y_2 are skew-adjoint.)

Next assume that the Hermitian operator $\mathrm{i}D$ on $\mathfrak{A}$ is moreover *Dunford scalar*. Then the Concluding remark (b) in [14] shows that the spectrum of $\mathrm{i}A$ is finite, which in turn implies that the spectrum of $\mathrm{i}D$ is also finite. In particular, $\mathrm{i}D$ has closed range in $\mathfrak{A}$. Let $\sigma(\mathrm{i}A) = \{\lambda_1, \dots, \lambda_k\}$, with $\lambda_1 > \dots > \lambda_k$. We have

$$\mathcal{H} = \mathrm{Ker}\,(\mathrm{i}A - \lambda_1) \oplus \dots \oplus \mathrm{Ker}\,(\mathrm{i}A - \lambda_1).$$

Writing the operators on $\mathcal{H}$ as block operator matrices, we get

$$\mathfrak{A}(D) = \left\{ X = (X_{ij})_{1 \le i,j \le k} \in \mathcal{C}_p(\mathcal{H}) \mid \begin{pmatrix} 0 & X_{12} & \dots & X_{1k} \\ X_{21} & 0 & \ddots & \vdots \\ \vdots & \ddots & \ddots & X_{k-1,k} \\ X_{k1} & \dots & X_{k,k-1} & 0 \end{pmatrix} \in \mathcal{C}_q(\mathcal{H}) \right\}$$

(see also Remark 2.11). Now Theorem 2.10 says that

$$\mathfrak{p} = \left\{ X = \begin{pmatrix} X_{11} & & & 0 \\ X_{21} & X_{22} & & \\ \vdots & \ddots & \ddots & \\ X_{k1} & \dots & X_{k,k-1} & X_{kk} \end{pmatrix} \in \mathcal{C}_p(\mathcal{H}) \mid \begin{pmatrix} 0 & & & 0 \\ X_{21} & 0 & & \\ \vdots & \ddots & \ddots & \\ X_{k1} & \dots & X_{k,k-1} & 0 \end{pmatrix} \in \mathcal{C}_q(\mathcal{H}) \right\},$$

since $\mathrm{i}A$ is just the diagonal operator matrix

$$\begin{pmatrix} \lambda_1 & & 0 \\ & \ddots & \\ 0 & & \lambda_k \end{pmatrix}$$

with $\lambda_1 > \dots > \lambda_k$, while $D(X) = [A, X]$ for each operator matrix X. Further note that

$$\mathfrak{A}(D)/\mathfrak{p} \cong \left\{ X = \begin{pmatrix} 0 & X_{12} & \dots & X_{1k} \\ & 0 & \ddots & \vdots \\ & & \ddots & X_{k-1,k} \\ 0 & & & 0 \end{pmatrix} \in \mathcal{C}_q(\mathcal{H}) \right\}$$

and for every such $X \in \mathfrak{A}(D)/\mathfrak{p}$ we have

$$\mathrm{i}\omega(X, X^*) = \sum_{1 \le i < j \le k} (\lambda_i - \lambda_j)\|X_{ij}\|^2_{\mathcal{C}_2(\mathcal{H})}.$$

This formula encodes the difference between the topology of $\mathfrak{A}(D)/\mathfrak{p}$ and the one defined by the scalar product constructed by means of ω. □

Acknowledgement

The present paper was at several places influenced by my conversations with Karl-Hermann Neeb during a research stage in Darmstadt. To mention just three of these places, it was he who essentially made Remark 1.5, who suggested that a result like Corollary 2.5 should be true, and who suggested that one should construct polarizations of the "restricted" algebra $\mathfrak{g}(D)$ (see Theorem 2.10 below), not of the algebra $\mathfrak{g}$ itself, as I had done in a previous version of the paper. It is a pleasure to thank him for this help, as well as for his kindness that made possible my pleasant and most instructive stay in Darmstadt. I also thank to the anonymous referee for a number of useful remarks, notably the last part of Remark 2.6, or the present shorter proofs of Proposition 2.3 and Lemma 2.8.

This research has been supported from contract ICA1-CT–2000–70022 with the European Commission.

References

[1] C. Apostol, *Inner Derivations with Closed Range.* Rev. Roumaine Math. Pures Appl. **21** (1976), 249–265.

[2] V.K. Balachandran, *Real L^*-Algebras.* Indian J. Pure Appl. Math. **3** (1972), 1224–1246.

[3] D. Beltiţă, *Equivariant Monotone Operators and Infinite-Dimensional Complex Homogeneous Spaces.* Submitted for publication.

[4] D. Beltiţă, *Complex Homogeneous Spaces of Pseudo-Restricted Groups.* Math. Research Letters **10** (2003), no. 4, 459–467.

[5] D. Beltiţă, M. Şabac, *Lie Algebras of Bounded Operators.* Operator Theory: Advances and Applications, vol. **120**. Birkhäuser Verlag, Basel, 2001.

[6] E. Berkson, R.J. Fleming, J.A. Goldstein, J. Jamison, *One-Parameter Groups of Isometries on C_p.* Rev. Roumaine Math. Pures Appl. **24** (1979), 863–868.

[7] R.P. Boyer, *Representation Theory of the Hilbert-Lie Group $U(\mathfrak{H})_2$.* Duke Math. J. **47** (1980), 325–344.

[8] A.L. Carey, *Some Homogeneous Spaces and Representations of the Hilbert Lie Group $\mathcal{U}(H)_2$.* Rev. Roum. Math. Pures Appl. **30** (1985), 505–520.

[9] I. Colojoară, *Elements of Spectral Theory* (in Romanian). Ed. Acad., Bucharest, 1968.

[10] N. Dunford, J.T. Schwartz, *Linear Operators. Part III: Spectral operators.* With the assistance of William G. Bade and Robert G. Bartle. Pure and Applied Math., vol. **VII**. Interscience Publ. [John Wiley & Sons, Inc.], New York–London–Sydney, 1971.

[11] J. Eschmeier, *Analytische Dualität und Tensorprodukte in der mehrdimensionalen Spektraltheorie.* Schriftenr. Math. Inst. Univ. Münster, 2. Ser. **42** (1987), 1–132.

[12] J. Eschmeier, M. Putinar, *Spectral Decompositions and Analytic Sheaves.* London Mathematical Society Monographs. New Series, vol. **10**. Oxford Univ. Press, Oxford, 1996.

[13] P. de la Harpe, *Classical Banach-Lie Algebras and Banach-Lie Groups of Operators in Hilbert Space.* Lecture Notes in Math., vol. **285**. Springer-Verlag, Berlin–Heidelberg–New York, 1972.

[14] M. Hladnik, *Spectrality of Elementary Operators.* J. Aus. Math. Soc. Ser. A **49** (1990), No. 2, 327–346.

[15] A.A. Kirillov, *Elements of the Theory of Representations.* Grundlehren der Mathematischen Wissenschaften, Band **220**. Springer-Verlag, Berlin–New York, 1976.

[16] N.P. Landsman, *Representations of the Infinite Unitary Group from Constrained Quantization.* J. Nonlinear Math. Physics **6** (1999), no. 2, 161–180.

[17] K. Mattila, *Normal Operators and Proper Boundary Points of the Spectra of Operators on a Banach Space.* Ann. Acad. Sci. Fenn., Ser. A I, Diss. **19** (1978).

[18] K.-H. Neeb, *Holomorphic Highest Weight Representations of Infinite-Dimensional Complex Classical Groups.* J. Reine Angew. Math. **497** (1998), 171–222.

[19] K.-H. Neeb, *Holomorphy and Convexity in Lie Theory.* de Gruyter Expositions in Mathematics, vol. **28**. Walter de Gruyter, Berlin, 2000.

[20] K.-H. Neeb, *Infinite-Dimensional Groups and their Representations.* Lie Theory, Progr. Math., vol. 228, Birkhäuser Verlag, Boston, 2004, 213–328.

[21] K.-H. Neeb, *Highest Weight Representations and Infinite-Dimensional Kähler Manifolds.* Recent Advances in Lie Theory (Vigo, 2000), vol. 25, Res. Exp. Math., Heldermann, Lemgo, 2002, pp. 367–392.

[22] K.-H. Neeb, *Classical Hilbert-Lie Groups, their Extensions and their Homotopy Groups.* in *Analysis and Geometry on Finite- and Infinite-Dimensional Lie Groups.* A. Strasburger, J. Hilgert, K.-H. Neeb, W. Wojtyński (eds.), Banach Center Publ. **55**, Warszawa, 2002, pp. 87–151.

[23] J.R. Schue, *Hilbert Space Methods in the Theory of Lie Algebras.* Trans. Amer. Math. Soc. **95** (1960), 69–80.

[24] J.R. Schue, *Cartan Decompositions for L^*-Algebras.* Trans. Amer. Math. Soc. **98** (1961), 334–349.

[25] H. Upmeier, *Symmetric Banach Manifolds and Jordan C^*-Algebras.* North-Holland Mathematics Studies, vol. **104**. Notas de Matemática, vol. **96**. North-Holland, Amsterdam-New York-Oxford, 1985.

[26] F.-H. Vasilescu, *Analytic Functional Calculus and Spectral Decompositions.* Mathematics and Its Applications (East European Series), vol. **1**. D. Reidel Publishing Company, Dordrecht–Boston–London; Editura Academiei, Bucharest, 1982.

Daniel Beltiţă
Institute of Mathematics "Simion Stoilow" of the Romanian Academy
P.O. Box 1–764
Bucharest, RO 014700
Romania
e-mail: `Daniel.Beltita@imar.ro`

Operator Theory:
Advances and Applications, Vol. 153, 39–59

Commuting Triples of Subnormal Operators and Related Moment Problems

Tudor Bînzar and Doru Păunescu

Abstract. Using a new characterization for subnormality of a commuting triple of operators (due to T. Trent, [12], for single operators, extended by P. Găvruţă and N. Suciu, [4], for commuting pairs) we shall give necessary and sufficient conditions on a triple indexed sequence of vectors from a Hilbert space such that it can be expressed as moments of an appropriate triple of commuting operators. We obtain an extension of Z. Sebestyén's similar result ([9], [10]) for simple sequences and D. Păunescu result ([7], [8]) for double sequences. The analogue problem for triple indexed sequences of operators acting on a Hilbert space is also obtained.

Mathematics Subject Classification (2000). 47A20, 44A66.

Keywords. Subnormal triple of operators, Positive definite kernel, Moment problem.

Let $\mathcal{H}$ be a complex Hilbert space and let $\mathcal{L}(\mathcal{H})$ be the C^*-algebra of all bounded linear operators on $\mathcal{H}$; an element S in $\mathcal{L}(\mathcal{H})$ will be called subnormal if there exists a normal operator N acting on a larger Hilbert space that extends S. If $S_1, S_2, S_3 \in \mathcal{L}(\mathcal{H})$ commute, then the triple (S_1, S_2, S_3) is *subnormal* if it has a *commuting normal extension*, i.e., there exist a larger Hilbert space $\mathcal{K} \supset \mathcal{H}$ and three normal operators N_1, N_2, N_3 such that:

- $\mathcal{H}$ is an invariant subspace under N_1, N_2 and N_3;
- N_1, N_2 and N_3 commute, $N_iN_j = N_jN_i$ $\left(i, j = \overline{1,3}\right)$;
- N_1, N_2 and N_3 extends respectively S_1, S_2 and S_3, $N_i|_{\mathcal{H}} = S_i$ $\left(i = \overline{1,3}\right)$.

J. Bram characterizes subnormality via positivity ([1]) and this works also for subnormal triples.

Theorem 1. *Let $S_1, S_2, S_3 \in \mathcal{L}(\mathcal{H})$ be commuting operators. The following affirmations are equivalent:*

(i) *(S_1, S_2, S_3) is a subnormal triple;*

(ii) *There exist a Hilbert space $\mathcal{K} \supset \mathcal{H}$ and the commuting unitary operators $U_1, U_2, U_3 \in \mathcal{L}(\mathcal{K})$, such that for $m, n, p \in \mathbb{N}$*

$$\begin{aligned}
S_1^{*m} S_2^{*n} S_3^{*p} &= P_{\mathcal{H}} U_1^m U_2^n U_3^p S_1^m S_2^n S_3^p \,, \\
S_1^{*m} S_2^{*n} S_3^{p} &= S_3^{*p} P_{\mathcal{H}} U_1^m U_2^n U_3^{*p} S_1^m S_2^n \,, \\
S_1^{*m} S_3^{*p} S_2^{n} &= S_2^{*n} P_{\mathcal{H}} U_1^m U_2^{*n} U_3^p S_1^m S_3^p \,, \\
S_1^{*m} S_2^{n} S_3^{p} &= S_2^{*n} S_3^{*p} P_{\mathcal{H}} U_1^m U_2^{*n} U_3^{*p} S_1^m;
\end{aligned}$$

(iii) *There exist the triple indexed sequences of operators in $\mathcal{L}(\mathcal{H})$ denoted $(A_{m\ n\ p})$, $(B_{m\ n\ p})$, $(C_{m\ n\ p})$, $(D_{m\ n\ p})$ with $(m, n, p) \in \mathbb{N} \times \mathbb{N} \times \mathbb{N}$, satisfying:*

$$\begin{aligned}
S_1^{*m} S_2^{*n} S_3^{*p} &= A_{m\ n\ p} S_1^m S_2^n S_3^p \,, \\
S_1^{*m} S_2^{*n} S_3^{p} &= S_3^{*p} B_{m\ n\ p} S_1^m S_2^n \,, \\
S_1^{*m} S_3^{*p} S_2^{n} &= S_2^{*n} C_{m\ n\ p} S_1^m S_3^p \,, \\
S_1^{*m} S_2^{n} S_3^{p} &= S_2^{*n} S_3^{*p} D_{m\ n\ p} S_1^m
\end{aligned}$$

such that the function

$$F_{m\ n\ p} = \begin{cases}
A_{m\ n\ p}, & m, n, p \geq 0; \\
A^*_{-m\ -n\ -p}, & m, n, p < 0; \\
B_{m\ n\ -p}, & m, n \geq 0,\ p < 0; \\
B^*_{-m\ -n\ p}, & m, n < 0,\ p \geq 0; \\
C_{m\ -n\ p}, & m, p \geq 0,\ n < 0; \\
C^*_{-m\ n\ -p}, & m, p < 0,\ n \geq 0; \\
D_{m\ -n\ -p}, & m \geq 0,\ n, p < 0; \\
D^*_{-m\ n\ p}, & m < 0,\ n, p \geq 0;
\end{cases}$$

is of positive type, that is

$$\sum_{\substack{j_1, j_2, j_3 \\ k_1, k_2, k_3}} \langle F_{j_1 - k_1\ j_2 - k_2\ j_3 - k_3} \mathbf{x}_{j_1\ j_2\ j_3}, \mathbf{x}_{k_1\ k_2\ k_3} \rangle \geq 0$$

for every finite system $\{\mathbf{x}_{j_1\ j_2\ j_3}\}_{j_1, j_2, j_3 \geq 0}$ from $\mathcal{H}$.

Proof. (i) $\Rightarrow$ (ii) Let $N_1, N_2, N_3 \in \mathcal{L}(\mathcal{K})$ three commuting normal operators on $\mathcal{K} \supset \mathcal{H}$ such that $N_i|_{\mathcal{H}} = S_i$, $i \in \overline{1,3}$. From a result of T. Furuta [2], it follows that

$$N_i = V_i P_i, \quad i \in \overline{1,3} \,,$$

where V_i $\left(i \in \overline{1,3}\right)$ are unitary operators and P_i $\left(i \in \overline{1,3}\right)$ are positive operators on $\mathcal{K}$ such that

$$\begin{cases} V_i V_j = V_j V_i \\ V_i P_j = P_j V_i \end{cases} \quad \text{for any } i, j \in \overline{1,3} \,.$$

Let us define $U_i = V_i^{*2}$, $\left(i \in \overline{1,3}\right)$; then for $m, n, p \in \mathbb{N}$, we have

$$\begin{aligned}
N_1^{*m} N_2^{*n} N_3^{*p} &= V_1^{*m} P_1^m V_2^{*n} P_2^n V_3^{*p} P_3^p = V_1^{*2m} V_1^m P_1^m V_2^{*2n} V_2^n P_2^n V_3^{*2p} V_3^p P_3^p \\
&= V_1^{*2m} V_2^{*2n} V_3^{*2p} V_1^m P_1^m V_2^n P_2^n V_3^p P_3^p = U_1^m U_2^n U_3^p N_1^m N_2^n N_3^p, \\
N_1^{*m} N_2^{*n} N_3^p &= N_3^p N_1^{*m} N_2^{*n} = V_3^p P_3^p V_1^{*m} P_1^m V_2^{*n} P_2^n \\
&= V_3^{*p} V_3^{2p} P_3^p V_1^{*2m} V_1^m P_1^m V_2^{*2n} V_2^n P_2^n \\
&= V_3^{*p} P_3^p V_1^{*2m} V_2^{*2n} V_3^{2p} V_1^m P_1^m V_2^n P_2^n = N_3^{*p} U_1^m U_2^n U_3^{*p} N_1^m N_2^n , \\
N_1^{*m} N_3^{*p} N_2^n &= N_2^n N_1^{*m} N_3^{*p} = V_2^n P_2^n V_1^{*m} P_1^m V_3^{*p} P_3^p \\
&= V_2^{*n} V_2^{2n} P_2^n V_1^{*2m} V_1^m P_1^m V_3^{*2p} V_3^p P_3^p \\
&= V_2^{*n} P_2^n V_1^{*2m} V_2^{2n} V_3^{*2p} V_1^m P_1^m V_3^p P_3^p = N_2^{*n} U_1^m U_2^{*n} U_3^p N_1^m N_3^p , \\
N_1^{*m} N_2^n N_3^p &= N_2^n N_3^p N_1^{*m} = V_2^n P_2^n V_3^p P_3^p V_1^{*m} P_1^m \\
&= V_2^{*n} V_2^{2n} P_2^n V_3^{*p} V_3^{2p} P_3^p V_1^{*2m} V_1^m P_1^m \\
&= V_2^{*n} P_2^n V_3^{*p} P_3^p V_1^{*2m} V_2^{2n} V_3^{2p} V_1^m P_1^m = N_2^{*n} N_3^{*p} U_1^m U_2^{*n} U_3^{*p} N_1^m.
\end{aligned}$$

Now the relations of (ii) follow easily from

$$P_{\mathcal{H}} N_i^* = S_i^* P_{\mathcal{H}} \text{ for any } i \in \overline{1,3} .$$

Indeed, if $\mathbf{k} \in \mathcal{K}$, $\mathbf{k} = \mathbf{x} + \mathbf{x}'$ with $\mathbf{x} \in H$ and $\mathbf{x}' \in \mathcal{K} \ominus \mathcal{H}$, then, for any $i \in \overline{1,3}$, we have

$$P_{\mathcal{H}} N_i^* \mathbf{k} = P_{\mathcal{H}} N_i^* \mathbf{x} + P_{\mathcal{H}} N_i^* \mathbf{x}' = P_{\mathcal{H}} N_i^* \mathbf{x} = S_i^* \mathbf{x} = S_i^* P_{\mathcal{H}} \mathbf{k} .$$

Now, let $\mathbf{x} \in \mathcal{H}$ and m, n, p be positive integers, then

$$\begin{aligned}
S_1^{*m} S_2^{*n} S_3^{*p} \mathbf{x} &= S_1^{*m} S_2^{*n} P_{\mathcal{H}} N_3^{*p} \mathbf{x} = S_1^{*m} P_{\mathcal{H}} N_2^{*n} N_3^{*p} \mathbf{x} = P_{\mathcal{H}} N_1^{*m} N_2^{*n} N_3^{*p} \mathbf{x} \\
&= P_{\mathcal{H}} U_1^m U_2^n U_3^p N_1^m N_2^n N_3^p \mathbf{x} = P_{\mathcal{H}} U_1^m U_2^n U_3^p S_1^m S_2^n S_3^p \mathbf{x} , \\
S_1^{*m} S_2^{*n} S_3^p \mathbf{x} &= S_1^{*m} S_2^{*n} N_3^p \mathbf{x} = S_1^{*m} P_{\mathcal{H}} N_2^{*n} N_3^p \mathbf{x} = P_{\mathcal{H}} N_1^{*m} N_2^{*n} N_3^p \mathbf{x} \\
&= P_{\mathcal{H}} N_3^{*p} U_1^m U_2^n U_3^{*p} N_1^m N_2^n \mathbf{x} = S_3^{*p} P_{\mathcal{H}} U_1^m U_2^n U_3^{*p} S_1^m S_2^n \mathbf{x} , \\
S_1^{*m} S_3^{*p} S_2^n \mathbf{x} &= S_1^{*m} S_3^{*p} N_2^n \mathbf{x} = S_1^{*m} P_{\mathcal{H}} N_3^{*p} N_2^n \mathbf{x} = P_{\mathcal{H}} N_1^{*m} N_3^{*p} N_2^n \mathbf{x} \\
&= P_{\mathcal{H}} N_2^{*n} U_1^m U_2^{*n} U_3^p N_1^m N_3^p \mathbf{x} = S_2^{*n} P_{\mathcal{H}} U_1^m U_2^{*n} U_3^p S_1^m S_3^p \mathbf{x}
\end{aligned}$$

and finally

$$\begin{aligned}
S_1^{*m} S_2^n S_3^p \mathbf{x} &= S_1^{*m} N_2^n N_3^p \mathbf{x} = P_{\mathcal{H}} N_1^{*m} N_2^n N_3^p \mathbf{x} \\
&= P_{\mathcal{H}} N_2^{*n} N_3^{*p} U_1^m U_2^{*n} U_3^{*p} N_1^m \mathbf{x} \\
&= S_2^{*n} P_{\mathcal{H}} N_3^{*p} U_1^m U_2^{*n} U_3^{*p} N_1^m \mathbf{x} = S_2^{*n} S_3^{*p} P_{\mathcal{H}} U_1^m U_2^{*n} U_3^{*p} S_1^m \mathbf{x}.
\end{aligned}$$

(ii) $\Rightarrow$ (iii) We denote

$$\begin{aligned}
A_{m\,n\,p} &= P_{\mathcal{H}} U_1^m U_2^n U_3^p|_{\mathcal{H}}, & B_{m\,n\,p} &= P_{\mathcal{H}} U_1^m U_2^n U_3^{*p}|_{\mathcal{H}} , \\
C_{m\,n\,p} &= P_{\mathcal{H}} U_1^m U_2^{*n} U_3^p|_{\mathcal{H}}, & D_{m\,n\,p} &= P_{\mathcal{H}} U_1^m U_2^{*n} U_3^{*p}|_{\mathcal{H}}
\end{aligned}$$

for each $m, n, p \in \mathbb{N}$. We shall establish that the function $F_{m\ n\ p}$ is of positive type. We have

$$\sum_{\substack{j_1,j_2,j_3\\k_1,k_2,k_3}} \langle F_{j_1-k_1\ j_2-k_2\ j_3-k_3}\mathbf{x}_{j_1\ j_2\ j_3}, \mathbf{x}_{k_1\ k_2\ k_3}\rangle$$

$$= \sum_{\substack{j_1\geq k_1\\j_2\geq k_2\\j_3\geq k_3}} \langle A_{j_1-k_1\ j_2-k_2\ j_3-k_3}\mathbf{x}_{j_1\ j_2\ j_3}, \mathbf{x}_{k_1\ k_2\ k_3}\rangle$$

$$+ \sum_{\substack{j_1\geq k_1\\j_2\geq k_2\\j_3<k_3}} \langle B_{j_1-k_1\ j_2-k_2\ k_3-j_3}\mathbf{x}_{j_1\ j_2\ j_3}, \mathbf{x}_{k_1\ k_2\ k_3}\rangle$$

$$+ \sum_{\substack{j_1\geq k_1\\j_2<k_2\\j_3\geq k_3}} \langle C_{j_1-k_1\ k_2-j_2\ j_3-k_3}\mathbf{x}_{j_1\ j_2\ j_3}, \mathbf{x}_{k_1\ k_2\ k_3}\rangle$$

$$+ \sum_{\substack{j_1\geq k_1\\j_2<k_2\\j_3<k_3}} \langle D_{j_1-k_1\ k_2-j_2\ k_3-j_3}\mathbf{x}_{j_1\ j_2\ j_3}, \mathbf{x}_{k_1\ k_2\ k_3}\rangle$$

$$+ \sum_{\substack{j_1<k_1\\j_2\geq k_2\\j_3\geq k_3}} \langle D^*_{k_1-j_1\ j_2-k_2\ j_3-k_3}\mathbf{x}_{j_1\ j_2\ j_3}, \mathbf{x}_{k_1\ k_2\ k_3}\rangle$$

$$+ \sum_{\substack{j_1<k_1\\j_2\geq k_2\\j_3<k_3}} \langle C^*_{k_1-j_1\ j_2-k_2\ k_3-j_3}\mathbf{x}_{j_1\ j_2\ j_3}, \mathbf{x}_{k_1\ k_2\ k_3}\rangle$$

$$+ \sum_{\substack{j_1<k_1\\j_2<k_2\\j_3\geq k_3}} \langle B^*_{k_1-j_1\ k_2-j_2\ j_3-k_3}\mathbf{x}_{j_1\ j_2\ j_3}, \mathbf{x}_{k_1\ k_2\ k_3}\rangle$$

$$+ \sum_{\substack{j_1<k_1\\j_2<k_2\\j_3<k_3}} \langle A^*_{k_1-j_1\ k_2-j_2\ k_3-j_3}\mathbf{x}_{j_1\ j_2\ j_3}, \mathbf{x}_{k_1\ k_2\ k_3}\rangle$$

$$= \sum_{\substack{j_1\geq k_1\\j_2\geq k_2\\j_3\geq k_3}} \left\langle P_{\mathcal{H}}U_1^{j_1-k_1}U_2^{j_2-k_2}U_3^{j_3-k_3}\mathbf{x}_{j_1\ j_2\ j_3}, \mathbf{x}_{k_1\ k_2\ k_3}\right\rangle$$

$$+ \sum_{\substack{j_1\geq k_1\\j_2\geq k_2\\j_3<k_3}} \left\langle P_{\mathcal{H}}U_1^{j_1-k_1}U_2^{j_2-k_2}U_3^{*k_3-j_3}\mathbf{x}_{j_1\ j_2\ j_3}, \mathbf{x}_{k_1\ k_2\ k_3}\right\rangle$$

$$+\sum_{\substack{j_1\ge k_1\\ j_2<k_2\\ j_3\ge k_3}}\left\langle P_{\mathcal H}U_1^{j_1-k_1}U_2^{*k_2-j_2}U_3^{j_3-k_3}\mathbf{x}_{j_1\ j_2\ j_3},\mathbf{x}_{k_1\ k_2\ k_3}\right\rangle$$

$$+\sum_{\substack{j_1\ge k_1\\ j_2<k_2\\ j_3<k_3}}\left\langle P_{\mathcal H}U_1^{j_1-k_1}U_2^{*k_2-j_2}U_3^{*k_3-j_3}\mathbf{x}_{j_1\ j_2\ j_3},\mathbf{x}_{k_1\ k_2\ k_3}\right\rangle$$

$$+\sum_{\substack{j_1<k_1\\ j_2\ge k_2\\ j_3\ge k_3}}\left\langle P_{\mathcal H}U_1^{*k_1-j_1}U_2^{j_2-k_2}U_3^{j_3-k_3}\mathbf{x}_{j_1\ j_2\ j_3},\mathbf{x}_{k_1\ k_2\ k_3}\right\rangle$$

$$+\sum_{\substack{j_1<k_1\\ j_2\ge k_2\\ j_3<k_3}}\left\langle P_{\mathcal H}U_1^{*k_1-j_1}U_2^{j_2-k_2}U_3^{*k_3-j_3}\mathbf{x}_{j_1\ j_2\ j_3},\mathbf{x}_{k_1\ k_2\ k_3}\right\rangle$$

$$+\sum_{\substack{j_1<k_1\\ j_2<k_2\\ j_3\ge k_3}}\left\langle P_{\mathcal H}U_1^{*k_1-j_1}U_2^{*k_2-j_2}U_3^{j_3-k_3}\mathbf{x}_{j_1\ j_2\ j_3},\mathbf{x}_{k_1\ k_2\ k_3}\right\rangle$$

$$+\sum_{\substack{j_1<k_1\\ j_2<k_2\\ j_3<k_3}}\left\langle P_{\mathcal H}U_1^{*k_1-j_1}U_2^{*k_2-j_2}U_3^{*k_3-j_3}\mathbf{x}_{j_1\ j_2\ j_3},\mathbf{x}_{k_1\ k_2\ k_3}\right\rangle$$

$$=\left\|\sum_{j_1,j_2,j_3\ge 0}U_1^{j_1}U_2^{j_2}U_3^{j_3}\mathbf{x}_{j_1\ j_2\ j_3}\right\|^2\ge 0\ .$$

(iii) $\Rightarrow$ (i) We apply Itô-Masani test ([5], [6]). Thus we have

$$\sum_{\substack{j_1,j_2,j_3\\ k_1,k_2,k_3}}\left\langle S_1^{k_1}S_2^{k_2}S_3^{k_3}\mathbf{x}_{j_1\ j_2\ j_3},S_1^{j_1}S_2^{j_2}S_3^{j_3}\mathbf{x}_{k_1\ k_2\ k_3}\right\rangle$$

$$=\sum_{\substack{j_1\ge k_1\\ j_2\ge k_2\\ j_3\ge k_3}}\left\langle S_1^{*j_1-k_1}S_2^{*j_2-k_2}S_3^{*j_3-k_3}S_1^{k_1}S_2^{k_2}S_3^{k_3}\mathbf{x}_{j_1\ j_2\ j_3},S_1^{k_1}S_2^{k_2}S_3^{k_3}\mathbf{x}_{k_1\ k_2\ k_3}\right\rangle$$

$$+\sum_{\substack{j_1\ge k_1\\ j_2\ge k_2\\ j_3<k_3}}\left\langle S_1^{*j_1-k_1}S_2^{*j_2-k_2}S_3^{k_3-j_3}S_1^{k_1}S_2^{k_2}S_3^{j_3}\mathbf{x}_{j_1\ j_2\ j_3},S_1^{k_1}S_2^{k_2}S_3^{j_3}\mathbf{x}_{k_1\ k_2\ k_3}\right\rangle$$

$$+\sum_{\substack{j_1\ge k_1\\ j_2<k_2\\ j_3\ge k_3}}\left\langle S_1^{*j_1-k_1}S_3^{*j_3-k_3}S_2^{k_2-j_2}S_1^{k_1}S_2^{j_2}S_3^{k_3}\mathbf{x}_{j_1\ j_2\ j_3},S_1^{k_1}S_2^{j_2}S_3^{k_3}\mathbf{x}_{k_1\ k_2\ k_3}\right\rangle$$

$$+\sum_{\substack{j_1\ge k_1\\ j_2<k_2\\ j_3<k_3}}\left\langle S_1^{*j_1-k_1}S_2^{k_2-j_2}S_3^{k_3-j_3}S_1^{k_1}S_2^{j_2}S_3^{j_3}\mathbf{x}_{j_1\ j_2\ j_3},S_1^{k_1}S_2^{j_2}S_3^{j_3}\mathbf{x}_{k_1\ k_2\ k_3}\right\rangle$$

$$+\sum_{\substack{j_1<k_1\\ j_2\ge k_2\\ j_3\ge k_3}}\left\langle S_1^{j_1}S_2^{k_2}S_3^{k_3}\mathbf{x}_{j_1\ j_2\ j_3}, S_1^{*k_1-j_1}S_2^{j_2-k_2}S_3^{j_3-k_3}S_1^{j_1}S_2^{k_2}S_3^{k_3}\mathbf{x}_{k_1\ k_2\ k_3}\right\rangle$$

$$+\sum_{\substack{j_1<k_1\\ j_2\ge k_2\\ j_3<k_3}}\left\langle S_1^{j_1}S_2^{k_2}S_3^{j_3}\mathbf{x}_{j_1\ j_2\ j_3}, S_1^{*k_1-j_1}S_3^{*k_3-j_3}S_2^{j_2-k_2}S_1^{j_1}S_2^{k_2}S_3^{j_3}\mathbf{x}_{k_1\ k_2\ k_3}\right\rangle$$

$$+\sum_{\substack{j_1<k_1\\ j_2<k_2\\ j_3\ge k_3}}\left\langle S_1^{j_1}S_2^{j_2}S_3^{k_3}\mathbf{x}_{j_1\ j_2\ j_3}, S_1^{*k_1-j_1}S_2^{*k_2-j_2}S_3^{j_3-k_3}S_1^{j_1}S_2^{j_2}S_3^{k_3}\mathbf{x}_{k_1\ k_2\ k_3}\right\rangle$$

$$+\sum_{\substack{j_1<k_1\\ j_2<k_2\\ j_3<k_3}}\left\langle S_1^{j_1}S_2^{j_2}S_3^{j_3}\mathbf{x}_{j_1\ j_2\ j_3}, S_1^{*k_1-j_1}S_2^{*k_2-j_2}S_3^{*k_3-j_3}S_1^{j_1}S_2^{j_2}S_3^{j_3}\mathbf{x}_{k_1\ k_2\ k_3}\right\rangle$$

$$=\sum_{\substack{j_1\ge k_1\\ j_2\ge k_2\\ j_3\ge k_3}}\Big\langle A_{j_1-k_1\ j_2-k_2\ j_3-k_3}S_1^{j_1-k_1}S_2^{j_2-k_2}S_3^{j_3-k_3}S_1^{k_1}S_2^{k_2}S_3^{k_3}\mathbf{x}_{j_1\ j_2\ j_3},$$
$$S_1^{k_1}S_2^{k_2}S_3^{k_3}\mathbf{x}_{k_1\ k_2\ k_3}\Big\rangle$$

$$+\sum_{\substack{j_1\ge k_1\\ j_2\ge k_2\\ j_3<k_3}}\Big\langle S_3^{*k_3-j_3}B_{j_1-k_1\ j_2-k_2\ k_3-j_3}S_1^{j_1-k_1}S_2^{j_2-k_2}S_1^{k_1}S_2^{k_2}S_3^{j_3}\mathbf{x}_{j_1\ j_2\ j_3},$$
$$S_1^{k_1}S_2^{k_2}S_3^{j_3}\mathbf{x}_{k_1\ k_2\ k_3}\Big\rangle$$

$$+\sum_{\substack{j_1\ge k_1\\ j_2<k_2\\ j_3\ge k_3}}\Big\langle S_2^{*k_2-j_2}C_{j_1-k_1\ k_2-j_2\ j_3-k_3}S_1^{j_1-k_1}S_3^{j_3-k_3}S_1^{k_1}S_2^{j_2}S_3^{k_3}\mathbf{x}_{j_1\ j_2\ j_3},$$
$$S_1^{k_1}S_2^{j_2}S_3^{k_3}\mathbf{x}_{k_1\ k_2\ k_3}\Big\rangle$$

$$+\sum_{\substack{j_1\ge k_1\\ j_2<k_2\\ j_3<k_3}}\Big\langle S_2^{*k_2-j_2}S_3^{*k_3-j_3}D_{j_1-k_1\ k_2-j_2\ k_3-j_3}S_1^{j_1-k_1}S_1^{k_1}S_2^{j_2}S_3^{j_3}\mathbf{x}_{j_1\ j_2\ j_3},$$
$$S_1^{k_1}S_2^{j_2}S_3^{j_3}\mathbf{x}_{k_1\ k_2\ k_3}\Big\rangle$$

$$+\sum_{\substack{j_1<k_1\\ j_2\ge k_2\\ j_3\ge k_3}}\Big\langle S_1^{j_1}S_2^{k_2}S_3^{k_3}\mathbf{x}_{j_1\ j_2\ j_3},$$
$$S_2^{*j_2-k_2}S_3^{*j_3-k_3}D_{k_1-j_1\ j_2-k_2\ j_3-k_3}S_1^{k_1-j_1}S_1^{j_1}S_2^{k_2}S_3^{k_3}\mathbf{x}_{k_1\ k_2\ k_3}\Big\rangle$$

$$+\sum_{\substack{j_1<k_1\\ j_2\geq k_2\\ j_3<k_3}} \Big\langle S_1^{j_1}S_2^{k_2}S_3^{j_3}\mathbf{x}_{j_1\ j_2\ j_3},$$

$$S_2^{*j_2-k_2}C_{k_1-j_1\ j_2-k_2\ k_3-j_3}S_1^{k_1-j_1}S_3^{k_3-j_3}S_1^{j_1}S_2^{k_2}S_3^{j_3}\mathbf{x}_{k_1\ k_2\ k_3}\Big\rangle$$

$$+\sum_{\substack{j_1<k_1\\ j_2<k_2\\ j_3\geq k_3}} \Big\langle S_1^{j_1}S_2^{j_2}S_3^{k_3}\mathbf{x}_{j_1\ j_2\ j_3},$$

$$S_3^{*j_3-k_3}B_{k_1-j_1\ k_2-j_2\ j_3-k_3}S_1^{k_1-j_1}S_2^{k_2-j_2}S_1^{j_1}S_2^{j_2}S_3^{k_3}\mathbf{x}_{k_1\ k_2\ k_3}\Big\rangle$$

$$+\sum_{\substack{j_1<k_1\\ j_2<k_2\\ j_3<k_3}} \Big\langle S_1^{j_1}S_2^{j_2}S_3^{j_3}\mathbf{x}_{j_1\ j_2\ j_3},$$

$$A_{k_1-j_1\ k_2-j_2\ k_3-j_3}S_1^{k_1-j_1}S_2^{k_2-j_2}S_3^{k_3-j_3}S_1^{j_1}S_2^{j_2}S_3^{j_3}\mathbf{x}_{k_1\ k_2\ k_3}\Big\rangle$$

$$=\sum_{\substack{j_1\geq k_1\\ j_2\geq k_2\\ j_3\geq k_3}} \Big\langle A_{j_1-k_1\ j_2-k_2\ j_3-k_3}S_1^{j_1}S_2^{j_2}S_3^{j_3}\mathbf{x}_{j_1\ j_2\ j_3}, S_1^{k_1}S_2^{k_2}S_3^{k_3}\mathbf{x}_{k_1\ k_2\ k_3}\Big\rangle$$

$$+\sum_{\substack{j_1\geq k_1\\ j_2\geq k_2\\ j_3<k_3}} \Big\langle B_{j_1-k_1\ j_2-k_2\ k_3-j_3}S_1^{j_1}S_2^{j_2}S_3^{j_3}\mathbf{x}_{j_1\ j_2\ j_3}, S_1^{k_1}S_2^{k_2}S_3^{k_3}\mathbf{x}_{k_1\ k_2\ k_3}\Big\rangle$$

$$+\sum_{\substack{j_1\geq k_1\\ j_2<k_2\\ j_3\geq k_3}} \Big\langle C_{j_1-k_1\ k_2-j_2\ j_3-k_3}S_1^{j_1}S_2^{j_2}S_3^{j_3}\mathbf{x}_{j_1\ j_2\ j_3}, S_1^{k_1}S_2^{k_2}S_3^{k_3}\mathbf{x}_{k_1\ k_2\ k_3}\Big\rangle$$

$$+\sum_{\substack{j_1\geq k_1\\ j_2<k_2\\ j_3<k_3}} \Big\langle D_{j_1-k_1\ k_2-j_2\ k_3-j_3}S_1^{j_1}S_2^{j_2}S_3^{j_3}\mathbf{x}_{j_1\ j_2\ j_3}, S_1^{k_1}S_2^{k_2}S_3^{k_3}\mathbf{x}_{k_1\ k_2\ k_3}\Big\rangle$$

$$+\sum_{\substack{j_1<k_1\\ j_2\geq k_2\\ j_3\geq k_3}} \Big\langle D^*_{k_1-j_1\ j_2-k_2\ j_3-k_3}S_1^{j_1}S_2^{j_2}S_3^{j_3}\mathbf{x}_{j_1\ j_2\ j_3}, S_1^{k_1}S_2^{k_2}S_3^{k_3}\mathbf{x}_{k_1\ k_2\ k_3}\Big\rangle$$

$$+\sum_{\substack{j_1<k_1\\ j_2\geq k_2\\ j_3<k_3}} \Big\langle C^*_{k_1-j_1\ j_2-k_2\ k_3-j_3}S_1^{j_1}S_2^{j_2}S_3^{j_3}\mathbf{x}_{j_1\ j_2\ j_3}, S_1^{k_1}S_2^{k_2}S_3^{k_3}\mathbf{x}_{k_1\ k_2\ k_3}\Big\rangle$$

$$+\sum_{\substack{j_1<k_1\\ j_2<k_2\\ j_3\geq k_3}} \Big\langle B^*_{k_1-j_1\ k_2-j_2\ j_3-k_3}S_1^{j_1}S_2^{j_2}S_3^{j_3}\mathbf{x}_{j_1\ j_2\ j_3}, S_1^{k_1}S_2^{k_2}S_3^{k_3}\mathbf{x}_{k_1\ k_2\ k_3}\Big\rangle$$

$$+\sum_{\substack{j_1<k_1\\ j_2<k_2\\ j_3<k_3}} \left\langle A^*_{k_1-j_1\ k_2-j_2\ k_3-j_3} S_1^{j_1} S_2^{j_2} S_3^{j_3} \mathbf{x}_{j_1\ j_2\ j_3}, S_1^{k_1} S_2^{k_2} S_3^{k_3} \mathbf{x}_{k_1\ k_2\ k_3} \right\rangle$$

$$= \sum_{\substack{j_1,j_2,j_3\\ k_1,k_2,k_3}} \left\langle F_{j_1-k_1,j_2-k_2,j_3-k_3} S_1^{j_1} S_2^{j_2} S_3^{j_3} \mathbf{x}_{j_1\ j_2\ j_3}, S_1^{k_1} S_2^{k_2} S_3^{k_3} \mathbf{x}_{k_1\ k_2\ k_3} \right\rangle \geq 0,$$

hence (S_1, S_2, S_3) is a subnormal triple and the proof of Theorem 1 is complete. □

The next theorem extends to triple indexed sequences the moment problem solved in [8] for subnormal pairs. The analogue for commuting pairs of contractions has been solved by P. Găvruţă and D. Păunescu in [3] using regular dilations; see B. Sz.-Nagy and C. Foiaş's book ([11]).

Theorem 2. *Let* $(\mathbf{h}_{m\ n\ p})_{m,n,p\in\mathbb{N}}$ *be a triple indexed sequence of the Hilbert space* $\mathcal{H}$ *such that:*

- $(\mathbf{h}_{m\ n\ p})_{m,n,p\in\mathbb{N}}$ *spans the entire space* $\mathcal{H}$;
- $\|\mathbf{h}_{m\ n\ p}\| \leq \Lambda^{m+n+p}$ $(m, n, p \in \mathbb{N})$ *for some positive number* Λ.

The following affirmations are equivalent:

(i) *There exists a subnormal triple* (S_1, S_2, S_3) *such that*

$$\mathbf{h}_{m\ n\ p} = S_1^m S_2^n S_3^p \mathbf{h}_{0\ 0\ 0} \text{ for } m, n, p \in \mathbb{N}.$$

(ii) *There exists* $\left(\mathbf{h}_{m\ n\ p}^{m'n'p'}\right)$ *a family of vectors in* $\mathcal{H}$ *with the following properties:*

1. $\mathbf{h}_{mnp}^{0\,0\,0} = \mathbf{h}_{m\ n\ p}$ *for* $m, n, p \in \mathbb{N}$;
2. $\left\langle \mathbf{h}_{m\ n\ p}^{m'n'p'}, \mathbf{h}_{i\ j\ k} \right\rangle = \langle \mathbf{h}_{m\ n\ p}, \mathbf{h}_{i+m'\ j+n'\ k+p'} \rangle$ *for* m, m', n, n', p, p', i, j, $k \in \mathbb{N}$;
3. *the inequality*

$$\left\| \sum_{\substack{m'n'p'\\ m\ n\ p}} c_{m\ n\ p}^{m'n'p'} \mathbf{h}_{m\ n\ p}^{m'n'p'} \right\|^2$$

$$\leq \sum_{\substack{m'n'p'\\ m\ n\ p}} \sum_{\substack{i'j'k'\\ i\ j\ k}} c_{m\ n\ p}^{m'n'p'} \overline{c_{i\ j\ k}^{i'j'k'}} \langle \mathbf{h}_{m+i'\ n+j'\ p+k'}, \mathbf{h}_{i+m'\ j+n'\ k+p'} \rangle$$

holds for each finite system of complex numbers $\left\{c_{m\ n\ p}^{m'n'p'}\right\}_{m,m',n,n',p,p'\in\mathbb{N}}$.

Proof. *(i)* ⇒ *(ii)* Let (S_1, S_2, S_3) be a commuting subnormal triple of operators in $\mathcal{L}(\mathcal{H})$ and U_1, U_2, U_3 the commuting unitary operators from the characterization Theorem 1. Then

$$S_1^{*m} S_2^{*n} S_3^{*p} = P_{\mathcal{H}} U_1^m U_2^n U_3^p S_1^m S_2^n S_3^p, \tag{1}$$

$$S_1^{*m} S_2^{*n} S_3^{p} = S_3^{*p} P_{\mathcal{H}} U_1^m U_2^n U_3^{*p} S_1^m S_2^n, \tag{2}$$

$$S_1^{*m} S_3^{*p} S_2^{n} = S_2^{*n} P_{\mathcal{H}} U_1^m U_2^{*n} U_3^p S_1^m S_3^p, \tag{3}$$

$$S_1^{*m} S_2^{n} S_3^{p} = S_2^{*n} S_3^{*p} P_{\mathcal{H}} U_1^m U_2^{*n} U_3^{*p} S_1^m \tag{4}$$

(recall that the commuting unitary operators U_1, U_2, U_3 are in the same space $\mathcal{L}(\mathcal{K})$ with the normal extension of the triple (S_1, S_2, S_3) and they are defined using the Furuta unitary operators. We define

$$\mathbf{h}_{m\ n\ p}^{m'n'p'} := S_1^{*m'} S_2^{*n'} S_3^{*p'} S_1^m S_2^n S_3^p \mathbf{h}_{0\ 0\ 0}\ .$$

The identities 1 and 2 are immediate since

$$\mathbf{h}_{m\ n\ p}^{0\ 0\ 0} = S_1^{*0} S_2^{*0} S_3^{*0} S_1^m S_2^n S_3^p \mathbf{h}_{0\ 0\ 0} = S_1^m S_2^n S_3^p \mathbf{h}_{0\ 0\ 0} = \mathbf{h}_{m\ n\ p}$$

respectively

$$\left\langle \mathbf{h}_{m\ n\ p}^{m'n'p'}, \mathbf{h}_{i\ j\ k} \right\rangle = \left\langle S_1^{*m'} S_2^{*n'} S_3^{*p'} S_1^m S_2^n S_3^p \mathbf{h}_{0\ 0\ 0}, \mathbf{h}_{i\ j\ k} \right\rangle$$

$$= \left\langle S_1^m S_2^n S_3^p \mathbf{h}_{0\ 0\ 0}, S_3^{p'} S_2^{n'} S_1^{m'} \mathbf{h}_{i\ j\ k} \right\rangle = \left\langle \mathbf{h}_{m\ n\ p}, S_1^{i+m'} S_2^{j+n'} S_3^{k+p'} \mathbf{h}_{0\ 0\ 0} \right\rangle$$

$$= \left\langle \mathbf{h}_{m\ n\ p}, \mathbf{h}_{i+m'\ j+n'\ k+p'} \right\rangle \quad \text{hold.}$$

Inequality 3 follows from the characterization of subnormal triples of operators.

$$\left\| \sum_{\substack{m'n'p' \\ m\ n\ p}} c_{m\ n\ p}^{m'n'p'} \mathbf{h}_{m\ n\ p}^{m'n'p'} \right\|_{\mathcal{H}}^2$$

$$= \left\| \sum_{\substack{m'n'p' \\ m\ n\ p}} c_{m\ n\ p}^{m'n'p'} \underbrace{S_1^{*m'} S_2^{*n'} S_3^{*p'}} S_1^m S_2^n S_3^p \mathbf{h}_{0\ 0\ 0} \right\|_{\mathcal{H}}^2$$

$$= \left\| \sum_{\substack{m'n'p' \\ m\ n\ p}} c_{m\ n\ p}^{m'n'p'} \underbrace{P_{\mathcal{H}} U_1^{m'} U_2^{n'} U_3^{p'} S_1^{m'} S_2^{n'} S_3^{p'}} S_1^m S_2^n S_3^p \mathbf{h}_{0\ 0\ 0} \right\|_{\mathcal{H}}^2$$

$$\leq \left\| \sum_{\substack{m'n'p' \\ m\ n\ p}} c_{m\ n\ p}^{m'n'p'} U_1^{m'} U_2^{n'} U_3^{p'} S_1^{m+m'} S_2^{n+n'} S_3^{p+p'} \mathbf{h}_{0\ 0\ 0} \right\|_{\mathcal{K}}^2$$

$$= \sum_{\substack{m'n'm' \\ m\ n\ p}} \sum_{\substack{i'j'k' \\ i\ j\ k}} c_{m\ n\ p}^{m'n'p'} \cdot \overline{c_{i\ j\ k}^{i'j'k'}}$$

$$\cdot \left\langle U_1^{m'} U_2^{n'} U_3^{p'} S_1^{m+m'} S_2^{n+n'} S_3^{p+p'} \mathbf{h}_{0\ 0\ 0}, U_1^{i'} U_2^{j'} U_3^{k'} S_1^{i+i'} S_2^{j+j'} S_3^{k+k'} \mathbf{h}_{0\ 0\ 0} \right\rangle_{\mathcal{K}}.$$

In order to compute the inner product

$$(\mathbf{P}) = \left\langle U_1^{m'} U_2^{n'} U_3^{p'} S_1^{m+m'} S_2^{n+n'} S_3^{p+p'} \mathbf{h}_{0\ 0\ 0}, U_1^{i'} U_2^{j'} U_3^{k'} S_1^{i+i'} S_2^{j+j'} S_3^{k+k'} \mathbf{h}_{0\ 0\ 0} \right\rangle_{\mathcal{K}},$$

this sum splits for every indices m, n, p, i, j, k in 2^3 summands as follows:

- if $m' \geq i'$, $n' \geq j'$ and $p' \geq k'$

$$(\mathbf{P}) = \left\langle P_{\mathcal{H}} U_1^{m'-i'} U_2^{n'-j'} U_3^{p'-k'} S_1^{m+m'} S_2^{n+n'} S_3^{p+p'} \mathbf{h}_{0\,0\,0}, S_1^{i+i'} S_2^{j+j'} S_3^{k+k'} \mathbf{h}_{0\,0\,0} \right\rangle_{\mathcal{H}}$$

$$= \Big\langle \underbrace{P_{\mathcal{H}} U_1^{m'-i'} U_2^{n'-j'} U_3^{p'-k'} S_1^{m'-i'} S_2^{n'-j'} S_3^{p'-k'}} S_1^{m+i'} S_2^{n+j'} S_3^{p+k'} \mathbf{h}_{0\,0\,0}, S_1^{i+i'} S_2^{j+j'} S_3^{k+k'} \mathbf{h}_{0\,0\,0} \Big\rangle_{\mathcal{H}}$$

$$\overset{(1)}{=} \Big\langle \underbrace{S_1^{*(m'-i')} S_2^{*(n'-j')} S_3^{*(p'-k')}} S_1^{m+i'} S_2^{n+j'} S_3^{p+k'} \mathbf{h}_{0\,0\,0}, S_1^{i+i'} S_2^{j+j'} S_3^{k+k'} \mathbf{h}_{0\,0\,0} \Big\rangle_{\mathcal{H}}$$

$$= \left\langle S_1^{m+i'} S_2^{n+j'} S_3^{p+k'} \mathbf{h}_{0\,0\,0}, S_1^{i+m'} S_2^{j+n'} S_3^{k+p'} \mathbf{h}_{0\,0\,0} \right\rangle_{\mathcal{H}} ;$$

- if $m' < i'$, $n' \geq j'$ and $p' \geq k'$

$$(\mathbf{P}) = \Big\langle S_1^{m+m'} S_2^{n+n'} S_3^{p+p'} \mathbf{h}_{0\,0\,0}, P_{\mathcal{H}} U_1^{i'-m'} U_2^{*(n'-j')} U_3^{*(p'-k')} S_1^{i+i'} S_2^{j+j'} S_3^{k+k'} \mathbf{h}_{0\,0\,0} \Big\rangle_{\mathcal{H}}$$

$$= \Big\langle S_1^{m+m'} S_2^{n+j'} S_3^{p+k'} \mathbf{h}_{0\,0\,0}, \underbrace{S_2^{*(n'-j')} S_3^{*(p'-k')} P_{\mathcal{H}} U_1^{i'-m'} U_2^{*(n'-j')} U_3^{*(p'-k')} S_1^{i'-m'}} \cdot S_1^{i+m'} S_2^{j+j'} S_3^{k+k'} \mathbf{h}_{0\,0\,0} \Big\rangle_{\mathcal{H}}$$

$$\overset{(4)}{=} \Big\langle S_1^{m+m'} S_2^{n+j'} S_3^{p+k'} \mathbf{h}_{0\,0\,0}, \underbrace{S_1^{*(i'-m')} S_2^{n'-j'} S_3^{p'-k'}} S_1^{i+m'} S_2^{j+j'} S_3^{k+k'} \mathbf{h}_{0\,0\,0} \Big\rangle_{\mathcal{H}}$$

$$= \left\langle S_1^{m+i'} S_2^{n+j'} S_3^{p+k'} \mathbf{h}_{0\,0\,0}, S_1^{i+m'} S_2^{j+n'} S_3^{k+p'} \mathbf{h}_{0\,0\,0} \right\rangle_{\mathcal{H}} ;$$

- if $m' \geq i'$, $n' < j'$ and $p' \geq k'$

$$(\mathbf{P}) = \Big\langle P_{\mathcal{H}} U_1^{m'-i'} U_2^{*(j'-n')} U_3^{p'-k'} S_1^{m+m'} S_2^{n+n'} S_3^{p+p'} \mathbf{h}_{0\,0\,0}, S_1^{i+i'} S_2^{j+j'} S_3^{k+k'} \mathbf{h}_{0\,0\,0} \Big\rangle_{\mathcal{H}}$$

$$= \Big\langle \underbrace{S_2^{*(j'-n')} P_{\mathcal{H}} U_1^{m'-i'} U_2^{*(j'-n')} U_3^{p'-k'} S_1^{m'-i'} S_3^{p'-k'}} S_1^{m+i'} S_2^{n+n'} S_3^{p+k'} \mathbf{h}_{0\,0\,0}, S_1^{i+i'} S_2^{j+n'} S_3^{k+k'} \mathbf{h}_{0\,0\,0} \Big\rangle_{\mathcal{H}}$$

$$\overset{(3)}{=} \Big\langle \underbrace{S_1^{*(m'-i')} S_3^{*(p'-k')} S_2^{j'-n'}} S_1^{m+i'} S_2^{n+n'} S_3^{p+k'} \mathbf{h}_{0\,0\,0}, S_1^{i+i'} S_2^{j+n'} S_3^{k+k'} \mathbf{h}_{0\,0\,0} \Big\rangle_{\mathcal{H}}$$

$$= \left\langle S_1^{m+i'} S_2^{n+j'} S_3^{p+k'} \mathbf{h}_{0\,0\,0}, S_1^{i+m'} S_2^{j+n'} S_3^{k+p'} \mathbf{h}_{0\,0\,0} \right\rangle_{\mathcal{H}} ;$$

- if $m' < i'$, $n' < j'$ and $p' \geq k'$

$$(\mathbf{P}) = \Big\langle S_1^{m+m'} S_2^{n+n'} S_3^{p+p'} \mathbf{h}_{0\,0\,0}, P_{\mathcal{H}} U_1^{i'-m'} U_2^{j'-n'} U_3^{*(p'-k')} S_1^{i+i'} S_2^{j+j'} S_3^{k+k'} \mathbf{h}_{0\,0\,0} \Big\rangle_{\mathcal{H}}$$

$$= \Big\langle S_1^{m+m'} S_2^{n+n'} S_3^{p+k'} \mathbf{h}_{0\,0\,0}, \underbrace{S_3^{*(p'-k')} P_{\mathcal{H}} U_1^{i'-m'} U_2^{j'-n'} U_3^{*(p'-k')} S_1^{i'-m'} S_2^{j'-n'}} S_1^{i+m'} S_2^{j+n'} S_3^{k+k'} \mathbf{h}_{0\,0\,0} \Big\rangle_{\mathcal{H}}$$

$$\overset{(2)}{=} \Big\langle S_1^{m+m'} S_2^{n+n'} S_3^{p+k'} \mathbf{h}_{0\,0\,0}, \underbrace{S_1^{*(i'-m')} S_2^{*(j'-n')} S_3^{p'-k'}} S_1^{i+m'} S_2^{j+n'} S_3^{k+k'} \mathbf{h}_{0\,0\,0} \Big\rangle_{\mathcal{H}}$$

$$= \Big\langle S_1^{m+i'} S_2^{n+j'} S_3^{p+k'} \mathbf{h}_{0\,0\,0}, S_1^{i+m'} S_2^{j+n'} S_3^{k+p'} \mathbf{h}_{0\,0\,0} \Big\rangle_{\mathcal{H}} ;$$

- if $m' < i'$, $n' \geq j'$ and $p' < k'$

$$(\mathbf{P}) = \Big\langle S_1^{m+m'} S_2^{n+n'} S_3^{p+p'} \mathbf{h}_{0\,0\,0}, P_{\mathcal{H}} U_1^{i'-m'} U_2^{*(n'-j')} U_3^{k'-p'} S_1^{i+i'} S_2^{j+j'} S_3^{k+k'} \mathbf{h}_{0\,0\,0} \Big\rangle_{\mathcal{H}}$$

$$= \Big\langle S_1^{m+m'} S_2^{n+j'} S_3^{p+p'} \mathbf{h}_{0\,0\,0}, \underbrace{S_2^{*(n'-j')} P_{\mathcal{H}} U_1^{i'-m'} U_2^{*(n'-j')} U_3^{k'-p'} S_1^{i'-m'} S_3^{k'-p'}} S_1^{i+m'} S_2^{j+j'} S_3^{k+p'} \mathbf{h}_{0\,0\,0} \Big\rangle_{\mathcal{H}}$$

$$\overset{(3)}{=} \Big\langle S_1^{m+m'} S_2^{n+j'} S_3^{p+p'} \mathbf{h}_{0\,0\,0}, \underbrace{S_1^{*(i'-m')} S_3^{*(k'-p')} S_2^{n'-j'}} S_1^{i+m'} S_2^{j+j'} S_3^{k+p'} \mathbf{h}_{0\,0\,0} \Big\rangle_{\mathcal{H}}$$

$$= \Big\langle S_1^{m+i'} S_2^{n+j'} S_3^{p+k'} \mathbf{h}_{0\,0\,0}, S_1^{i+m'} S_2^{j+n'} S_3^{k+p'} \mathbf{h}_{0\,0\,0} \Big\rangle_{\mathcal{H}} ;$$

- if $m' \geq i'$, $n' < j'$ and $p' < k'$

$$(\mathbf{P}) = \Big\langle P_{\mathcal{H}} U_1^{m'-i'} U_2^{*(j'-n')} U_3^{*(k'-p')} S_1^{m+m'} S_2^{n+n'} S_3^{p+p'} \mathbf{h}_{0\,0\,0}, S_1^{i+i'} S_2^{j+j'} S_3^{k+k'} \mathbf{h}_{0\,0\,0} \Big\rangle_{\mathcal{H}}$$

$$= \Big\langle \underbrace{S_2^{*(j'-n')} S_3^{*(k'-p')} P_{\mathcal{H}} U_1^{m'-i'} U_2^{*(j'-n')} U_3^{*(k'-p')} S_1^{m'-i'}} S_1^{m+i'} S_2^{n+n'} S_3^{p+p'} \mathbf{h}_{0\,0\,0}, S_1^{i+i'} S_2^{j+n'} S_3^{k+p'} \mathbf{h}_{0\,0\,0} \Big\rangle_{\mathcal{H}}$$

$$\overset{(4)}{=} \Big\langle \underbrace{S_1^{*(m'-i')} S_2^{j'-n'} S_3^{k'-p'}} S_1^{m+i'} S_2^{n+n'} S_3^{p+p'} \mathbf{h}_{0\,0\,0}, S_1^{i+i'} S_2^{j+n'} S_3^{k+p'} \mathbf{h}_{0\,0\,0} \Big\rangle_{\mathcal{H}}$$

$$= \Big\langle S_1^{m+i'} S_2^{n+j'} S_3^{p+k'} \mathbf{h}_{0\,0\,0}, S_1^{i+m'} S_2^{j+n'} S_3^{k+p'} \mathbf{h}_{0\,0\,0} \Big\rangle_{\mathcal{H}} ;$$

- if $m' < i'$, $n' < j'$ and $p' < k'$

$$(\mathbf{P}) = \left\langle S_1^{m+m'} S_2^{n+n'} S_3^{p+p'} \mathbf{h}_{0\ 0\ 0}, P_{\mathcal{H}} U_1^{i'-m'} U_2^{j'-n'} U_3^{k'-p'} S_1^{i+i'} S_2^{j+j'} S_3^{k+k'} \mathbf{h}_{0\ 0\ 0} \right\rangle_{\mathcal{H}}$$

$$= \Big\langle S_1^{m+m'} S_2^{n+n'} S_3^{p+p'} \mathbf{h}_{0\ 0\ 0}, \underbrace{P_{\mathcal{H}} U_1^{i'-m'} U_2^{j'-n'} U_3^{k'-p'} S_1^{i'-m'} S_2^{j'-n'} S_3^{k'-p'}} S_1^{i+m'} S_2^{j+n'} S_3^{k+p'} \mathbf{h}_{0\ 0\ 0} \Big\rangle_{\mathcal{H}}$$

$$\stackrel{(1)}{=} \Big\langle S_1^{m+m'} S_2^{n+n'} S_3^{p+p'} \mathbf{h}_{0\ 0\ 0}, \underbrace{S_1^{*(i'-m')} S_2^{*(j'-n')} S_3^{*(k'-p')}} S_1^{i+m'} S_2^{j+n'} S_3^{k+p'} \mathbf{h}_{0\ 0\ 0} \Big\rangle_{\mathcal{H}}$$

$$= \left\langle S_1^{m+i'} S_2^{n+j'} S_3^{p+k'} \mathbf{h}_{0\ 0\ 0}, S_1^{i+m'} S_2^{j+n'} S_3^{k+p'} \mathbf{h}_{0\ 0\ 0} \right\rangle_{\mathcal{H}} .$$

By adding the previous equalities, we obtain:

$$\Big\| \sum_{\substack{m'n'p' \\ m\ n\ p}} c_{m\ n\ p}^{m'n'p'} \mathbf{h}_{m\ n\ p}^{m'n'p'} \Big\|_{\mathcal{H}}^2$$

$$\leq \sum_{\substack{m'n'p' \\ m\ n\ p}} \sum_{\substack{i'j'k' \\ i\ j\ k}} c_{m\ n\ p}^{m'n'p'} \overline{c_{i\ j\ k}^{i'j'k'}} \left\langle S_1^{m+i'} S_2^{n+j'} S_3^{p+k'} \mathbf{h}_{0\ 0\ 0}, S_1^{i+m'} S_2^{j+n'} S_3^{k+p'} \mathbf{h}_{0\ 0\ 0} \right\rangle_{\mathcal{H}}$$

$$= \sum_{\substack{m'n'p' \\ m\ n\ p}} \sum_{\substack{i'j'k' \\ i\ j\ k}} c_{m\ n\ p}^{m'n'p'} \overline{c_{i\ j\ k}^{i'j'k'}} \left\langle \mathbf{h}_{m+i'\ n+j'\ p+k'}, \mathbf{h}_{i+m'\ j+n'\ k+p'} \right\rangle_{\mathcal{H}} .$$

(ii) ⇒ (i) follows from the classical Kolmogorov-Aronszajn-Pedrick kernel theorem. Let $\left(\mathbf{h}_{m\ n\ p}^{m'n'p'}\right)_{m,m',n,n',p,p'}$ be a six indexed sequence of vectors in $\mathcal{H}$ such that:

$$\begin{aligned} \mathbf{h}_{m\ n\ p}^{0\ 0\ 0} &= \mathbf{h}_{m\ n\ p} \\ \left\langle \mathbf{h}_{m\ n\ p}^{m'n'p'}, \mathbf{h}_{i\ j\ k} \right\rangle &= \left\langle \mathbf{h}_{m\ n\ p}, \mathbf{h}_{i+m'\ j+n'\ k+p'} \right\rangle \end{aligned} \qquad (m, m', n, n', p, p', i, j, k \in \mathbb{N})$$

and

$$\Big\| \sum_{\substack{m'n'p' \\ m\ n\ p}} c_{m\ n\ p}^{m'n'p'} \mathbf{h}_{m\ n\ p}^{m'n'p'} \Big\|_{\mathcal{H}}^2$$

$$\leq \sum_{\substack{m'n'p' \\ m\ n\ p}} \sum_{\substack{i'j'k' \\ i\ j\ k}} c_{m\ n\ p}^{m'n'p'} \overline{c_{i\ j\ k}^{i'j'k'}} \left\langle \mathbf{h}_{m+i'\ n+j'\ p+k'}, \mathbf{h}_{i+m'\ j+n'\ k+p'} \right\rangle$$

holds for each finite system of complex numbers $\left\{c_{m\ n\ p}^{m'n'p'}\right\}_{m,m',n,n',p,p' \in \mathbb{N}}$.

We define the kernel

$$\Phi = \left(\varphi^{(m',i')(n',j')(p',k')}_{(m,\,i)\,(n,\,j)\,(p,\,k)}\right)_{(m,i),(n,j),(p,k),(m',i'),(n',j'),(p',k')\in\mathbb{N}\times\mathbb{N}},$$

where

$$\varphi^{(m',i')(n',j')(p',k')}_{(m,\,i)\,(n,\,j)\,(p,\,k)} : \mathbb{C}\times\mathbb{C}\to\mathbb{C}\,,$$
$$\varphi^{(m',i')(n',j')(p',k')}_{(m,\,i)\,(n,\,j)\,(p,\,k)}(x,y) = x\overline{y}\,\langle \mathbf{h}_{m+i'\;n+j'\;p+k'}, \mathbf{h}_{i+m'\;j+n'\;k+p'}\rangle\,.$$

It follows immediately that Φ is a positive definite kernel:

$$\begin{aligned}
&\sum \varphi^{(m',i')(n',j')(p',k')}_{(m,\,i)\,(n,\,j)\,(p,\,k)}\left(x^{m'n'p'}_{m\,n\,p}, x^{i'j'k'}_{i\,j\,k}\right)\\
&= \sum_{\substack{m'n'p'\\ m\,n\,p}}\sum_{\substack{i'j'k'\\ i\,j\,k}} x^{m'n'p'}_{m\,n\,p}\overline{x^{i'j'k'}_{i\,j\,k}}\left(\mathbf{h}_{m+i',n+j',p+k'}, \mathbf{h}_{i+m',j+n',k+p'}\right)\\
&\geq \left\|\sum_{\substack{m'n'p'\\ m\,n\,p}} x^{m'n'p'}_{m\,n\,p}\mathbf{h}^{m'n'p'}_{m\,n\,p}\right\|^2 \geq 0
\end{aligned}$$

Then there exist a complex Hilbert space $\mathbf{H}$ and a family of linear operators $E^{m'n'p'}_{m\,n\,p} : \mathbb{C}\to\mathbf{H}$ such that

$$\varphi^{(m',i')(n',j')(p',k')}_{(m,\,i)\,(n,\,j)\,(p,\,k)}(\cdot,\cdot) = \left[\left(E^{i'j'k'}_{i\,j\,k}\right)^* E^{m'n'p'}_{m\,n\,p}(\cdot)\right](\cdot)$$

$$\mathbf{H} = \overline{\mathbf{H}}_0 \text{where} \mathbf{H}_0 = Sp\left\{E^{m'n'p'}_{m\,n\,p}(x) \mid m,m',n,n',p,p'\in\mathbb{N},\ x\in\mathbb{C}\right\}.$$

The inner product on $\mathbf{H}_0$ is given by

$$\begin{aligned}
&\left\langle \sum_{\substack{m'n'p'\\ m\,n\,p}}\sum_{\alpha,\beta,\gamma} E^{m'n'p'}_{m\,n\,p}(x_{\alpha\;\beta\;\gamma}), \sum_{\substack{i'j'k'\\ i\,j\,k}}\sum_{\rho,\theta,\sigma} E^{i'j'k'}_{i\,j\,k}(y_{\rho\;\theta\;\sigma})\right\rangle\\
&= \sum_{\substack{m'n'p'\\ m\,n\,p}}\sum_{\substack{i'j'k'\\ i\,j\,k}}\sum_{\alpha,\beta,\gamma}\sum_{\rho,\theta,\sigma} x_{\alpha\;\beta\;\gamma}\overline{y_{\rho\;\theta\;\sigma}}\langle \mathbf{h}_{m+i'\;n+j'\;p+k'}, \mathbf{h}_{i+m'\;j+n'\;k+p'}\rangle\,.
\end{aligned}$$

We define the operators

$$\begin{aligned}
N_{01} &: \mathbf{H}_0\to\mathbf{H}_0\,,\quad N_{01}\left(E^{m'n'p'}_{m\,n\,p}(x)\right) = E^{m'\quad n'p'}_{m+1\,n\,p}(x)\\
N_{02} &: \mathbf{H}_0\to\mathbf{H}_0\,,\quad N_{02}\left(E^{m'n'p'}_{m\,n\,p}(x)\right) = E^{m'\;n'\quad p'}_{m\,n+1\,p}(x)\\
N_{03} &: \mathbf{H}_0\to\mathbf{H}_0\,,\quad N_{03}\left(E^{m'n'p'}_{m\,n\,p}(x)\right) = E^{m'n'\;p'}_{m\,n\,p+1}(x)
\end{aligned}$$

and

$$V_0 : \mathbf{H}_0\to\mathcal{H}\,,\quad V_0\left(E^{m'n'p'}_{m\,n\,p}(x)\right) = x\mathbf{h}^{m'n'p'}_{m\,n\,p}\,.$$

More generally, we remark that N_{01} (and in the same manner N_{02} and N_{03}), respectively V_0 has the form

$$N_{01}\left(\sum_{\substack{m'n'p'\\ m\,n\,p}}\sum_{\alpha,\beta,\gamma} E^{m'n'p'}_{m\,n\,p}(x_{\alpha\;\beta\;\gamma})\right) = \sum_{\substack{m'n'p'\\ m\,n\,p}}\sum_{\alpha,\beta,\gamma} E^{m'\quad n'p'}_{m+1\,n\,p}(x_{\alpha\;\beta\;\gamma})\,,$$

respectively

$$V_0\Bigg(\sum_{\substack{m'n'p'\\ m\,n\,p}}\sum_{\alpha,\beta,\gamma} E^{m'n'p'}_{m\,n\,p}(x_{\alpha\ \beta\ \gamma})\Bigg) = \sum_{\substack{m'n'p'\\ m\,n\,p}}\sum_{\alpha,\beta,\gamma} x_{\alpha\ \beta\ \gamma}\mathbf{h}^{m'n'p'}_{m\,n\,p}\,.$$

We will prove that N_{01} is bounded.

$$\Bigg\|N_{01}\Bigg(\sum_{\substack{m'n'p'\\ m\,n\,p}}\sum_{\alpha,\beta,\gamma} E^{m'n'p'}_{m\,n\,p}(x_{\alpha\ \beta\ \gamma})\Bigg)\Bigg\|^2$$

$$= \Bigg\|\sum_{\substack{m'n'p'\\ m\,n\,p}}\sum_{\alpha,\beta,\gamma} E^{m'\quad n'p'}_{m+1\,n\,p}(x_{\alpha\ \beta\ \gamma})\Bigg\|^2$$

$$= \sum_{\substack{m'n'p'\\ m\,n\,p}}\sum_{\substack{i'j'k'\\ i\,j\,k}}\sum_{\alpha,\beta,\gamma}\sum_{\rho,\theta,\sigma} x_{\alpha\ \beta\ \gamma}\overline{x_{\rho\ \theta\ \sigma}}\langle \mathbf{h}_{m+1+i'\ n+j'\ p+k'}, \mathbf{h}_{i+1+m'\ j+n'\ k+p'}\rangle$$

Now, we fix ν, ν' in $\mathbb{N}$ and we remark that the Cauchy-Schwarz-Buniakowski inequality implies:

$$\Bigg\|\sum_{\substack{m'n'p'\\ m\,n\,p}}\sum_{\alpha,\beta,\gamma} E^{m'+\nu'n'p'}_{m+\nu\ n\,p}(x_{\alpha\ \beta\ \gamma})\Bigg\|^2$$

$$= \sum_{\substack{m'n'p'\\ m\,n\,p}}\sum_{\substack{i'j'k'\\ i\,j\,k}}\sum_{\alpha,\beta,\gamma}\sum_{\rho,\theta,\sigma} x_{\alpha\ \beta\ \gamma}\overline{x_{\rho\ \theta\ \sigma}}\langle \mathbf{h}_{m+\nu+i'+\nu'\ n+j'\ p+k'}, \mathbf{h}_{i+\nu+m'+\nu'\ j+n'\ k+p'}\rangle$$

$$= \Bigg\langle\sum_{\substack{m'n'p'\\ m\,n\,p}}\sum_{\alpha,\beta,\gamma} E^{m'+\nu'+\nu\ n'p'}_{m+\nu+\nu'\quad n\,p}(x_{\alpha\ \beta\ \gamma}), \sum_{\substack{i'j'k'\\ i\,j\,k}}\sum_{\rho,\theta,\sigma} E^{i'j'k'}_{i\,j\,k}(x_{\rho\ \theta\ \sigma})\Bigg\rangle$$

$$\overset{CSB}{\leq} \Bigg\|\sum_{\substack{m'n'p'\\ m\,n\,p}}\sum_{\alpha,\beta,\gamma} E^{m'+\nu'+\nu\ n'p'}_{m+\nu+\nu'\quad n\,p}(x_{\alpha\ \beta\ \gamma})\Bigg\|\Bigg\|\sum_{\substack{i'j'k'\\ i\,j\,k}}\sum_{\rho,\theta,\sigma} E^{i'j'k'}_{i\,j\,k}(x_{\rho\ \theta\ \sigma})\Bigg\|.$$

Evaluating the norm of N_{01}, we obtain:

- for $\nu = 1$ and $\nu' = 0$:

$$\Bigg\|N_{01}\Bigg(\sum_{\substack{m'n'p'\\ m\,n\,p}}\sum_{\alpha,\beta,\gamma} E^{m'n'p'}_{m\,n\,p}(x_{\alpha\ \beta\ \gamma})\Bigg)\Bigg\|^2$$

$$= \Bigg\|\sum_{\substack{m'n'p'\\ m\,n\,p}}\sum_{\alpha,\beta,\gamma} E^{m'\quad n'p'}_{m+1\,n\,p}(x_{\alpha\ \beta\ \gamma})\Bigg\|^2$$

$$\overset{\substack{\nu=1\\ \nu'=0}}{\leq} \Bigg\|\sum_{\substack{m'n'p'\\ m\,n\,p}}\sum_{\alpha,\beta,\gamma} E^{m'+1n'p'}_{m+1\ \,n\,p}(x_{\alpha\ \beta\ \gamma})\Bigg\|\cdot\Bigg\|\sum_{\substack{i'j'k'\\ i\,j\,k}}\sum_{\rho,\theta,\sigma} E^{i'j'k'}_{i\,j\,k}(x_{\rho\ \theta\ \sigma})\Bigg\|;$$

- for $\nu = 1$ and $\nu' = 1$:

$$\left\| N_{01}\Big(\sum_{\substack{m'n'p' \\ m\ n\ p}} \sum_{\alpha,\beta,\gamma} E^{m'n'p'}_{m\ n\ p}(x_{\alpha\ \beta\ \gamma}) \Big) \right\|^{2^2}$$

$$\leq \left\| \sum_{\substack{m'n'p' \\ m\ n\ p}} \sum_{\alpha,\beta,\gamma} E^{m'+1n'p'}_{m+1\ n\ p}(x_{\alpha\ \beta\ \gamma}) \right\|^2 \cdot \left\| \sum_{\substack{i'j'k' \\ i\ j\ k}} \sum_{\rho,\theta,\sigma} E^{i'j'k'}_{i\ j\ k}(x_{\rho\ \theta\ \sigma}) \right\|^2$$

$$\overset{\substack{\nu=1 \\ \nu'=1}}{\leq} \left\| \sum_{\substack{m'n'p' \\ m\ n\ p}} \sum_{\alpha,\beta,\gamma} E^{m'+2n'p'}_{m+2\ n\ p}(x_{\alpha\ \beta\ \gamma}) \right\| \cdot \left\| \sum_{\substack{i'j'k' \\ i\ j\ k}} \sum_{\rho,\theta,\sigma} E^{i'j'k'}_{i\ j\ k}(x_{\rho\ \theta\ \sigma}) \right\|^{1+2};$$

- for $\nu = 2$ and $\nu' = 2$:

$$\left\| N_{01}\Big(\sum_{\substack{m'n'p' \\ m\ n\ p}} \sum_{\alpha,\beta,\gamma} E^{m'n'p'}_{m\ n\ p}(x_{\alpha\ \beta\ \gamma}) \Big) \right\|^{2^3}$$

$$\leq \left\| \sum_{\substack{m'n'p' \\ m\ n\ p}} \sum_{\alpha,\beta,\gamma} E^{m'+2n'p'}_{m+2\ n\ p}(x_{\alpha\ \beta\ \gamma}) \right\|^2$$

$$\cdot \left\| \sum_{\substack{i'j'k' \\ i\ j\ k}} \sum_{\rho,\theta,\sigma} E^{i'j'k'}_{i\ j\ k}(x_{\rho\ \theta\ \sigma}) \right\|^{2+2^2}$$

$$\overset{\substack{\nu=2 \\ \nu'=2}}{\leq} \left\| \sum_{\substack{m'n'p' \\ m\ n\ p}} \sum_{\alpha,\beta,\gamma} E^{m'+2^2n'p'}_{m+2^2\ n\ p}(x_{\alpha\ \beta\ \gamma}) \right\|$$

$$\cdot \left\| \sum_{\substack{i'j'k' \\ i\ j\ k}} \sum_{\rho,\theta,\sigma} E^{i'j'k'}_{i\ j\ k}(x_{\rho\ \theta\ \sigma}) \right\|^{1+2+2^2}.$$

- In the general case, for $\nu = 2^K$ and $\nu' = 2^K$ it results that:

$$\left\| N_{01}\Big(\sum_{\substack{m'n'p' \\ m\ n\ p}} \sum_{\alpha,\beta,\gamma} E^{m'n'p'}_{m\ n\ p}(x_{\alpha\ \beta\ \gamma}) \Big) \right\|^{2^{K+1}}$$

$$\leq \left\| \sum_{\substack{m'n'p' \\ m\ n\ p}} \sum_{\alpha,\beta,\gamma} E^{m'+2^Kn'p'}_{m+2^K\ n\ p}(x_{\alpha\ \beta\ \gamma}) \right\|$$

$$\cdot \left\| \sum_{\substack{i'j'k' \\ i\ j\ k}} \sum_{\rho,\theta,\sigma} E^{i'j'k'}_{i\ j\ k}(x_{\rho\ \theta\ \sigma}) \right\|^{1+2+2^2+\cdots+2^K}.$$

The given sequence $(\mathbf{h}_{m\ n\ p})$ is power bounded, i.e., $\|\mathbf{h}_{m\ n\ p}\| \leq \Lambda^{m+n+p}$ (for some positive number Λ), hence:

$$\left\| \sum_{\substack{m'n'p'\\ m\,n\,p}} \sum_{\alpha,\beta,\gamma} E_{m+2^K\ \ n\ p}^{m'+2^K n' p'} (x_{\alpha\ \beta\ \gamma}) \right\|^2$$

$$= \sum_{\substack{m'n'p'\\ m\,n\,p}} \sum_{\substack{i'j'k'\\ i\,j\,k}} \sum_{\alpha,\beta,\gamma} \sum_{\rho,\theta,\sigma} x_{\alpha\ \beta\ \gamma} \overline{x_{\rho\ \theta\ \sigma}}$$

$$\cdot \left\langle \mathbf{h}_{m+i'+2^{K+1}\ n+j'\ p+k'}, \mathbf{h}_{i+m'+2^{K+1}\ j+n'\ k+p'} \right\rangle$$

$$\leq \sum_{\substack{m'n'p'\\ m\,n\,p}} \sum_{\substack{i'j'k'\\ i\,j\,k}} \sum_{\alpha,\beta,\gamma} \sum_{\rho,\theta,\sigma} |x_{\alpha\ \beta\ \gamma}| \left|\overline{x_{\rho\ \theta\ \sigma}}\right|$$

$$\cdot \left| \left\langle \mathbf{h}_{m+i'+2^{K+1}\ n+j'\ p+k'}, \mathbf{h}_{i+m'+2^{K+1}\ j+n'\ k+p'} \right\rangle \right|$$

$$\leq \sum_{\substack{m'n'p'\\ m\,n\,p}} \sum_{\substack{i'j'k'\\ i\,j\,k}} \sum_{\alpha,\beta,\gamma} \sum_{\rho,\theta,\sigma} |x_{\alpha\ \beta\ \gamma}| \left|\overline{x_{\rho\ \theta\ \sigma}}\right| \left\| \mathbf{h}_{m+i'+2^{K+1}\ n+j'\ p+k'} \right\|$$

$$\cdot \left\| \mathbf{h}_{i+m'+2^{K+1}\ j+n'\ k+p'} \right\|$$

$$\leq \sum_{\alpha,\beta,\gamma} \sum_{\rho,\theta,\sigma} |x_{\alpha\ \beta\ \gamma}| \left|\overline{x_{\rho\ \theta\ \sigma}}\right|$$

$$\cdot \sum_{\substack{m'n'p'\\ m\,n\,p}} \sum_{\substack{i'j'k'\\ i\,j\,k}} \Lambda^{m+i'+2^{K+1}+n+j'+p+k'} \Lambda^{i+m'+2^{K+1}+j+n'+k+p'}$$

$$= \left(\Lambda^{2^{K+1}}\right)^2$$

$$\underbrace{\cdot \sum_{\alpha,\beta,\gamma} \sum_{\rho,\theta,\sigma} |x_{\alpha\ \beta\ \gamma}| \left|\overline{x_{\rho\ \theta\ \sigma}}\right| \sum_{\substack{m'n'p'\\ m\,n\,p}} \sum_{\substack{i'j'k'\\ i\,j\,k}} \Lambda^{m+i'+n+j'+p+k'} \Lambda^{i+m'+j+n'+k+p'}}_{\overset{not}{=}\ r}.$$

Using the last evaluation for the norm of N_{01}, we will obtain the desired result. Indeed, the inequality

$$\left\| N_{01}\left(\sum_{\substack{m'n'p'\\ m\,n\,p}} \sum_{\alpha,\beta,\gamma} E_{m\ n\ p}^{m'n'p'} (x_{\alpha\ \beta\ \gamma}) \right) \right\|^{2^{K+1}}$$

$$\leq \Lambda^{2^{K+1}} r \left\| \sum_{\substack{m'n'p'\\ m\,n\,p}} \sum_{\alpha,\beta,\gamma} E_{m\ n\ p}^{m'n'p'} (x_{\alpha\ \beta\ \gamma}) \right\|^{2^{K+1}-1}$$

is equivalent to

$$\left\| N_{01}\Bigg(\sum_{\substack{m'n'p' \\ m\,n\,p}} \sum_{\alpha,\beta,\gamma} E^{m'n'p'}_{m\,n\,p}(x_{\alpha\ \beta\ \gamma}) \Bigg) \right\|$$

$$\leq \Lambda r^{2^{-K-1}} \left\| \sum_{\substack{m'n'p' \\ m\,n\,p}} \sum_{\alpha,\beta,\gamma} E^{m'n'p'}_{m\,n\,p}(x_{\alpha\ \beta\ \gamma}) \right\|^{1-2^{-K-1}},$$

for every positive integer K, hence, for $K \to \infty$ it results

$$\left\| N_{01}\Bigg(\sum_{\substack{m'n'p' \\ m\,n\,p}} \sum_{\alpha,\beta,\gamma} E^{m'n'p'}_{m\,n\,p}(x_{\alpha\ \beta\ \gamma}) \Bigg) \right\| \leq \Lambda \cdot \left\| \sum_{\substack{m'n'p' \\ m\,n\,p}} \sum_{\alpha,\beta,\gamma} E^{m'n'p'}_{m\,n\,p}(x_{\alpha\ \beta\ \gamma}) \right\|$$

In the same manner, we obtain the boundedness of N_{02} and N_{03}.

$$\left\| N_{02}\Bigg(\sum_{\substack{m'n'p' \\ m\,n\,p}} \sum_{\alpha,\beta,\gamma} E^{m'n'p'}_{m\,n\,p}(x_{\alpha\ \beta\ \gamma}) \Bigg) \right\| \leq \Lambda \left\| \sum_{\substack{m'n'p' \\ m\,n\,p}} \sum_{\alpha,\beta,\gamma} E^{m'n'p'}_{m\,n\,p}(x_{\alpha\ \beta\ \gamma}) \right\|$$

$$\left\| N_{03}\Bigg(\sum_{\substack{m'n'p' \\ m\,n\,p}} \sum_{\alpha,\beta,\gamma} E^{m'n'p'}_{m\,n\,p}(x_{\alpha\ \beta\ \gamma}) \Bigg) \right\| \leq \Lambda \left\| \sum_{\substack{m'n'p' \\ m\,n\,p}} \sum_{\alpha,\beta,\gamma} E^{m'n'p'}_{m\,n\,p}(x_{\alpha\ \beta\ \gamma}) \right\|.$$

Each operator of the triple (N_{01}, N_{02}, N_{03}) commutes with the others. Indeed,

$$N_{01}N_{02}\Bigg(\sum_{\substack{m'n'p' \\ m\,n\,p}} \sum_{\alpha,\beta,\gamma} E^{m'n'p'}_{m\,n\,p}(x_{\alpha\ \beta\ \gamma}) \Bigg)$$

$$= N_{01}\Bigg(\sum_{\substack{m'n'p' \\ m\,n\,p}} \sum_{\alpha,\beta,\gamma} E^{m'\ n'\ p'}_{m n+1 p}(x_{\alpha\ \beta\ \gamma}) \Bigg)$$

$$= \sum_{\substack{m'n'p' \\ m\,n\,p}} \sum_{\alpha,\beta,\gamma} E^{\ m'\ \ n'\ p'}_{m+1 n+1 p}(x_{\alpha\ \beta\ \gamma})$$

$$= N_{02}\Bigg(\sum_{\substack{m'n'p' \\ m\,n\,p}} \sum_{\alpha,\beta,\gamma} E^{\ m'\ n'p'}_{m+1 n\ p}(x_{\alpha\ \beta\ \gamma}) \Bigg)$$

$$= N_{02}N_{01}\Bigg(\sum_{\substack{m'n'p' \\ m\,n\,p}} \sum_{\alpha,\beta,\gamma} E^{m'n'p'}_{m\,n\,p}(x_{\alpha\ \beta\ \gamma}) \Bigg).$$

The other identities follow in the same manner.

The operator V_0 is a contraction:

$$\left\| V_0\Big(\sum_{\substack{m'n'p'\\ m\,n\,p}} \sum_{\alpha,\beta,\gamma} E^{m'n'p'}_{m\,n\,p}(x_{\alpha\,\beta\,\gamma})\Big)\right\| = \left\| \sum_{\substack{m'n'p'\\ m\,n\,p}} \sum_{\alpha,\beta,\gamma} x_{\alpha\,\beta\,\gamma}\mathbf{h}^{m'n'p'}_{m\,n\,p}\right\|$$

$$\leq \sqrt{\sum_{\substack{m'n'p'\\ m\,n\,p}} \sum_{\substack{i'j'k'\\ i\,j\,k}} \sum_{\alpha,\,\beta,\,\gamma} \sum_{\rho,\,\theta,\,\sigma} x_{\alpha\,\beta\,\gamma}\overline{x_{\rho\,\theta\,\sigma}}\,\langle \mathbf{h}_{m+i'\;n+j'\;p+k'}, \mathbf{h}_{i+m'\;j+n'\;k+p'}\rangle}$$

$$= \sqrt{\left\langle \sum_{\substack{m'n'p'\\ m\,n\,p}} \sum_{\alpha,\,\beta,\,\gamma} E^{m'n'p'}_{m\,n\,p}(x_{\alpha\,\beta\,\gamma}), \sum_{\substack{m'n'p'\\ m\,n\,p}} \sum_{\alpha,\,\beta,\,\gamma} E^{m'n'p'}_{m\,n\,p}(x_{\alpha\,\beta\,\gamma})\right\rangle}$$

$$= \left\| \sum_{\substack{m'n'p'\\ m\,n\,p}} \sum_{\alpha,\,\beta,\,\gamma} E^{m'n'p'}_{m\,n\,p}(x_{\alpha\,\beta\,\gamma})\right\| .$$

These linear bounded operators on $\mathbf{H}_0$ extend to $\mathbf{H}$ with the same norm; let us denote N_1, N_2, N_3 and V these extensions.

Now we shall study the adjoint of V; to do this it is enough to study the behavior of V^* on the sequence $(\mathbf{h}_{m\,n\,p})$:

$$\left\langle \sum_{\substack{m'n'p'\\ m\,n\,p}} \sum_{\alpha,\,\beta,\,\gamma} E^{m'n'p'}_{m\,n\,p}(x_{\alpha\,\beta\,\gamma}), V^*\mathbf{h}_{i\;j\;k}\right\rangle$$

$$= \left\langle V\Big(\sum_{\substack{m'n'p'\\ m\,n\,p}} \sum_{\alpha,\,\beta,\,\gamma} E^{m'n'p'}_{m\,n\,p}(x_{\alpha\,\beta\,\gamma})\Big), \mathbf{h}_{i\;j\;k}\right\rangle$$

$$= \left\langle \sum_{\substack{m'n'p'\\ m\,n\,p}} \sum_{\alpha,\,\beta,\,\gamma} x_{\alpha\,\beta\,\gamma}\,\mathbf{h}^{\,m'n'p'}_{\,m\,n\,p}, \mathbf{h}_{\,i\;j\;k}\right\rangle$$

$$= \sum_{\substack{m'n'p'\\ m\,n\,p}} \sum_{\alpha,\,\beta,\,\gamma} x_{\alpha\,\beta\,\gamma}\,\langle \mathbf{h}_{\,m\;n\;p}, \mathbf{h}_{\,i+m'\;j+n'\;k+p'}\rangle .$$

On the other hand:

$$\left\langle \sum_{\substack{m'n'p'\\ m\,n\,p}} \sum_{\alpha,\,\beta,\,\gamma} E^{m'n'p'}_{m\,n\,p}(x_{\alpha\,\beta\,\gamma}), E^{0\,0\,0}_{i\,j\,k}(1)\right\rangle$$

$$\sum_{\substack{m'n'p'\\ m\,n\,p}} \sum_{\alpha,\,\beta,\,\gamma} x_{\alpha\,\beta\,\gamma}\,\langle \mathbf{h}_{\,m\;n\;p}, \mathbf{h}_{\,i+m'\;j+n'\;k+p'}\rangle ,$$

hence we have:

$$V^*\mathbf{h}_{i\;j\;k} = E^{0\,0\,0}_{i\,j\,k}(1) .$$

Finally, we define S_1, S_2 and S_3 :

$$S_1 = VN_1V^*,\ S_2 = VN_2V^* \text{ and } S_3 = VN_3V^*.$$

The operators S_1 and S_2 commute:

$$\begin{aligned} S_1S_2\mathbf{h}_{i\ j\ k} &= VN_1V^*VN_2V^*\mathbf{h}_{i\ j\ k} = VN_1V^*VN_2E_{i\ j\ k}^{0\ 0\ 0}(1) \\ &= VN_1V^*VE_{ij+1k}^{0\ \ 0\ \ 0}(1) = VN_1V^*\mathbf{h}_{ij+1k}^{0\ \ 0\ \ 0} \\ &= VN_1V^*\mathbf{h}_{i\,j+1\,k} = VN_1E_{ij+1k}^{0\ \ 0\ \ 0}(1) \\ &= VE_{i+1j+1k}^{0\ \ \ 0\ \ 0} = (1)\,\mathbf{h}_{i+1j+1k}^{0\ \ \ 0\ \ 0} = \mathbf{h}_{i+1\ j+1\ k}\ ; \end{aligned}$$

$$\begin{aligned} S_2S_1\mathbf{h}_{i\ j\ k} &= VN_2V^*VN_1V^*\mathbf{h}_{i\ j\ k} = VN_2V^*VN_1E_{i\ j\ k}^{0\ 0\ 0}(1) \\ &= VN_2V^*VE_{i+1jk}^{0\ \ 0\ 0}(1) = VN_2V^*\mathbf{h}_{i+1jk}^{0\ \ 0\ 0} \\ &= VN_2V^*\mathbf{h}_{i+1\ j\ k} = VN_2E_{i+1jk}^{0\ \ 0\ 0}(1) \\ &= VE_{i+1j+1k}^{0\ \ \ 0\ \ 0}(1) = \mathbf{h}_{i+1j+1k}^{0\ \ \ 0\ \ 0} = \mathbf{h}_{i+1\,j+1\,k}\,. \end{aligned}$$

In the same manner, it follows the other two relations $S_1S_3 = S_3S_1$ and $S_2S_3 = S_3S_2$.

We notice that

$$S_1^mS_2^nS_3^p\mathbf{h}_{i\ j\ k} = \mathbf{h}_{m+i\ n+j\ p+k} \text{ for } m,n,p,i,j,k \in \mathbb{N}$$

thus $S_1^mS_2^nS_3^p\mathbf{h}_{0\ 0\ 0} = \mathbf{h}_{m\ n\ p}$.

We prove that (S_1, S_2, S_3) is a subnormal triple verifying Itô's condition. Let

$$\{\mathbf{g}_{m'\ n'\ p'}\}_{m',n',p'\in\mathbb{N}}\ ,\quad \mathbf{g}_{m'\ n'\ p'} = \sum_{m,n,p} x_{m\ n\ p}^{m'n'p'}\mathbf{h}_{m\ n\ p},$$

be a finite system of vectors in $\mathcal{H}$, then we have

$$\begin{aligned} &\sum_{\substack{i'j'k'\\m'n'p'}} \left\langle S_1^{i'}S_2^{j'}S_3^{k'}\mathbf{g}_{m'\ n'\ p'}, S_1^{m'}S_2^{n'}S_3^{p'}\mathbf{g}_{i'\ j'\ k'}\right\rangle \\ &= \sum_{\substack{i'j'k'\\m'n'p'}} \left\langle S_1^{i'}S_2^{j'}S_3^{k'}\left(\sum_{m,n,p} x_{m\ n\ p}^{m'n'p'}\mathbf{h}_{m\ n\ p}\right), S_1^{m'}S_2^{n'}S_3^{p'}\left(\sum_{i,j,k} x_{i\ j\ k}^{i'j'k'}\mathbf{h}_{i\ j\ k}\right)\right\rangle \\ &= \sum_{\substack{i'\ j'k'\\m'n'p'}}\sum_{\substack{m\ n\ p\\i\ j\ k}} x_{m\ n\ p}^{m'n'p'}\overline{x_{i\ j\ k}^{i'j'k'}}\left\langle S_1^{i'}S_2^{j'}S_3^{k'}\mathbf{h}_{m\ n\ p}, S_1^{m'}S_2^{n'}S_3^{p'}\mathbf{h}_{i\ j\ k}\right\rangle \\ &= \sum_{\substack{i'j'k'\\m'n'p'}}\sum_{\substack{m\ n\ p\\i\ j\ k}} x_{m\ n\ p}^{m'n'p'}\overline{x_{i\ j\ k}^{i'j'k'}}\left\langle \mathbf{h}_{m+i'\ n+j'\ p+k'}, \mathbf{h}_{i+m'\ j+n'\ k+p'}\right\rangle \\ &\geq \left\| \sum_{\substack{m'n'p'\\m\ n\ p}} x_{m\ n\ p}^{m'n'p'}\mathbf{h}_{m\ n\ p}^{m'n'p'}\right\|^2 \geq 0\,. \end{aligned}$$

□

Using similar arguments, we are able to prove the following analogue of Theorem 2 for operators:

Theorem 3. *Let* $(A_{m\ n\ p})_{m,n,p\in\mathbb{N}}$ *be a triple indexed sequence of operators in* $\mathcal{L}(\mathcal{H})$ *such that:*

- *the ranges of* $A_{m\ n\ p}$ $(m,n,p\in\mathbb{N})$ *span the entire space* $\mathcal{H}$;
- $\|A_{m\ n\ p}\| \le \Lambda^{m+n+p}$ $(m,n,p\in\mathbb{N})$ *for some positive number* Λ.

The following affirmations are equivalent:

(i) *There exists a subnormal triple* (S_1, S_2, S_3) *such that*

$$A_{m\ n\ p} = S_1^m S_2^n S_3^p A_{0\ 0\ 0} \text{ for } m,n,p\in\mathbb{N}.$$

(ii) *There exists* $\left(A_{m\ n\ p}^{m'n'p'}\right)$ *a family of operators in* $\mathcal{L}(\mathcal{H})$ *with the following properties:*

1. $A_{m\ n\ p}^{0\ 0\ 0} = A_{m\ n\ p}$ *for* $m,n,p\in\mathbb{N}$.
2. $(A_{i\ j\ k})^* A_{m\ n\ p}^{m'n'p'} = (A_{i+m'\ j+n'\ k+p'})^* A_{m\ n\ p}$ $(m,\ m',\ n,\ n',\ p,\ p',\ i,\ j,\ k\in\mathbb{N})$;
3. *the inequality*

$$\left\| \sum_{\substack{m'n'p'\\ m\ n\ p}} A_{m\ n\ p}^{m'n'p'} \mathbf{x}_{m\ n\ p}^{m'n'p'} \right\|^2$$

$$\le \sum_{\substack{m'n'p'\\ m\ n\ p}} \sum_{\substack{i'j'k'\\ i\ j\ k}} \left\langle A_{m+i'\ n+j'\ p+k'} \mathbf{x}_{m\ n\ p}^{m'n'p'}, A_{i+m'\ j+n'\ k+p'} \mathbf{x}_{i\ j\ k}^{i'j'k'} \right\rangle,$$

holds for each finite system of vectors $\left\{\mathbf{x}_{m\ n\ p}^{m'n'p'}\right\}_{m,m',n,n',p,p'\in\mathbb{N}} \subset \mathcal{H}$.

References

[1] J. Bram, *Subnormal Operators.* Duke Math. J. **22** (1955), 75–94.

[2] T. Furuta, *On the Polar Decomposition of an Operator.* Acta Sci. Math. (Szeged) **46** (1983), 261–268.

[3] P. Găvruţă, D. Păunescu, *Sebestyén's Moment Problem and Regular Dilations*, Acta. Math. Hung. **94**(3) (2002), 223–232.

[4] P. Găvruţă, N. Suciu, *A Characterization of Subnormal Pair.* Proceedings of 16th OT Conference, Theta 1997, 149–157.

[5] T. Itô, *On the Commutative Family of Subnormal Operators.* J. Fac. Sci. Hokkaido Univ. Ser. I **14** (1958), 1–15.

[6] P. Masani, *Dilations as Propagators of Hilbertian Varieties.* SIAM J. Math. Anal. **9**(3) (1978), 414–456.

[7] D. Păunescu, *Some Moment Problems in Hilbert Spaces.* I. Proceedings of the 9th Symposium of Math. and Appl., Timişoara 2001, 110–117.

[8] D. Păunescu, *Some Moment Problems in Hilbert Spaces.* II. (to be published in Bull. Appl. Comp. Math., Tech. Univ. of Budapesta, 2002).

[9] Z. Sebestyén, *Moment Theorems for Operators on Hilbert Spaces*, Acta Sci. Math. (Szeged) **44** (1982), 165–171.

[10] Z. Sebestyén, *Moment Theorems for Operators on Hilbert Spaces*. II. Acta Sci. Math. (Szeged) **47** (1984), 101–106.

[11] B. Sz.-Nagy, C. Foiaş, *Analyse harmonique des opérateurs de l'espace de Hilbert*. Masson et Cie, Akadémiai Kiadó, Paris, Budapest, 1967.

[12] T.T. Trent, *New Conditions for Subnormalit*. Pacific J. Math. **93** (1981), 459–464.

Tudor Bînzar
University "Politehnica" of Timişoara
Piaţa Regina Maria Nr.1
1900 Timişoara
Romania
e-mail: `Tudor_Binzar@hotmail.com`

Doru Păunescu
University "Politehnica" of Timişoara
Piaţa Regina Maria Nr.1
1900 Timişoara
Romania
e-mail: `dpaunesc@etv.utt.ro`

Operator Theory:
Advances and Applications, Vol. 153, 61–78

The Equality of the Reduced and the Full C^*-Algebras and the Amenability of a Topological Groupoid

Mădălina Buneci

Abstract. C. Anantharaman-Delaroche and J. Renault have proved that the amenability of a topological locally compact groupoid implies the equality of the reduced and the full C^*-algebras. In this paper we shall prove the converse assertion under a technical hypothesis. We shall prove that if G is a locally compact second countable groupoid endowed with a Haar system having "a bounded decomposition over the principal groupoid associated to G", then the equality $C^*_{\text{red}}(G) = C^*(G)$ implies the amenability of all quasi-invariant measures. In order to prove this we shall see that the inequality $\|II_\mu(f)\| \leq \|\text{Reg}_\mu(f)\|$ for all $f \in C_c(G)$ implies a similar inequality for all $f \in I(G, \nu, \mu)$ (where Reg_μ is the left regular representation of $C_c(G)$ on a quasi invariant measure μ, and II_μ is the trivial representation on μ).

Mathematics Subject Classification (2000). 22A22, 43A07, 46L52, 43A65.

Keywords. Locally compact groupoid, Amenable quasi-invariant measure, Full C^*-algebra, Reduced C^*-algebra.

1. Introduction

For establishing notation, we include some definitions that can be found in several places (e.g., [5], [7], [9], [12]). A groupoid is a set G endowed with a product map

$$(x, y) \to xy \left[: G^{(2)} \to G \right]$$

where $G^{(2)}$ is a subset of $G \times G$ called the set of composable pairs, and an inverse map

$$x \to x^{-1} \ [: G \to G]$$

such that the following conditions hold:

(1) If $(x, y) \in G^{(2)}$ and $(y, z) \in G^{(2)}$, then $(xy, z) \in G^{(2)}$, $(x, yz) \in G^{(2)}$ and $(xy)\,z = x\,(yz)$.
(2) $\left(x^{-1}\right)^{-1} = x$ for all $x \in G$.
(3) For all $x \in G$, $\left(x, x^{-1}\right) \in G^{(2)}$, and if $(z, x) \in G^{(2)}$, then $(zx)\,x^{-1} = z$.
(4) For all $x \in G$, $\left(x^{-1}, x\right) \in G^{(2)}$, and if $(x, y) \in G^{(2)}$, then $x^{-1}\,(xy) = y$.

The maps r and d on G, defined by the formulae $r(x) = xx^{-1}$ and $d(x) = x^{-1}x$, are called the range and the source maps. It follows easily from the definition that they have a common image called the unit space of G, which is denoted $G^{(0)}$. Its elements are units in the sense that $xd(x) = r(x)\,x = x$. Units will usually be denoted by letters as u, v, w while arbitrary elements will be denoted by x, y, z. It is useful to note that a pair (x, y) lies in $G^{(2)}$ precisely when $d(x) = r(y)$, and that the cancellation laws hold (e.g., $xy = xz$ iff $y = z$). The fibres of the range and the source maps are denoted $G^u = r^{-1}(\{u\})$ and $G_v = d^{-1}(\{v\})$, respectively. More generally, given the subsets A, $B \subset G^{(0)}$, we define $G^A = r^{-1}(A)$, $G_B = d^{-1}(B)$ and $G_B^A = r^{-1}(A) \cap d^{-1}(B)$. The reduction of G to $A \subset G^{(0)}$ is $G|A = G_A^A$. The relation $u\tilde{\ }v$ iff $G_v^u \neq \phi$ is an equivalence relation on $G^{(0)}$. Its equivalence classes are called orbits and the orbit of a unit u is denoted $[u]$. A groupoid is called transitive iff it has a single orbit. The quotient space for this equivalence relation is called the orbit space of G and denoted $G^{(0)}/G$. $\pi : G^{(0)} \to G^{(0)}/G, \pi(u) = \dot{u}$ is the canonical projection. A subset of $G^{(0)}$ is said saturated if it contains the orbits of its elements.

A topological groupoid consists of a groupoid G and a topology compatible with the groupoid structure. This means that:

(1) $x \to x^{-1}$ $[: G \to G]$ is continuous.
(2) (x, y) $\left[: G^{(2)} \to G\right]$ is continuous where $G^{(2)}$ has the induced topology from $G \times G$.

We are exclusively concerned with topological groupoids which are second countable, locally compact Hausdorff. It was shown in [10] that measured groupoids (in the sense of Definition 2.3./p. 6 [5]) may be assume to have locally compact topologies, with no loss in generality. A subset U of G is said conditionally compact if for every compact subset K of $G^{(0)}$, $U \cap r^{-1}(K)$ and $U \cap d^{-1}(K)$ is compact in G. If G is locally compact and $G^{(0)}$ is paracompact then $G^{(0)}$ has a fundamental system of conditionally compact neighborhoods (Proposition II.1.9/p.56 [12]). If X is a locally compact space, $C_c(X)$ denotes the space of complex-valuated continuous functions with compact support. The Borel sets of a topological space are taken to be the σ-algebra generated by the open sets.

Let G be a locally compact second countable groupoid equipped with a Haar system, i.e., a family of positive Radon measures on G, $\left\{\nu^u,\, u \in G^{(0)}\right\}$, such that

1) For all $u \in G^{(0)}$, $\mathrm{supp}(\nu^u) = G^u$.
2) For all $f : G \to \mathbf{C}$ continuous with compact support,

$$u \to \int f(x)\, d\nu^u(x) \quad \left[: G^{(0)} \to \mathbf{C}\right] \quad \text{is continuous.}$$

3) For all $f : G \to \mathbf{C}$ continuous with compact support, and all $x \in G$,

$$\int f(y)\, d\nu^{r(x)}(y) = \int f(xy)\, d\nu^{d(x)}(y)$$

As a consequence of the existence of continuous Haar systems, $r, d : G \to G^{(0)}$ are open maps ([14]).

The construction of the C^*-algebra of a groupoid extends the well-known case of a group. The space of continuous functions with compact support on groupoid is made into a $*$-algebra and endowed with the smallest C^*-norm making its representations continuous. Let $C_c(G)$ be the space of continuous functions with compact support on the groupoid G. For $f,\, g \in C_c(G)$ the convolution is defined by:

$$f * g(x) = \int f(xy)\, g\left(y^{-1}\right) d\nu^{d(x)}(y)$$

and the involution by

$$f^*(x) = \overline{f\left(x^{-1}\right)}.$$

Under these operations, $C_c(G)$ becomes a topological $*$-algebra. A representation of $C_c(G)$ is a $*$-homomorphism from $C_c(G)$ into $\mathcal{B}(H)$, for some Hilbert space H, that is continuous with respect to the inductive limit topology on $C_c(G)$ and the weak operator topology on $\mathcal{B}(H)$. The full C^*-algebra $C^*(G)$ is defined as the completion of the involutive algebra $C_c(G)$ with respect to the full C^*-norm

$$\|f\| = \sup \|L(f)\|$$

where runs over all non-degenerate representation of $C_c(G)$ which are continuous for the inductive limit topology. Let us single out a special class of representations of $C_c(G)$ that serve as analogues of the regular representation of a group. If μ is a measure on $G^{(0)}$, then the measure $\nu = \int \nu^u d\mu(u)$, defined by

$$\int f(y)\, d\nu(y) = \int \left(\int f(y)\, d\nu^u(y) \right) d\mu(u), \quad f \geq 0 \quad \text{Borel}$$

is called the measure on G induced by μ. The image of ν by the inverse map $x \to x^{-1}$ is denoted ν^{-1}. μ is said quasi-invariant if its induced measure ν is equivalent to its inverse ν^{-1}. A measure belongings to the class of a quasi-invariant measure is also quasi-invariant. We say that the class is invariant. If μ quasi-invariant measure, then $\mathrm{Ind}\,\mu(f)$ is the operator on $L^2\left(G, \nu^{-1}\right)$ defined by formula

$$\mathrm{Ind}\,\mu(f)\,\xi(x) = f * \xi(x)$$

and $\mathrm{Ind}_u(f)$ is the operator on $L^2\left(G, (\nu^u)^{-1}\right)$ defined by formula

$$\mathrm{Ind}_u(f)\,\xi(x) = f * \xi(x) = \int f(xy)\,\xi\left(y^{-1}\right) d\nu^u(y).$$

The reduced C^*-algebra $C^*_{\mathrm{red}}(G)$ is defined as the completion of the involutive algebra $C_c(G)$ with respect to the reduced C^*-norm

$$\|f\| = \sup \|\mathrm{Ind}_u(f)\|.$$

If μ is a quasi-invariant measure with $\mathrm{supp}(\mu) = G^{(0)}$, then

$$\|f\|_{\mathrm{red}} = \|\mathrm{Ind}\,\mu(f)\|$$

If μ is a quasi-invariant measure on $G^{(0)}$ and ν is the measure induced on G, then the Radon-Nikodym derivative $\Delta = \frac{d\nu}{d\nu^{-1}}$ is called the modular function of μ. Let $\nu_0 = \Delta^{-\frac{1}{2}}\nu$.

For $f \in L^1(G, \nu_0)$ define

$$\|f\|_{I,\mu} = \max\left\{\left\|u \to \int |f| d\nu^u\right\|_{\infty}, \left\|u \to \int |f^*| d\nu^u\right\|_{\infty}\right\}.$$

Let $I(G, \nu, \mu) = \left\{f \in L^1(G, \nu_0), \|f\|_{I,\mu} < \infty\right\}$.

Under the convolution and the involution, $I(G, \nu, \mu)$ becomes a Banach $*$-algebra.

Every representation $(\mu, G^{(0)} * \mathcal{H}, L)$(see Definition 3.20/p.68 [7]) of G can be integrated into a representation, still denoted by L, of $I(G, \nu, \mu)$. The relation between the two representation is:

$$\langle L(f)\xi_1, \xi_2\rangle = \int f(x) \langle L(x)\xi_1(d(x)), \xi_2(r(x))\rangle \Delta^{-\frac{1}{2}}(x)\, d\nu^u(x)\, d\mu_1(u)$$

where $f \in I(G, \nu, \mu)$, $\xi_1, \xi_2 \in \int_{G^{(0)}}^{\oplus} \mathcal{H}(u)\, d\mu(u)$.

Conversely, every non-degenerate $*$-representation of any suitably large $*$-algebra of $I(G, \nu, \mu)$ is obtained in this fashion (see [6], [12]).

We denote by $C^*(G, \nu, \mu)$ the completion of $I(G, \nu, \mu)$ with respect to the norm

$$\|f\| = \sup \|L(f)\|$$

where L ranges over all representation of (G, ν, μ).

The (left) regular representation of G on μ is the representation

$$\left(\mu, G^{(0)} * L^{(2)}(\nu), \mathrm{Reg}\,\mu\right)$$

where $\mathrm{Reg}(x) : L^2\left(\nu^{d(x)}\right) \to L^2\left(\nu^{r(x)}\right)$ is defined by the formula

$$\mathrm{Reg}(x)\xi(d(x))(y) = \xi\left(x^{-1}y\right).$$

The integrated form of this representation is called the (left) regular representation of $C_c(G)$ (also, of $C^*(G)$ on μ). The map

$$W : L^2(\nu) \to L^2\left(\nu^{-1}\right), W\xi = \xi\Delta^{\frac{1}{2}}$$

is a Hilbert space isomorphism that implements a unitary equivalence between Reg_μ and $\mathrm{Ind}\,\mu$ (II.1.10 [12]). Therefore, if μ is a quasi invariant measure with $\mathrm{supp}(\mu) = G^{(0)}$ and Reg_μ is the left regular representation of $C_c(G)$ on μ, then

$$\|f\|_{\mathrm{red}} = \left\|\mathrm{Reg}_\mu(f)\right\|.$$

In [1], C. Anantharaman-Delaroche and J. Renault have proved the following theorem:

Theorem 1.1 (Proposition 6.1.8/p. 146 [1]). *Let G be a locally compact groupoid equipped with a continuous Haar system $\{\nu^u, u \in G^{(0)}\}$. If G is measurewise amenable (amenable with respect to all quasi invariant measures), then $C^*(G) = C^*_{\text{red}}(G)$.*

We shall prove that under a technical hypothesis the converse assertion is true. We shall use the following result:

Theorem 1.2. (Theorem 6.1.4, p.142 [1]). *The following conditions are equivalent:*

(1) *(G, ν, μ) is amenable.*
(2) *The trivial representation of $C^*(G, \nu, \mu)$ is weakly contained in the regular representation.*
(3) *The regular representation is faithful on $C^*(G, \nu, \mu)$.*

We shall also use the decomposition

$$\nu^u = \int \nu_{u,v} d\eta_u(v) \quad \text{for all } u \in G^{(0)}$$

of the Haar system for G over the principal groupoid associated to G (see Section1 [13], $\nu_{u,v}$ is supported on G^u_v for all $u\tilde{\ }v$, η_u is supported on $[u]$ for all u and $\eta_u = \eta_v$ for all $u\tilde{\ }v$). We shall assume that the decomposition of the Haar system over the principal groupoid is bounded. This means that there is a positive Radon measure η_0 on $G^{(0)}$ such that the family $\{\eta_u\}_{u \in G^{(0)}}$ is dominated by η_0, i.e.,

$$\eta_u(f) \le \eta_0(f), \text{ for all positive function } f \in C_c\left(G^{(0)}\right).$$

Obviously, locally compact second countable transitive groupoids and locally compact groupoids for which the applications $d_u : G^u \to G^{(0)}, d_u(x) = d(x)$ are open (in particular, locally trivial groupoids) satisfy the hypothesis.

We shall prove that if the decomposition of the Haar system over the principal groupoid is bounded, then the equality $C^*_{\text{red}}(G) = C^*(G)$ implies the amenability of all quasi-invariant measures. In [3] we have proved the same result for transitive locally compact second countable groupoids. In order to prove the result for groupoids endowed with Haar systems having bounded decompositions over the associated principal groupoids, we shall show that the inequality $\|II\mu(f)\| \le \left\|\text{Reg}_\mu(f)\right\|$ for all $f \in C_c(G)$ implies a similar inequality for all $f \in I(G, \nu, \mu)$ (where Reg_μ is the left regular representation of $C_c(G)$ on a quasi invariant measure μ, and II_μ is the trivial representation on μ).

2. Functions approximated by continuous functions in the norm of $C^*(G, \nu, \mu)$

First we present some results on the structure of the Haar systems, as developed by J. Renault in Section 1 of [13] and also by A. Ramsay and M.E. Walter in Section 2 of [11]. Let $\{\nu^u, u \in G^{(0)}\}$ be a continuous Haar system on G.

In Section 1 of [13] Jean Renault constructs a Borel Haar system for G'. One way to do this is to choose a function F_0 continuous with conditionally support which is nonnegative and equal to 1 at each $u \in G^{(0)}$. Then for each $u \in G^{(0)}$ choose a left Haar measure $\nu_{u,u}$ on G_u^u so the integral of F_0 with respect to $\nu_{u,u}$ is 1.

Renault defines $\nu_{u,v} = x\nu_{v,v}$ if $x \in G_v^u$ (where $x\nu_{u,u}(f) = \int f(xy)\, d\nu_{v,v}(y)$ as usual). If z is another element in G_v^u, then $x^{-1}z \in G_v^v$, and since $\nu_{v,v}$ is a left Haar measure on G_v^v, it follows that $\nu_{u,v}$ is independent of the choice of x. If K is a compact subset of G, then $\sup\limits_{u,v} \nu_{u,v}(K) < \infty$. Renault also defines a 1-cocycle δ on G such that for every $u \in G^{(0)}$, $\delta|_{G_u^u}$ is the modular function for $\nu_{u,u}$. δ and $\delta^{-1} = 1/\delta$ are bounded on compact sets in G.

Let

$$R = (r,d)(G) = \{(r(x), d(x)),\ x \in G\}$$

be the graph of the equivalence relation induced on $G^{(0)}$. This R is the image of G under the homomorphism (r,d), so it is a σ-compact groupoid. With this apparatus in place, Renault describes a decomposition of the Haar system $\{\nu^u, u \in G^{(0)}\}$ for G over the equivalence relation R (the principal groupoid associated to G). He proves that there is a unique Borel Haar system α for R with the property that

$$\nu^u = \int \nu_{s,t}\, d\alpha^u(s,t) \quad \text{for all } u \in G^{(0)}.$$

In Section 2 of [11] A. Ramsay and M.E. Walter prove that

$$\sup_u \alpha^u((r,d)(K)) < \infty, \text{ for all compact } K \subset G$$

If μ is a quasi-invariant measure for $\{\nu^u, u \in G^{(0)}\}$, then μ is a quasi-invariant measure for $\{\alpha^u, u \in G^{(0)}\}$. Also if Δ_R is the modular function associated to $\{\alpha^u, u \in G^{(0)}\}$ and μ, then $\Delta = \delta\Delta_R \circ (r,d)$ can serve as the modular function associated to $\{\nu^u, u \in G^{(0)}\}$ and μ.

For each $u \in G^{(0)}$ the measure α^u is concentrated on $\{u\} \times [u]$. Therefore there is a measure η_u concentrated on $[u]$ such that $\alpha^u = \varepsilon_u \times \eta_u$, where ε_u is the unit point mass at u. Since $\{\alpha^u, u \in G^{(0)}\}$ is a Haar system, we have $\eta_u = \eta_v$ for all $(u,v) \in R$, and the function

$$u \to \int f(s)\, \eta_u(s)$$

is Borel for all $f \geq 0$ Borel on $G^{(0)}$. For each u the measure η_u is quasi-invariant (Section 2, [11]) Therefore η_u is equivalent to $d_*(v^u)$ (Lemma 4.5, p. 277, [9]).

Let μ be a quasi-invariant measure and let $\mu_1 = \int \eta_u d\mu(u)$. Then μ_1 is equivalent to μ. Indeed, let $f \geq 0$ Borel on $G^{(0)}$ such that $\mu(f) = 0$.Since μ is quasi-invariant, it follows that for $\mu\, a.a.\ u$ $\nu^u(f \circ d) = 0$, and since η_u is equivalent to $d_*(v^u)$, it results $\eta_u(f) = 0$ for μ $a.a.$ u. Conversely if $\mu_1(f) = 0$, then $\eta_u(f) = 0$ for μ $a.a.$ u, and therefore $\nu^u(f \circ d) = 0$. Thus the quasi-invariance of μ implies $\mu(f) = 0$.

Let α the measure induced by μ_1 on R, and let Δ_R be the modular function of μ_1. Then $\Delta_R = \frac{d\alpha}{d\alpha^{-1}}$. It is easy to note that α is symmetric. Hence $\Delta_R = 1$. If Δ is the modular function associated to $\{\nu^u, u \in G^{(0)}\}$ and μ_1, then $\Delta = \delta \Delta_R \circ (r, d) = \delta$.

The next lemma contains the properties of the decomposition of a Haar system. For its proof see Section 1 of [13], Section 2 of [11], Theorem 4.4. of [5], or [2].

Lemma 2.1. *Let $\{\nu^u, u \in G^{(0)}\}$ be a continuous Haar system on G. Let*

$$\nu^u = \int \nu_{u,v} d\eta_u(v) \quad \text{for all } u \in G^{(0)}$$

be the decomposition of the Haar system for G over the equivalence relation R. Let μ be a quasi-invariant measure and $\mu_1 = \int \eta_u d\mu(u)$. Let Δ be the modular function associated to $\{\nu^u, u \in G^{(0)}\}$ and μ_1. Then

1) *$\nu_{u,v}$ is concentrated on G^u_v, and $\nu_{u,v} \neq 0$, for all $(u,v) \in R$.*
2) *For all $f \geq 0$ Borel on G,*

$$(u,v) \mapsto \int f(y)\, d\nu_{u,v}(y) \ \left[: R \to \overline{\mathbf{R}}\right]$$

is an extended real-valued Borel function.

3) $\sup\limits_{u,v} \nu_{u,v}(K) < \infty$, *for all compact $K \subset G$.*
4) *For all $f \geq 0$ Borel on G,*

$$\int f(xy)\, d\nu_{d(x),v}(y) = \int f(y)\, d\nu_{r(x),v}(y) \quad \text{for all } x \in G,\ v \in [d(x)]$$

5) *For all $f \geq 0$ Borel on G,*

$$\Delta(x) \int f(yx)\, d\nu_{u,r(x)}(y) = \int f(y)\, d\nu_{u,d(x)}(y) \quad \text{for all } x \in G_0,\ u \in [d(x)]$$

6) *$\Delta : G \to \mathbf{R}^*_+$ is a homomorphism.*
7) *Δ and $\Delta^{-1} = 1/\Delta$ are bounded on compact sets in G.*
8) *For all $f \geq 0$ Borel on G,*

$$\int f(y)\, d\nu_{u,v}(y) = \int f(y^{-1})\, \Delta(y^{-1})\, d\nu_{v,u}(y) \quad \text{for all } (u,v) \in R$$

9) $\sup\limits_{u} \varepsilon_u \times \eta_u((r,d)(K)) < \infty$, *for all compact $K \subset G$*

Throughout this section the Haar system $\{\nu^u, u \in G^{(0)}\}$ and the systems of measures in its decomposition (as in preceding lemma) will be considered fixed.

We shall need that $\sup\limits_u \eta_u(K) < \infty$, for all compact $K \subset G^{(0)}$. If there is a positive Radon measure η_0 on $G^{(0)}$ such that the family $\{\eta_u\}_{u \in G^{(0)}}$ is dominated by η_0, i.e.,

$$\eta_u(f) \leq \eta_0(f), \quad \text{for all positive function } f \in C_c\left(G^{(0)}\right),$$

we shall say that *the decomposition of the Haar system over the principal groupoid is bounded.*

Notation 2.2. Let B be a set with the property that it intersects each orbit in exactly one element. Let $e : G^{(0)} \to G^{(0)}$ be the function defined by $e(u)$ is the only element in $B \cap [u]$. Let d_B be the function

$$d_B : G^B \to G^{(0)}, \quad d_B(x) = d(x).$$

Lemma 2.3. *If the function d_B is open, then*

$$\sup_u \eta_u(K) < \infty \quad \textit{for all compact } K \subset G^{(0)}.$$

Proof. Let K be a compact subset of $G^{(0)}$. Since G is locally compact and d_B is continuous and open from G^B onto $G^{(0)}$, there is a compact $L \subset G$, such that $K \subset d_B\left(L \cap G^B\right)$. We have

$$\begin{aligned}\int 1_K(v)\, d\eta_u(v) &= \int 1_K(v)\, d\eta_{e(u)}(v)\\ &\leq \int 1_{d_B(L\cap G^B)}(v)\, d\eta_{e(u)}(v)\\ &= \int 1_{(r,d)(L\cap G^B)}(e(u), v)\, d\eta_{e(u)}(v)\end{aligned}$$

Hence $\sup\limits_u \eta_u(K) < \infty$. □

Example 2.4. Groupoids which satisfy the hypothesis (d_B is open):

1) locally compact transitive groupoids
2) locally compact groupoids for which the applications $d_u : G^u \to G^{(0)}$, $d_u(x) = d(x)$ are open. In particular, locally trivial groupoids satisfy the hypothesis.
3) locally compact group bundles having the bundle maps open.

Remark 2.5. If the orbit space of the groupoid G is discrete with respect to the quotient topology, then any Haar system on G has a bounded decomposition over the principal groupoid associated to G.

Notation 2.6. Let μ be a quasi-invariant probability measure and let $\mu_1 = \int \eta_u d\mu(u)$. Then μ_1 is equivalent to μ. We replace μ with μ_1 (see [4]). Let Δ be the modular function associated to $\left\{\nu^u, u \in G^{(0)}\right\}$ and μ_1 Let ν be the measure induced by μ_1 on G, and $\nu_0 = \Delta^{-\frac{1}{2}}\nu$.

For $f \in L^1(G, \nu_0)$ define f^* by $f^*(y) = \overline{f(y^{-1})}$ and

$$\|f\|_{I,\mu} = \max\left\{\left\|u \to \textstyle\int |f| d\nu^u\right\|_\infty, \left\|u \to \textstyle\int |f^*| d\nu^u\right\|_\infty\right\}$$

$$\|f\|_{II,\mu} = \sup\{\textstyle\int |f(y)\, j(d(y)\, k(r(y)))| \Delta(y)^{-\frac{1}{2}} d\nu(y) :$$
$$j, k \in L^2\left(G^{(0)}, \mu_1\right), \textstyle\int |j|^2 d\mu_1 = \int |k|^2 d\mu_1 = 1\}.$$

It is easy to see that $\|f\|_{L^1(G,\nu_0)} = \|f^*\|_{L^1(G,\nu_0)} \leq \|f\|_{II,\mu} = \|f^*\|_{II,\mu} \leq \|f\|_{I,\mu} = \|f^*\|_{I,\mu}$.

Let $I(G,\nu,\mu)=\left\{f\in L^1(G,\nu_0),\|f\|_{I,\mu}<\infty\right\}$
Let $II(G,\nu,\mu)=\left\{f\in L^1(G,\nu),\|f\|_{II,\mu}<\infty\right\}$

Remark 2.7. If L is the integrated form of a representation, $\left(\mu,G^{(0)}*\mathcal{H},L\right)$, of the groupoid G, then

$$|\langle L(f)\xi,\eta\rangle|\leq\left\langle II_\mu(|f|)\widetilde{\xi},\widetilde{\eta}\right\rangle$$

where $\widetilde{\xi}(u)=\|\xi(u)\|$. Therefore $\|L(f)\|\leq\|II_\mu(|f|)\|=\|f\|_{II,\mu}\leq\|f\|_{I,\mu}$.

Lemma 2.8. *Let us assume that the decomposition of the Haar system over the principal groupoid is bounded. Let $f\in L^1(G,\nu_0)$ such that $f_\Delta\in L^\infty\left(G^{(0)},\mu\right)$, where for μ-a.a $w\in G^{(0)}$*

$$f_\Delta(w)=\iint\left(\int|f(x)|\Delta(x)^{-\frac{1}{2}}d\nu_{u,v}(x)\right)^2d\eta_{e(w)}(u)\,d\eta_{e(w)}(v).$$

Then there is a sequence $(f_n)_n$ in $C_c(G)$ such that

$$\lim_n\|f-f_n\|_{II,\mu}=0.$$

Proof. Let $g\in L^1(G,\nu_0)$ such that $g_\Delta\in L^\infty\left(G^{(0)},\mu\right)$, where

$$g_\Delta(w)=\iint\left(\int|g(x)|\Delta(x)^{-\frac{1}{2}}d\nu_{u,v}(x)\right)^2d\eta_{e(w)}(v)\,d\eta_{e(w)}(v),\quad w\in G^{(0)}.$$

We claim that

$$\|g\|_{II,\mu}^2\leq\|g_\Delta\|_\infty\quad\left(\text{in }L^\infty\left(G^{(0)},\mu\right)\right)$$

Indeed, let $j,k\in L^2\left(G^{(0)},\mu_1\right)$ with $\int|j|^2d\mu_1=\int|k|^2d\mu_1=1$. We have

$$\begin{aligned}
&\int|g(x)\,j(d(x))\,k(r(x))|\,\Delta(x)^{-\frac{1}{2}}d\nu(x)\\
&=\iiiint|g(x)|\,\Delta(x)^{-\frac{1}{2}}d\nu_{u,v}(x)\,|j(v)|\,|k(u)|\,d\eta_{e(w)}(u)\,d\eta_{e(w)}(v)\,d\mu(w)\\
&\leq\int\left(\left(\iint\left(\int|g(x)|\,\Delta(x)^{-\frac{1}{2}}d\nu_{u,v}(x)\right)^2d\eta_{e(w)}(u)\,d\eta_{e(w)}(v)\right)^{\frac{1}{2}}\right.\\
&\qquad\left.\cdot\left(\iint|j(v)|^2|k(u)|^2d\eta_{e(w)}(u)\,d\eta_{e(w)}(v)\right)^{\frac{1}{2}}\right)d\mu(w)\\
&\leq\sqrt{\|g_\Delta\|_\infty}\cdot\left(\iint|j(v)|^2d\eta_{e(w)}(v)\,d\mu(w)\right)^{\frac{1}{2}}\left(\iint|k(u)|^2\eta_{e(w)}(u)\,d\mu(w)\right)^{\frac{1}{2}}\\
&=\sqrt{\|g_\Delta\|_\infty}.
\end{aligned}$$

Consequently,

$$\|g\|_{II,\mu}\leq\|g_\Delta\|_\infty$$

Let $B_c(G)$ be the space of bounded Borel functions on G, each with compact support. Let η_0 be a dominant for $\{\eta_u\}$. If $f \in B_c(G)$, then f is the limit almost everywhere with respect to $\int \nu_{u,v} d(\eta_0 \times \eta_0)(u,v)$, of a sequence, $(f_n)_n$, in $C_c(G)$ that is uniformly bounded and supported on some compact set supporting f. Let K be the support of f. Since

$$\iint |f(x) - f_n(x)| \Delta(x)^{-\frac{1}{2}} d\nu_{u,v}(x) d(\eta_0 \times \eta_0)(u,v) \to 0 \ (n \to \infty)$$

it follows that there is a subsequence of $(f_n)_n$ such that

$$\int |f(x) - f_{n_k}(x)| \Delta(x)^{-\frac{1}{2}} d\nu_{u,v}(x) \to 0 \ (k \to \infty) \text{ a.e}$$

Because of the boundedness properties of f, $(f_n)_n$ and the boundedness of the system $\{\nu_{u,v}, (u,v) \in R\}$ and Δ on compact sets, it results that there exists $M > 0$ such that

$$\left(\int |f(x) - f_{n_k}(x)| \Delta(x)^{-\frac{1}{2}} d\nu_{u,v}(x) \right)^2 \leq M 1_{r(K)}(u) 1_{d(K)}(v).$$

Since

$$\sup_w \iint \left(\int |f(x) - f_{n_k}(x)| \Delta(x)^{-\frac{1}{2}} d\nu_{u,v}(x) \right)^2 d\eta_{e(w)}(v) d\eta_{e(w)}(u)$$
$$\leq \iint \left(\int |f(x) - f_{n_k}(x)| \Delta(x)^{-\frac{1}{2}} d\nu_{u,v}(x) \right)^2 d\eta_0(v) d\eta_0(u)$$

which converges to zero, by the Dominated Convergence Theorem, it follows that

$$\lim_k \|f - f_{n_k}\|_{II,\mu} = 0.$$

Let $f \in L^1(G, \nu_0)$ such that

$$\|f_\Delta\|_\infty < \infty$$

Let $g_n(x) = f(x)$ if $|f(x)| \leq n$, $g_n(x) = 0$ otherwise. Let $(K_n)_n$ be an increasing sequence of compact sets with $\bigcup_n K_n = G$. Let $f_n = g_n 1_{K_n}$. $|f - f_n|$ converges pointwise to zero, dominated by $|f|$. Hence

$$\left\| w \to \iint \left(\int |f(x) - f_n(x)| \Delta(x)^{-\frac{1}{2}} d\nu_{u,v}(x) \right)^2 d\eta_{e(w)}(v) d\eta_{e(w)}(u) \right\|_\infty$$
$$\leq \iint \left(\int |f(x) - f_n(x)| \Delta(x)^{-\frac{1}{2}} d\nu_{u,v}(x) \right)^2 d\eta_0(v) d\eta_0(u)$$

which converges to zero, by the Dominated Convergence Theorem. It follows that there is a sequence in $B_c(G)$ such that

$$\lim_n \|f - f_n\|_{II,\mu} = 0. \qquad \square$$

Proposition 2.9. *Let us assume that the decomposition of the Haar system over the principal groupoid is bounded. If $f \in I(G, \nu, \mu)$ and $g \in C_c(G)$, then the function $f * g \in L^1(G, \nu_0)$ and the function*

$$w \to \iint \left(\int |f * g(x)| \Delta(x)^{-\frac{1}{2}} d\nu_{u,v}(x) \right)^2 d\eta_{e(w)}(v) \, d\eta_{e(w)}(v), \; w \in G^{(0)}$$

is in $L^\infty\left(G^{(0)}, \mu\right)$.

Proof. Let K be the support of g. Let $M = \sup_{v,w} \left(\int |g(y)| \Delta(y)^{-\frac{1}{2}} d\nu_{w,v}(y) \right)$.

$$\begin{aligned}
&\nu_{u,v}\left(|f * g| \Delta^{-\frac{1}{2}} \right) \\
&= \int \left| \iint f(xy) g\left(y^{-1}\right) d\nu_{d(x),w}(y) \, d\eta_{e(d(x))}(w) \right| \Delta(x)^{-\frac{1}{2}} d\nu_{u,v}(x) \\
&\leq \iint \int \left| f(xy) g\left(y^{-1}\right) \right| d\nu_{d(x),w}(y) \, d\eta_{e(u)}(w) \, \Delta(x)^{-\frac{1}{2}} d\nu_{u,v}(x) \\
&= \iint \int \left| f(xy) g\left(y^{-1}\right) \right| \Delta(x)^{-\frac{1}{2}} d\nu_{v,w}(y) \, d\nu_{u,v}(x) \, d\eta_{e(u)}(w) \\
&= \iint \int \left| f(xy) g\left(y^{-1}\right) \right| \Delta(x)^{-\frac{1}{2}} d\nu_{u,v}(x) \, d\nu_{v,w}(y) \, d\eta_{e(u)}(w) \\
&= \iint \int \left| f(x) g\left(y^{-1}\right) \right| \Delta\left(y^{-1}\right) \Delta\left(xy^{-1}\right)^{-\frac{1}{2}} d\nu_{u,d(y)}(x) \, d\nu_{v,w}(y) \, d\eta_{e(u)}(w) \\
&= \iint \int \left| f(x) g\left(y^{-1}\right) \right| \Delta(x)^{-\frac{1}{2}} \Delta(y)^{-\frac{1}{2}} d\nu_{u,w}(x) \, d\nu_{v,w}(y) \, d\eta_{e(u)}(w) \\
&= \int \left(\int |f(x)| \Delta(x)^{-\frac{1}{2}} d\nu_{u,w}(x) \int g\left(y^{-1}\right) \Delta(y)^{-\frac{1}{2}} d\nu_{v,w}(y) \right) d\eta_{e(u)}(w) \\
&= \int \left(\int |f(x)| \Delta(x)^{-\frac{1}{2}} d\nu_{u,w}(x) \int |g(y)| \Delta(y)^{-\frac{1}{2}} d\nu_{w,v}(y) \right) d\eta_{e(u)}(w) \\
&= \int \left(\int |f(x)| \Delta(x)^{-\frac{1}{2}} d\nu_{u,w}(x) \int |g(y)| \Delta(y)^{-\frac{1}{2}} d\nu_{w,v}(y) \right) \\
&\qquad \cdot 1_{r(K)}(w) \, d\eta_{e(u)}(w) \, 1_{d(K)}(v) \\
&\leq \sup_{v,w} \left(\int |g(y)| \Delta(y)^{-\frac{1}{2}} d\nu_{w,v}(y) \right) \iint |f(x)| \Delta(x)^{-\frac{1}{2}} d\nu_{u,w}(x) \\
&\qquad \cdot 1_{r(K)}(w) \, d\eta_{e(u)}(w) \, 1_{d(K)}(v) \\
&= M 1_{d(K)}(v) \iint |f(x)| \Delta(x)^{-\frac{1}{2}} d\nu_{u,w}(x) \, 1_{r(K)}(w) \, d\eta_{e(u)}(w) \\
&\leq M 1_{d(K)}(v) \int \left(\int |f(x)| d\nu_{u,w}(x) \right)^{\frac{1}{2}} \left(\int |f(x)| \Delta(x)^{-1} d\nu_{u,w}(x) \right)^{\frac{1}{2}} \\
&\qquad \cdot 1_{r(K)}(w) \, d\eta_{e(u)}(w)
\end{aligned}$$

$$\leq M1_{d(K)}(v)\int\left(\int|f(x)|\,d\nu_{u,w}(x)\right)^{\frac{1}{2}}\left(\int\left|f\left(x^{-1}\right)\right|d\nu_{w,u}(x)\right)^{\frac{1}{2}}\cdot 1_{r(K)}(w)\,d\eta_{e(u)}(w)$$

$$\leq M1_{d(K)}(v)\left(\iint|f(x)|\,d\nu_{u,w}(x)\,d\eta_{e(u)}(w)\right)^{\frac{1}{2}}\cdot\left(\iint\left|f\left(x^{-1}\right)\right|d\nu_{w,u}(x)\,1_{d(K)}(w)\,d\eta_{e(u)}(w)\right)^{\frac{1}{2}}$$

$$\leq M1_{d(K)}(v)\left(\iint|f(x)|\,d\nu^{u}(x\right)^{\frac{1}{2}}\cdot\left(\iint\left|f\left(x^{-1}\right)\right|d\nu_{w,u}(x)\,1_{r(K)}(w)\,d\eta_{e(u)}(w)\right)^{\frac{1}{2}}$$

$$\leq M1_{d(K)}(v)\left(\|f\|_I\right)^{\frac{1}{2}}\left(\iint\left|f\left(x^{-1}\right)\right|d\nu_{w,u}(x)\,1_{r(K)}(w)\,d\eta_{e(u)}(w)\right)^{\frac{1}{2}}\quad\mu\text{-a.e.}$$

Then

$$\iint\left(\int|f*g(x)|\,\Delta(x)^{-\frac{1}{2}}\,d\nu_{u,v}(x)\right)^{2}d\eta_{e(s)}(v)\,d\eta_{e(s)}(u)$$

$$=\iint M^2\|f\|_I\,1_{d(K)}(v)\cdot\left(\int\left|f\left(x^{-1}\right)\right|d\nu^{w}(x)\right)1_{d(K)}(w)\,d\eta_{e(s)}(v)\,d\eta_{e(s)}(w)$$

$$=M^2\|f\|_I\iint 1_{d(K)}(v)\cdot\left(\int\left|f\left(x^{-1}\right)\right|d\nu^{w}(x)\right)1_{d(K)}(w)\,d\eta_{e(s)}(v)\,d\eta_{e(s)}(w)$$

$$=M^2\|f\|_I\iint 1_{d(K)}(v)\cdot\int\left|f\left(x^{-1}\right)\right|d\nu_{w,u}(x)\,d\eta_{e(s)}(u)\,d\eta_{e(s)}(v)\,1_{r(K)}(w)\,d\eta_{e(s)}(w)$$

$$\leq M^2\|f\|_I^2\iint 1_{d(K)}(v)\,1_{d(K)}(w)\,d\eta_{e(s)}(v)\,d\eta_{e(s)}(w)$$

$$<\infty,\quad\mu\text{-a.e.}$$

□

3. The inequality $\|II_\mu(\cdot)\|\leq\left\|\mathrm{Reg}_\mu(\cdot)\right\|$

Proposition 3.1. *Let G be a locally compact groupoid equipped with a continuous Haar system. Assume that the decomposition of the Haar system over the principal groupoid is bounded and that $C^*_{\mathrm{red}}(G)=C^*(G)$. Let μ be a quasi-invariant* *probability measure with* $\operatorname{supp}(\mu) = G^{(0)}$. *Then*

$$\|II_\mu(f)\| \leq \|\operatorname{Reg}_\mu(f)\|$$

for all $f \in I(G, \nu, \mu)$. *(*Reg_μ *is the left regular representation of* $C_c(G)$ *on* μ.*)*

Proof. Let

$$\nu^u = \int \nu_{u,v} d\eta_u(v) \quad \text{for all } u \in G^{(0)}$$

be the decomposition of the Haar system for G over the principal groupoid associated to G Let $\mu_1 = \int \eta_u d\mu(u)$. Then μ_1 is equivalent to μ. Let Δ be the modular function associated to $\{\nu^u, u \in G^{(0)}\}$ and μ_1. Let ν be the measure induced by μ_1 on G, and $\nu_0 = \Delta^{-\frac{1}{2}}\nu$.

If $f \in C_c(G)$, then

$$\|f\|_{\mathrm{red}} = \|\operatorname{Ind}\mu(f)\|$$

where $\operatorname{Ind}\mu(f)$ is the operator on $L^2(G, \nu^{-1})$ defined by formula

$$\operatorname{Ind}\mu(f)\xi(x) = f * \xi(x)$$

([7], p. 50). The map $W : L^2(G, \nu) \to L^2(G, \nu^{-1})$ defined by the formula, $W\xi = \xi\Delta^{\frac{1}{2}}$ is a Hilbert space isomorphism that implements a unitary equivalence between Reg_μ (the left regular representation of $C_c(G)$ on μ) and $\operatorname{Ind}\mu$.([12], II.1.10). Therefore, if μ is a quasi-invariant measure with $\operatorname{supp}(\mu) = G^{(0)}$, then

$$\|f\|_{\mathrm{red}} = \|\operatorname{Reg}_\mu(f)\|. \tag{3.1}$$

Since $\mu_1 = \iint \eta_{e(u)}(w)\, d\mu(u)$ is equivalent to μ, it follows that $\|\operatorname{Reg}_\mu(f)\| = \|\operatorname{Reg}_{\mu_1}(f)\|$ and $\|\operatorname{Ind}\mu(f)\| = \|\operatorname{Ind}\mu_1(f)\|$.

Since $C^*_{\mathrm{red}}(G) = C^*(G)$, for all $f \in C_c(G)$

$$\|II_\mu(f)\| \leq \|f\|_{\mathrm{red}}. \tag{3.2}$$

From (3.1) and (3.2) it follows that

$$\|II_\mu(f)\| \leq \|\operatorname{Reg}_\mu(f)\| = \|\operatorname{Ind}\mu(f)\| \tag{3.3}$$

for all $f \in C_c(G)$. If $f \in C_c(G)$, $f \geq 0$, then $\|II_\mu(f)\| = \|\operatorname{Reg}_\mu(f)\| = \|\operatorname{Ind}\mu(f)\|$.

If $f \in L^1(G, \nu_0)$ and

$$\iiint \left(\int |f(x)|\, \Delta(x)^{-\frac{1}{2}}\, d\nu_{u,v}(x)\right)^2 d\eta_{e(w)}(v)\, d\eta_{e(w)}(v)\, d\mu(w) < \infty,$$

then there is a sequence a sequence $(f_n)_n$, in $C_c(G)$ such that

$$\lim_n \|f - f_n\|_{II,\mu} = 0.$$

Hence

$$\lim_n \|L(f - f_n)\| = 0$$

for any representation L of G. From (3.3) it follows that $\|II\mu(f)\| \leq \|\operatorname{Reg}_\mu(f)\|$ for all f with the above property.

Let $f \in I(G,\nu,\mu)$ and $g \in C_c(G)$. By the preceding lemma, the function $f * g$ has the property that

$$\iiint \left(\int |f * g(x)| \, \Delta(x)^{-\frac{1}{2}} \, d\nu_{u,v}(x) \right)^2 d\eta_{e(w)}(v) \, d\eta_{e(w)}(v) \, d\mu(w) < \infty.$$

It follows that

$$\| II_\mu(f * g) \| \leq \| \operatorname{Reg}_\mu(f * g) \|.$$

Let $(K_n)_n$ be a sequence of compact subsets of $G^{(0)}$ with $\cup K_n = G^{(0)}$. By Corollary 2.11 of [8] (or Lemma 5.48 of [7]), the $*$-algebra $C_c(G)$ has a two-sided approximate identity with respect to the inductive limit topology, $(e_n)_n$, with the following properties:

1) $e_n(x) \geq 0$, for all $x \in G$
2) $\left| \int e_n(x) \, d\nu^u(x) - 1 \right| < \frac{1}{n}$, for all $u \in K_n$
3) $e_n(x) = e_n(x^{-1})$, for all $x \in G$.

Let $\xi \in L^2(G,\nu^{-1})$. Since $\| \operatorname{Ind}\mu(e_n)\xi - \xi \|_2 = \| e_n * \xi - \xi \|_2 \to 0$ $(n \to \infty)$, the Banach-Steinhaus Theorem implies that $(\| \operatorname{Ind}\mu(e_n) \|)_n$ is bounded. Thus there is $M > 0$, such that $\| e_n \|_{II,\mu} = \| II_\mu(e_n) \| = \| \operatorname{Ind}\mu(e_n) \| \leq M$ for all n.

Let $f \in I(G,\nu,\mu)$. First, we shall assume that $d(\operatorname{supp}(f))$ is a compact set in $G^{(0)}$.

For each n, $e_n \in L^2(G,\nu^{-1})$ because $e_n \in C_c(G)$. Hence $f * e_n \in L^2(G,\nu^{-1})$ $(\| f * e_n \|_2 \leq \| f \|_I \| e_n \|_2)$.

Since $f \in I(G,\nu,\mu) \subset L^1(G,\nu)$, it follows that for each ε, there is $g \in C_c(G)$, such that $\int |f - g| \, d\nu < \varepsilon$. We have

$$\int |f - f * e_n| \, d\nu \leq \int |f - g| \, d\nu + \int |g - g * e_n| \, d\nu + \int |(f - g) * e_n| \, d\nu$$

and

$$\begin{aligned}
\int |(f - g) * e_n| \, d\nu &= \int \left| \int (f - g)(y) \, e_n(y^{-1}x) \, d\nu^{r(x)}(y) \right| d\nu(x) \\
&\leq \iint |f - g|(y) \, e_n(y^{-1}x) \, d\nu^{r(x)}(y) \, d\nu(x) \\
&\leq \iiint |f - g|(y) \, e_n(y^{-1}x) \, d\nu^{r(x)}(y) \, d\nu^u(x) \, d\mu(u) \\
&= \iiint |f - g|(y) \, e_n(y^{-1}x) \, d\nu^u(x) \, d\nu^u(y) \, d\mu(u) \\
&= \iiint |f - g|(y) \, e_n(x) \, d\nu^{d(y)}(x) \, d\nu^u(y) \, d\mu(u) \\
&\leq \left(1 + \frac{1}{n}\right) \int |f - g| \, d\nu
\end{aligned}$$

for n with the property that $d(\operatorname{supp}(f) \cup \operatorname{supp}(g)) \subset K_n$.

Hence $\lim_n \int |f - f * e_n| \, d\nu = 0$. Since $\lim_n \iint |f - f * e_n| \, d\nu^u d\mu_1(u) = 0$, passing to a subsequence we may assume that $\lim_n \int |f - f * e_n| \, d\nu^u = 0$ *a.e.*, dominated by $3 \|f\|_I$ for all n such that $d(\operatorname{supp}(f)) \subset K_n$.

If $a \in C_c\left(G^{(0)}\right)$, then

$$\begin{aligned}
&\int |f * e_n|(x) \, |a(d(x))|^2 \Delta^{-1}(x) \, d\nu(x) \\
&= \int |f * e_n| \left(x^{-1}\right) |a(r(x))|^2 \, d\nu(x) \\
&= \iint \left|e_n * \overline{f^*}\right|(x) \, |a(r(x))|^2 \, d\nu^u(x) \, d\mu_1(u) \\
&\leq \iiint e_n(y) \left|f^*\left(y^{-1}x\right)\right| d\nu^{r(x)}(y) \, |a(r(x))|^2 \, d\nu^u(x) \, d\mu_1(u) \\
&= \iiint e_n(y) \left|f^*\left(y^{-1}x\right)\right| \, |a(u)|^2 \, d\nu^u(y) \, d\nu^u(x) \, d\mu_1(u) \\
&= \iint e_n(y) \left(\int \left|f^*\left(y^{-1}x\right)\right| d\nu^u(x) \right) |a(u)|^2 \, d\nu^u(y) \, d\mu_1(u) \\
&= \iint e_n(y) \left(\int |f^*(x)| \, \nu^{d(y)}(x) \right) \, |a(u)|^2 \, d\nu^u(y) \, d\mu_1(u) \\
&\leq \|f\|_I \iint e_n(y) \, d\nu^u(y) \, |a(u)|^2 \, d\mu_1(u) \\
&\leq \left(1 + \frac{1}{n}\right) \|f\|_I \, \|a\|_2^2
\end{aligned}$$

for all n such that $\operatorname{supp}(a) \subset K_n$.

If $a \in C_c\left(G^{(0)}\right)$ and $b \in L^2\left(G^{(0)}, \mu\right)$, then $\lim_n |\langle II(f - f * e_n) a, b\rangle| = 0$. Indeed,

$$\begin{aligned}
|\langle II_\mu(f - f * e_n) a, b\rangle| &\leq \int |f(x) - f * e_n(x)| \, |a(d(x))| \, |b(r(x))| \, d\nu_0(x) \\
&\leq \left(\int |f(x) - f * e_n(x)| \, |a(d(x))|^2 \Delta^{-1}(x) \, d\nu(x) \right)^{\frac{1}{2}} \\
&\qquad \cdot \left(\int |f(x) - f * e_n(x)| \, |b(r(x))|^2 \, d\nu(x) \right)^{\frac{1}{2}} \\
&\leq \left(\iint |f(x) - f * e_n(x)| \, d\nu_u(x) \, |a(u)|^2 \, d\mu(u) \right)^{\frac{1}{2}} \\
&\qquad \cdot \left(\int |f(x) - f * e_n(x)| \, d\nu^u(x) \, |b(u)|^2 \, d\mu(u) \right)^{\frac{1}{2}} \\
&\leq \|a\|_2 \left(3 \|f\|_I\right)^{\frac{1}{2}} \left(\int |f(x) - f * e_n(x)| \, d\nu^u(x) \, |b(u)|^2 \, d\mu_1(u) \right)^{\frac{1}{2}}.
\end{aligned}$$

Since $C_c\left(G^{(0)}\right)$ is dense in $L^2\left(G^{(0)},\mu_1\right)$ and $\left(\|II_\mu\left(f-f*e_n\right)\|\right)_n$ is bounded, it follows that $\lim_n \left|\left\langle II_\mu\left(f-f*e_n\right)a,b\right\rangle\right|=0$ for all $a,b\in L^2\left(G^{(0)},\mu_1\right)$.

Finally, we have

$$\begin{aligned}
\left|\left\langle II_\mu((f)\,a,b\right\rangle\right| &\leq \lim_n \left|\left\langle II_\mu((f*e_n-f)\,a,b\right\rangle\right| + \lim_n \left|\left\langle II_\mu\left((f*e_n)\,a,b\right\rangle\right| \\
&\leq \lim_n \|II_\mu\left((f*e_n)\right)\| \, \|a\|_2 \, \|b\|_2 \\
&\leq \lim_n \|\mathrm{Ind}\,\mu\left((f*e_n)\right)\| \, \|a\|_2 \, \|b\|_2 \\
&\leq \lim_n \|\mathrm{Ind}\,\mu\left(f\right)\mathrm{Ind}\,\mu\left(e_n\right)\| \, \|a\|_2 \, \|b\|_2 \\
&\leq \lim_n \|\mathrm{Ind}\,\mu\left(f\right)\| \, \|\mathrm{Ind}\,\mu\left(e_n\right)\| \, \|a\|_2 \, \|b\|_2 .
\end{aligned}$$

Thus $\|II_\mu\left(f\right)\| \leq \|\mathrm{Ind}\,\mu((f)\|$.

Let $f\in I\left(G,\nu,\mu\right)$. Let $\left(K_n\right)_n$ be a sequence of compact subsets of $G^{(0)}$ with $\cup K_n = G^{(0)}$. If $f_n = f1_{K_n}\circ d$, then $|f_n-f|$ converges to zero dominated by $|f|$. Hence, if $a,b\in L^2\left(G^{(0)},\mu_1\right)$, then $\lim_n\left|\left\langle II_\mu\left(f-f_n\right)a,b\right\rangle\right|=0$. On the other hand, if $\xi\in L^2\left(G,\nu^{-1}\right)$,

$$\begin{aligned}
\|\mathrm{Ind}\,\mu\left(f_n\right)\xi\|_2 &= \|f_n*\xi\|_2 = \|(f1_{K_n}\circ d)*\xi\|_2 \\
&= \|f*(\xi 1_{K_n}\circ r)\|_2 \leq \|\mathrm{Ind}\,\mu\left(f\right)(\xi 1_{K_n}\circ r)\|_2 \\
&\leq \|\mathrm{Ind}\,\mu\left(f\right)\| \, \|(\xi 1_{K_n}\circ r)\|_2 \leq \|\mathrm{Ind}\,\mu\left(f\right)\| \, \|\xi\|_2 .
\end{aligned}$$

Thus $\|\mathrm{Ind}\,\mu\left(f_n\right)\| \leq \|\mathrm{Ind}\,\mu\left(f\right)\|$.

For $a,b\in L^2\left(G^{(0)},\mu\right)$, we have

$$\begin{aligned}
\left|\left\langle II_\mu\left(f\right)a,b\right\rangle\right| &\leq \lim_n \left|\left\langle II_\mu(f_n-fa,b\right\rangle\right| + \lim_n \left|\left\langle II_\mu\left(f_n\right)a,b\right\rangle\right| \\
&\leq \lim_n \|II_\mu\left(f_n\right)\| \, \|a\|_2 \, \|b\|_2 \\
&\leq \lim_n \|\mathrm{Ind}\,\mu\left((f_n)\right)\| \, \|a\|_2 \, \|b\|_2 \\
&\leq \lim_n \|\mathrm{Ind}\,\mu\left(f\right)\| \, \|a\|_2 \, \|b\|_2 .
\end{aligned}$$

It follows that

$$\|II_\mu\left(f\right)\| \leq \|\mathrm{Ind}\,\mu\left(f\right)\| = \|\,\mathrm{Reg}_\mu\left(f\right)\|$$

for all $f\in I\left(G,\nu,\mu\right)$. □

Proposition 3.2. *Let G be a locally compact groupoid equipped with a continuous Haar system $\{\nu^u, u\in G^{(0)}\}$. Assume that the decomposition of the Haar system over the principal groupoid is bounded and that $C^*_{\mathrm{red}}\left(G\right)=C^*\left(G\right)$. Then G is measurewise amenable.*

Proof. Let μ be a quasi-invariant probability measure. First, we shall assume that μ is a quasi-invariant probability measure with $\mathrm{supp}\left(\mu\right)=G^{(0)}$. From the preceding propositions it follows that the trivial representation of $C^*\left(G,\nu,\mu\right)$ is weakly contained in the regular representation. Therefore (G, ν, μ) is amenable (Theorem 6.1.4, p. 142, [1]).

Let μ be an arbitrary quasi-invariant measure, μ_0 be a quasi-invariant measure with $\operatorname{supp}(\mu_0) = G^{(0)}$, and $\mu_1 = \mu_0 + \mu$. Then μ_1 is quasi-invariant measure with $\operatorname{supp}(\mu_1) = G^{(0)}$, and therefore (G, ν, μ_1) is amenable. Proposition 3.2.14 (iii) of [1] shows that there exists a sequence (g_n) in $B_b(G, \nu)^+$ normalized and such that

$$\lim_n \int h(u) \left[\int |f * g_n - (\nu(f) \circ r) g_n| \, d\nu^u \right] d\mu_1(u) = 0$$

for all $f \in B_b(G, \nu)$ and $h \in L^1(\mu_1)$.

Since $\mu << \mu_1$, there exists a Borel function g such that $\mu = g\mu_1$. If $h \in L^1(\mu)$, then $hg \in L^1(\mu_1)$ and

$$\lim_n \int h(u) g(u) \left[\int |f * g_n - (\nu(f) \circ r) g_n| \, d\nu^u \right] d\mu_1(u) = 0$$

$$\lim_n \int h(u) \left[\int |f * g_n - (\nu(f) \circ r) g_n| \, d\nu^u \right] d\mu(u) = 0.$$

Therefore (G, ν, μ) is amenable. □

References

[1] C. Anantharaman-Delaroche, J. Renault, *Amenable Groupoids.* Monographie de L'Enseignement Mathématique No **36**, Genève, 2000.

[2] M. Buneci, *Consequences of Hahn Structure Theorem for the Haar Measure.* Math. Reports, **4**(54), 4(2002), 321–334.

[3] M. Buneci, *C^*-Algebras Associated to the Transitive Groupoids.* An. Univ. Craiova Ser. Mat. Inform, Vol. XXVIII, 2001, 79–92.

[4] M. Buneci, *The Structure of the Haar Systems on Locally Compact Groupoids.* Math. Phys. Electron. J, Vol. **8** (4) (2002), 1–11.

[5] P. Hahn, *Haar Measure for Measure Groupoids.* Trans. Amer. Math. Soc. **242** (1978), 1–33.

[6] P. Hahn, *The Regular Representations of Measure Groupoids.* Trans. Amer. Math. Soc. **242** (1978), 34–72.

[7] P. Muhly, *Coordinates in Operator Algebra.* Book in preparation.

[8] P. Muhly, J. Renault and D. Williams, *Equivalence and Isomorphism for Groupoid C^*-Algebras.* J. Operator Theory **17**(1987), 3–22

[9] A. Ramsay, *Virtual Groups and Groups Actions.* Adv. in Math. **6**(1971), 253–322.

[10] A. Ramsay, *Topologies on Measured Groupoids.* J. Funct. Anal. **47** (1982), 314–343.

[11] A. Ramsay and M. E. Walter, *Fourier-Stieltjes Algebras of Locally Compact Groupoids*, J. Funct. Anal. **148** (1997), 314–367.

[12] J. Renault, *A Groupoid Approach to C^*-Algebras.* Lecture Notes in Math., Springer-Verlag, **793**, 1980.

[13] J. Renault, *The Ideal Structure of Groupoid Crossed Product Algebras*. J. Operator Theory, **25** (1991), 3–36.

[14] J. Westman, *Nontransitive Groupoid Algebras*. Univ. of California at Irvine, 1967.

Mădălina Buneci
University Constantin Brâncuşi of Târgu-Jiu
Bulevardul Republicii, Nr. 1
210152 Târgu-Jiu
Gorj
Romania
e-mail: `ada@utgjiu.ro`

Operator Theory:
Advances and Applications, Vol. 153, 79–101

ρ-Numerical Radius in Banach Spaces

Laurent Carrot

Abstract. In this paper, we extend the definition of the ρ-numerical radius to Banach spaces. In order to do so, we use one of the classical characterizations of the $\mathcal{C}_\rho$ classes, which can be naturally extended. Then, we give a first study of this concept. Some of the properties given in the Banach case are original, but most of them are a generalization of the Hilbert case, even if the proofs have to be done in completely different ways.

Mathematics Subject Classification (2000). Primary: 47A12, 46B25.

Keywords. Operator on Banach space, Numerical radius.

1. Introduction

The ρ-numerical radius was first introduced by Sz.-Nagy and Foias [13] on Hilbert spaces as the operator radius associated with the classes $\mathcal{C}_\rho$, where an operator T is in $\mathcal{C}_\rho$ ($\rho > 0$) if and only if T is in $\mathcal{B}(H)$ and there exists a unitary operator U on some Hilbert space K such that K contains H as a subspace and such that

$$T^n h = \rho P_H U^n h, \quad \text{for all } h \in H,\ n = 1, 2, \ldots.$$

This concept has been studied a lot (see especially [1, 10, 11, 13]), and, as I was giving a kind of survey on this subject for the working party "Théorie des opérateurs et analyse complexe" in Lyon, G. Cassier has given me the idea of extending the definition of classes $\mathcal{C}_\rho$ to Banach spaces. Of course, we can't apply the definition of Sz.-Nagy and Foias. We use one of the multiple characterizations of these classes (pointed out by Davis [7]):

$$T \in \mathcal{C}_\rho \Leftrightarrow r(T) \leq 1 \quad \text{and} \quad \|zT[(\rho-1)zT - \rho I]^{-1}\| \leq 1, \quad \text{for each } z \in \mathbb{D}.$$

where $r(T)$ denotes the spectral radius of T. This characterization can be immediately extended to Banach spaces. So, from now on, if $(E,\ \|\cdot\|)$ is a Banach space, we define:

$$\mathcal{C}_{\rho,\ \|\cdot\|} = \{T \in \mathcal{B}(E):\ r(T) \leq 1 \text{ and } \|zT[\rho-1)zT - \rho I]^{-1}\| \leq 1, \text{ for each } z \in \mathbb{D}\},$$

and

$$w_{\rho,\ \|\cdot\|}(T) = \inf\{1/r : r > 0 \text{ and } rT \in \mathcal{C}_{\rho,\ \|\cdot\|}\}\ ; \quad T \in \mathcal{B}(E).$$

Here, we indicate the norm because one can see that the $\mathcal{C}_\rho$ classes depend much on it. Similarity problems show that we should indicate which is the considerated norm even in Hilbert spaces because an operator which is in $\mathcal{C}_{\rho,\ \|\cdot\|_1}$ can be in $\mathcal{C}_{1,\ \|\cdot\|_2}$ for another Hilbert norm. Of course, in case of no ambiguity, we write $\mathcal{C}_\rho$ instead of $\mathcal{C}_{\rho,\ \|\cdot\|}$.

In this paper, we are interested in listing some hilbertian properties which still hold in Banach spaces, and sometimes, we give counter-examples to some unverified properties. We will not (and cannot) be exhaustive. This study has mainly two interests:

- It allows us to distinguish norm properties from (typically) hilbertian ones.
- the own interest of the functionals $w_{\rho,\ \|\cdot\|}(\,\cdot\,)$ which include the norm ($w_1(\,\cdot\,)$) and the spectral radius $r(\,\cdot\,)$.

This paper will be divided in four sections

- First, we give some "elementary" properties of $\mathcal{C}_{\rho,\ \|\cdot\|}$ and $w_{\rho,\ \|\cdot\|}(\,\cdot\,)$, and we do a first comparison with existing "tools" (the norm $\|\cdot\|$, the spectral radius $r(\,\cdot\,)$ and the numerical radius introduced by Bauer [3] $v(\,\cdot\,)$)
- Then, we study the ρ-numerical radius of direct sum of operators and of quotient operators.
- In the third part, we construct a power bounded operator which is in no $\mathcal{C}_{\rho,\ \|\cdot\|}$, $\rho > 0$.
- Finally, we give a look to nilpotent-operators case, which is a source of counter-examples. It will also allow us to obtain the optimality of some constants.

From now on, we denote by $(E,\ \|\cdot\|)$ a Banach space.

2. Elementary properties

First, we have to verify that $w_\rho(\,\cdot\,)$ is well defined:

Proposition 2.1. *We have:*

$$w_\rho(T) < \infty$$

for any operator $T \in \mathcal{B}(E)$.

Proof. The mapping

$$f : z \mapsto \|zT[(\rho-1)zT - \rho I]^{-1}\|$$

is defined at least in a neighborhood of 0, and it is continuous on this neighborhood. Moreover, as $f(0) = 0$, there exists $r_0 > 0$ such that:

$$r(r_0 T) \leq 1 \quad \text{and} \quad f(z) \leq 1, \quad \text{for each complex } z \text{ such that } |z| < r_0.$$

Finally, if $r < r_0$, then $rT \in \mathcal{C}_\rho$. Hence we have:

$$w_\rho(T) \leq 1/r_0 < \infty. \qquad \square$$

Remarks.

1) This proof also allows us to write without ambiguity:

$$w_\rho(T) = 1/\sup\{r > 0 : \; rT \in \mathcal{C}_\rho\},$$

a notation that we shall use in some proofs.

2) From the definition, one can immediately see that an operator T is in $\mathcal{C}_1$ if and only if T is a contraction, and so $w_1(\,\cdot\,) = \|\cdot\|$.

In Hilbert spaces, several proofs (see [7, 11]) lead to the redundancy of the condition $r(T) \leq 1$ in the $\mathcal{C}_\rho$ classes definition. We give a new proof of this redundancy, based on the following lemma:

Lemma 2.2. *Let T be an operator in $\mathcal{B}(E)$ and $\rho > 0$ such that:*

$$\|zTh\| \leq \|((\rho-1)zTh - \rho h\| \; ; \quad z \in \mathbb{D}, \; h \in E.$$

Then we have $r(T) \leq 1$.

Proof. As, for $\rho = 1$, the condition become

$$\|zTh\| \leq \|h\| \; ; \quad z \in \mathbb{D}, \; h \in E.$$

T is a contraction, and its spectral radius is smaller than 1. In order to prove this lemma for $\rho \neq 1$, we chose λ an element of the spectrum of T such that $|\lambda| = r(T)$. With this choice, λ is an approximate eigenvalue for T, i.e.,

$$\exists h_n : \; \|h_n\| = 1 \quad \text{and} \quad Th_n = \lambda h_n + g_n, \text{ with } g_n \xrightarrow[n\to\infty]{} 0.$$

Then, if z is in $\mathbb{D} \setminus \{0\}$, we have:

$$\begin{aligned} &\big|\|[(\rho-1)z - \rho I]h_n\| - |(\rho-1)z\lambda - \rho|\,\big| \\ &= \big|\|[(\rho-1)z - \rho I]h_n\| - \|[(\rho-1)z\lambda - \rho]h_n\|\big| \\ &\leq \|[(\rho-1)zT - (\rho-1)z\lambda]h_n\| \\ &= \|(\rho-1)zg_n\| \to 0 \text{ when } n \to \infty. \end{aligned}$$

Hence we deduce:

$$\lim_{n\to\infty} \|zTh_n\| = |z|\,|\lambda| \leq |(\rho-1)z\lambda - \rho|, \quad \text{for all } z \in \mathbb{D} \setminus \{0\}.$$

In order to be able to conclude, we must first show that $r(T) < \rho/|\rho-1|$. To do so, let us suppose that $r(T) = |\lambda| \geq \rho/|\rho-1|$. Then, by dividing the preceding inequality by $|z|$ ($\neq 0$), we obtain:

$$r(T) \leq |(\rho-1)\lambda - \rho z| \quad (z \notin \bar{\mathbb{D}}),$$

and so

$$r(T) \leq \inf\{|(\rho-1)\lambda - \rho z|, \; z \notin \bar{\mathbb{D}}\}.$$

Now, in that case ($|\lambda| \geq \rho/|\rho-1|$), this infimum is equal to 0, therefore we get $r(T) = 0$, which is impossible. We have proved that

$$|\lambda| = r(T) < \rho/|\rho-1|.$$

We can now conclude. Let ϵ be in $]0,1[$ and take z in $\mathbb{D}$ such that $|z| = 1-\epsilon$ and $arg(z) = -arg((\rho-1)\lambda)$. Then, the inequality $|z||\lambda| \leq |(\rho-1)z\lambda - \rho|$ become:

$$(1-\epsilon)|\lambda| \leq |(1-\epsilon)|\rho-1||\lambda| - \rho| = \rho - (1-\epsilon)|\rho-1||\lambda|.$$

Let ϵ tend to 0, then we get

$$|\lambda| \leq \rho - |\rho-1||\lambda|,$$

and finally

$$1 \geq |\lambda| = r(T).$$

□

Remark. A more precise end of proof leads to the inequality

$$r(T) \leq \rho/(2-\rho),$$

which holds only for $\rho \leq 1$.

Now, if T is an operator in $\mathcal{B}(E)$ such that

$$\|zT[(\rho-1)zT - \rho I]^{-1}\| \leq 1 \quad (z \in \mathbb{D}),$$

then we have:

$$\|zTh\| \leq \|(\rho-1)zTh - \rho h\| \; ; \quad z \in \mathbb{D},\ h \in E.$$

Hence we get $r(T) \leq 1$, from the preceding lemma.

Reciprocally, let T be an operator such that:

$$\|zTh\| \leq \|(\rho-1)zTh - \rho h\| \; ; \quad z \in \mathbb{D},\ h \in E.$$

Then, fix z in $\mathbb{D}$,

- if $\rho \geq 1$, we have $r(T) \leq 1$, and so $((\rho-1)zT - \rho I)$ is invertible. Hence we get:
$$\|zT[(\rho-1)zT - \rho I]^{-1}\| \leq 1.$$
- if $0 < \rho \leq 1$, we have, from the preceding remark,
$$r(T) \leq \frac{\rho}{2-\rho} < \frac{\rho}{1-\rho},$$
and so $((\rho-1)zT - \rho I)$ is invertible. Hence we also get:
$$\|zT[(\rho-1)zT - \rho I]^{-1}\| \leq 1.$$

Now, we can give the new definition of the classes $\mathcal{C}_\rho$, in which the redundant condition $r(T) \leq 1$ does not appear:

Definition 2.3. Let T be an operator in $\mathcal{B}(E)$,

$$\begin{aligned} T \in \mathcal{C}_\rho &\Leftrightarrow \|zT[(\rho-1)zT - \rho I]^{-1}\| \leq 1 \; ; \quad z \in \mathbb{D} \\ &\Leftrightarrow \|zTh\| \leq \|(\rho-1)zTh - \rho h\| \; ; \quad z \in \mathbb{D},\ h \in E. \end{aligned}$$

The next characterization for $w_\rho(\cdot)$ was pointed out by Ando and Nishio ([1]) in Hilbert spaces:

Proposition 2.4. *Let T be an operator in $\mathcal{B}(E)$. Then*

$$w_\rho(T) \leq \alpha \Leftrightarrow \|T[(\rho-1)T - \rho\alpha zT]^{-1}\| \leq 1, \quad z \notin \bar{\mathbb{D}},$$

Proof. Ando's and Nishio's proof, already based on norm characterizations, still holds in Banach spaces. □

With this characterization, we immediately obtain, as in Hibert case ([1]), the Ando's and Nishio's law:

Proposition 2.5. *Let T be an operator in $\mathcal{B}(E)$ and $\rho \in]0,2[$. We have:*

$$\rho w_\rho(T) = (2-\rho) w_{2-\rho}(T).$$

The growth of the classes $\mathcal{C}_\rho$, as ρ increases, is still true, but the proof is quite different:

Proposition 2.6. *The classes $\mathcal{C}_\rho$ are growing as ρ increases.*

Proof. Using Ando's and Nishio's law, we only have to prove the growth for $\rho \geq 1$. So, let $\rho \geq 1$, T be an operator in $\mathcal{C}_\rho$ and $\rho' > \rho$. Now, with such assumptions, we can compute, if $z \in \mathbb{D}$ and $h \in E$,

$$\begin{aligned}
&\rho'\Big\|\frac{\rho'-1}{\rho'}zTh - h\Big\| - \rho\Big\|\frac{\rho-1}{\rho}zTh - h\Big\| \\
&\geq \rho'\Big(\Big\|\frac{\rho-1}{\rho}zTh - h\Big\| - \Big\|(\frac{\rho'-1}{\rho'} - \frac{\rho-1}{\rho})zTh\Big\|\Big) - \rho\Big\|\frac{\rho-1}{\rho}zTh - h\Big\| \\
&= \rho'\Big\|\frac{\rho-1}{\rho}zTh - h\Big\| - \rho'\frac{\rho'-\rho}{\rho\rho'}\|zTh\| - \rho\Big\|\frac{\rho-1}{\rho}zTh - h\Big\| \\
&= (\rho'-\rho)\Big(\Big\|\frac{\rho-1}{\rho}zTh - h\Big\| - \|zTh\|/\rho\Big) \\
&\geq 0, \text{ because } T \in \mathcal{C}_\rho.
\end{aligned}$$

Therefore, we have:

$$\|zTh\| \leq \rho\Big\|\frac{\rho-1}{\rho}zTh - h\Big\| \leq \rho'\Big\|\frac{\rho'-1}{\rho'}zTh - h\Big\| \; ; \quad z \in \mathbb{D}, \; h \in E,$$

which is a sufficient condition(see Definition 2.3) for T to be in $\mathcal{C}_{\rho'}$. The proof is complete. □

This growth and the definition of the ρ-numerical radius lead immediately to the following corollary:

Corollary 2.7. *If T is any fixed operator in $\mathcal{B}(E)$, then the mapping $\rho \mapsto w_\rho(T)$ is decreasing.*

Now, we can define:

Definition 2.8. For any operator T in $\mathcal{B}(E)$, we write:

$$w_\infty(T) = \lim_{\rho \to \infty} w_\rho(T).$$

As in Hilbert spaces ([10]), we have $w_\infty(T) = r(T)$. In order to prove this equality, we need the following lemma:

Lemma 2.9. *Let T be an operator in $\mathcal{B}(E)$ such that $r(T) < 1$. Then there exists $\rho > 0$ such that $T \in \mathcal{C}_\rho$.*

Proof. As $r(T) < 1$, there exists $s > 1$ such that $r(sT) < 1$. Hence there exists $B > 0$ such that

$$s^n \|T^n\| \leq B, \text{ for any } n \geq 0.$$

Let $\rho > 1$ and $z \in \mathbb{D}$. Then,

$$\begin{aligned}
&\left\|\left(\frac{\rho-1}{\rho}zT - I\right)^{-1}\right\| \\
&\qquad \leq \|I\| + \sum_{n=1}^{\infty} \left\|\left(\frac{\rho-1}{\rho}zT\right)^n\right\| \\
&\qquad \leq 1 + \sum_{n=1}^{\infty} \|T^n\| \leq 1 + \sum_{n=1}^{\infty} B/s^n \leq M, \text{ which does not depend on } \rho.
\end{aligned}$$

Therefore, if $\rho > 1$ and $z \in \mathbb{D}$, we have:

$$\begin{aligned}
\|zT((\rho-1)zT - \rho I)^{-1}\| &= \left\|\frac{z}{\rho}T\left(\frac{\rho-1}{\rho}zT - I\right)^{-1}\right\| \\
&\leq \left\|z\frac{T}{\rho}\right\| \cdot \left\|\left(\frac{\rho-1}{\rho}zT - I\right)^{-1}\right\| \leq \frac{\|T\|}{\rho}M.
\end{aligned}$$

One only has to chose ρ big enough ($\rho \geq \|T\|M$) to get $T \in \mathcal{C}_\rho$. The proof is complete. □

Now, the same proof as in Hilbert spaces [10] entails the desired equality:

Proposition 2.10. *Let T be an operator of $\mathcal{B}(E)$, we have:*

$$w_\infty(T) = r(T).$$

We also obtain (same proof as in [10]):

Proposition 2.11. *If we denote by $PWB(E)$ the set of power bounded operators on E, then*

$$(PWB(E)) \cap \Big(\bigcup_{\rho>0} \mathcal{C}_\rho\Big) \text{ is dense in } PWB(E) \text{ for the norm of } \mathcal{B}(E).$$

In Hilbert spaces, Ando and Nishio ([1]) obtained the log-convexity of $\rho \mapsto w_\rho(T)$. This question seems much more difficult in Banach spaces. However, we still have a convexity property:

Proposition 2.12. *Let T be an operator in $\mathcal{B}(E)$. The mapping*

$$f : x \mapsto log((e^x + 1)w_{e^x+1}(T))$$

is convex on $\mathbb{R}$.

Proof. The proof of Ando and Nishio ([1]), based on a modified version of Hadamard's three circles theorem ([14]), still works in Banach spaces. □

Then, as a corollary, we have:

Corollary 2.13. *Let T be an operator in $\mathcal{B}(E)$, then we have:*

$$\|T\| \geq w_\rho(T) \geq \frac{\|T\|}{\rho} \quad (\rho > 1).$$

Proof. We have already proved the left hand side of this inequality in Corollary 2.7. This is obvious for $T = 0$. Now, for $T \neq 0$, the mapping of the previous proposition is convex, and

$$\lim_{x \to -\infty} f(x) = log(\|T\|).$$

Therefore, we obtain:

$$\log((e^x + 1)w_{e^x+1}(T)) \geq \log(\|T\|), \quad \text{for any real number } x.$$

Finally, we get:

$$\rho\, w_\rho(T) \geq \|T\| \quad (\rho > 1).$$

□

Remark. this inequality should lead us to think that $w_\rho(\,\cdot\,)$ is a norm equivalent to $\|\cdot\|$ on $\mathcal{B}(E)$. But this is surely not always the case: of course, for $\rho = 1$ ($w_1(T) = \|T\|$), this is true, but even in Hilbert spaces [10], $w_\rho(\,\cdot\,)$ fails to be a norm for $\rho > 2$. This question is more difficult in Banach spaces.

As a second corollary, we can give the following:

Corollary 2.14. *Let T be a fixed operator in $\mathcal{B}(E)$. The mapping*

$$\rho \mapsto w_\rho(T) \quad (\rho > 0)$$

is continuous.

Proof. As the mapping f of Proposition 2.12 is convex on $\mathbb{R}$, it is continuous on $\mathbb{R}$. And then, we easily obtain the announced corollary. □

Now, we are interested in obtaining, after a few steps, the continuity of the mapping:

$$T \mapsto w_\rho(T).$$

The first step is the following lemma, which we shall give without proof. It gives a formula for the resolvent of a perturbated operator:

Lemma 2.15. *Let T and R be two operators in $\mathcal{B}(E)$. We denote by $\rho(T)$ the resolvent set of T, and by $R_z(T)$ the following operator: $(zI - T)^{-1}$. If z is an element of $\rho(T)$ such that $r(R.R_z(T)) < 1$, then $z \in \rho(T + R)$ and*

$$\begin{aligned} R_z(T+R) &= R_z(T) + \sum_{n=1}^{\infty} R_z(T).(R.R_z(T))^n \\ &= R_z(T) + \sum_{n=1}^{\infty} (R.R_z(T))^n.R_z(T). \end{aligned}$$

The second lemma gives an idea of the behavior of $\|zT((\rho - 1)zT - \rho I)^{-1}\|$ as we perturb T:

Lemma 2.16. *Let T be an operator in $\mathcal{B}(E)$ such that $r(T) \leq 1$, $\rho \neq 1$ and β be such that $\beta < \rho/|\rho - 1|$. Then, for any $\epsilon > 0$, there exists $\eta > 0$ such that:*

$$\forall R,\ \|R\| \leq \eta \Rightarrow \forall z:\ |z| < \beta,$$
$$\big|\|z(T+R)((\rho-1)z(T+R)-\rho I)^{-1}\| - \|zT((\rho-1)zT-\rho I)^{-1}\|\big| \leq \epsilon.$$

Proof. The inequality is always verified for $z = 0$. So we only have to consider the case $z \neq 0$. In the proof, R may be chosen small enough that it satisfies the conditions of the previous lemma. Hence, using the notations of this lemma, if z is an element of $\rho(T)$, we have:

$$z \in \rho(T+R) \quad \text{and} \quad R_z(T+R) = R_z(T) + R_z(T) \cdot \sum_{n=1}^{\infty} (R.R_z(T))^n.$$

For $z \neq 0$, let note $\alpha = \rho/((\rho-1)z)$, then

$$\begin{aligned}
&\|z(T+R)((\rho-1)z(T+R)-\rho I)^{-1}\| \\
&= \|\frac{z}{\rho}\alpha(T+R)R_\alpha(T+R)\| \\
&= \Big\|\frac{z}{\rho}\alpha T \cdot R_\alpha(T) + \frac{z}{\rho}\alpha R \cdot R_\alpha(T) + \frac{z}{\rho}\alpha(T+R) \cdot R_\alpha(T) \cdot \sum_{n=1}^{\infty}(R \cdot R_\alpha(T))^n\Big\|.
\end{aligned}$$

Therefore,

$$\begin{aligned}
&\Big|\Big\|\frac{z}{\rho}\alpha R_\alpha(T+R)\Big\| - \Big\|\frac{z}{\rho}\alpha T \cdot R_\alpha(T)\Big\|\Big| \\
&\leq \Big\|\frac{z}{\rho}\alpha R \cdot R_\alpha(T) + \frac{z}{\rho}\alpha(T+R) \cdot R_\alpha(T) \cdot \sum_{n=1}^{\infty}(R \cdot R_\alpha(T))^n\Big\| \\
&\leq \frac{\|R\|}{\rho-1} \cdot (\|R_\alpha(T)\| + \|R_\alpha(T)\| \cdot \|(T+R)R_\alpha(T)\| \cdot \|(I - R \cdot R_\alpha(T))^{-1}\|).
\end{aligned}$$

Now, as $\beta < \rho/|\rho-1|$, the term into brackets is uniformely bounded on $\{z:\ |z| \leq \beta\}$, and therefore, there exist $K \geq 0$ and $\eta > 0$ such that, if R is an operator with $\|R\| \leq \eta$, then

$$1 + \|(T+R)R_\alpha(T)\| \cdot \|(I - R \cdot R_\alpha(T))^{-1}\| \leq K, \quad \text{for all } z \text{ with } |z| < \beta.$$

Then, choosing R small enough, we have the announced lemma. □

The third lemma is concerned with a growth property of a mapping:

Lemma 2.17. *Let $T \neq 0$ be an operator in $\mathcal{B}(E)$. Then the mapping*

$$\Phi : r \mapsto \max\{\|zT((\rho-1)zT-\rho I)^{-1}\|,\ |z| \leq r\}$$

is strictly growing.

Proof. The proof, based on the maximum modulus principle for analytic functions, is easy. □

Now, we have all the necessary tools to prove:

Proposition 2.18. *Let $\rho > 0$. The mapping $T \mapsto w_\rho(T)$ is continuous on $\mathcal{B}(E)$.*

Proof. We only consider the case $\rho \neq 1$, because it is trivial for $\rho = 1$ (remember that $w_1(X) = \|X\|$). Using Ando's and Nishio's law, we can suppose for the proof that $\rho > 1$. As, for any operator R in $\mathcal{B}(E)$, $0 \leq w_\rho(R) \leq \|R\|$, the continuity at $T = 0$ is verified. Let T be a fixed operator of $\mathcal{B}(E) \setminus \{0\}$. By homogeneity, we can suppose that $w_\rho(T) = 1$, and therefore $r(T) \leq 1$. We prove the continuity at T in 2 steps: first, we prove upper-semi-continuity, and then lower-semi-continuity.

- ⋆ Fix $\epsilon > 0$ and let γ be such that $1 + \epsilon = 1/(1 - \gamma)$. Then, from Lemma 2.17, there exists $\delta > 0$ such that, for any $z \neq 0$ with modulus smaller than $1 - \gamma$,

$$\left\| \frac{z}{\rho} \alpha T R_\alpha(T) \right\| < 1 - \delta,$$

where $R_z(T) = (zI - T)^{-1}$ and, as in Lemma 2.16, $\alpha = \rho/((\rho - 1)z$.

Now, from Lemma 2.16, as $1 - \gamma < 1 < \rho/(\rho - 1)$, there exists $\eta > 0$ such that:

$$\left| \left\| \frac{z}{\rho} \alpha (T + R) R_\alpha (T + R) \right\| - \left\| \frac{z}{\rho} \alpha T R_\alpha(T) \right\| \right| < \delta,$$

for any $z \neq 0$ with modulus smaller than $1 - \gamma$ and any R in $\mathcal{B}(E)$ with $\|R\| < \eta$, and consequently, we have:

$$\left\| \frac{z}{\rho} \alpha (T + R) R_\alpha (T + R) \right\| \leq 1.$$

This means that, for any R in a neighborhood of 0,

$$\sup\{x : \ x.(T + R) \in \mathcal{C}_\rho\} \geq 1 - \alpha,$$

and therefore

$$w_\rho(T + R) = \inf \left\{ \frac{1}{x} : x.(T + R) \in \mathcal{C}_\rho \right\} \leq \frac{1}{1 - \gamma} = 1 + \epsilon.$$

- ⋆ To obtain lower-semi-continuity, we can suppose that ϵ is small enough to ensure $0 < 1/(1 - \epsilon) < \rho/(\rho - 1)$ and chose γ such that $1 - \epsilon = 1/(1 + \gamma)$. Then, from Lemma 2.17, there exist $\beta > 0$ and z_0 such that,

$$|z_0| = 1 + \gamma \quad \text{and} \quad \left\| \frac{z_0}{\rho} \alpha_0 R_{\alpha_0}(T) \right\| = 1 + \beta,$$

where $\alpha_0 = \rho/((\rho - 1)z_0)$.

Now, from Lemma 2.16, as $|z_0| = 1 + \gamma = 1/(1 - \epsilon) < \rho/(\rho - 1)$, there exists $\eta > 0$ such that, for any R in $\mathcal{B}(E)$ with $\|R\| \leq \eta$, we have:

$$\left| \left\| \frac{z_0}{\rho} \alpha_0 (T + R) R_{\alpha_0}(T + R) \right\| - \left\| \frac{z_0}{\rho} \alpha_0 T R_{\alpha_0}(T) \right\| \right| < \frac{\beta}{2},$$

and consequently:

$$\left\| \frac{z_0}{\rho} \alpha_0 (T + R) R_{\alpha_0}(T + R) \right\| \geq 1 + \frac{\beta}{2} > 1.$$

This means that: $\sup\{x : x \cdot (T+R) \in \mathcal{C}_\rho\} \leq 1+\gamma$, for any R in a neighborhood of 0, and therefore

$$w_\rho(T+R) = \inf\left\{\frac{1}{x} : x \cdot (T+R) \in \mathcal{C}_\rho\right\} \geq \frac{1}{1+\gamma} = 1-\epsilon.$$

⋆ Combining these two points, we obtain the continuity. □

Now, we can prove the following theorem:

Theorem 2.19. *The mapping* $W : (\mathbb{R}^+ \times \mathcal{B}(E)) \to \mathbb{R}^+$ *defined by* $W(\rho, T) = w_\rho(T)$ *is continuous.*

Proof. As the mapping $T \mapsto w_1(T) = \|T\|$ is of course continuous, $T \mapsto w_\rho(T)$ is continuous for all $\rho > 0$. Then, combining with Corollary 2.14, a standard argument gives the theorem. □

Finally, we can wonder whether there is a link between $w_\rho(\cdot)$ and $v(\cdot)$, the numerical radius on Banach spaces introduced by Bauer ([3]; but also, in a different way by Lumer, [12]); a survey about this subject can be found in [4, 5]). We shall give, for the convenience of the reader, the definition of the numerical radius introduced by Bauer:

Definition 2.20. If T is an operator in $\mathcal{B}(E)$, the numerical range, $V(T)$, and the numerical radius $v(T)$ of T are defined by

$$V(T) = \{f(Th) : \ h \in E, \ f \in E', \ f(h) = \|f\| = \|h\| = 1\},$$
$$v(T) = \sup\{|\lambda| : \ \lambda \in V(T)\},$$

where E' denotes the Banach space of continuous linear functionals on E.

We can immediately observe that $v(\cdot)$ and $w_2(\cdot)$ coincide on Hilbert spaces, and, as we shall see later, we can think that this equality only occurs in this case. In general Banach spaces, we can only give the following inequality:

Proposition 2.21. *Let* T *be an operator in* $\mathcal{B}(E)$. *Then, we have:*

$$\frac{\rho}{e+\rho-1}\, w_\rho(T) \leq v(T) \leq \rho\, w_\rho(T) \quad (\rho \geq 1),$$

where e *denotes* $\exp(1)$.

Proof. First, we give an inequality (for a proof, see for example [4] or [8]) verified by Bauer's numerical radius:

$$\frac{\|T\|}{e} \leq v(T) \leq \|T\|.$$

Combining this inequality with Corollary 2.13, we have:

$$\frac{w_\rho(T)}{e} \leq v(T) \leq \rho w_\rho(T).$$

The right-hand side of the inequality is proved. In order to obtain the left-hand side, let $\lambda > 0$, $z \in \mathbb{D}$, $\rho \geq 1$ and $h \in E$ such that $\|h\| = 1$. Then, if we denote by f_h an element of E' (dual of E) such that $\|f_h\| = |f_h(h)| = 1$ (such an element always exists), we have:

$$\begin{aligned}\left\|\left(\rho-1\right)z\frac{T}{\lambda v(T)}h-\rho h\right\| &\geq \left|f_h((\rho-1)z\frac{T}{\lambda v(T)}-\rho h)\right| \\ &= \left|(\rho-1)z\frac{f_h(Th)}{\lambda v(T)}-\rho f_h(h)\right| \\ &\geq \rho-(\rho-1)\frac{v(T)}{\lambda v(T)} = \rho\left(1-\frac{1}{\lambda}\right)+\frac{1}{\lambda}\end{aligned}$$

and

$$\left\|\frac{z}{\lambda v(T)}Th\right\| \leq \frac{1}{\lambda v(T)}\|T\| \leq \frac{ev(T)}{\lambda v(T)} = \frac{e}{\lambda}.$$

Therefore, if λ is such that

$$\frac{e}{\lambda} \leq \rho\left(1-\frac{1}{\lambda}\right)+\frac{1}{\lambda},$$

i.e.,

$$\lambda \geq \frac{e+\rho-1}{\rho},$$

then

$$\left\|z\frac{T}{\lambda v(T)}h\right\| \leq \left\|(\rho-1)z\frac{T}{\lambda v(T)}h-\rho h\right\| ; \quad z \in \mathbb{D},\ h \in E,$$

and therefore $\frac{T}{\lambda v(T)}$ is in $\mathcal{C}_\rho$, i.e.,

$$w_\rho\left(\frac{T}{\lambda v(T)}\right) \leq 1.$$

Finally, by homogeneity, we have

$$w_\rho(T) \leq \lambda v(T),$$

and, by choosing $\lambda = \frac{e+\rho-1}{\rho}$, we obtain the desired inequality:

$$w_\rho(T) \leq \frac{e+\rho-1}{\rho}v(T). \qquad \square$$

Remarks.

1) Proposition 2.21 is a generalization of the following inequality:

$$\frac{\|T\|}{e} \leq v(T) \leq \|T\|.$$

2) We shall see later (Corollary 5.3), that the constant ρ in this inequality is the best possible.

3. ρ-numerical radius of "evoluated" operators

In this part, at first, we compute what the ρ-numerical radius of a direct sum of operators is. Then, we give an estimate of the ρ-numerical radius of a quotient operator, and finally, an estimate of the ρ-numerical radius of a restriction.

Proposition 3.1. *Let $T_n \in \mathcal{B}(E_n)$, $n \in \mathbb{N}^*$, a family of operators such that $\sup\{\|T_n\|,\ n \geq 1\} < \infty$. Then, we have:*

$$w_\rho(\oplus T_n) = \sup\{w_\rho(T_n),\ n \geq 1\}, \quad \rho > 0,\ \rho \neq \infty.$$

This is true as soon as the norm defined on the direct sum of the spaces E_n verifies:

for any $(x_1, \ldots, x_n, \ldots)$ and $(y_1, \ldots, y_n, \ldots)$ in $\oplus E_n$ such that

$$\|x_i\|_{E_i} = \|y_i\|_{E_i} \quad (i \geq 1),$$

we have:

$$\|(x_1, \ldots, x_n, \ldots)\| = \|(y_1, \ldots, y_n, \ldots)\|.$$

Proof. First, as $\sup\{\|T_n\|,\ n \geq 1\} < \infty$, we have $\sup\{w_\rho(T_n),\ n \geq 1\} < \infty$ for any $\rho \geq 1$ (use Corollary 2.13). Now, using Ando's and Nishio's law, we obtain the existence of this supremum for any $\rho > 0$. Let $\rho > 0$ and denote by a this supremum. Then,

$$\frac{T_n}{a} \in \mathcal{C}_\rho, \text{ for all } n \geq 1,$$

i.e., by definition,

$$\left\|\frac{z}{a}T_n((\rho-1)\frac{z}{a}T_n - \rho I_n)^{-1}\right\| \leq 1\ ; \quad z \in \mathbb{D},\ n \geq 1.$$

Hence, we have:

$$\left\| \oplus \left(\frac{z}{a}T_n((\rho-1)\frac{z}{a}T_n - \rho I_n)^{-1}\right)\right\| \leq 1\ ; \quad z \in \mathbb{D},$$

and consequently

$$\left\|\frac{z}{a} \oplus T_n((\rho-1)\frac{z}{a} \oplus T_n - \rho \oplus I_n)^{-1}\right\| \leq 1\ ; \quad z \in \mathbb{D}.$$

This means that

$$\frac{1}{a} \oplus T_n \in \mathcal{C}_\rho,$$

and so, by homogeneity,

$$w_\rho(\oplus T_n) \leq a = \sup\{w_\rho(T_n)\}.$$

Now, if we note $b = w_\rho(\oplus T_n) \leq a < \infty$, then

$$\left\|\frac{z}{b} \oplus T_n((\rho-1)\frac{z}{b} \oplus T_n - \rho \oplus I_n)^{-1}\right\| \leq 1\ ; \quad z \in \mathbb{D}.$$

We immediately obtain:

$$\left\| \oplus \left(\frac{z}{b}T_n((\rho-1)\frac{z}{b}T_n - \rho I_n)^{-1}\right)\right\| \leq 1\ ; \quad z \in \mathbb{D},$$

and consequently

$$\left\| \frac{z}{b} T_n ((\rho - 1) \frac{z}{b} T_n - \rho I_n)^{-1} \right\| \leq 1 \; ; \quad z \in \mathbb{D}, \; n \geq 1.$$

This means that:

$$w_\rho(T_n) \leq b, \quad \text{for all } n \geq 1,$$

and finally

$$\sup\{w_\rho(T_n)\} \leq b = w_\rho(\oplus T_n). \qquad \square$$

Remark. We have proved (Proposition 2.10) that $w_\infty(\cdot) = r(\cdot)$, the spectral radius. Therefore, we cannot have an equality in the previous proposition for $\rho = \infty$ (see for example [11]), but we always have:

$$r(\oplus T_n) \geq \sup\{r(T_n), \; n \geq 1\}.$$

Now, we investigate the case of operators acting on a quotient space:

Proposition 3.2. *Let T be an operator in $\mathcal{B}(E)$ and F a closed subspace of $ker(T)$. We define $T_p \in \mathcal{B}(E/F)$ by:*

$$h + F \mapsto Th + F.$$

Then

$$w_{\rho,\, \|\cdot\|_E}(T) \geq w_{\rho,\, \|\cdot\|_{E/F}}(T_p) \quad (\rho > 0).$$

Proof. Let z in $\mathbb{C}$ be such that:

$$\|zTh\|_E \leq \|(\rho - 1)zTh - \rho h\|_E, \text{ for any } h \in E.$$

As we have:

$$\|(\rho - 1)zT_p h - \rho h\|_{E/F} = \inf\{\|(\rho - 1)zTh - \rho h + f\|_E, \; f \in F\},$$

and, if f is in F,

$$\begin{aligned}
\|(\rho - 1)zTh - \rho h + f\|_E &= \|((\rho - 1)zT - \rho I)(h + f/\rho)\|_E && \text{(because } f \in kerT) \\
&\geq \|zT(h + f/\rho)\|_E && \text{(cf. choice of } z) \\
&= \|zTh\|_E && \text{(because } f \in kerT) \\
&\geq \inf\{\|zTh + f\|_E, \; f \in F\} \\
&= \|zT_p h\|_{E/F},
\end{aligned}$$

we get:

$$\|(\rho - 1)zT_p h - \rho h\|_{E/F} \geq \|zT_p h\|_{E/F}.$$

Using Definition 2.3, we obtain:

$$\{r : \; rT \in \mathcal{C}_\rho\} \subset \{r : \; rT_p \in \mathcal{C}_{\rho, \|\cdot\|_{E/F}}\},$$

and therefore the desired inequality. $\square$

Finally, we give an estimate of the ρ-numerical radius of a restriction of an operator to any closed invariant subspace:

Proposition 3.3. *Let T be an operator in $\mathcal{B}(E)$ and F a closed subspace of E, invariant by T. Then*

$$w_\rho(T\mid_F) \leq w_\rho(T).$$

Proof. By homogeneity, we only have to prove that

$$T \in \mathcal{C}_\rho \Rightarrow T\mid_F \in \mathcal{C}_\rho.$$

Now, to prove this, using Definition 2.3, one only has to see that if we have:

$$\|zTh\| \leq \|(\rho-1)zTh - \rho h\|, \text{ for any } z \in \mathbb{D} \text{ and } h \in E,$$

then

$$\|zT\mid_F f\| = \|zTf\| \leq \|(\rho-1)zTf - \rho f\| = \|(\rho-1)zT\mid_F f - \rho f\|,$$

for any $z \in \mathbb{D}$ and $f \in F \subset E$. □

4. Example of operator power bounded, but not in any $\mathcal{C}_\rho$

In this part, we are interested in producing such an operator. In Hilbert spaces, the first example was given by Sz.-Nagy and Foias ([13]). This operator was involutive. Our example will also be involutive.

First, we compute an underestimate of the ρ-numerical radius for some involutive operators:

Lemma 4.1. *Let T be an operator in $\mathcal{B}(E)$ such that $T^2 = I$ and $\|T\| \geq 5$. Then*

$$w_\rho(T) \geq \frac{\rho+1}{\rho} \quad (\rho \geq 1).$$

Proof. As $T^2 = I$, we immediately have $r(T) = 1$. Therefore, if $z \in \mathbb{D}$, we can compute:

$$\Big(I - \frac{\rho-1}{\rho}zT\Big)^{-1} = \Big(\sum_{k=0}^{\infty}\Big(\frac{\rho-1}{\rho}z\Big)^{2k}\Big)I + \Big(\sum_{k=0}^{\infty}\Big(\frac{\rho-1}{\rho}z\Big)^{2k}\Big)\frac{\rho-1}{\rho}zT$$

and then

$$\begin{aligned}
&\Big\|z\frac{T}{\rho}\Big(I - \frac{\rho-1}{\rho}zT\Big)^{-1}\Big\| \\
&= \Big\|z\frac{T}{\rho}\Big(\Big(\sum_{k=0}^{\infty}\Big(\frac{\rho-1}{\rho}z\Big)^{2k}\Big)I + \Big(\sum_{k=0}^{\infty}\Big(\frac{\rho-1}{\rho}z\Big)^{2k}\Big)\frac{\rho-1}{\rho}zT\Big)\Big\| \\
&= \Big|\frac{z}{\rho}\Big| \cdot \Big|\sum_{k=0}^{\infty}\Big(\frac{\rho-1}{\rho}z\Big)^{2k}\Big| \cdot \Big\|T\Big(I + \frac{\rho-1}{\rho}zT\Big)\Big\| \\
&= \Big|\frac{z}{\rho}\Big| \cdot \frac{1}{\Big|1 - \Big(\frac{\rho-1}{\rho}z\Big)^2\Big|} \cdot \Big\|T + \frac{\rho-1}{\rho}zI\Big\|.
\end{aligned}$$

As $\|T\| \geq 5$, we have,

$$\left\|T - \frac{\rho-1}{\rho}z\right\| \geq \|T\| - \left\|\frac{\rho-1}{\rho}zI\right\| \geq 4,$$

and

$$2r\Big(T + \frac{\rho-1}{\rho}zI\Big) \leq 2\Big(r(T) + r\Big(\frac{\rho-1}{\rho}zI\Big)\Big) \leq 4.$$

Therefore,

$$\left\|z\frac{T}{\rho}\Big(I - \frac{\rho-1}{\rho}zT\Big)^{-1}\right\| \geq 2\left|\frac{z}{\rho}\right| \cdot \frac{1}{\left|1 - \left(\frac{\rho-1}{\rho}z\right)^2\right|} \cdot r\Big(T + \frac{\rho-1}{\rho}zI\Big).$$

Now, as $T \neq \lambda I$ (because $T^2 = I$ and $\|T\| \geq 5 > 1$), we obtain:

$$r\Big(T + \frac{\rho-1}{\rho}zI\Big) = \max\Big(|1 + \frac{\rho-1}{\rho}z|,\ |1 - \frac{\rho-1}{\rho}z|\Big),$$

and, consequently,

$$\begin{aligned}\left\|z\frac{T}{\rho}\Big(I - \frac{\rho-1}{\rho}zT\Big)^{-1}\right\| &\geq 2\left|\frac{z}{\rho}\right| \cdot \frac{1}{\left|1 - \left(\frac{\rho-1}{\rho}z\right)^2\right|} \cdot \left|1 + \frac{\rho-1}{\rho}z\right| \\ &= 2\left|\frac{z}{\rho}\right| \cdot \frac{1}{\left|1 - \frac{\rho-1}{\rho}z\right|}, \qquad \text{for any } z \in \mathbb{D}.\end{aligned}$$

Therefore, we have the following inclusion:

$$[0,\ 1] \cap \{r : \forall |z| \leq r,\ 2\left|\frac{z}{\rho}\right| \cdot \frac{1}{\left|1 - \frac{\rho-1}{\rho}z\right|} \leq 1\} \supset [0,\ 1] \cap \{r : rT \in \mathcal{C}_\rho\}.$$

But a quick study of the left set shows that it is in fact $[0,\ \rho/(\rho+1)]$, which finally leads to the desired inequality. □

Remarks.

1) The constant 5 is not the best possible to get this underestimate.
2) It can be shown that, if T is an operator such that $T^2 = 1$ and $T \in \mathcal{C}_\rho$, then $\|T\| \leq 2$, where 2 is the best possible constant.

Proposition 4.2. *Let T be an operator in $\mathcal{B}(E)$ such that $T^2 = I$ and $\|T\| \geq 5$. Then T is not contained in any of the classes $\mathcal{C}_\rho$ $(\rho > 0)$.*

Proof. The previous lemma shows that every involutive operator T of norm greater than 5 is not in any $\mathcal{C}_\rho$ $(\rho \geq 1)$, but as $\|T\| > 1$, T will not be in $\mathcal{C}_\rho$ for $\rho < 1$. □

Remark. We can prove that if the norm of T is strictly greater than 2, then T is not in any $\mathcal{C}_\rho$ $(\rho > 0)$.

The second lemma, which we shall not prove, will enable us to build an involutive operator on E from another on a subspace of dimension 2.

Lemma 4.3. *In a Banach space, any finite-dimensional subspace is complemented (ie there exists a continuous projection on this subspace).*

Finally, we can produce our example:

Theorem 4.4. *Let E be a Banach space of dimension greater than 2, then there exists a power bounded operator on E which does not belong to any of the classes $\mathcal{C}_\rho$ ($\rho > 0$).*

Proof. Using Proposition 4.2, we only have to produce an involutive operator with norm greater than 5, which will be automatically power bounded. Using Lemma 4.3, we take two closed subspaces F_1 and F_2 such that $\dim(F_1) = 2$ and $E = F_1 \oplus F_2$. Then, we define T on E by:

- $T\,|_{F_2} = I\,|_{F_2}$;
- $Te_1 = 5e_2$ and $Te_2 = e_1/5$, where $(e_1,\ e_2)$ is a normed basis of F_1.

Then, T is well defined, $\|T\| \geq 5$ and $T^2 = I$. The proof is complete. □

This example also entails the following corollary:

Corollary 4.5. *Let E be a Banach space, and $\|.\|_1$ and $\|.\|_2$ be two (Banach) norms on E. We cannot affirm*

$$\bigcup_{\rho>0} \mathcal{C}_{\rho,\ \|\cdot\|_1} = \bigcup_{\rho>0} \mathcal{C}_{\rho,\ \|\cdot\|_2},$$

even if these norms are equivalent.

Proof. One only has to consider the Banach space $\mathbb{C}^2$, the canonical basis (e_1, e_2), and the two equivalent norms:

$$\|xe_1 + ye_2\|_1 = |x| + |y| \text{ and } \|xe_1 + ye_2\|_\alpha = |x| + 5|y|.$$

Then, let T be the operator defined by

$$T(e_1) = e_2 \text{ and } T(e_2) = e_1.$$

This operator verifies $\|T\|_1 = 1$, and therefore $T \in \mathcal{C}_{1,\|\cdot\|_1}$; but as $\|T\|_\alpha \geq 5$, T is not in $\mathcal{C}_{\rho,\|\cdot\|_\alpha}$ ($\rho > 0$). □

5. Study of nilpotent operators

Here, we will focus on the study of nilpotent operators which, in Hilbert spaces, produce a lot of examples and counter-examples. The study of 2-nilpotent operators (ie such that $T^2 = 0$) have many consequences. First we compute the ρ-numerical radius for 2-nilpotent operators:

Proposition 5.1. *Let T be a 2-nilpotent operator in $\mathcal{B}(E)$. Then*

$$w_\rho(T) = \frac{\|T\|}{\rho}.$$

Proof. For any $\alpha > 0$, we obtain, from the characterization given in Proposition 2.4

$$\begin{aligned} w_\rho(T) \leq \alpha &\Leftrightarrow \forall z \notin \bar{\mathbb{D}},\ \|T((\rho-1)T - \rho\alpha z I)^{-1}\| \leq 1 \\ &\Leftrightarrow \forall z \notin \bar{\mathbb{D}},\ \left\|\frac{T}{\rho\alpha z}(\frac{\rho-1}{\rho\alpha z}T - I)^{-1}\right\| \leq 1 \\ &\Leftrightarrow \forall z \notin \bar{\mathbb{D}},\ \left\|\frac{T}{\rho\alpha z}(\frac{\rho-1}{\rho\alpha z}T + I)\right\| \leq 1 \\ &\Leftrightarrow \forall z \notin \bar{\mathbb{D}},\ \left\|\frac{T}{\rho\alpha z}\right\| \leq 1 \\ &\Leftrightarrow \forall z \notin \bar{\mathbb{D}},\ \|T\| \leq \rho\alpha|z| \\ &\Leftrightarrow \alpha \geq \frac{\|T\|}{\rho}. \end{aligned}$$

□

As a first corollary, we have, as in Hilbert spaces,

Corollary 5.2. *The $\mathcal{C}_\rho$ classes are strictly growing, as ρ increases, as soon as the dimension is greater than 2.*

Proof. Using Lemma 4.3, we obtain two closed subspaces F and G such that $E = F \oplus G$ and $\dim F = 2$. Then, we define T on E by $T\ |_F$ is a 2-nilpotent operator of norm 1, and $T\ |_G = 0$. T is a 2-nilpotent operator of norm 1 on E. Now, if we denote by $T_\rho = \rho T$, then $T_\rho \in \mathcal{C}_\rho$, and, as we have:

$$w_{\rho'}(T_\rho) = \frac{\rho}{\rho'}, \quad \text{for any } \rho' > 0,$$

we obtain:

$$T_\rho \notin \mathcal{C}_{\rho'}, \quad \text{for any } 0 < \rho' < \rho.$$

□

Using this computation, we also obtain:

Corollary 5.3. *In general, the constant ρ in the following inequality:*

$$v(T) \leq \rho w_\rho(T)$$

is optimum.

Proof. Just consider the space l_1^2 ($= \mathbb{C}^2$, with norm 1). If we denote by S_2 the truncated shift on this space, then $v(S_2) = 1$ and $w_\rho(S_2) = 1/\rho$. The proof is complete. □

Remark. An easy computation shows that:

$$v(S_2, \|\cdot\|_p) = \left(\frac{p-1}{p}\right)^{\frac{p-1}{p}} \left(\frac{1}{p}\right)^{\frac{1}{p}} \quad (p > 1).$$

when $w_{2,\|\cdot\|_p}(S_2) = 1/2$. The graph of $v(S_2, \|\cdot\|_p)$, as p varies, leads us to think that the following is true:

$$\text{if } v(T) = w_2(T) \text{ for every } T \in \mathcal{B}(E), \quad \text{then } E \text{ is a Hilbert space.}$$

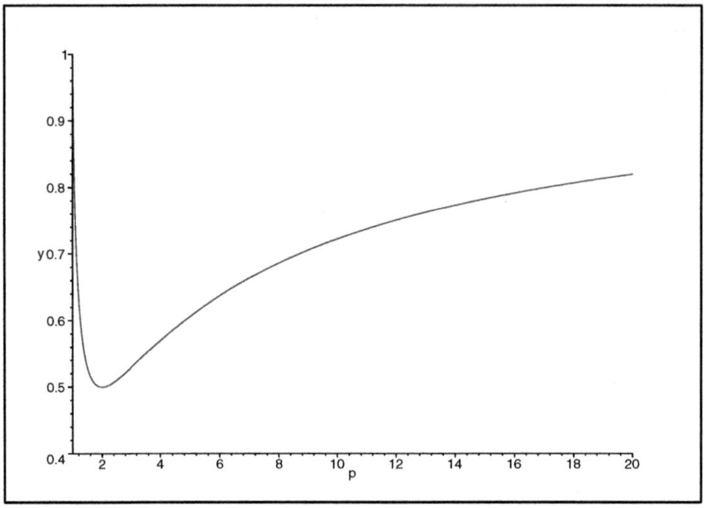

FIGURE 1. Graph of $v(S_2, \|\cdot\|_p)$, as p varies

As a matter of fact, this is true when $E = l_p$ $(p > 1)$, and, in this spaces, one only has to see what happens with S_2. Moreover, further computations, even if they do not allow us to give a definitive answer, seem to support this idea.

Studying 3-nilpotent operators also leads to many consequences. First, let us give an estimate of the ρ-numerical radius for such operators:

Proposition 5.4. *Let T be an operator such that $T^3 = 0$ and $\|T\| = \|T^2\| = 1$. Then, we have:*

$$\frac{2(\rho-1)}{\rho(\sqrt{1+4(\rho-1)}-1)} \geq w_\rho(T) \geq \frac{2(\rho-1)}{\rho(\sqrt{1+4(\rho-1)}+1)} \quad (\rho > 1).$$

Remark. The overestimate remains true, even if $\|T^2\| \neq 1$.

Proof. Let T be a 3-nilpotent operator, $\rho > 1$ and $r > 0$. We have:

$$\begin{aligned} rT \in \mathcal{C}_\rho &\Leftrightarrow \forall z \in r\mathbb{D},\ \|zT((\rho-1)zT - \rho I)^{-1}\| \leq 1 \\ &\Leftrightarrow \forall z \in r\mathbb{D},\ \left|\frac{z}{\rho}\right| \cdot \left\|T\Big(\frac{\rho-1}{\rho}zT - I\Big)^{-1}\right\| \leq 1 \\ &\Leftrightarrow \forall z \in r\mathbb{D},\ \left|\frac{z}{\rho}\right| \cdot \left\|T\Big(I + \Big(\frac{\rho-1}{\rho}zT\Big) + \Big(\frac{\rho-1}{\rho}zT\Big)^2\Big)\right\| \leq 1 \\ &\Leftrightarrow \forall z \in r\mathbb{D},\ \left|\frac{z}{\rho}\right| \cdot \left\|T + \frac{\rho-1}{\rho}zT^2\right\| \leq 1 \end{aligned}$$

Now, we have:

$$\frac{\rho-1}{\rho}|z|\|T^2\| - \|T\| \leq \left\|T + \frac{\rho-1}{\rho}zT^2\right\| \leq \left\|T\right\| + \frac{\rho-1}{\rho}|z|\left\|T^2\right\|,$$

and, as $\|T\| = \|T^2\| = 1$, we obtain:

$$\Big\{r>0 : \frac{r}{\rho}\Big(\frac{\rho-1}{\rho}r-1\Big) \leq 1\Big\} \supset \{r>0 : rT \in \mathcal{C}_\rho\} \supset \Big\{r>0 : \frac{r}{\rho}\Big(\frac{\rho-1}{\rho}r+1\Big) \leq 1\Big\}.$$

A quick study of the left and right sets shows that

$$\Big\{r>0 : \frac{r}{\rho}\Big(\frac{\rho-1}{\rho}r-1\Big) \leq 1\Big\} = \Big]0, \frac{\rho+\rho\sqrt{1+4(\rho-1)}}{2(\rho-1)}\Big]$$

and

$$\Big\{r>0 : \frac{r}{\rho}\Big(\frac{\rho-1}{\rho}r+1\Big) \leq 1\Big\} = \Big]0, \frac{-\rho+\rho\sqrt{1+4(\rho-1)}}{2(\rho-1)}\Big].$$

So we obtain the desired inequality. □

Corollary 5.5. *Let S_3 denotes the truncated shift on $\mathbb{C}^3$ and T be an operator acting on $(E, \|\cdot\|)$ such that $T^3 = 0$. Then, we have:*

$$w_{\rho,\ \|\cdot\|}(T) \leq \|T\| w_{\rho,\ \|\cdot\|_1}(S_3) \quad (\rho > 0).$$

Proof. Using Ando's and Nishio's law, we can suppose $\rho > 1$. Now, in that case, if $z \in \mathbb{C}$, we have:

$$\Big\|S_3 + \frac{\rho-1}{\rho} z S_3^2\Big\| = 1 + \frac{\rho-1}{\rho}|z|,$$

and therefore

$$w_{\rho,\|\cdot\|_1}(S_3) = \frac{2(\rho-1)}{\rho(\sqrt{1+4(\rho-1)}-1)}.$$ □

A precise re-reading of the proof of Proposition 5.4 leads to a generalization of the previous corollary:

Proposition 5.6. *Let $n \geq 2$, S_n denote the truncated shift on $\mathbb{C}^n$, and T be an operator acting on $(E, \|\cdot\|)$ such that $T^n = 0$. Then, we have:*

$$w_{\rho,\ \|\cdot\|}(T) \leq \|T\|.w_{\rho,\|\cdot\|_1}(S_n) \quad (\rho > 0).$$

Proof. By homogeneity, we can assume that $\|T\| = 1$, and, for such an operator,

$$\begin{aligned}
&\|zT((\rho-1)zT-\rho I)^{-1}\| \\
&\quad \leq \Big|\frac{z}{\rho}\Big|\Big(\|T\| + \cdots + \Big(\frac{\rho-1}{\rho}\Big)^{n-2}\|T^{n-1}\|\Big) \\
&\quad \leq \Big|\frac{z}{\rho}\Big|\Big(1 + \cdots + \Big(\frac{\rho-1}{\rho}\Big)^{n-2}\Big) \\
&\quad \leq \Big|\frac{z}{\rho}\Big|\Big(\|S_n\|_1 + \cdots + \Big(\frac{\rho-1}{\rho}\Big)^{n-2}\|S_n^{n-1}\|_1\Big) \\
&\quad = \|zS_n((\rho-1)zS_n-\rho I)^{-1}\|_1.
\end{aligned}$$

The end of the proof is easy. □

Remark. In Hilbert spaces, Haagerup and de la Harpe [9] gave an estimate for the numerical radius of nilpotent operators:

$$\text{if } T^n = 0, \text{ then } w_2(T) \leq \|T\| w_2(S_n).$$

Later, in their study of constrained von Neumann inequalities [2], Badea and Cassier gave, as a consequence, a generalization of this result:

$$\text{if } T \text{ is a contraction such that } T^n = 0, \text{ then}$$
$$w_\rho(p(T)) \leq w_\rho(p(S_n)) \text{ for any polynomial } p \text{ in } \mathbb{C}[X].$$

The previous result is a kind of analogous to these results in Banach spaces.

Finally, we give a last theorem, which is a generalization of a computation for truncated shifts in Hilbert spaces [6]:

Theorem 5.7. *Let T be an operator such that $\|T\| = \|T^{n-1}\| = 1$ and $T^n = 0$, then*

$$w_\rho(T) \underset{\rho\to\infty}{\sim} \rho^{-\frac{1}{n-1}}.$$

Proof. As $T^n = 0$, we have:

$$\left(I - \frac{\rho-1}{\rho} zT\right)^{-1} = \sum_{j=0}^{n-1} \left(\frac{\rho-1}{\rho} zT\right)^j, \text{ for any } z \in \mathbb{C}.$$

Moreover,

$$r(T) = 0 \Rightarrow w_\rho(T) \xrightarrow[\rho\to\infty]{} 0,$$

so we can, for any $k \geq 2$, by considering $\rho \geq \rho_k > 2$, suppose that $w_\rho(T) \leq 1/k$, and so $]0,\ k] \subset \{r > 0 : rT \in \mathcal{C}_\rho\}$.

We will prove this theorem using two asymptotic estimates. First, we give an overestimate. We compute an overestimate of:

$$\begin{aligned}
\left\|\frac{z}{\rho} T(I - \frac{\rho-1}{\rho} zT)^{-1}\right\| &= \left\|\frac{z}{\rho} T(\sum_{j=0}^{n-1} (\frac{\rho-1}{\rho} zT)^j)\right\| \\
&= \left\|\sum_{j=1}^{n-1} \frac{(\rho-1)^{j-1}}{\rho^j} z^j T^j\right\| \\
&\leq \sum_{j=1}^{n-1} \left\|\frac{(\rho-1)^{j-1}}{\rho^j} z^j T^j\right\| \\
&\leq \sum_{j=1}^{n-1} \frac{(\rho-1)^{j-1}}{\rho^j} |z|^j \\
&= \frac{|z|}{\rho} \cdot \frac{1 - \left(\frac{\rho-1}{\rho}|z|\right)^{n-1}}{1 - \left(\frac{\rho-1}{\rho}|z|\right)} \quad (\text{if } \frac{\rho-1}{\rho}|z| \neq 1).
\end{aligned}$$

Therefore we obtain the following inclusion:

$$\Big\{r \geq k : \frac{r}{\rho} \cdot \frac{1-(\frac{\rho-1}{\rho}r)^{n-1}}{1-(\frac{\rho-1}{\rho}r)} \leq 1\Big\} \subset \{r \geq k : rT \in \mathcal{C}_\rho\}.$$

Now, as $r \geq k \geq 2$ and $\rho \geq \rho_k \geq 2$, we have $r > \rho/(\rho-1)$, and so $\rho-(\rho-1)r < 0$. Then we deduce:

$$\begin{aligned}
& r \geq k \text{ and } \frac{r}{\rho}\frac{1-(\frac{\rho-1}{\rho}r)^{n-1}}{1-(\frac{\rho-1}{\rho}r)} \leq 1 \\
& \Leftrightarrow r \geq k \text{ and } r - r\Big(\frac{\rho-1}{\rho}r\Big)^{n-1} \geq \rho - \rho r + r \\
& \Leftrightarrow r \geq k \text{ and } r\Big(\frac{\rho-1}{\rho}r\Big)^{n-1} \leq \rho r - \rho \\
& \Leftrightarrow r \geq k \text{ and } \Big(\frac{\rho-1}{\rho}r\Big)^{n-1} \leq \rho(1-1/r) \\
& \text{it is enough to verify that } r \geq k \text{ and } \Big(\frac{\rho-1}{\rho}r\Big)^{n-1} \leq \rho\Big(1-\frac{1}{k}\Big) \\
& \Leftrightarrow r \geq k \text{ and } r \leq \rho^{1/(n-1)}\Big(1-\frac{1}{k}\Big)^{1/(n-1)}\frac{\rho}{\rho-1}.
\end{aligned}$$

As the right member of the previous inequality tends to $+\infty$ as $\rho \to \infty$, we deduce the existence of $\rho_k' \geq \rho_k$ such that:

$$\rho^{1/(n-1)}\Big(1-\frac{1}{k}\Big)^{1/(n-1)}\frac{\rho}{\rho-1} \geq k, \qquad \text{for any } \rho \geq \rho_k'.$$

Consequently, for any $k \geq 2$, there exists $\rho_k' > 2$ such that:

$$\begin{aligned}
\{r > 0 : rT \in \mathcal{C}_\rho\} &=]0,\ k] \cup \{r \geq k : rT \in \mathcal{C}_\rho\} \\
&\supset]0,\ k] \cup \Big[k,\ \rho^{1/(n-1)}\Big(1-\frac{1}{k}\Big)^{1/(n-1)}\frac{\rho}{\rho-1}\Big] \\
&= \Big]0,\ \rho^{1/(n-1)}\Big(1-\frac{1}{k}\Big)^{1/(n-1)}\frac{\rho}{\rho-1}\Big], \qquad \text{for any } \rho \geq \rho_k',
\end{aligned}$$

and finally:

$$w_\rho(T) \leq \frac{\rho-1}{\rho} \cdot \frac{1}{\rho^{1/(n-1)}(1-1/k)^{1/(n-1)}}.$$

Now, we look for an underestimate. We compute an underestimate of:

$$\begin{aligned}
\Big\|\frac{z}{\rho}T\Big(I-\frac{\rho-1}{\rho}zT\Big)^{-1}\Big\| &\geq \Big\|\frac{(\rho-1)^{n-2}}{\rho^{n-1}}z^{n-1}T^{n-1}\Big\| - \Big\|\sum_{j=1}^{n-2}\frac{(\rho-1)^{j-1}}{\rho^j}z^jT^j\Big\| \\
&\geq \frac{(\rho-1)^{n-2}}{\rho^{n-1}}|z|^{n-1} - \frac{|z|}{\rho} \cdot \frac{1-\Big(\frac{\rho-1}{\rho}|z|\Big)^{n-2}}{1-\frac{\rho-1}{\rho}|z|}.
\end{aligned}$$

Therefore we have the following inclusion:

$$\{r \geq k : rT \in \mathcal{C}_\rho\} \subset \Big\{r \geq k : \frac{(\rho-1)^{n-2}}{\rho^{n-1}} r^{n-1} - \frac{r}{\rho} \cdot \frac{1-\left(\frac{\rho-1}{\rho}r\right)^{n-2}}{1-\left(\frac{\rho-1}{\rho}r\right)} \leq 1\Big\}.$$

Now,

$$r \geq k \text{ and } \frac{(\rho-1)^{n-2}}{\rho^{n-1}} r^{n-1} - \frac{r}{\rho} \cdot \frac{1-(\frac{\rho-1}{\rho}r)^{n-2}}{1-(\frac{\rho-1}{\rho}r)} \leq 1$$

$$\Leftrightarrow r \geq k \text{ and } -r\Big(1-\Big(\frac{\rho-1}{\rho}r\Big)^{n-2}\Big) \geq (\rho-(\rho-1)r)\Big(1-\frac{(\rho-1)^{n-2}}{\rho^{n-1}}r^{n-1}\Big)$$

$$\Leftrightarrow r \geq k \text{ and } -r + r\Big(\frac{\rho-1}{\rho}r\Big)^{n-2} \geq \rho - r\rho + r - r\Big(\frac{\rho-1}{\rho}r\Big)^{n-2} + r\Big(\frac{\rho-1}{\rho}r\Big)^{n-1}$$

$$\Leftrightarrow r \geq k \text{ and } r\rho - \rho - 2r \geq r\Big(\frac{\rho-1}{\rho}r\Big)^{n-2}(\frac{\rho-1}{\rho}r - 2)$$

$$\Rightarrow r \geq k \text{ and } r\rho \geq r\Big(\frac{\rho-1}{\rho}r\Big)^{n-2}\Big(\frac{\rho-1}{\rho}r - 2\Big)$$

$$\Rightarrow r \geq k \text{ and } \rho \geq (\frac{\rho-1}{\rho}r)^{n-2}\frac{\rho-1}{\rho}r\Big(1-2\frac{\rho}{r(\rho-1)}\Big)$$

$$\Rightarrow r \geq k \text{ and } \rho \geq \Big(\frac{\rho-1}{\rho}r\Big)^{n-2}\frac{\rho-1}{\rho}r\Big(1-\frac{4}{k}\Big)$$

$$\text{(because } \rho \geq 2 \text{ and } r \geq k, \text{ and so } -2\frac{\rho}{r(\rho-1)} \geq -\frac{4}{k})$$

$$\Rightarrow r \geq k \text{ and } \frac{\rho}{1-4/k} \geq \Big(\frac{\rho-1}{\rho}r\Big)^{n-1} \quad \text{(for all } k > 4)$$

$$\Rightarrow r \geq k \text{ and } r \leq \Big(\frac{\rho}{1-\frac{4}{k}}\Big)^{1/(n-1)}\frac{\rho}{\rho-1}$$

As the right member of the previous inequality tends to $+\infty$ as $\rho \to \infty$, we deduce the existence of $\rho''_k \geq \rho_k$ such that:

$$\Big(\frac{\rho}{1-4/k}\Big)^{1/(n-1)}\frac{\rho}{\rho-1} \geq k, \qquad \text{for any } \rho \geq \rho''_k.$$

Consequently, for any $k \geq 2$, there exists $\rho''_k > 2$ such that:

$$\begin{aligned}\{r > 0 : rT \in \mathcal{C}_\rho\} &=]0,\ k] \cup \{r \geq k : rT \in \mathcal{C}_\rho\} \\ &\subset]0,\ k] \cup \Big[k,\ \Big(\frac{\rho}{1-4/k}\Big)^{1/(n-1)}\frac{\rho}{\rho-1}\Big] \\ &= \Big]0,\ \Big(\frac{\rho}{1-4/k}\Big)^{1/(n-1)}\frac{\rho}{\rho-1}\Big], \quad \text{for any } \rho \geq \rho''_k,\end{aligned}$$

and finally

$$w_\rho(T) \geq \frac{\rho-1}{\rho}\Big(\frac{1-4/k}{\rho}\Big)^{1/(n-1)}.$$

In order to conclude, one only has to combine the two estimates to obtain:
$\forall\epsilon>0,\ \exists\rho_\epsilon:\forall\rho\geq\rho_\epsilon,\ 1-\epsilon\leq\rho^{1/(n-1)}w_\rho(T)\leq 1+\epsilon,$ i.e. $|\rho^{1/(n-1)}w_\rho(T)-1|\leq\epsilon,$
where ρ_ϵ is computed using ρ'_k and ρ''_k. We have the desired asymptotic result. □

Remark. A precise re-reading of the proof shows that the overestimate remains true event if $\|T^{n-1}\| < 1$, but not the underestimate.

Acknowledgments
The author wishes to express his gratitude to Gilles Cassier for his help.

References

[1] T. Ando, K. Nishio, *Convexity Properties of Operator Radii Associated with Unitary ρ-Dilations.* Michigan Math. J. **20** (1973), 303–307.

[2] C. Badea, G. Cassier, *Constrained von Neumann Inequalities.* Adv. Math. **166** (2002), no 2, 260–297.

[3] F.L. Bauer, *On the Field of Values Subordinate to a Norm.* Numer. Math. **4** (1962), 103–111.

[4] F.F. Bonsall, J. Duncan, *Numerical Ranges of Operators on Normed Spaces and of Elements of Normed Algebras.* London Math. Soc. Lect. Note. Series **2** (1971).

[5] F.F. Bonsall, J. Duncan, *Numerical Ranges.* II, London Math. Soc. Lect. Note. Series, No. 10 (1973).

[6] L. Carrot, *Computation of the ρ-Numerical Radius for Truncated Shifts.* to appear in Math. Sz.

[7] C. Davis, *The Shell of a Hilbert-Space Operator.* Acta Sci. Math. (Szeged) **29** (1968), 69–86.

[8] B.W. Glickfeld, *On an Inequality of Banach Algebra Geometry and Semi-Inner Product Space Theory.* Ill. J. Math. **14** (1970), 76–81.

[9] U. Haagerup, P. de la Harpe, *The Numerical Radius of a Nilpotent Operator on a Hilbert Space.* Proc. Amer. Math. Soc. **115** (1992), 371–379.

[10] J.A.R. Holbrook, *On the Power-Bounded Operators of Sz.-Nagy and Foias.* Acta Sci. Math. (Szeged) **29** (1968), 299–310.

[11] J.A.R. Holbrook, *Inequalities Governing the Operator Radii Associated with Unitary ρ-Dilations.* Michigan Math. J. **18** (1971).

[12] G. Lumer, *Semi Inner-Product Spaces.* Trans. Amer. Math. Soc. **100** (1961), 29–43.

[13] B. Sz.-Nagy, C. Foias, *On Certain Classes of Power-Bounded Operators in Hilbert Space.* Acta Sci. Math. **27** (1966), 17–25.

[14] W. Rudin, *Real and Complex Analysis.* third edition, McGraw-Hill Book Co., New York, 1987.

Laurent Carrot
Institut G. Desargues
Université de Lyon I
F-69622 Villeurbanne, France

Operator Theory:
Advances and Applications, Vol. 153, 103–122

Generalized Toeplitz Operators and Cyclic Vectors

G. Cassier, H. Mahzouli and E.H. Zerouali

Abstract. We give in this paper some asymptotic von Neumann inequalities for power bounded operators in the class $C_\rho \cap C_{1,\cdot}$ and some spacial von Neumann inequalities associated with non zero elements of the point spectrum, when it is non void, of generalized Toeplitz operators. Introducing perturbed kernel, we consider classes C_R which extend the classical classes C_ρ. We give results about absolute continuity with respect to the Haar measure for operators in class $C_R \cap C_{1,\cdot}$. This allows us to give new results on cyclic vectors for such operators and provides invariant subspaces for their powers. Relationships between cyclic vectors for T and T^* involving generalized Toeplitz operators are given and the commutativity of $\{T\}'$, the commutant of T is discussed.

Mathematics Subject Classification (2000). 47B35, 47B99.

Keywords. Generalized Toeplitz operator, von Neumann inequality, Invariant subspaces.

1. Introduction

Throughout of this paper, we denote H^2 the Hardy space of analytic functions on the unit disc D, S the unilateral unweighted shift on H^2 and S^* its adjoint operator. Toeplitz operators are defined as solutions of the operator equation $S^*XS = X$. They have been intensively treated during the last years and turn out to be a rich source of examples and counter examples in operator theory. Generalized Toeplitz operators were introduced as follows. Let H be a complex Hilbert space, let $B(H)$ be the algebra of all bounded operators on H, and suppose $T \in B(H)$. The set of T-Toeplitz operators, $C(T)$ is defined to contain operators satisfying the operator equation $T^*XT = X$. We denote $C_+(T)$ as the set of positive T-Toeplitz operators.

Recall first the following known result ([5]). Let $\rho > 0$, and $T \in B(H)$ whose spectrum is included in the closed unit disc. The operator T is in C_ρ if and only if $K^\rho_{r,t}(T) \geq 0$ for every $(r,t) \in]0,1] \times \mathbf{R}$. This leads to the next definition.

Definition 1.1. Let R be a polynomial and let T be an operator acting on a Hilbert space, we say that T is in class C_R if and only if

$$\sigma(T) \subset \bar{\mathbf{D}} \qquad \text{and} \qquad K^R_{r,t}(T) = K_{r,t}(T) + R(re^{-it}T) + R(re^{it}T)^* - I \geq 0.$$

Remark. When $R = \rho/2 > 0$ the class C_R is exactly the usual class C_ρ.

G. Cassier and T. Fack showed, in [5], that if T is a contraction in $C_{1,.}$ which has a positive generalized Toeplitz operator with eigenvalues, then T has a non zero noncyclic element, that is a non zero vector $y \in H$ such that $\operatorname{span}\{T^n y : n \geq 0\} \neq H$. In particular T has a non trivial invariant subspace. To this aim, they provided some spacial von Neumann inequalities for T.

In this paper we study the structure of cyclic vectors for powers of power bounded operators in the class $C_R \cap C_{1,.}$. See Section 2 for definitions.

In Section 2, basic background and some known results are given. We recall the notion of almost convergence and Banach limits. We also relate these notions to generalized Toeplitz operators and define the asymptotic kernel for power bounded operators.

In Section 3, we establish the von Neumann spacial inequality for operator in the class $C_\rho \cap C_{1,.}$. We refine this inequality in the case where T has a generalized Toeplitz operator with non void point spectrum.

We investigate in Section 4 the problem of cyclic vectors for power bounded operators in $C_R \cap C_{1,.}$. We show in Theorem 4.4, that completely non unitary operators in $C_R \cap C_{1,.}$ are intertwined with the unilateral shift on some Hilbert space in a satisfactory way. This leads to an extension of a result of Nagy-Foias ([12]) "*there exists N such that T^n has no cyclic vector for $n \geq N$*". Other results from [11] are generalized. The use of generalized Toeplitz operators allows to give more precise results in the general case of power bounded operators. Namely, we link cyclic vectors for T and T^* through generalized Toeplitz operators. We finally study the commutativity of the commutant of completely non unitary power bounded operators in the class $C_{1,.}$.

In Section 5, we provide spacial von Neumann inequalities associated to perturbed kernel. We give explicit computations in the case where $R(T) = \beta T + \bar{\beta} T^* + I$. This shed light on the role of the existence of eigenvalues of a generalized Toeplitz vector associated with T.

2. Preliminaries

2.1. Generalized Toeplitz operators

A complex sequence $\xi \in l^\infty(\mathbb{N}, \mathbb{C})$ almost converges to the complex number c if

$$\lim_{k \to \infty} \sup_{n \in \mathbb{N}} \left| c - k^{-1} \sum_{j=n}^{n+k-1} \xi(j) \right| = 0.$$

The sequence ξ is said to be almost convergent is the strong sense to c if $|\xi - c \cdot \mathbf{1}|$ is almost convergent to 0, where $\mathbf{1}$ is the constant sequence of value 1.

A Banach limit L is a positive functional on l^∞ satisfying

1. $L((u_n)_{n\geq 0}) = 1$ for $u_n \equiv 1$
2. $L((u_n)_{n\geq 0}) = L((u_{n+1})_{n\geq 0})$
3. If $(u_n)_{n\geq 0}$ converges, then $L((u_n)_{n\geq 0}) = \lim_{n\to+\infty} u_n$.

We denote by $\mathcal{B}$ the set of all Banach limits. We refer to [8, 9, 10] for further details.

The following classes of operators were considered in [11, 12, 13, 14].

$$C_{1,\cdot} = \{T \in B(H)/ \ \forall x \in H \setminus \{0\},\ \|T^n x\| \text{ does not converges to zero}\}$$

$$C_{\cdot,1} = \{T \in B(H)/ \ \forall x \in H \setminus \{0\},\ \|(T^*)^n x\| \text{ does not converges to zero}\}$$

and

$$C_{1,1} = C_{\cdot,1} \cap C_{1,\cdot}.$$

Let $PWB(H) := \{T \in B(H) : \sup_{n\geq 0} \|T^n\| < +\infty\}$ be the set of power bounded operators. Simple computations permit to show that if $T \in PWB(H)$ is such that $\lim_{n\geq 0} \|T^n\| \neq 0$ then the spectral radius $r(T)$ of T equals 1. This last affirmation will be assumed to be true for all our operators.

Let T be a power bounded operator and L be a Banach limit. Consider the functional

$$y \mapsto (x \mid y) := L((\langle T^n x \mid T^n y\rangle)_{n\geq 0}).$$

Using positivity of Banach limits, we see that

$$|(x \mid y)| \leq M^2 \|x\| \|y\|,$$

where $M = \sup_{n\geq 0} \|T^n\|$. In particular $(\cdot \mid \cdot)$ is continuous and by Reisz representation theorem, there exists a positive bounded operator X_L such that $(x \mid y) = \langle X_L x \mid y\rangle$. Following ([4]) such operators are called canonical generalized Toeplitz operators and their set is denoted $\tau(T)$. Recall that, $\tau(T)$ is reduced to a singleton if and only if the sequence $(T^*)^n T^n$ is weakly almost convergent. In every finite von Neumann algebra, the sequence $(T^*)^n T^n$ is always weakly almost convergent ([4]). If T is in the class C_ρ, then $(T^*)^n T^n$ converges in the strong topology of operators ([6]). In particular, in both previous cases $\tau(T)$ is reduced to a singleton.

We collect in the following proposition some properties of canonical Toeplitz operators (for the proof see [4]).

Proposition 2.1. *Let T be a Power Bounded Operator and $X \in \tau(T)$. Then*

a. $X \in C_+(T)$.
b. *$\tau(T)$ contains an injective element if an only if X is injective for any $X \in \tau(T)$.*
c. *$\{x \in H$ such that $T^n x \to 0\} \subset \ker(X)$ for any $X \in \tau(T)$. In particular, $T \in C_{1,\cdot}$ when $\tau(T)$ contains an injective element.*

We assume for the following theorem that $T \in C_{1,.}$ and let $X \in \tau(T)$. The associated mapping $(\cdot \mid \cdot)$ induces an inner product on H. Denote by $\widetilde{H}$ the Hilbert completion of H with respect to the norm $|||\cdot|||$, induced by the inner product $(\cdot|\cdot)$ and denote by $\widetilde{T}$ the canonical extension of T to $\widetilde{H}$. For every $x \in H$ we have

$$|||\widetilde{T}x|||^2 = (\widetilde{T}x \,|\, \widetilde{T}x) = L((\langle T^{n+1}x \,|\, T^{n+1}x\rangle)_{n\geq 0}) = L((\langle T^n x \,|\, T^n x\rangle)_{n\geq 0}) = |||x|||^2.$$

Since H is a dense subspace of $\widetilde{H}$, we conclude that $\widetilde{T}$ is an isometry.

In the invariant subspace problem, since $\overline{\mathrm{Im}(T)}$ and $\ker(T)$, if non trivial, provide invariant subspaces. It is always assumed (and we will do) that T has a dense range. This forces $\widetilde{T}$ to be dense range, and so $\widetilde{T}$ became a unitary transformation. Let μ_0 be the spectral measure on $S^1 = \{e^{it}/t \in \mathbb{R}\}$ associated with $\widetilde{T}$. Consider for $f \in L^\infty(S^1, \mu_0)$ the invariant subspace introduced by B. Beauzamy in [1]:

$$E_f = H \cap Ker(f(\widetilde{T})). \tag{2.1}$$

In [5], it is shown, when T is a contraction, that $E_f = KerX_f$ for some positive generalized Toeplitz operator X_f associated with T. Outlining the proof of this result, we generalize as follows

Theorem 2.2. ([5], Theorem 7) *Let T be a cyclic power bounded operator acting on Hilbert space H with dense range, assume that T is completely non unitary (c.n.u for short) in the class $C_{1,.}$. For $f \in L^\infty(S^1, \mu_0)$, let E_f be the closed invariant subspace generated by f. Then, there exists $X_f \in C_+(T)$ such that $E_f = Ker(X_f)$.*

We end this section by recovering some properties of generalized Toeplitz operators associated with power bounded operators. Denote $p_M =: M$ the constant gauge, we have

Proposition 2.3. *Let T be a power bounded operator, $\|T^n\| \leq M$ and let $X_L \in C_+(T)$ be associated with a Banach limit L_{p_M}. Then for $X \in C_+(T)$ we have*

$$0 \leq X \leq M^2 \|X\| X_L.$$

In particular $ker(X) \subset ker(X_L)$ and hence if X_L is one to one, then every $X \in C_+(T)$ is one to one.

Proof. Let $x \in H$,

$$|\langle Xx \mid x\rangle| = |\langle XT^n x \mid T^n x\rangle| \leq \|X\| \, \|T^n x\|^2.$$

By using the p-Banach limit we obtain

$$\langle Xx \mid x\rangle \leq M^2 \|X\| \langle X_L x \mid x\rangle. \qquad \square$$

3. Asymptotic spacial von Neumann inequalities for operators in C_ρ

3.1. Asymptotic kernels

Let $T \in PWB(H)$ be such that $r(T) = 1$ and $X \in B(H)$ be an arbitrary operator. The asymptotic kernel associated with X is the operator-valued function,

$$K_{r,t}^{\infty}(T, X) = X(I - re^{-it}T)^{-1} + (I - re^{it}T^*)^{-1}X - X$$

and the kernel associated with T is

$$K_{r,t}(T) = K_{r,t}^{\infty}(T, I) = (I - re^{-it}T)^{-1} + (I - re^{it}T^*)^{-1} - I$$

for $(r, t) \in [0, 1[\times[0, 2\pi[$.

These kernels were studied in [5] and applied to obtain asymptotic spacial von Neumann inequalities which allowed the authors to provide invariant subspaces for some classes of contractions.

The following proposition is inspired from [5] assemble some properties of the asymptotic kernel and the proof runs similarly.

Proposition 3.1. ([5]) *Let T be a power bounded operator, $X \in C_+(T)$ and let $K_{r,t}^{\infty}(T, X)$ be the asymptotic kernel associated with X defined on $(r, t) \in [0, 1[\times[0, 2\pi[$. Then*

1. *$K_{r,t}^{\infty}(T, X) \in C_+(T)$ and satisfies*
$$K_{r,t}^{\infty}(T, X) = (1 - r^2)(I - re^{it}T^*)^{-1}X(I - re^{-it}T)^{-1}$$
2. *For f an analytic function on neighborhood of the unit closed disc, we have*
$$Xf(rT) = \int_0^{2\pi} f(e^{it})K_{r,t}^{\infty}(T, X)\frac{dt}{2\pi}$$
and
$$f(rT^*)X = \int_0^{2\pi} f(e^{-it})K_{r,t}^{\infty}(T, X)\frac{dt}{2\pi}$$
for every $0 \le r < 1$.

Using Proposition 2.3, We get

Proposition 3.2. *Let T be a power bounded operator, then for every $X \in C_+(T)$ and $(r, t) \in [0, 1[\times[0, 2\pi[$, we have $K_{r,t}^{\infty}(T, X) \in C_+(T)$ and*

$$K_{r,t}^{\infty}(T, X) \le \|X\| K_{r,t}^{\infty}(T, X_L),$$

where $X_L \in C_+(T)$ is associated with a Banach limit L.

Proof.

$$\begin{aligned} K_{r,t}^{\infty}(T, X) &= (1 - r^2)(I - re^{it}T^*)^{-1}X(I - re^{-it}T)^{-1} \\ &\le (1 - r^2)\|X\|(I - re^{it}T^*)^{-1}X_L(I - re^{-it}T)^{-1} \\ &\le \|X\| K_{r,t}^{\infty}(T, X_L). \end{aligned}$$

□

3.2. Asymptotic spacial von Neumann inequalities and point spectrum of generalized Toeplitz operators

We fit the generalized von Neumann inequality given in [3] to our situation. The use of asymptotic kernels in the proof was crucial. For $X \in C_+(T)$, $r \in [0,1[$ and $x \in H$, denote $d\mu^X_{r,x}(t) = \langle K^\infty_{r,t}(T,X)x \mid x\rangle \frac{dt}{2\pi}$, then

Proposition 3.3. *Let T be a power bounded operator and $X \in C_+(T)$. Then for any pair (f,g) of analytic functions on a neighborhood of the closed unit disc, $x \in H$ and $0 \le r \le 1$, we have*

$$\|[f(rT^*)X + Xg(rT)]\,x\|^2 \le \|X\|^2 \int_0^{2\pi} \left|f(e^{-it}) + g(e^{it})\right|^2 d\mu^X_{r,x}(t).$$

We develop in the rest of this section asymptotic spacial von Neumann inequalities for power bounded operators in C_ρ.

Proposition 3.4. *Let T be a power bounded operator and $X \in C_+(T)$. Suppose that X has a nonzero eigenvalue α and denote y a corresponding normalized eigenvector. Let P, Q be polynomials, then*

$$|\langle Q(T)P(T)y \mid y\rangle| \le \frac{1}{\alpha}\|\sqrt{X}P(T)y\|\,\|\sqrt{X}Q(T)y\|, \tag{3.1}$$

Proof. Simple computations give

$$\langle K^\infty_{r,t}(T,X)y \mid y\rangle = \alpha\langle K_{r,t}(T)y \mid y\rangle.$$

For $(r,t) \in [0,1[\times[0,2\pi[$ and for every polynomials P, Q, we obtain

$$\begin{aligned}|\langle Q(rT)P(rT)y \mid y\rangle| &= \left|\int_0^{2\pi} P(e^{it})Q(e^{it})\langle K_{r,t}(T)y \mid y\rangle \frac{dt}{2\pi}\right|\\ &= \frac{1}{\alpha}\left|\int_0^{2\pi} P(e^{it})Q(e^{it})d\mu^X_{r,y}(t)\right|.\end{aligned}$$

Since $K^\infty_{r,t}(T,X)$ is a positive operator, by applying the Cauchy-Schwarz inequality we obtain

$$\begin{aligned}&|\langle Q(rT)P(rT)y \mid y\rangle|\\ &\le \frac{1}{\alpha}\int_0^{2\pi} |P(e^{it})Q(e^{it})|\left\|\sqrt{K^\infty_{r,t}(T,X)}y\right\|\,\left\|\sqrt{K^\infty_{r,t}(T,X)}y\right\|\frac{dt}{2\pi}.\end{aligned}$$

Denote I the left side of the inequality above and apply the Cauchy-Schwarz inequality again

$$\begin{aligned}I &\le \frac{1}{\alpha}\left(\int_0^{2\pi} |P(e^{it})|^2 d\mu^X_{r,y}(t)\right)^{\frac{1}{2}}\left(\int_0^{2\pi} |Q(e^{it})|^2 d\mu^X_{r,x}(t)\right)^{\frac{1}{2}}\\ &= \frac{1}{\alpha}\left\|\sqrt{X}P(rT)y\right\|\,\left\|\sqrt{X}Q(rT)y\right\|.\end{aligned}$$

Equation 3.1 is then obtained by taking $r \to 1$. The proposition is proved. □

As in [5] we associate with an operator T in C_ρ the kernel given by

$$K^\rho_{r,t}(T) = (I - re^{-it}T)^{-1} + (I - re^{it}T^*)^{-1} + (\rho - 2)I = K_{r,t}(T) + (\rho - 1)I.$$

Note that *ρ-kernels* allow to reformulate the definition of the class C_ρ. More precisely an operator T belongs to C_ρ if the associated ρ-asymptotic kernel $(K^\rho_{r,t})$ is positive and if ρ is minimal in this sense. See [5] for further information.

Let P_1, P_2 be polynomials. Fix $I(y) = \|(P_1(rT^*) + P_2(rT))y\|\|^2$ and set $J(x) = \langle (P_1(rT^*) + P_2(rT))y \mid x\rangle$ for $x \in H$. The following provides asymptotic spacial von Neumann inequalities for operators in the class C_ρ.

Proposition 3.5. *Under the assumptions above, we have*

$$\begin{aligned} I(y) \leq \frac{1}{\alpha}\int_0^{2\pi} &|(1-\rho)(P_1(0) + P_2(0)) + P_1(e^{-it}) + P_2(e^{it})|^2 d\mu^X_{r,y} \\ &+ \rho(\rho - 1)\{\|P_1 - P_1(0)\|_2^2 + \|P_2 - P_2(0)\|_2^2\}\|y\|^2 \\ &+ \frac{\rho - 1}{\rho}|P_1(0) + P_2(0)|^2\|y\|^2. \end{aligned} \tag{3.2}$$

Proof. Since

$$J(x) = \int_0^{2\pi} \Big[\frac{1-\rho}{\rho}(P_1(0) + P_2(0)) + P_1(e^{-it}) + P_2(e^{it})\Big]\langle K^\rho_{r,t}(T)y \mid x\rangle \frac{dt}{2\pi}$$

and since $K^\rho_{r,t}(T)$ is a positive operator ($T \in C_\rho$), by applying the Cauchy-Schwarz inequality twice, we obtain

$$\begin{aligned} |J(x)|^2 \leq &\int_0^{2\pi} \Big|\Big(\frac{1-\rho}{\rho}\Big)(P_1(0) + P_2(0)) + P_1(e^{-it}) + P_2(e^{it})\Big|^2 \langle K^\rho_{r,t}(T)y \mid y\rangle \frac{dt}{2\pi} \\ &\times \int_0^{2\pi} \langle K^\rho_{r,t}(T)x \mid x\rangle \frac{dt}{2\pi}. \end{aligned}$$

This leads to the inequality,

$$|J(x)|^2 \leq \rho\|x\|^2 \int_0^{2\pi} \Big|\frac{1-\rho}{\rho}(P_1 + P_2)(0) + P_1(e^{-it}) + P_2(e^{it})\Big|^2 \langle K^\rho_{r,t}(T)y \mid y\rangle \frac{dt}{2\pi}.$$

Using the identity $\langle K^\infty_{r,t}(T, X)y \mid y\rangle = \alpha\langle K_{r,t}(T)y \mid y\rangle$, we get

$$\langle K^\infty_{r,t}(T, X)y \mid y\rangle = \alpha(\langle K^\rho_{r,t}(T)y \mid y\rangle - (\rho - 1)\|y\|^2),$$

thus

$$\langle K^\rho_{r,t}(T)y \mid y\rangle = \frac{1}{\alpha}\langle K^\infty_{r,t}(T, X)y \mid y\rangle + (\rho - 1)\|y\|^2.$$

Now taking the supremum for $\|x\| \leq 1$ gives

$$\begin{aligned} I(y) &\leq \rho \int_0^{2\pi} \left| \frac{1-\rho}{\rho}(P_1(0)+P_2(0)) + P_1(e^{-it}) + P_2(e^{it}) \right|^2 \\ &\quad \times \left\{ \frac{1}{\alpha} \langle K_{r,t}^{\infty}(T,X)y \mid y \rangle + (\rho - 1)\|y\|^2 \right\} \frac{dt}{2\pi} \\ &= \frac{\rho}{\alpha} \int_0^{2\pi} \left| \frac{1-\rho}{\rho}(P_1(0)+P_2(0)) + P_1(e^{-it}) + P_2(e^{it}) \right|^2 \langle K_{r,t}^{\infty}(T,X)y \mid y \rangle \\ &\quad + \rho(\rho-1)\|y\|^2 \left\| \frac{1-\rho}{\rho}(P_1(0)+P_2(0)) + P_1(e^{-it}) + P_2(e^{it}) \right\|_2^2. \end{aligned}$$

The last term in the right side of the inequality above equals

$$\frac{\rho-1}{\rho}|P_1(0)+P_2(0)|^2\|y\|^2 + \rho(\rho-1)\{\|P_1 - P_1(0)\|_2^2 + \|P_2 - P_2(0)\|_2^2\}\|y\|^2.$$

Thus

$$\begin{aligned} I(y) &\leq \frac{\rho}{\alpha} \int_0^{2\pi} \left| \frac{1-\rho}{\rho}(P_1+P_2)(0) + P_1(e^{-it}) + P_2(e^{it}) \right|^2 d\mu_{r,x}^X(t) \\ &\quad + \frac{\rho-1}{\rho}|P_1(0)+P_2(0)|^2\|y\|^2 + \rho(\rho-1) \\ &\quad \times \{[\|P_1 - P_1(0)\|_2^2 + \|P_2 - P_2(0)\|_2^2]\|y\|^2\}. \end{aligned}$$

The proof is complete. □

Remark. In the case where $\rho = 1$ we obtain for every polynomial Q,

$$\|(Q(T))y\|^2 \leq \frac{1}{\alpha}\|\sqrt{X}Q(T)y\|^2 \leq \frac{1}{\alpha}\|\sqrt{X}\|^2\|Q(T)y\|^2$$

thus, T is similar to an isometry on $E = \text{span}\{T^n y, n \geq 0\}$. The existence of non trivial invariant subspaces is derived directly from [5].

Proposition 3.6. *Let T be a power bounded operator in the class C_ρ and suppose that $\mu_{y,y}^X$ is absolutely continuous with respect to the Haar measure $dm = \frac{dt}{2\pi}$. Define Φ on polynomials by $\Phi(P) = P(T)y$. Then there exists a positive measure ν such that Φ is extendible to $H^2(d\nu)$ and intertwines T with the unilateral shift on $H^2(d\nu)$.*

Proof. Before starting the proof, note that every vector, in the singular invariant subspace associated with the singular part of $\mu_{y,y}^X$, is non cyclic thus we can assume without loss of generality that "*$\mu_{y,y}^X$ is absolutely continuous with respect to the Haar measure $dm = \frac{dt}{2\pi}$*".

For $n \geq 1$, we have $X = T^{*n}XT^n$, it follows that

$$\begin{aligned} \|T^n Q(T)y\|^2 &\leq \frac{\rho}{\alpha}\|\sqrt{X}T^n Q(T)y\|^2 + \rho(\rho-1)\|Q(e^{it})\|_2^2\|y\|^2 \\ &= \frac{\rho}{\alpha}\|\sqrt{X}Q(T)y\|^2 + \rho(\rho-1)\|Q(e^{it})\|_2^2\|y\|^2. \end{aligned}$$

If Y is a canonical Toeplitz operator, there exists a Banach limit L such that $Y = L(T^{*n}T^n)$. Thus,

$$\begin{aligned}\|\sqrt{Y}Q(T)y\|^2 &= \langle YQ(T)y \mid Q(T)y\rangle \\ &\leq \frac{\rho}{\alpha}\|\sqrt{X}Q(T)y\|^2 + (\rho)(\rho-1)\|Q(e^{it})\|_2^2\|y\|^2 \\ &= \int_0^{2\pi} |Q(e^{it})|^2 d\nu\end{aligned}$$

with $d\nu = \frac{\rho}{\alpha}\mu^X_{y,y} + \rho(\rho-1)dm$. Define $\Phi : H^2(d\nu) \to H$ by $\Phi(Q) = Q(T)y$, then Φ can be extended to $H^2(d\nu)$. Let S_ν be the unweighted shift on $H^2(d\nu)$, we clearly have,

$$T \circ \Phi = \Phi \circ S_\nu.$$

□

4. Cyclic vectors for operators in $C_R \cap C_{1,\cdot}$

Recall that $T \in B(H)$ is said to be a cyclic operator if $E_T(x) := \overline{\text{span}}\{T^n x,\ n \geq 0\} = H$ for some $x \in H$. Such a vector is called a cyclic vector and we denote $Cyc(T)$ the set of all cyclic vector for T. It is clear that T has a non trivial invariant subspace if and only if $Cyc(T) \neq H$. We also see that the lattice of invariant subspaces is large if $Cyc(T) = \emptyset$

Let $R = \sum_{k=0}^n a_k z^k$ be a polynomial and T be an operator in the class C_R. Since $K^R_{r,t}(T) \geq 0$, when $r \to 1^-$, the positive quasi spectral measures $K^R_{r,t}(T)dt/2\pi$ converge in the weak measure topology to a positive quasi-spectral measure γ. Consequently, the quasi-spectral measure $K_{r,t}(T)dt/2\pi$ weakly converges to a quasi-spectral measure μ_T (not necessarily positive), and we have

$$\begin{aligned}\gamma_{x,y} = \mu_{T,x,y} + \Big[\sum_{k=1}^n a_k e^{-ikt}\langle T^k x, y\rangle + \sum_{k=1}^n \overline{a_k} e^{ikt}\langle T^{*k}x, y\rangle \\ + (2\mathcal{R}e(a_0) - 1)\langle x, y\rangle\Big]\frac{dt}{2\pi}.\end{aligned} \tag{4.1}$$

for any $(x, y) \in H \times H$. As in the proof of the classical von Neumann inequality given in [2], we can prove a spacial von Neumann inequality associated to the kernel $K^R_{r,t}(T)$

$$\begin{aligned}&\left\|\left[f(rT^*)T^{*\,n+1} + T^{n+1}g(rT)\right]x\right\|^2 \\ &\leq \int_0^{2\pi} \left|f(e^{-it}) + g(e^{it})\right|^2 \langle K^R_{r,t}(T)x, x\rangle \frac{dt}{2\pi}.\end{aligned} \tag{4.2}$$

We investigate in this section the nature of cyclic vectors for operators in $C_R \cap C_{1,\cdot}$. We show that if T is an operator in $C_R \cap C_{1,\cdot}$, then T^N fails to have cyclic vectors for some N, and hence the lattice of invariant subspaces of T^N is very large. We assume throughout of this section that T is injective with dense range. If such assumption is not satisfied the existence of non zero non cyclic vectors is trivial. We first give some preliminary results.

4.1. Perturbed kernel and absolute continuity

The following result has independent interest and will be used in our proof.

Proposition 4.1. *Let T be an injective operator with dense range and for any positive integer p, set $E_p = \{x, \forall\, n \geq p,\ T^{*n}T^n x = T^n T^{*n} x = x\}$. Then $E_p = E_0$ for any p, furthermore $T_{|E_0}$ is a unitary operator and $T_{|E_0^\perp}$ is completely non-unitary.*

Proof. Let us first prove that E reduces T^p (that is invariant for T^p and for $(T^p)^*$), we claim that if $x \in E$, we have $T^q x \in E$ for every $q \geq p$. Indeed

$$T^{*q}(T^{*n}T^nT^qx - T^qx) = T^{(n+q)*}T^{n+q}x - T^{*q}T^qx = x - x = 0,$$

thus, $T^{*n}T^nT^qx - T^qx \in KerT^* = \overline{R(T)}^\perp = \{0\}$ and so, $T^{*n}T^nT^qx = T^qx$ for $n \geq p$. Similarly $T^nT^{*n}T^qx = T^qx$, the claim is proved. E_p is invariant for T^{*q} in a same way. For every $x \in E_p$ and $y \in E_p^\perp$, we have

$$\langle Tx \mid y\rangle = \langle TT^pT^{*p}x \mid y\rangle = \langle T^{*p}x \mid T^{*(p+1)}y\rangle = 0$$

and since $E_p^\perp$ is invariant for T^p and for $T^{*(p+1)}$, we obtain E_p is invariant for T. Likewise we see that E_p is invariant for T^*.
To see that $E_p = E_0$ for $p \geq 0$, let x be in E_p and note that

$$\|Tx\|^2 = \|TT^pT^{*p}x\|^2 = \left\|T^{p+1}(T^{*p}x)\right\|^2 = \|T^{*p}x\|^2 = \|x\|^2 .$$

Similar arguments show that $\|T^*x\|^2 = \|x\|^2$, for any $x \in E_p$. It follows that $T_{|E_p}$ is a unitary operator, consequently $E_p = E_0$.

To complete the proof, note that if E is a reducing space for T such that $T_{|E'}$ is a unitary operator, then necessarily $T^{*n}T^nx = T^nT^{*n}x = x$ and hence $E \subset E_0$. In particular $T_{|E_0^\perp}$ is completely non unitary. □

Remark.

1. The set E_0 considered here was introduced when T is a contraction under the form $\{x, \|T^nx\| = \|T^{*n}x\| = \|x\|\}$, ([11], Chapter 1, Theorem 3.2. for example).
2. The proposition holds for some class of operators without the assumption injective with dense range namely normal operators, contractions…

Proposition 4.2. *Let $T \in C_R$ be an injective operator with dense range. If T is completely non unitary, then T is absolutely continuous. Moreover, for every $x \in H\backslash\{0\}$ there exists $h_{x,x} \in L^1(dm)$ such that $\ln(h_{x,x}) \in L^1(dm)$ and that satisfies*

$$\gamma_{x,y} = \lim_{w^*}\langle K^R_{r,t}(T)x \mid x\rangle dm(t) = h_{x,x}dm.$$

Proof. For x, y in H we have

$$\langle K^R_{r,t}(T)x \mid y\rangle dm(t) \rightarrow_{w^*} \gamma_{x,y}$$

and by the Jordan decomposition we get $\gamma_{x,y} = h_{x,y}dm + \nu_{x,y}$ with $h_{x,y} \in L^1(dm)$ and $\nu_{x,y}$ a singular measure with respect to the Haar measure. Applying the Cauchy-Schwarz inequality to the bilinear application

$$\begin{array}{ccc} H \times H & \rightarrow & L^1(dm) \\ (x, y) & \mapsto & h_{x,y} \end{array}$$

gives

$$\left|\int f(e^{it})h_{x,y}dm(t)\right| \leq \sqrt{\int |f(e^{it})|h_{x,x}dm(t)}\sqrt{\int |f(e^{it})|h_{y,y}dm(t)}$$

for all positive functions. Then it is clear that $F = \{x \in H : h_{x,y} = 0 \forall y \in H\}$ is a closed subspace of H. Indeed more is given, $F = \{x \in H \backslash h_{x,x} = 0\}$ and $F^\perp = \{x \in H : \nu_{x,x} = 0\}$. Let Ω be a compact subset of the support of $\nu_{x,y}$ such that $m(\Omega) = 0$ (since $\nu_{x,y}$ is singular with respect to the Haar measure). There exists a sequence $(f_k)_{k\in\mathbb{N}}$ in the disc algebra A(D)

(i) $|f_k| \leq 1$
(ii) $f_k(z) \rightarrow 1_\Omega(z)$, $|\tilde{\nu}|$-almost everywhere, where $\tilde{\nu}$ is an associated basic measure,
(iii) $f_k(e^{it}) \rightarrow 0$ m-almost everywhere (see [7] for more details).

We deduce that $f_k(T) \rightarrow \mu_T(\Omega)$ in the strong convergence of operators topology with $\langle \mu_T(\Omega)x \mid y\rangle = \int_\Omega d\mu_{T,x,y}$. But from equation (4.1), we see that

$$\begin{aligned}\gamma_{x,y} = \mu_{T,x,y} + \Big[\sum a_k e^{-ikt}\langle T^k x, y\rangle + \sum \overline{a_k} e^{ikt}\langle T^{*k}x, y\rangle \\ + (2\mathcal{R}e(a_0) - 1)\langle x, y\rangle\Big] dm.\end{aligned}$$

Thus $\gamma(\Omega) = \mu_T(\Omega) = \mu_T(\Omega)^*$ and $\mu_T(\Omega) \in \{T\}'$ is an orthogonal projection. In particular $T_{|F}$ is unitary. Since T is completely non unitary, we get $F = \{0\}$ and hence T is absolutely continuous.

To prove the second assertion, suppose that $ln(h_{x,x})$ is not integrable. By Szegő's formula, we have

$$\inf_{p\in\mathbb{C}[e^{i\theta}]} \int |1 - e^{i\theta}p(e^{i\theta})|^2 d\gamma_{x,x} = exp\Big[\int ln(h_{x,x})dm(e^{i\theta})\Big] = 0.$$

We derive that for every $k \geq p = d^\circ R + 1$, there exists a sequence $(p_n)_{n\in\mathbb{N}} \subseteq \mathbb{C}[e^{i\theta}]$ such that

$$\int |1 - e^{i(k+p)\theta}p_n(e^{i\theta})|^2 d\gamma_{x,x}(\theta) \rightarrow 0$$

Using spacial von Neumann's inequality (equation (4.2)), we get

$$\|T^p x - T^{k+2p}p_n(T)x\|^2 \leq \int |1 - e^{i(k+p)\theta}p_n(e^{i\theta})|^2 d\gamma_{x,x}(\theta) \rightarrow 0$$

and

$$\|T^{*k}x - T^p p_n(T)x\|^2 \leq \int |e^{-ik\theta} - e^{ip\theta}p_n(e^{i\theta})|^2 d\gamma_{x,x}(\theta) \to 0$$

we deduce that $T^p x = \lim_{n\to+\infty} T^{k+2p}p_n(T)x = T^{k+p}T^{*k}x$ and since $KerT = \{0\}$, we obtain $T^kT^{*k}x = x$. Using the same arguments, we have $T^{*k}T^kx = x$, thus $x \in E_p$. Using Proposition 3.2, we derive that $T_{|E_0}$ is unitary, this contradicts, T is *c.n.u.* Finally $ln(h^{\rho}_{x,x})$ is integrable. □

Proposition 4.3. *Let $X \in C_+(T)$. For every $x, y \in H$ the measure $\mu^X_{x,y}$ is absolutely continuous with respect to the Haar measure on the unit circle.*

Proof. Let p be a polynomial. For $x, y \in H$, we have

$$\langle Xp(T)x \mid y\rangle = \int_0^{2\pi} p(e^{it})d\mu^X_{x,y}(t)$$

and

$$\begin{aligned}\int_0^{2\pi} & p(e^{it})d\gamma_{x,Xy}(t)\\ &= \sum_{k=1}^n a_k\widehat{p}(k)\langle T^kx, Xy\rangle + (2\mathcal{R}e(a_0) - 1)\langle x, Xy\rangle + \int_0^{2\pi} p(e^{it})d\mu_{T,x,Xy}(t)\\ &= \sum_{k=1}^n a_k\widehat{p}(k)\langle T^kx, Xy\rangle + (2\mathcal{R}e(a_0) - 1)\langle x, Xy\rangle + \langle XP(T)x, y\rangle\\ &= \sum_{k=1}^n a_k\widehat{p}(k)\langle T^kx, Xy\rangle + (2\mathcal{R}e(a_0) - 1)\langle x, Xy\rangle + \int_0^{2\pi} p(e^{it})d\mu^X_{x,y}(t).\end{aligned}$$

If we set

$$\nu = \mu^X_{x,y} + \Big[\sum_{k=1}^n a_k e^{-ikt}\langle T^kx, Xy\rangle + (2\mathcal{R}e(a_0) - 1)\langle x, Xy\rangle\Big]dm - \gamma_{x,Xy},$$

we see that $\widehat{\nu}(n) = 0$ for $n \geq 0$ then ν is absolutely continuous with respect to the Haar measure. The preceding lemma asserts that $\gamma_{x,y}$ is absolutely continuous and hence $\mu^X_{x,y}$ is absolutely continuous with respect to the Haar measure. □

4.2. Cyclic vectors

We give and prove now one of the two main theorems of this section.

Theorem 4.4. *Let $T \in C_R \cap C_{1,.}$ and $x \in H$*

1. *There exists $N_x \in \mathbb{N}$ such that $x \notin Cyc(T^N)$ for arbitrary $N \geq N_x$.*
2. *There exists $N_0 \in \mathbb{N}$ such that $Cyc(T^N) = \emptyset$ for any $N \geq N_0$.*

Proof. **1.** Let $x \in H$ be a nonzero element. If $x \notin Cyc(T)$, then it is trivial that $x \notin Cyc(T^n)$ for every $n \geq 0$. Hence, we can assume that $x \in Cyc(T)$. Denote $\mu = \mu^X_{x,x} = \varphi dm$ with $\varphi \in L^1(dm)$. We have

$$\|p\|^2_{L^2(\mu)} = \int_0^{2\pi} |p(e^{it})|^2 d\mu(t) = \|\sqrt{X}p(T)x\|^2 \leq \|X\| \, \|p(T)x\|^2$$

and hence the operator $L : p(T)x \in E_T(x) \to p \in H^2_\mu$ is extendible to a bounded operator on H (that we also denote L). The following diagram is commutative

$$\begin{array}{ccc} H & \xrightarrow{L} & H^2_\mu \\ T \downarrow & & \downarrow S_\mu \\ H & \xrightarrow{L} & H^2_\mu \end{array} \tag{4.3}$$

where S_μ is the usual shift on H^2_μ.

Seeking a contradiction, let u be arbitrary in H and suppose that $u \in E_{T^{n_k}}(x)$ for some sequence $n_k \geq 0$ that goes to infinity. Thus there exist p_n polynomials, such that $u = \lim_{n\to\infty} p_n(T^{n_k})x$. Using the diagram (4.3),

$$L(u) = \lim_{n\to\infty} L(p_n(T^{n_k})x) = \lim_{n\to\infty} p_n(S^{n_k}_\mu)Lx$$

this leads to the inclusion $Im(L) \subset \text{span}\{(S^{n_k}_\mu)^n Lx \ : n \geq 0\}$ and since L is dense range, we deduce that $\psi := L(x) \in Cyc(S^{n_k}_\mu)$ for every $k \geq 0$. We show now that this fact is impossible. □

Lemma 4.5. *Let S_μ be the shift operator on H^2_μ and $\psi \in H^2_\mu$. Then there exists N_0 such that $\psi \notin Cyc(S^N_\mu)$ for every $N \geq N_0$.*

Proof. Suppose that $\psi \in Cyc(S^{n_k}_\mu)$ for some given sequence $(n_k)_{k\geq 0}$ tending to infinity. As $ln(t)$, $(t \in]0, 2\pi])$ is integrable, there exists $h \in H^\infty(\subset H^2_\mu)$ such that $|h(e^{it})| = t$, *a.e* with respect to Lebesgue measure. Thus for some sequence $(q_n)_{n\geq 0}$ of polynomials, we have $\lim_{n\to\infty} \|q_n\psi - h\psi\|_{H^2_\mu} = 0$. We derive that for $N \geq 0$,

$$\int_0^{2\pi} |t - |q_n(e^{iNt})||^2 |\psi|^2 d\mu \leq \int_0^{2\pi} |h(e^{iNt}) - q_n(e^{iNt})| \, |\psi|^2 d\mu$$
$$= \|q_n\psi - h\psi\|_{H^2_\mu} \to 0 \quad \text{when } n \to \infty.$$

Let Ω be the support of ψ, there exists a subsequence of q_n that we denote q_{n_k} such that

$$|q_{n_k}(e^{iNt})| \to t \ \ \textit{a.e. with respect to Lebesgue measure.} \tag{4.4}$$

If we set $\Omega_k = \Omega - \frac{2k\pi}{N}$ for $0 \leq k \leq N-1$, then $\Omega_k \cap \Omega_p = \emptyset$ for $0 \leq k, p \leq N-1$. To see this consider $t \in \Omega_k \cap \Omega_p$, and write $t = t_1 + \frac{2k\pi}{N} = t_2 + \frac{2p\pi}{N}$. Then by equation (4.4), and since $0 \leq p, k \leq N-1$, we obtain $t_1 = t_2$ and $p = k$. To get our contradiction, note that

$$\begin{aligned} 1 = m(T) &\geq \sum_{k=0}^{N-1} m(\Omega_k) \\ &= Nm(\Omega) \qquad\qquad \textit{(Invariance by translations)} \end{aligned}$$

for every N, hence $m(\Omega) = 0$. Which is impossible.

2. In a similar way *2.* is a direct consequence of the following lemma. □

Lemma 4.6. *Let S_μ be the shift operator on H^2_μ and $\Omega = supp(\mu)$. Then for every $N > \frac{1}{m(\Omega)}$, the operator S_μ^N has no cyclic vector.*

Proof. It outlines the proof of Lemma 4.5 by observing that $\Omega \subset supp(\phi)$ for every cyclic vector ϕ for S^n_μ. □

Remark. Theorem 4.4 gives a rich set of invariant subspaces for some power of T but seems not to give any information about invariant subspaces for T that arise directly from non cyclic vectors. If S is the usual shift on the Hardy space H^2, then an easy proof shows that S^2 has no cyclic vectors (a fact that is also a consequence of Lemma 4.6) while S has a large set of cyclic vectors. Namely all outer functions.

In [11] B. Nagy and C. Foias were interested in the following problem. Let T be a contraction in the class $C_{1,.}$, under which assumption we have, T is cyclic implies T^* is cyclic.

The rest of this section deals with this problem for power bounded operators in the class $C_{1,.}$. We generalize their result to such operators by relying generalized Toeplitz operators, cyclic vectors for T and cyclic vectors for T^*.

Theorem 4.7. *Let T be a power bounded cyclic operator of the class $C_{1,.}$, and let $X \in C_+(T)$ be injective.*

1. *If $D \cap (\sigma(T) \setminus \sigma_p(T)) \neq \emptyset$, then $X(Cyc(T)) \subset Cyc(T^*)$. In particular T^* is cyclic.*

Moreover, if T is a completely non unitary the assumption $D \cap (\sigma(T) \setminus \sigma_p(T)) \neq \emptyset$ is no more needed to say that T^* is cyclic and we have

2. *If the function $\ln(\mu^X_{a,a}/dm)$ is not Lebesgue integrable, then $X(Cyc(T)) \subset Cyc(T^*)$.*
3. *If the function $\ln(\mu^X_{a,a}/dm)$ is Lebesgue integrable, then $Im(\sqrt{X}) \cap Cyc(T^*) \neq \emptyset$*

Proof. Fix $a \in Cyc(T)$ and write $\sqrt{X}T = U\sqrt{X}$ and suppose that $T^* - \lambda$ is injective for some $\lambda \in \sigma(T)$ such that $|\lambda| < 1$. From the equality $(T^* - \lambda)\sqrt{X} = \sqrt{X}(U^* - \lambda)$, it follows that $\lambda \notin \sigma_p(U^*)$. In particular $U - \bar{\lambda}$ is dense range, and since it is closed range we get $U - \bar{\lambda}$ is onto. We conclude, because U is an isometry, that U is a unitary transformation. On the other hand, observing that

$\sqrt{X}P(T)a = P(U)\sqrt{X}a$ and using the fact that X is dense range, we have $b = Xa$ is a cyclic vector for U. But for unitary transformation (in fact for normals) we have $\|q(U)b - \bar{q}(U^*)b\| = \|\bar{q}(U^*)b - q(U)b\|$ for every polynomials P, Q. This lead to $\sqrt{X}a$ is a cyclic vector for U^* and it follows that Xa is a cyclic vector for T^*.

By Proposition 4.1, we know that $\mu^X_{a,a}$ is absolutely continuous with respect to m, thus there exists $\varphi \in L^1(dm)$ such that $\mu^X_{a,a} = \varphi dm$. We have the following alternative:

1°) The function $\ln h_{a,a} = \ln \varphi$ is not Lebesgue integrable. Using Szegő's theorem, we see that for every positive integer p there exists a sequence $(p_n)_{n\geq 0} \subseteq \mathbb{C}[e^{i\theta}]$ such that $\int |1 - e^{i(p+1)\theta} p_n(e^{i\theta})|^2 \varphi(e^{i\theta}) dm(\theta) \to 0$. If $p(z) = \sum_{k=1}^n \widehat{p}(k) z^k$ is a polynomial, we denote by $\overline{p}$ the polynomial given by $\overline{p}(z) = \sum_{k=1}^n \overline{\widehat{p}(k)} z^k$. By Proposition 3.3, we have

$$\|\overline{p}(T^*)T^*Xa - XT^p a\|^2 \leq \|X\|^2 \int_0^{2\pi} \left|\overline{p(e^{it})}e^{-it} - e^{ipt}\right|^2 \varphi(e^{it}) dm(t) \to 0.$$

Let us denote by $F = \bigvee_{k\geq 0} T^{*k}(Xa)$ the closed subspace generated by the sequence $(T^{*k}(Xa))_{k\geq 0}$. The previous inequality implies that $T^p(a) \in F$ for any $p \geq 0$. Therefore, we get that

$$X(H) = X\Big(\bigvee_{k\geq 0} T^k(a)\Big) \subseteq F.$$

Since X is an injective positive operator, we have $H = \overline{X(H)} \subseteq F$, thus $F = H$ and Xa is cyclic for T^*.

2°) The function $\ln h_{a,a} = \ln \varphi$ is Lebesgue integrable. Then, there exists ψ in H^2 such that $\varphi(e^{it}) = \left|\psi(e^{it})\right|^2$. Since X is an injective operator which belongs to $C_+(H)$, by the polar decomposition there exists an isometry V such that $\sqrt{X}T = V\sqrt{X}$. Observe that

$$\left\|p(V)\sqrt{X}a\right\|^2 = \int_0^{2\pi} \left|p(e^{it})\right|^2 \left|\psi(e^{it})\right|^2 dm(t).$$

and it follows that $b = \sqrt{X}a$ is cyclic for V. On the other hand, the previous formula shows that the operator $W : p(V)b \in H \to p(e^{it}) \in H^2_\mu$ is extendible to a unitary operator (that we also denote W). Let S_γ be the usual shift on the space $H^2(\gamma)$, where γ is the measure given by $\gamma = \left|\psi(e^{it})\right|^2 dm(t)$. Since we clearly have $W \circ V = S_\gamma \circ W$, we get that V is unitarily similar to S_γ. The operator S^*_γ clearly admit cyclic vectors (for instance, every vector ϕ/ψ where ϕ is a cyclic vector for the usual backward shift). It follows immediately that V^* admit cyclic vectors, let b_0 be such a cyclic vector. Since X is a positive operator with a dense range, the relation $T^*\sqrt{X} = \sqrt{X}V^*$ ensures that $b = \sqrt{X}b_0$ is cyclic for T^*. This ends the proof. □

As a corollary we retrieve the first result in [9],

Corollary 4.8. *Let T be a cyclic contraction in the class $C_{1,\cdot}$. Then T^* is cyclic in each of the following cases*

1. *If $D \cap (\sigma(T) \setminus \sigma_p(T)) \neq \emptyset$.*
2. *T is a completely non unitary.*

4.3. Cyclic vectors and the commutativity of the commutant

The second problem treated in [13] is the following. Let T be a contraction in the class $C_{1,\cdot}$. Under which conditions is the commutant $\{T\}'$ commutative?

In this section, we describe the commutant of a power bounded operator and deduce its commutativity under more general assumptions.

We first give a generalization of a theorem in [5]. The proof runs in a similar way and will be omitted.

Theorem 4.9. *Let $T \in PWB(H)$ and $X \in \tau(T)$. Assume that T is cyclic. Then the map*

$$f \to \mu^X(f)$$

is a linear surjection from $L^\infty(\mu^X)$ onto $C(T)$, which induces a faithful order preserving surjection from $L^\infty(\mu^X)_+$ onto $C_+(T)$.

We have

Theorem 4.10. *Let $T \in PWB(H)$ be in the class $C_{1,\cdot}$ and $a \in Cyc(T)$. Then $\{T\}'$ is commutative. Moreover, for any $X \in \tau(T)$, we have*

$$\overline{X\{T\}'} = \mu^X(H^\infty(\mu^X_{a,a}))$$

where $H^\infty(\mu^X_{a,a})$ is the weak closure of polynomials is $L^\infty(\mu^X_{a,a})$.

Proof. The proof lies on the two following ingredients.
Ingredient 1. For any $R \in \{T\}'$ and p_n such that $p_n(T)a \to Ra$, the sequence p_n converges in $L^2(\mu^X)$ to the symbol associated with $XR = \mu^X(\phi)$ by the previous theorem. Indeed, since

$$\|Xp_n(T)a - Xp_m(T)a\|^2 = \int_0^{2\pi} |p_n(e^{it}) - p_m(e^{it})|^2 d\mu^X_{a,a}$$

the sequence p_n converges to a function $\psi \in L^2(\mu^X)$. Observe that for any polynomials p, q

$$\int_0^{2\pi} \phi(e^{it})p(e^{it})\overline{q}(e^{it})d\mu^X_{a,a} = \langle\mu^X(\phi)p(T)a, q(T)a\rangle = \langle XRp(T)a, q(T)a\rangle$$

$$= \lim_{n\to+\infty} \langle Xp_n(T)p(T)a, q(T)a\rangle = \lim_{n\to+\infty} \int_0^{2\pi} \phi(e^{it})p_n(e^{it})p(e^{it})\overline{q(e^{it})}d\mu^X_{a,a}$$

$$= \int_0^{2\pi} \psi(e^{it})p(e^{it})\overline{q(e^{it})}d\mu^X_{a,a}.$$

Hence $\psi = \phi\ d\mu^X_{a,a}$ *a.e* and since $a \in Cyc(T)$, it follows that $\psi = \phi\ d\mu^X$ *a.e.*

Ingredient 2. Let R_1 and R_2 be in the commutant of T and write $R_1 a = \lim_{n\to+\infty} p_n(a)$ and $R_2 a = \lim_{n\to+\infty} q_n(a)$. Fix $f_1 = \lim p_n$ and $f_2 = \lim q_n$ obtained by *Ingredient 1.* Since $X_1 = XR_1$ and $X_2 = XR_2$ belong to $C(T)$, using again the previous theorem, there exists f_1, and f_2 in $H^\infty(\mu^X)$ such that $X_1 = \mu^X(f_1)$ and $X_2 = \mu^X(f_2)$. For any r, s polynomials, we have

$$\begin{aligned}\langle X_1 R_2 r(T)a, s(T)a\rangle &= \lim_{n\to+\infty} \langle X q_n(T) r(T)a, s(T)a\rangle \\ &= \lim_{n\to+\infty} \int_0^{2\pi} q_n(e^{it}) r(e^{it}) \overline{s(e^{it})} f_1(e^{it}) d\mu^X_{a,a} \\ &= \int_0^{2\pi} f_2(e^{it}) r(e^{it}) \overline{s(e^{it})} f_1(e^{it}) d\mu^X_{a,a}.\end{aligned}$$

Similarly, $\langle X_2 R_1 r(T)a, s(T)a\rangle = \int_0^{2\pi} f_2(e^{it}) r(e^{it}) \overline{s}(e^{it}) f_1(e^{it}) d\mu^X_{a,a}$.

Thus $XR_1R_2 = X_2R_1 = X_1R_2 = XR_2R_1$. The injectivity of X leads to $R_1R_2 = R_2R_1$.

The commutativity of $\{T\}'$ is obtained.

From the proof of the first assumption, we see that $\overline{X\{T\}'} \subset H^\infty(\mu^X)$. To show the reverse inclusion, it suffices to use again Theorem 4.9. □

5. The perturbed kernel and spacial von Neumann inequalities

The aim of this section is to provide asymptotic von Neumann inequality associated with perturbed kernels. We give complete computations in the case where the perturbation is $R(T) = \overline{\beta}T + \beta T^*$. The general case of trigonometric perturbation follows immediately.

Let $T \in L(H)$ and $\beta = re^{it} \in \mathbb{C}$. We denote the perturbed kernel associated with T, $L_\beta(T) = K_\beta(T) + \overline{\beta}T + \beta T^*$, the asymptotic perturbed kernel associated with T, $L^\infty_\beta(T,X) = K^\infty_\beta(T,X) + \overline{\beta}XT + \beta T^*X$, and the ρ-perturbed kernel associated with T $L^\rho_\beta(T) = K^\rho_\beta(T) + \overline{\beta}T + \beta T^*$. Then, direct computations give

1. $\langle L_\beta(T,X)y \mid y\rangle = \alpha\langle L_\beta(T)y \mid y\rangle$.
2. $f(rT) = \int_0^{2\pi} f(e^{it}) L_\beta(T) \frac{dt}{2\pi} - |\beta| f'(0)T$.
3. $f(rT^*) = \int_0^{2\pi} f(e^{-it}) L_\beta(T) \frac{dt}{2\pi} - |\beta| f'(0)T^*$.

As before, we set

$$J(x) = \langle (P_1(rT^*) + P_2(rT))y \mid x\rangle.$$

Then

$$\begin{aligned}
J(x) &= \int_0^{2\pi} [(\frac{1-\rho}{\rho})(P_1(0)+P_2(0)) + P_1(e^{-it}) + P_2(e^{it})]\langle K^{\rho}_{r,t}(T)y \mid x\rangle \frac{dt}{2\pi} \\
&= \int_0^{2\pi} \Big[\Big(\frac{1-\rho}{\rho}\Big)(P_1(0)+P_2(0)) + P_1(e^{-it}) + P_2(e^{it})\Big]\langle L^{\rho}_{\beta}(T)y \mid x\rangle \\
&\quad - \langle \overline{\beta}T + \beta T^* y \mid x\rangle \frac{dt}{2\pi} \\
&= \int_0^{2\pi} \Big[\Big(\frac{1-\rho}{\rho}\Big)(P_1(0)+P_2(0)) + P_1(e^{-it}) + P_2(e^{it})\Big] \\
&\quad \times \langle L^{\rho}_{\beta}(T)y \mid x\rangle \frac{dt}{2\pi} - |\beta| P_1'(0)\langle Ty \mid x\rangle - |\beta| P_2'(0)\langle T^* y \mid x\rangle.
\end{aligned}$$

By the Cauchy-Schwarz inequality

$$\begin{aligned}
|J(x)| \leq &\Big(\int_0^{2\pi} \Big[\Big|\Big(\frac{1-\rho}{\rho}\Big)(P_1(0)+P_2(0)) + P_1(e^{-it}) + P_2(e^{it})\Big|^2\Big] \\
&\times \langle L^{\rho}_{r,t}(T)y \mid y\rangle \frac{dt}{2\pi} \int_0^{2\pi} \langle L^{\rho}_{r,t}(T)x \mid x\rangle \frac{dt}{2\pi}\Big)^{\frac{1}{2}} \\
&+ |\beta|\,|P_1'(0)|\,|\langle Ty \mid x\rangle| + |\beta|\,|P_2'(0)|\,|\langle T^* y \mid x\rangle|,
\end{aligned}$$

we obtain

$$\begin{aligned}
|J(x)| \leq \rho^{\frac{1}{2}} \|x\| &\Big\{ \int_0^{2\pi} \Big[\Big| \Big(\frac{1-\rho}{\rho}\Big)(P_1(0)+P_2(0)) + P_1(e^{-it}) + P_2(e^{it})\Big|^2\Big] \\
&\times \langle L^{\rho}_{\beta}(T)y \mid y\rangle \frac{dt}{2\pi}\Big\}^{\frac{1}{2}} + |\beta|\,|P_1'(0)|\,\|Ty\|\,\|x\| + |\beta|\,|P_2'(0)|\,\|T^*y\|\,\|x\|.
\end{aligned}$$

As $\langle L^{\rho}_{\beta}(T)y \mid y\rangle = \frac{1}{\alpha}\langle L_{\beta}(T,X)y \mid y\rangle + (\rho-1)\|y\|^2$, by the preceding inequality and taking the supremum for $\|x\| \leq 1$, we get

$$\begin{aligned}
I(y)^{1/2} \leq \frac{\rho^{\frac{1}{2}}}{\alpha} &\Big[\int_0^{2\pi} \Big|\Big(\frac{1-\rho}{\rho}\Big)(P_1(0)+P_2(0)) + P_1(e^{-it}) + P_2(e^{it})\Big|^2 \\
&\times \langle L_{\beta}(T,X)y \mid y\rangle \frac{dt}{2\pi} \\
&+ \frac{\rho}{(\rho-1)}\Big[\|P_1+P_2\|_2^2 + \Big(\frac{1-\rho^2}{\rho^2}\Big)|(P_1(0)+P_2(0))|^2\Big]\|y\|^2)\Big]^{\frac{1}{2}} \\
&+ |\beta|\,|P_1'(0)|\,\|Ty\| + |\beta|\,|P_2'(0)|\,\|T^*y\|.
\end{aligned}$$

Remark. It is easy to see that if we take, $R(T) = \overline{\beta}T^n + \beta T^{n*}$, then

$$\int_0^{2\pi} f(e^{it})K_{\beta}(T)\frac{dt}{2\pi} = \int_0^{2\pi} f(e^{it})L_{\beta}(T)\frac{dt}{2\pi} - |\beta| f^{(n)}(0)T^n$$

the corresponding von Neumann asymptotic inequality is

$$\begin{aligned} I(y)^{1/2} \leq \frac{\rho^{\frac{1}{2}}}{\alpha} \Bigg[\int_0^{2\pi} \Big| \Big(\frac{1-\rho}{\rho}\Big)(P_1(0)+P_2(0)) + P_1(e^{-it}) + P_2(e^{it}) \Big|^2 \\ \times \langle L_{re^{it}}(T,X)y \mid y\rangle \frac{dt}{2\pi} \\ + \frac{\rho}{(\rho-1)} \Big[\|P_1+P_2\|_2^2 + \Big(\frac{1-\rho^2}{\rho^2}\Big) |(P_1(0)+P_2(0))|^2 \Big] \|y\|^2) \Bigg]^{\frac{1}{2}} \\ + |\beta|\,|P_1^{(n)}(0)|\,\|T^n y\| + |\beta|\,|P_2^{(n)}(0)|\,\|T^{*n}y\|. \end{aligned}$$

Now, if we consider $L_{re^{it}}(T) = K_{re^{it}}(T) + \sum_{j=1}^{j=n} \nu_j e^{ijt} T^j + \sum_{i=1}^{j=n} \mu_j e^{-ijt} T^{*j}$, then it is not hard to show that

$$\begin{aligned} I(y)^{1/2} \leq & \frac{\rho^{\frac{1}{2}}}{\alpha} \Bigg[\int_0^{2\pi} \Big| \Big(\frac{1-\rho}{\rho}\Big)(P_1(0)+P_2(0)) + (P_1(e^{-it}) + P_2(e^{it}) \Big|^2 \\ & \times \langle L_{re^{it}}(T,X)y \mid y\rangle \frac{dt}{2\pi} \\ & + \frac{\rho}{(\rho-1)} \Big[\|P_1+P_2\|_2^2 + \Big(\frac{1-\rho^2}{\rho^2}\Big) |(P_1(0)+P_2(0))|^2 \Big] \|y\|^2) \Bigg]^{\frac{1}{2}} \\ & + \sum_{j=1}^{j=inf(n,degP_1)} |\nu_j|\,|P_1^{(j)}(0)|\,\|T^j y\| \\ & + \sum_{i=1}^{j=inf(n,degP_2)} |\mu_j|\,|P_2^{(j)}(0)|\,\|T^{*j} y\|. \end{aligned}$$

Acknowledgment

The authors thank Abdus Salam ICTP for the kind support provided during the preparation of a part of this paper. The two last authors acknowledge the support of the Mathematics Department of University Lyon 1, where this work has been initiated.

References

[1] B. Beauzamy, *Sous-espaces de type fonctionnel dans les espaces de Banach.* Acta. Math. 144 (1980) 65–82.

[2] G. Cassier, *Ensemble K spectraux et algèbres duales d'opérateurs.* prepublication LAFP, no. 2 (1991), 1–13.

[3] G. Cassier, *Autour de quelques interactions récentes entre l'analyse complexe et la théorie des opérateurs.* In Operator Theory and Banach Algebras, Proceedings of the International Conference in Analysis Rabat (Marocco), April 12–14, 1999, Theta, Bucharest 2003.

[4] G. Cassier, *Generalized Toeplitz operators, restriction to invariant subspaces and similarity problems.* J. Operator Theory, to appear.

[5] G. Cassier, T. Fack, *Contractions in von Neumann Algebras.* J. Funct. Anal. **135**(2) (1996), 297–338.

[6] G. Eckstein, *Sur les opérateurs de classe C_ρ.* Acta. Sci. Math. **33** (1972), 354–352.

[7] K. Hoffman, *Banach Spaces of Analytic Functions.* Reprint of the 1962 original, Dover Publications, Inc., New York, 1988.

[8] L. Kérchy, *Operators with Regular Norm-Sequences.* Acta. Sci. Math. (Szeged) **63** (1997), 571–605.

[9] L. Kérchy, *Hyperinvariant Subspaces of Operators with Non-Vanishing Orbits.* Proc. Amer. Math. Soc 127(5) (1999), 1363–1370.

[10] G.G. Lorentz, *A Contribution to the Theory of Divergent Sequences.* Acta Math. **80** (1948), 167–190.

[11] B. Sz.-Nagy, C. Foias, *Analyse harmonique des opérateurs de l'espace de Hilbert.* Masson et Akad. Kiadó, Paris 1967.

[12] B. Sz.-Nagy, C. Foias, *Contractions without Cyclic Vectors.* Proc. Amer. Math. Soc. **87**(4) (1983), 671–674.

[13] B. Sz.-Nagy, C. Foias, *Vecteurs cycliques et commutativité du commutant.* Acta Sci. Math., (1970) 177–183.

[14] B. Sz.-Nagy, C. Foias, *Harmonic Analysis of Operators on Hilbert Spaces.* North Holland, Amsterdam, 1970.

G. Cassier
Institut Gerard Desargues
Université Claude Bernard Lyon I
Batiment 101, La Doua, Villeurbanne
France
e-mail: `cassier@desargues.univ-lyon1.fr`

H. Mahzouli
Département de Mathématiques et Informatique
Faculté des Sciences de Rabat
BP 1014, Rabat
Maroc
e-mail: `mahzouli@desargues.univ-lyon1.fr`

E.H. Zerouali
Département de Mathématiques et Informatique
Faculté des Sciences de Rabat
BP 1014, Rabat
Maroc
e-mail: `zerouali@fsr.ac.ma`

Operator Theory:
Advances and Applications, Vol. 153, 123–141

Statistical Properties of Disordered Quantum Systems

Francesco Fidaleo and Carlangelo Liverani

Abstract. We discuss two different approaches to the study of the long-time behavior of some disordered quantum anharmonic chains.

Mathematics Subject Classification (2000). Primary 82C10, 47A35. Secondary 82B44, 46L55.

Keywords. Quantum dynamics; Non equilibrium statistical mechanics; Ergodic theory; Disordered systems; Non commutative dynamical systems.

1. Introduction

Notwithstanding the rising interest in the ergodic properties of infinite systems, only a limited amount of results are available for non linear quantum evolutions. This is particularly true if one wants to gain a knowledge of invariant states that are not necessarily KMS states for the system. The latter problem is of interest since, if there exist invariant non-KMS states which are stable with respect to a wide class of perturbations, then their physical relevance could be comparable with the one of the KMS states. Indeed, it is shown in [6], that this is the case for an infinite chain with one defect, in which only one spring is subject to a small anharmonic perturbation. More precisely, starting from the model studied in [4], it was considered in proper units, the Hamiltonian

$$\begin{aligned} H(q,p) = \frac{1}{2}\sum_{i\in\mathbb{Z}} p_i^2 + \frac{1}{2}\left(\frac{1}{M}-1\right)p_0^2 + \frac{1}{8}\sum_{i\in\mathbb{Z}}(q_{i+1}-q_i)^2 \\ + \sum_{i\in\mathbb{Z}}\frac{\kappa}{2}q_i^2 + \frac{K}{2}q_0^2 + V(q_0)\,, \end{aligned} \tag{1.1}$$

where the non linear part V is small and regular in some appropriate sense.

It is well-known (see, e.g., [3]) that the corresponding linear system (i.e., when $V \equiv 0$) exhibits a multitude of quasi-free states invariant with respect to the free dynamics α_t^0, due to the integrable nature of the Hamiltonian. It is proven in [6] that, for such a class of perturbations $V \neq 0$, the infinite dynamics α_t associated to (1.1) is well defined on an appropriate C^*-algebra $\mathfrak{M}$ and is described by a uniformly convergent series with respect to the time.

Recall that for a dynamical system $(\mathfrak{A}, \gamma_t, \varphi)$ with φ invariant for the dynamics, forward mixing (or equivalently strong clustering) means that

$$\lim_{t \to +\infty} \varphi(B^* \gamma_t(A) B) = \varphi(B^* B) \varphi(A) \tag{1.2}$$

for each $A, B \in \mathfrak{A}$.[1] Our definition of (forward) mixing slightly differs from the standard one

$$\lim_{t \to +\infty} \varphi(B \gamma_t(A)) = \varphi(A) \varphi(B) \,, \tag{1.3}$$

see, e.g., [9], see also [11] for very recent developments on ergodicity in quantum setting. We adopt the former as it is more suitable for applications to quantum mechanics (e.g., [8] where (1.2) is referred as the property of "return to equilibrium").

It is expected that (1.2) is stronger than (1.3). However, we have

Proposition 1.1. *Property* (1.2) *implies* (1.3). *Conversely, if the dynamical system* $(\mathfrak{A}, \gamma_t)$ *is asymptotically abelian in norm, or if* Ω_φ *is cyclic for* $\pi_\varphi(\mathfrak{A})'$, *then* (1.2) *and* (1.3) *are equivalent,* $(\pi_\varphi, \mathcal{H}_\varphi, \Omega_\varphi)$ *being the GNS triplet of* φ.

Proof. It is immediate to verify that (1.2) and (1.3) are equivalent if $(\mathfrak{A}, \gamma_t)$ is asymptotically abelian in norm.

As Ω_φ is cyclic for $\pi_\varphi(\mathfrak{A})$, (1.2) implies that, if $\Psi \in \mathcal{H}_\varphi$, then $\langle \Psi, \pi_\varphi(\gamma_t(A)) \Psi \rangle \longrightarrow \|\Psi\|^2 \varphi(A)$ which gives by polarization

$$\begin{aligned} \varphi(B \gamma_t(A)) &\equiv \langle \pi_\varphi(B^*) \Omega_\varphi, \pi_\varphi(\gamma_t(A)) \Omega_\varphi \rangle \\ &\longrightarrow \left(\frac{1}{4} \sum_{\omega^4 = 1} \omega \| (\pi_\varphi(B^*) + \omega I) \Omega_\varphi \|^2 \right) \varphi(A) \equiv \varphi(A) \varphi(B) \,, \end{aligned}$$

that is (1.2) $\Rightarrow$ (1.3).

Let Ω_φ be cyclic for $\pi_\varphi(\mathfrak{A})'$. As Ω_φ is cyclic for $\pi_\varphi(\mathfrak{A})$, we get $\langle \Phi, U_t \Psi \rangle \longrightarrow \langle \Phi, \Omega_\varphi \rangle \langle \Omega_\varphi, \Psi \rangle$, for each $\Phi, \Psi \in \mathcal{H}_\varphi$, where U_t is the unitary implementation of γ_t in the GNS representation. By applying this with $\Phi = T^* \pi_\varphi(B) \Omega_\varphi$, $\Psi = \pi_\varphi(A) \Omega_\varphi$ for $T \in \pi_\varphi(\mathfrak{A})'$, we obtain $\langle \Omega_\varphi, \pi_\varphi(B^*) \pi_\varphi(\gamma_t(A)) T \Omega_\varphi \rangle \longrightarrow \langle \Omega_\varphi, \pi_\varphi(B^*) T \Omega_\varphi \rangle \varphi(A)$. As Ω_φ is cyclic also for $\pi_\varphi(\mathfrak{A})'$, the last result is true for generic $\Psi \in \mathcal{H}_\varphi$ instead of $T\Omega_\varphi$. The assertion (1.3) $\Rightarrow$ (1.2) when Ω_φ is also cyclic for $\pi_\varphi(\mathfrak{A})'$ follows by computing the last limit for $\Psi = \pi_\varphi(B) \Omega_\varphi$. □

The results contained in [6] can be summarized as follows.

Theorem 1.2. *Suppose that the potential* V *in* (1.1) *is sufficiently small and regular. Then there exists a class of states on* $\mathfrak{M}$, *which is invariant and mixing w.r.t. the linear dynamics* α_t^0, *such that the following assertions hold true.*

---

[1] The backward mixing is defined analogously.

(i) *To any state ω in the class mentioned above, it corresponds a unique state ω_∞ invariant w.r.t. the non linear dynamics α_t, which is obtained as the following pointwise limit*

$$\omega_\infty(A) = \lim_{t\to\pm\infty} \omega(\alpha_t A) .$$

(ii) *The dynamical system $(\mathfrak{M}, \alpha_t, \omega_\infty)$ is mixing.*

In the situation described above, the perturbed dynamical system $(\mathfrak{M},\alpha_t,\omega_\infty)$ exhibits the same strongly ergodic properties as the unperturbed system $(\mathfrak{M},\alpha_t^0,\omega)$. Namely, there exists a non trivial class of states, non necessarily KMS, which are "stable" for suitable perturbations of the dynamics.

From the technical point of view, the results in [6] are based on a sort of L^1-asymptotic abelianess for the free dynamics, in accordance with the standard technique used in perturbation theory, see, e.g., [9] and the literature cited therein. Loosely speaking, it is shown that, calling $\{W(\lambda e_k)\}_{\lambda\in\mathbb{R}}$ the Weyl operators associated to the variables of the kth particle, it holds true for each fixed j,k in the chain,

$$\int_{\mathbb{R}+} \| [W(e_j), \alpha_t^0 W(e_k)] \| \, dt < +\infty .$$

Of course, one can criticize [6] by saying that it is limited to a bounded perturbation of a linear infinite system, and that such a small and local perturbation cannot suffice to alter the statistical properties of an infinite system. In other words, the allowed perturbations are unreasonably restricted from the physical point of view. Accordingly, the stability of non-KMS states with respect to such perturbations is really too limited to imply something about the realistic stability of infinite states.

It is then natural to address the possibility to investigate a translation invariant perturbation, that is a locally small perturbation, but present at each site of the chain. Of course, in such a case the perturbation would be unbounded. To treat such a model with a perturbative approach in the spirit of [6], it would be necessary to have a sort of L^1 space-time asymptotic abelianess, that is

$$\sup_{k\in\mathbb{Z}} \Big[\sum_{j\in\mathbb{Z}} \int_{\mathbb{R}+} \| [W(e_k), \alpha_t^0 W(e_j)] \| \, dt \Big] < +\infty . \tag{1.4}$$

Unfortunately, harmonic lattices, or harmonic lattices with finitely many defects, do not seem to enjoy such strong ergodic properties. Nevertheless, there exist translation invariant non linear dynamics for which (1.4) is satisfied. These are related to the attempt to make sense of the dynamics associated to the formal random Hamiltonian

$$H(q,p) = \frac{1}{2}\sum_{i\in\mathbb{Z}} p_i^2 + \kappa_i q_i^2 + \frac{1}{2}\sum_{|i-j|=1} (q_i - q_j)^2 + V(q_i - q_j) , \tag{1.5}$$

where the κ_i are random variables on a common probability space satisfying certain reasonable properties.[2] The model described by the Hamiltonian (1.5) can be thought as an anharmonic chain with infinitely many random defects subjected to neighbor interactions. The construction of such a dynamics is based on the perturbation of a quasi-free dynamics given by Bogoliubov transformations acting on a phase-space of a CCR algebra, defined by averaging the commutators with respect to the random environment. One then considers quasi-free states ω determined by the truncated quenched state. In such a context, it is proven in [7], the analogous version of Theorem 1.2, see the next sections for further details.

A natural criticism to the above result is that to average and then quantize it is not the same as to quantize and then average. In other words, it could be physically more relevant to study the dynamics associated to the (formal) Hamiltonian (1.5) for each fixed realization of the spring constants $\{\kappa_i\}_{i\in\mathbb{Z}}$, and then average with respect to the randomness. These two alternatives are better illustrated in Section 2. Unfortunately, it is not clear how to define the infinite dynamics for each spring configuration, since (1.4) does not hold pointwise. In addition, the C^*-algebra in such a case would inevitably have a non trivial center, and this also contributes to spoil the ergodic properties (again see Section 2 for more details). On such issues, no rigorous result relative to long-time behavior seems to be available in literature.

In order to pursue the program related to the latter approach, some new preliminary results on the long-time behavior can be obtained, limited to finite systems and to linear dynamics. This is done in Section 4.

2. Disordered harmonic chain. Two alternative constructions

Let us describe more precisely how the model associated to the Hamiltonian (1.5) can be investigated.

We start by considering the disordered system obtained by putting $V = 0$ in (1.5). By disordered we mean that all the elastic constants $\{\kappa_i\}_{i\in\mathbb{Z}}$ appearing in (1.5), are random variables defined on a probability space (X, P). Namely, let us consider the infinite block matrix

$$A = \begin{pmatrix} 0 & I \\ \Delta - K & 0 \end{pmatrix} \tag{2.1}$$

where Δ is the discrete Laplacian given by $(\Delta v)_i = v_{i+1} + v_{i-1} - 2v_i$, and $K = K(\xi)$ is a random multiplication operator given, for generic $\xi \in X$, by $(K(\xi)v)_i := \kappa_i(\xi)v_i$. The block matrix $A(\xi)$ acts componentwise on the double sequence $\{(q_i, p_i)\}_{i\in\mathbb{Z}}$. It is the generator of the linear flow at fixed environment $\{\kappa_i(\xi)\}_{i\in\mathbb{Z}}$. We assume that all κ_i are identically equidistributed independent random variables with law p supported in some interval $[a, b] \subset \mathbb{R}^+$. In such a way, we

[2] We have chosen the unity of measure such that the mass of all particles is equal to one, and the Planck constant $\hbar = 1$.

get for the sample space, $X \equiv [a,b]^{\mathbb{Z}}$, and for the law, $P \equiv \prod_{i\in\mathbb{Z}} p$. To insure that the results in the present section apply to non trivial cases, it suffices to assume that

$$p(\,dx) = f(x)\,dx$$

with $f \in \mathcal{D}((a,b))$; see [7], Section 3.

Such a system can be seen both as a collection of Hamiltonian systems, one for each fixed environment (that is for almost all the realizations of the random variables), as well as a single Hamiltonian system on an appropriate (huge) phase space. Let us briefly discuss the second alternative. We start with a classical disordered one-dimensional chain described by the random Hamiltonian

$$H_\xi(q,p) := \frac{1}{2}\sum_{i=1}^{n} p_i^2 + \kappa_i(\xi)q_i^2 + \frac{1}{2}\sum_{|i-j|=1}(q_i - q_j)^2$$

which takes into account the quadratic part of (1.5) and where, for simplicity, the system is a finite one. Consider the infinite-dimensional symplectic space $(\mathcal{V},\Omega)$ consisting of an appropriate linear subspace $\mathcal{V}$ of measurable functions on X with values in the phase-space $\mathbb{R}^{2n}$ equipped with the symplectic form Ω given by

$$\Omega((q,p),(Q,P)) := \int_X \sigma((q(\xi),p(\xi)),(Q(\xi),P(\xi)))\mu(\,d\xi)\,. \tag{2.2}$$

Here, σ is the canonical symplectic form on the phase-space $\mathbb{R}^{2n}$, and μ is any measure on X in the measure-class determined by P.

The equations of motion associated to the functional Hamiltonian

$$\mathcal{H}(q,p) := \int_X H_\xi(q(\xi),p(\xi))\mu(\,d\xi)\,, \tag{2.3}$$

read $(\dot q(\xi),\dot p(\xi)) = A(\xi)(q(\xi),p(\xi))$ and determine a Hamiltonian flow T_t^*. Such a flow determines uniquely, almost everywhere, the time evolution at fixed environment. In such a situation, *the above points of view are essentially equivalent.* Moreover, in the second case it is natural to quantize the symplectic space $(\mathcal{V},\Omega)$ by specifying the measure in Formulae (2.2), (2.3) as the probability P describing the randomness. In this way we have a genuine CCR-algebra: the commutation function between (linear combinations of) fields and conjugate momenta is a multiple of the identity operator determined by the symplectic form (2.2). Such an approach was firstly pursued in [7] in order to study the long-time behavior of anharmonic crystals in the setting of perturbation theory. The arising model exhibits good strongly ergodic properties which are stable for a wide class of perturbations.

This approach differs from the natural procedure in the alternative point of view: that is first quantize at fixed environment and then study the collection of the resulting systems. The algebra of observables in this case can be taken to be the "product" of the CCR-algebras corresponding to each fixed environment.[3]

[3]More precisely, the algebra of observables is modeled by a direct integral, see Section 4.

This is a different algebra than the previous one. In fact, it has a non trivial center and it is not a CCR.

To have a more concrete idea of the situation, consider the trivial case in which the system consists of a one-dimensional harmonic oscillator with two possible values of the spring. In this case, the largest possible $\mathcal{V}$ is isomorphic to $\mathbb{R}^4$. For such a choice, the CCR in our quantization scheme (i.e., the approach followed in [7]) is naturally represented on the Hilbert space $L^2(\mathbb{R}^4)$. The generated von Neumann algebra is $\mathcal{B}(L^2(\mathbb{R}^4))$, i.e., the algebra of all the bounded operators on $L^2(\mathbb{R}^4)$. The alternative quantization scheme yields the algebra $\mathcal{B}(L^2(\mathbb{R}^2)) \oplus \mathcal{B}(L^2(\mathbb{R}^2))$. The non trivial center consists of all the operators with two diagonal blocks proportional to the identity. Notice that there is no canonical way to compare these algebras, as well as the Hilbert spaces on which they canonically act. However, both algebras contain a common remarkable subalgebra, that is the algebra generated by the constant functions at time $t = 0$.

To insure strong ergodic properties in the case of the Hamiltonian (1.5), one must choose $\mathcal{V}$ rather small, this is precisely what is done in [7], where $\mathcal{V}$ consists of the smallest space containing the constant functions and invariant with respect to the linear evolution. Further, only quasi-free states are considered. Such states are defined on the Weyl operators as

$$\omega(W(v)) = e^{\mathbb{E}(-\frac{1}{2}B(v,v))}\,, \tag{2.4}$$

where B is a random two-point function which uniquely determines the quasi-free state, and $\mathbb{E}$ denotes the average w.r.t. the randomness, see Formulae (3.6), (3.7).

On the contrary, in the environment-by-environment scheme, it is natural to define the *quenched*, or eventually, the *annealed* states. For example, the quenched state ω_q and the annealed one ω_a at inverse temperature β are defined as

$$\omega_q(A) = \mathbb{E}\left(\frac{\mathrm{Tr}(e^{-\beta H}A)}{\mathrm{Tr}(e^{-\beta H})}\right), \quad \omega_a(A) = \frac{\mathbb{E}(\mathrm{Tr}(e^{-\beta H}A))}{\mathbb{E}(\mathrm{Tr}(e^{-\beta H}))}\,,$$

when the r.h.s. are well defined. In the case at hand, the quenched state can be defined in the environment-by-environment scheme as

$$\omega_q(W(v)) = \mathbb{E}\left(e^{-\frac{1}{2}B(v,v)}\right). \tag{2.5}$$

Notice that the quenched state (2.5) is not quasi-free. To compare the two states one must consider them on some common subalgebra, in our case the algebra generated by the constant functions. On such a subalgebra, the states described by (2.4) are the truncated functionals associated to the quenched states given in (2.5), see, e.g., [3].[4]

Let us see in more detail what can be said in the two approaches just outlined.

[4]In addition, their two point functions coincide even at different times, that is

$$\partial_\lambda\partial_\mu\omega(W(\lambda v)W(\mu T_t w))\big|_{\lambda=\mu=0} = \partial_\lambda\partial_\mu\omega_q(W(\lambda v)W(\mu T_t w))\big|_{\lambda=\mu=0}.$$

3. Disordered chains. The approach of [7]

In the present section, we report the main results of [7]. Consider the sequences $v = \{v_m^k\}$ where $k = 1, 2$ distinguishes the position from the momentum, and $m \in \mathbb{Z}$ is the index relative to the site. Define then the norm

$$\llbracket v \rrbracket := \sum_{m\in\mathbb{Z}} e^{\epsilon|m|}(|v_m^1| + |v_m^2|)$$

where $\epsilon > 0$ is a fixed constant determined by the linear part of the Hamiltonian (1.5). Let $L^2 := \{v \mid \llbracket v \rrbracket < \infty\}$. The space L^2 becomes in a natural way a phase-space when it is equipped with the complex conjugation C, and the commutation function

$$\theta(u, v) = \left\langle u, \begin{pmatrix} 0 & iI \\ -iI & 0 \end{pmatrix} v \right\rangle, \tag{3.1}$$

see, e.g., [1, 2]. Further, for each $\xi \in X$, let $A(\xi)$ be the random operator (2.1) on the phase-space (L^2, C, θ). The associated one parameter group of Bogoliubov automorphisms is given by

$$T_t(\xi)v := e^{tA(\xi)^*}v\,. \tag{3.2}$$

The group $T_t(\xi)$ is easily computed ([5], Formula (1.2)) obtaining

$$T_t(\xi) = \begin{pmatrix} \cos(H(\xi)^{1/2}t) & -H(\xi)^{1/2}\sin(H(\xi)^{1/2}t) \\ H(\xi)^{-1/2}\sin(H(\xi)^{1/2}t) & \cos(H(\xi)^{1/2}t) \end{pmatrix}. \tag{3.3}$$

As already mentioned, we will restrict our discussion to a rather small C^*-algebra. We start by considering the constant functions $v(\xi) = v$, and the functions $T_t(\xi)v$ generated by the dynamics. We call the collection $\mathbf{v} := (v, t)$ of such functions *deterministic variables.*

The space $\mathbf{D}(L)$, spanned of all deterministic variables, is a phase-space in a natural way, if one defines as $\mathbf{C}$ the usual complex conjugation on functions, and the "commutation function" $\boldsymbol{\theta}$ as

$$\boldsymbol{\theta}(\mathbf{u}, \mathbf{v}) := \int_X \theta(\mathbf{u}(\xi), \mathbf{v}(\xi))P(\,\mathrm{d}\xi)\,. \tag{3.4}$$

If $\mathbf{v} = (v, \tau)$, the map (3.2) again defines a one parameter group of Bogoliubov automorphisms $\mathbf{T}_t$ on all of $\mathbf{D}(L)$ by

$$\mathbf{T}_t\mathbf{v}(\xi) := T_{t+\tau}(\xi)v\,. \tag{3.5}$$

To define a state, we consider fiberwise the two-point function

$$B_F(u, v) = \left\langle u(\xi), \begin{pmatrix} F(H(\xi)) & \frac{i}{2}I \\ -\frac{i}{2}I & H(\xi)F(H(\xi)) \end{pmatrix} v(\xi) \right\rangle, \tag{3.6}$$

where F is a suitable bounded Borel function on the common spectrum (see, e.g., [10]) of almost all $H(\xi)$.[5] The two-point function $\mathbf{B}_F$ on the phase-space

[5] One must assume $F(x) > 0$, and $xF(x)^2 \geq \frac{1}{4}$ in order to insure that the form be positive defined.

$(\mathbf{D}(L), \mathbf{C}, \boldsymbol{\theta})$ is given by

$$\mathbf{B}_F(\mathbf{u}, \mathbf{v}) = \int_X B_F(\mathbf{u}(\xi), \mathbf{v}(\xi))P(\mathrm{d}\xi) \,. \tag{3.7}$$

The corresponding quasi-free state ω on the Weyl algebra associated to the real part of $(\mathbf{D}(L), \mathbf{C}, \boldsymbol{\theta})$ is obtained by

$$\omega(W(\mathbf{u})) = e^{-\frac{1}{2}\mathbf{B}_F(\mathbf{u},\mathbf{u})} \,.$$

Note that ω is invariant for the one parameter group of Bogoliubov automorphisms (3.5). The commutation rule (3.4) translates on the Weyl operators as

$$W(\mathbf{u})W(\mathbf{v}) = e^{-\frac{1}{2}\boldsymbol{\theta}(\mathbf{u},\mathbf{v})}W(\mathbf{u}+\mathbf{v}) \,.$$

Let $\mathfrak{W}$ be the C^*-algebra generated by all the Weyl operators associated to the GNS representation of the quasi-free state ω which is kept fixed during the analysis.

While the C^*-algebra $\mathfrak{W}$ is large enough to accommodate the linear dynamics, it cannot be invariant under the non linear time evolution. On the other hand, $\mathfrak{W}''$ is most likely too large for our purposes, and the dynamics on it may very well have poor ergodic properties. This problem can be solved by enlarging $\mathfrak{W}$ enough to accommodate the non linear dynamics, but not as much as to spoil its ergodic properties.

Let M be the collection of all the triples $\mathfrak{m} = (m, \mathbf{h}, \mu)$, where $m \in \mathbb{N}$, $\mathbf{h} : \mathbb{R}^m \to \mathbf{D}(L)$ and μ is an appropriate measure on $\mathbb{R}^m$, see [7], Section 6 for more details. For each $\mathfrak{m} \in M$ define

$$W(\mathfrak{m}) = \int_{\mathbb{R}^m} W(\mathbf{h}(x))\mu(\mathrm{d}x) \,,$$

where the integrals are meant w.r.t the weak (or, equivalently strong) operator topology of $\mathcal{B}(\mathcal{H})$, $\mathcal{H}$ being the Hilbert space of the GNS representation of the state under consideration. Next, we consider the linear set of operators

$$\mathcal{M} := \mathrm{span}\,\{W(\mathfrak{m}) \mid \mathfrak{m} \in M\} \,.$$

It is easy to check that $\mathcal{M}$ is a $*$-subalgebra of $\mathfrak{W}''$. Define $\mathfrak{M}$ as the C^*-algebra generated by $\mathcal{M}$. Notice that $\mathfrak{W} \subset \mathfrak{M} \subset \mathfrak{W}''$. Denoting again by ω and α_t^0 the natural extensions of the state and the linear dynamics to all of $\mathfrak{M}$, it is shown in [7] that the resulting dynamical system $(\mathfrak{M}, \alpha_t^0, \omega)$ is mixing.

In order to treat the non linear part V in (1.5), we consider perturbations associated to functions given by the (inverse) Fourier transform of "good" measures. Namely,

$$V(x) := \int_{\mathbb{R}^d} e^{i\lambda\cdot x}\nu(\mathrm{d}\lambda)$$

of a complex bounded Borel measure ν on $\mathbb{R}$ satisfying certain boundedness properties. The corresponding perturbation is obtained by an integral in the weak operator topology of $\mathcal{B}(\mathcal{H})$; see [7], Section 5 for further details. Let the dynamics

$\{\alpha_t^{\Lambda_N}\}_{\Lambda_N \subset \mathbb{Z}}$ restricted to an increasing sequence of finite boxes $\{\Lambda_N\}$ be defined by the Dyson expansion (see, e.g., [3, 9]).

The main results of [7] are summarized in the following

Theorem 3.1. *There exists a one parameter group of automorphisms $\{\alpha_t\}_{t\in\mathbb{R}}$ of $\mathfrak{M}$ such that, for each $A \in \mathfrak{M}$, the following pointwise norm-limit holds true*

$$\alpha_t A = \lim_{\Lambda_N \uparrow \mathbb{Z}} \alpha_t^{\Lambda_N} A .$$

In addition, the $\sigma(\mathfrak{M}^*, \mathfrak{M})$-limit

$$\lim_{t\to\pm\infty} \alpha_t^* \omega =: \omega_\infty$$

exists and defines an invariant state which is mixing for the dynamics α_t.

The proof of the above theorem is based on a uniform control in time of the power series defining the dynamics (see [6, 7]). In turns, this depends on the fact that the linear dynamics satisfies a kind of space-time asymptotic abelianess (1.4).

In conclusion, the model constructed in [7] and briefly outlined here, exhibits good strongly ergodic properties which are stable for a wide class of infinitely extended perturbations of the dynamics.

4. Disordered chains. The quenched approach

As already mentioned, there are serious technical problems to develop, in the case described in this section, a program similar to that of the previous one. To clearly illustrate the situation, we will limit ourselves to the study of finite linear systems. Although such results are limited, they are the first necessary step in order to understand ergodic properties exhibited by disordered systems. Further, they set the stage for the possible subsequent steps:

(1) take the thermodynamic limit,
(2) consider non linear perturbations.

In order to investigate the long-time behavior of the disordered harmonic chain, we start again with the Hamiltonian

$$H(q,p) := \frac{1}{2}\sum_{i=1}^{n} p_i^2 + k_i q_i^2 + \frac{1}{2}\sum_{i=1}^{n-1}(q_{i+1}-q_i)^2$$

describing the finite system of particles allocated to the sites $1, 2, \dots, n$ on the line, with free boundary conditions. The i-particle is subjected to the elastic force described by the elastic constant k_i, and a harmonic uniform nearest neighbor interaction. The infinitesimal generator of the Hamilton flow is given as usual by the truncation of the matrix (2.1).

In order to introduce the disorder, we fix a normalized L^1 function h on the hypercube $Q := [a,b]^n$ with $0 < a < b < +\infty$, and define on the hypercube

$$\mathrm{d}\mu_h(k) := h(k)\,\mathrm{d}^n k .$$

As already mentioned, to construct the appropriate C^*-algebra of observables, we consider the direct integral $\int_Q^\oplus \Gamma(\mathbb{C}^n)\,\mathrm{d}\mu_h(k)$ where $\Gamma(\mathbb{C}^n)$ is the Fock space for the n-dimensional CCR. Define $\mathfrak{A}$ as the C^*-algebra generated by

$$\left\{\int_Q^\oplus W(u(k))\,\mathrm{d}\mu_h(k)\,,\, u\in L^2_{\mathbb{R}}(Q,\,\mathrm{d}\mu_h;\mathbb{C}^{2n})\right\},$$

where $W(u(k))$ is the Weyl operator associated to $u(k)\in\mathbb{R}^{2n}$ acting on $\Gamma(\mathbb{C}^n)$. With an abuse of language, we call also

$$W(u):=\int_Q^\oplus W(u(k))\,\mathrm{d}\mu_h(k)$$

a *Weyl operator.*

It is easy to check that the Weyl operators satisfy the commutation relation

$$W(u)W(v)=\left(\int_Q^\oplus e^{-\frac{1}{2}\theta(u(k),v(k))}I_k\,\mathrm{d}\mu_h(k)\right)W(u+v) \tag{4.1}$$

where θ is given in (3.1), and I_k is the identity operator acting on $\Gamma(\mathbb{C}^n)$ relative to the fibre k.

On the C^*-algebra $\mathfrak{A}$, it is naturally acting the one parameter group of automorphisms $\{\alpha_t\}_{t\in\mathbb{R}}$, described on the Weyl operators by

$$\alpha_t W(u)=W(T_t u)$$

where T_t are made of Bogoliubov automorphisms acting fiberwise as

$$(T_t u)(k):=T_t(k)u(k)\,,$$

and $T_t(k)$ is given by (3.3).

Consider a two-point function B_F as in (3.6) based on the function F which we suppose to be continuous. For each environment k, consider the state φ_k given fiberwise by

$$\varphi_k(W(u(k))):=e^{-\frac{1}{2}B_F(u(k),u(k))}\,. \tag{4.2}$$

The state φ, defined on the Weyl operators by

$$\varphi(W(u)):=\int_Q \varphi_k(W(u(k))\,\mathrm{d}\mu_h(k)\,, \tag{4.3}$$

is uniquely extended on all of $\mathfrak{A}$. By construction, such a state is invariant w.r.t. the dynamics.

Let $B\in\mathfrak{A}$ such that $\varphi(B^*B)>0$, be fixed. As usual, one can consider the state

$$\varphi_B:=\frac{\varphi(B^*\,\cdot\,B)}{\varphi(B^*B)} \tag{4.4}$$

where φ is given in (4.3). Of course, in general φ_B will not be an invariant state, yet it is meaningful to ask if it has or not an asymptotic limit.

Theorem 4.1. *The pointwise limit*

$$\lim_{t\to+\infty} \varphi_B(\alpha_t(A)) =: \varphi_{B,\infty}(A)$$

exists and defines an invariant state on $\mathfrak{A}$.[6]

Proof. By compactness, the net $\{\varphi(B^*\alpha_t(\,\cdot\,)B)\}_{t\in\mathbb{R}}$ has weak*-cluster points. The assertion will follow by showing that there is only one cluster point. By a standard approximation argument, it suffices to investigate the three-point function $I(t) := \varphi(W(u)W(T_t v)W(w))$ where u, v, w are generic elements of $L^2(Q,\, \mathrm{d}\mu_h; \mathbb{R}^{2n})$. By the commutation rule (4.1), we compute

$$I(t) = \int_Q e^{-\frac{i}{2}\left\{\theta(u,w)+\theta(u-w,T_t v)\right\}} e^{-\frac{1}{2}B_F(u+T_t v+w,u+T_t v+w)}\,\mathrm{d}\mu_h\,.$$

Notice that the first exponent is purely imaginary, and the second one is negative. Further, the one parameter group

$$T_t : L^2(Q,\, \mathrm{d}\mu_h; \mathbb{C}^{2n}) \mapsto L^2(Q,\, \mathrm{d}\mu_h; \mathbb{C}^{2n})$$

is equibounded. Hence, for each fixed $\varepsilon > 0$, we can choose smooth functions h_ε, u_ε, v_ε, w_ε, F_ε such that for every $t \in \mathbb{R}$, we have $|I(t) - I_\varepsilon(t)| < \varepsilon$. Here, I_ε is the corresponding function constructed as above by using the mentioned list of smooth functions. We obtain

$$\begin{aligned}|I(t) - I(t')| &< |I(t) - I_\varepsilon(t)| + |I_\varepsilon(t) - I_\varepsilon(t')| + |I_\varepsilon(t') - I(t')|\\ &< 2\varepsilon + |I_\varepsilon(t) - I_\varepsilon(t')|\,.\end{aligned}$$

Namely, the above 3ε-trick allows us to reduce ourselves to the case when all the involved functions are smooth.

We write

$$I(t) = \int_Q h(k) e^{-\frac{i}{2}\mathcal{E}(k)} e^{-\frac{1}{2}E(k,t)}\,\mathrm{d}^n k$$

where

$$\begin{aligned}\mathcal{E}(k) &:= \theta(u(k), w(k)) + B_F(u(k) + w(k), u(k) + w(k)) + B_F(v(k), v(k))\\ E(k,t) &:= \theta(u(k) - w(k), T_t(k)v(k)) + 2B_F(u(k) + w(k), T_t(k)v(k))\,.\end{aligned}$$

After some computations, we obtain

$$\begin{aligned}E(k,t) = \sum_{l=1}^{n} \Big(&\cos(\sqrt{\lambda_l(k)}\,t) \sum_{\sigma=1,2} \langle \xi^\sigma(k) + 2\eta^\sigma(k) | e_l(k)\rangle\langle e_l(k)|v^\sigma(k)\rangle\\ &+ \sin(\sqrt{\lambda_l(k)}\,t) \sum_{\sigma=1,2} \langle \hat{\xi}^\sigma(k) + 2\hat{\eta}^\sigma(k) | e_l(k)\rangle\langle e_l(k)|v^\sigma(k)\rangle \Big)\end{aligned}$$

[6] The analogous limit at $-\infty$ exists as well, and defines another (in general different) invariant state on $\mathfrak{A}$.

where

$$\xi(k) := \begin{pmatrix} 0 & iI \\ -iI & 0 \end{pmatrix} (u(k) - w(k)),$$
$$\hat{\xi}(k) := \begin{pmatrix} -iH(k)^{-1/2} & 0 \\ 0 & -iH(k)^{1/2} \end{pmatrix} (u(k) - w(k)),$$
$$\eta(k) := \begin{pmatrix} F(H(k)) & 0 \\ 0 & H(k)F(H(k)) \end{pmatrix} (u(k) + w(k)),$$
$$\hat{\eta}(k) := \begin{pmatrix} 0 & H(k)^{1/2}F(H(k)) \\ -H(k)^{1/2}F(H(k)) & 0 \end{pmatrix} (u(k) + w(k)),$$

$\{\lambda_l(k)\}_{l=1}^n$ and $\{e_l(k)\}_{l=1}^n$ being the finite sequences of the eigenvalues and the corresponding eigenvectors of the matrix $H(k)$.

Notice that we have for such a matrix,

$$H(k) = \begin{pmatrix} (k_1+2) & -1 & 0 & \cdot & \cdot & \cdot & \cdot \\ -1 & (k_2+2) & -1 & 0 & \cdot & \cdot & \cdot \\ \cdot & \cdot & \cdot & \cdot & \cdot & \cdot & \cdot \\ \cdot & 0 & -1 & (k_l+2) & -1 & 0 & \cdot \\ \cdot & \cdot & \cdot & \cdot & \cdot & \cdot & \cdot \\ \cdot & \cdot & \cdot & 0 & -1 & (k_{n-1}+2) & -1 \\ \cdot & \cdot & \cdot & \cdot & 0 & -1 & (k_n+2) \end{pmatrix}.$$

This is exactly the situation described in Appendix A. Namely, the eigenvalues of $H(k)$ are the n distinct roots of the orthogonal polynomial $P_n(k+2,x)$ corresponding to the parameters $(k_1+2,\dots,k_n+2)$. The functions

$$\omega(k) := (\sqrt{\lambda_1(k)},\dots,\sqrt{\lambda_n(k)}) \tag{4.5}$$

are, locally, smooth functions of $(k_1,\dots,k_n)$. Further, as the eigenvalues $(\lambda_1(k), \dots, \lambda_n(k))$ are strictly positive under our assumptions, there exists local inverses for $\omega(k)$ apart from a closed null-set $N \subset \mathbb{R}^n$, see Proposition A.3.

Let now $\{O_m\}_{m=1}^\infty$ be an open covering of the (open) set $\overset{\circ}{Q} \setminus N$ of full Lebesgue measure, together with a partition of unity $\{\psi_m\}_{m=1}^\infty$ subordinate to the mentioned covering, such that for each m, the function (4.5) admits the inverse K_m in O_m. We have

$$I(t) = \sum_{m=1}^\infty \int_{K_m(\mathrm{supp}(\psi_m))} \psi_m(K_m(\omega))h(K_m(\omega)) \times e^{-\frac{1}{2}\mathcal{E}(K_m(\omega))}e^{-\frac{1}{2}E(K_m(\omega),t)}\big|\det \partial_\omega K_m(\omega)\big|\,\mathrm{d}^n\omega$$

where the last series is summable, uniformly in $t \in \mathbb{R}$; see, e.g., [12], Theorem 3.12.

As $\mathcal{E}(k)$ and $E(k,t)$ are smooth functions,[7] we can apply Proposition B.1 after exchanging the sum with the limit. This leads to the assertion. □

Notice that, taking into account (B.1), the limit of the three-point function $\lim_{t\to+\infty}\varphi(W(u)W(T_t v)W(w))$, can be explicitly computed.

The above theorem differs substantially from the mixing property (1.2). In fact, mixing would imply $\varphi_{B,\infty}=\varphi$. It is not hard to show that this is generically impossible. This is a consequence of the abundance of invariant states and shows that the asymptotic state depends on the initial conditions. However, Theorem 4.1 tells us that the disordered chain exhibit a "good" long-time behavior, contrary to the usual chain (non disordered) where the dynamics is quasi-periodic.

Equally well, one can consider the "quenched" state φ^B associated to an element B as above. It is given in a natural way on the Weyl operators by

$$\varphi^B(W(u)) := \int_Q \frac{\varphi_k(B(k)^*W(u(k))B(k))}{\varphi_k(B(k)^*B(k))}\,\mathrm{d}\mu_h(k)\,, \tag{4.6}$$

where φ_k is given in (4.2), and $\{B(k)\}_{k\in Q}$ is the measurable decomposition of B which always exists; see, e.g., [14], Section IV.8.

Suppose that the measurable function $k\mapsto\varphi_k(B(k)^*B(k))$ is essentially bounded from below by a constant $\delta>0$. Define fiberwise the operator $\widetilde{B}$ as

$$\widetilde{B}(k) := \varphi_k(B(k)^*B(k))^{-1/2}B(k)\,.$$

Clearly, $\widetilde{B}$ belongs to $\mathfrak{A}$. In addition, $\varphi^B=\varphi_{\widetilde{B}}$, that is most of the states described in (4.6) are particular cases of those defined in (4.4).

Appendix A. Note on orthogonal polynomials

Consider a sequence $\{a_n\}_{n\in\mathbb{N}}$ of real numbers, together with the associated sequence of polynomials uniquely defined by

$$\begin{aligned} &P_{-1}(x) := 0\,, \qquad P_0(x) := 1\,,\\ &P_j(x) := (x-a_j)P_{j-1}(x) - P_{j-2}(x)\,. \end{aligned} \tag{A.1}$$

It is well known that the P_n are the orthogonal polynomials associated to some probability measure on the real line. Hence, the P_n have n distinct roots with nice separation properties, see, e.g., [13].

Consider, for each $n\in\mathbb{N}$, the finite sequence of parameters $a:=(a_1,\dots,a_n)$, together with the corresponding polynomial P_n. Let $\bar{P}^n_j$ be the polynomial constructed by the relations (A.1) using the parameters $(a_{n-j+1},\dots,a_n)$. To simplify

[7] This follows as, for all k nearby $\bar{k}$,

$$|e_l(k)\rangle\langle e_l(k)| = \frac{1}{2\pi i}\oint_\gamma (zI-H(k))^{-1}\,\mathrm{d}z$$

where γ is a sufficiently small fixed circle surrounding counterclockwise the eigenvalue $\lambda_l(\bar{k})$ corresponding to the one-dimensional projector $|e_l(\bar{k})\rangle\langle e_l(\bar{k})|$.

notations, we omit for the moment to make explicit the dependence of the polynomials on the parameters.

Lemma A.1. *The relation*

$$Q_j^n(x) := \partial_{a_j} P_n(x) = -\bar{P}_{n-j}^n(x) P_{j-1}(x)$$

holds true for each $n \in \mathbb{N}$, $(a_1, \ldots, a_n) \in \mathbb{R}^n$ *and* $x \in \mathbb{R}$.

Proof. The proof is by induction over n. For $n = 1$ we have

$$\partial_{a_1} P_1(x) = \partial_{a_1}(x - a_1) = -1 = -\bar{P}_0^1(x) P_0(x) .$$

Let us suppose the lemma true for n, then

$$\begin{aligned}\partial_{a_{n+1}} P_{n+1}(x) &= -P_n(x) = -\bar{P}_0^{n+1}(x) P_n(x), \\ \partial_{a_n} P_{n+1}(x) &= (x - a_{n+1}) \partial_{a_n} P_n(x) = -(x - a_{n+1}) \bar{P}_0^n(x) P_{n-1}(x) \\ &= -\bar{P}_1^{n+1}(x) P_{n-1}(x) .\end{aligned}$$

For $j < n$ we obtain

$$\begin{aligned}\partial_{a_j} P_{n+1}(x) &= (x - a_{n+1}) \partial_{a_j} P_n(x) - \partial_{a_j} P_{n-1}(x) \\ &= -(x - a_{n+1}) \bar{P}_{n-j}^n(x) P_{j-1}(x) + \bar{P}_{n-j-1}^{n-1}(x) P_{j-1}(x) \\ &= -\bar{P}_{n+1-j}^{n+1}(x) P_{j-1}(x) ,\end{aligned}$$

which concludes the proof. □

Note that Lemma A.1 implies that the Q_j^n are all polynomials of degree $n - 1$ in x. We can thus set

$$Q_j^n(x) =: \sum_{k=0}^{n-1} \alpha_{j,k}^n x^k \tag{A.2}$$

where the coefficients depends on the parameters a. Define the $n \times n$ matrix A_n as that having the coefficients appearing in the r.h.s. of (A.2). Consider, for a finite sequence $\lambda := (\lambda_1, \ldots, \lambda_n)$ of unknowns, the $n \times n$ matrix $\Lambda \equiv \Lambda(a, \lambda)$ whose entries are given by

$$[\Lambda(a, \lambda)]_{i,j} := \partial_{a_j} P_n(\lambda_i) .$$

Lemma A.2. *The following holds true.*

$$\det \Lambda = \det \big(A_n\big) \det \big(1 \quad \lambda_i \quad \ldots \quad \lambda_i^{n-1}\big) . \tag{A.3}$$

Moreover, the polynomial $\det \Lambda$ *is not identically zero.*

Proof. Let $\mathbb{P}_n$ be the set of permutations of $\{0, \ldots, n-1\}$. Then, by the multilinearity of the determinant, we get

$$\det \Lambda = \sum_{\sigma \in \mathbb{P}_n} \prod_{j=1}^{n} \alpha_{j,\sigma(j-1)}^n \det \Big(\lambda_i^{\sigma(i-1)}\Big)$$

and the first half follows by reordering the determinants in the r.h.s..

As we have for the Vandermonde determinant,

$$\det\begin{pmatrix}1 & \lambda_i & \dots & \lambda_i^{n-1}\end{pmatrix} = \prod\nolimits_{\substack{i,j=1\\ i>j}}^{n} (\lambda_i - \lambda_j)\,,$$

we obtain that $\det \Lambda \equiv 0$ is equivalent to $\det A_n \equiv 0$. To rule out this last possibility, it suffices to show that it is different from zero for a specific choice of the parameters a. We compute it for $a_i = 0$ for each $i < n$, and $a_n = 1$. Let us call $\hat{P}, \hat{Q}$ and $\hat{\alpha}$ the polynomials and coefficients corresponding to parameters $a_i = 0$ for all i.

Clearly, $\bar{P}_j^i(x) = \hat{P}_j(x)$ for all $i < n$, while

$$\bar{P}_j^n(x) = (x-1)\hat{P}_{j-1}(x) - \hat{P}_{j-2}(x) = \hat{P}_j(x) - \hat{P}_{j-1}(x)\,.$$

Hence, for $j < n$,

$$Q_j^n(x) = -\hat{P}_{n-j}(x)\hat{P}_{j-1}(x) - \hat{P}_{n-j-1}(x)\hat{P}_{j-1}(x) = \hat{Q}_j^n(x) - \hat{Q}_j^{n-1}(x)\,.$$

Next, notice that $\hat{P}_j$ is even for j even and odd for j odd. Thus $\hat{Q}_j^n$ is even for n odd and vice versa. This means that the non zero coefficients $\alpha_{k,j}^n$ and $\alpha_{k,j}^{n-1}$ fall in different columns. In addition, for $j < n-1$,

$$\begin{aligned}\hat{Q}_j^n(x) &= -x\hat{P}_{n-j-1}(x)\hat{P}_{j-1}(x) + \hat{P}_{n-j-2}(x)P_{j-1}(x)\\ &= x\hat{Q}_j^{n-1}(x) - \hat{Q}_j^{n-2}(x)\,.\end{aligned}$$

Let us see what the above equations means in terms of the coefficients $\hat{\alpha}$. For $j < n-1$ we have

$$\begin{aligned}\hat{\alpha}_{j,0}^n &= -\hat{\alpha}_{j,0}^{n-2}\,,\\ \hat{\alpha}_{j,k}^n &= \hat{\alpha}_{j,k-1}^{n-1} - \hat{\alpha}_{j,k}^{n-2}\,, \qquad k = 1,\dots,n-3\,,\\ \hat{\alpha}_{j,n-2}^n &= \hat{\alpha}_{j,n-3}^{n-1}\,,\\ \hat{\alpha}_{j,n-1}^n &= \hat{\alpha}_{j,n-2}^{n-1}\,.\end{aligned} \tag{A.4}$$

The above considerations imply that A_n has the following form

$$A_n = \begin{pmatrix} \hat{\alpha}_{1,0}^n - \hat{\alpha}_{1,0}^{n-1} & \dots & \hat{\alpha}_{1,n-2}^n - \hat{\alpha}_{1,n-2}^{n-1} & \hat{\alpha}_{1,n-1}^n \\ \vdots & \vdots\vdots\vdots & \vdots & \vdots \\ \hat{\alpha}_{n-1,0}^n - \hat{\alpha}_{n-1,0}^{n-1} & \dots & \hat{\alpha}_{n-1,n-2}^n - \hat{\alpha}_{n-1,n-2}^{n-1} & \hat{\alpha}_{n-1,n-1}^n \\ \hat{\alpha}_{n,0}^n & \dots & \hat{\alpha}_{n,n-2}^n & 1 \end{pmatrix}.$$

We restrict ourselves to the even case, the other one being similar.

$$A_n = \begin{pmatrix} -\hat{\alpha}_{1,0}^{n-1} & \hat{\alpha}_{1,1}^n & -\hat{\alpha}_{1,2}^n & \dots & -\hat{\alpha}_{1,n-2}^{n-1} & \hat{\alpha}_{1,n-1}^n \\ \vdots & \vdots & \vdots & \vdots\vdots\vdots & \vdots & \vdots \\ -\hat{\alpha}_{n-1,0}^{n-1} & \hat{\alpha}_{n-1,1}^n & -\hat{\alpha}_{n-1,2}^n & \dots & -\hat{\alpha}_{n-1,n-2}^{n-1} & \hat{\alpha}_{n-1,n-1}^n \\ 0 & \hat{\alpha}_{n,1}^n & 0 & \dots & 0 & 1 \end{pmatrix}.$$

Now, we construct a new matrix by keeping the columns with $k = 2l$ and substituting the columns with $k = 2l+1$ with the sum of the same column $2l+1$

and the previous column $2l$. Clearly, the new matrix has the same determinant as the original one. Keeping in mind (A.4), and developing the determinant w.r.t. the last column (which has all zeroes but in the last entry), it yields

$$\begin{aligned}|\det A_n| &= \left|\det\begin{pmatrix} \hat{\alpha}^{n-1}_{1,0} & \hat{\alpha}^{n-2}_{1,1} & \hat{\alpha}^{n-1}_{1,2} & \cdots & \hat{\alpha}^{n-2}_{1,n-3} & \hat{\alpha}^{n-1}_{1,n-2} \\ \vdots & \vdots & \vdots & \vdots\vdots\vdots & \vdots & \vdots \\ \hat{\alpha}^{n-1}_{n-2,0} & \hat{\alpha}^{n-2}_{n-2,1} & \hat{\alpha}^{n-1}_{n-2,2} & \cdots & \hat{\alpha}^{n-2}_{n-2,n-3} & \hat{\alpha}^{n-1}_{n-1,n-1} \\ \hat{\alpha}^{n-1}_{n-1,0} & 0 & \hat{\alpha}^{n-1}_{n-1,2} & \cdots & 0 & 1 \end{pmatrix}\right| \\ &= |\det A_{n-1}|\,.\end{aligned}$$

By the same arguments for the odd case, we conclude that $|\det A_n| = |\det A_{n-1}|$ for each n, which means by induction, $|\det A_n| = 1$ for each n as $\det A_2 = 1$. □

Consider the polynomial $P_n \equiv P_n(a,x)$ in the unknown x, constructed by (A.1) relatively to the parameters $a \equiv (a_1,\dots,a_n)$. A trivial application of the Implicit Function Theorem tells us that the n distinct roots

$$\lambda(a) := (\lambda_1(a),\dots,\lambda_n(a)) \tag{A.5}$$

of $P_n(a,x)$ depends smoothly on the parameters a.[8] In other words, defining

$$F(a,\lambda) := (P_n(a,\lambda_1),\dots,P_n(a,\lambda_n))\,,$$

the equation

$$F(a,\lambda) = 0$$

defines at least locally, the n distinct roots $\lambda(a)$ as smooth functions of the parameters a. It is crucial for our aims to ask for local invertibility of the function (A.5).

Proposition A.3. *Apart from a negligible set in the parameters a, the map $a \in \mathbb{R}^n \mapsto \lambda(a) \in \mathbb{R}^n$ defines a local diffeomorphism.*

Proof. For the cases under consideration, we have for the Jacobian matrix

$$\frac{\partial\lambda}{\partial a} = \left(\frac{\partial F}{\partial\lambda}(a,\lambda(a))\right)^{-1}\frac{\partial F}{\partial a}(a,\lambda(a))\,,$$

at least locally. Hence, it is sufficient to investigate the function $(\alpha,\lambda) \in \mathbb{R}^{2n} \mapsto \det\frac{\partial F}{\partial a}(a,\lambda) \equiv \det\Lambda(a,\lambda) \in \mathbb{R}$ under the conditions $\lambda_i \neq \lambda_j$, $i \neq j$, see Footnote 8. Taking into account Lemma A.2, the last determinant factorizes into two pieces, where the first one depends only on the parameters, and the second one never vanishes in our context. Further, it is also shown in Lemma A.2, that such a determinant is not identically zero. The negligible set we are looking for is precisely the locus of zeroes of the non trivial polynomial $\det A_n$ in (A.3), which is well known to be negligible w.r.t. the n-dimensional Lebesgue measure. □

---

[8]This easily follows from the fact that $P_n(a,x)$ has n distinct roots which means that $\partial_x P_n(a,x)\big|_{x=\lambda_i(a)} \neq 0$, $i = 1,\dots,n$.

Appendix B. A convergence property

Consider a real-valued C^1 function h supported into $(0,L)^n$, sequences $\{a_i\}_{i=1,\dots,n}$, $\{b_i\}_{i=1,\dots,n}$ of complex-valued C^1 functions on $\mathbb{R}^n$, and finally a holomorphic function $f:\mathbb{C}\mapsto\mathbb{C}$.

Proposition B.1. *Under the above conditions, we have*

$$\lim_{t\to+\infty}\int_{[0,L]^n} \mathrm{d}^n\omega\, h(\omega) f\Big(\sum_{i=1}^n \big(a_i(\omega)\sin\omega_i t + b_i(\omega)\cos\omega_i t\big)\Big) \tag{B.1}$$
$$= \int_{[0,L]^n} \mathrm{d}^n\omega\, h(\omega)\frac{1}{(2\pi)^n}\int_{[0,2\pi]^n} \mathrm{d}^n x\, f\Big(\sum_{i=1}^n \big(a_i(\omega)\sin x_i + b_i(\omega)\cos x_i\big)\Big).$$

Proof. Taking into account the support of h, we can write

$$\Big|\int_{[0,L]^n} \mathrm{d}^n\omega\, h(\omega) f\Big(\sum_{i=1}^n \big(a_i(\omega)\sin\omega_i t + b_i(\omega)\cos\omega_i t\big)\Big)$$
$$-\int_{[0,L]^n} \mathrm{d}^n\omega\, h(\omega)\frac{1}{(2\pi)^n}\int_{[0,2\pi]^n} \mathrm{d}^n x\, f\Big(\sum_{i=1}^n \big(a_i(\omega)\sin x_i + b_i(\omega)\cos x_i\big)\Big)\Big|$$
$$\leq \sum_{j_1,\dots,j_n=0}^{\left[\frac{tL}{2\pi}\right]} \int_{Q_j} \mathrm{d}^n\omega\, \Big|h(\omega)-h\big(\tfrac{2\pi}{t}j\big)\Big|\,\Big|f\Big(\sum_{i=1}^n \big(a_i(\omega)\sin\omega_i t + b_i(\omega)\cos\omega_i t\big)\Big)\Big|$$
$$+\sum_{j_1,\dots,j_n=0}^{\left[\frac{tL}{2\pi}\right]} \Big|h\big(\tfrac{2\pi}{t}j\big)\Big| \int_{Q_j} \mathrm{d}^n\omega\, \Big|f\Big(\sum_{i=1}^n \big(a_i(\omega)\sin\omega_i t + b_i(\omega)\cos\omega_i t\big)\Big)$$
$$-f\Big(\sum_{i=1}^n \big(a_i\big(\tfrac{2\pi}{t}j\big)\sin\omega_i t + b_i\big(\tfrac{2\pi}{t}j\big)\cos\omega_i t\big)\Big)\Big|$$
$$+\Big|\sum_{j_1,\dots,j_n=0}^{\left[\frac{tL}{2\pi}\right]} \Big[\big(\tfrac{2\pi}{t}\big)^n h\big(\tfrac{2\pi}{t}j\big)\frac{1}{(2\pi)^n}\int_{[0,2\pi]^n} \mathrm{d}^n x\, f\Big(\sum_{i=1}^n \big(a_i(\omega)\sin x_i + b_i(\omega)\cos x_i\big)\Big)$$
$$-\int_{Q_j} \mathrm{d}^n\omega\, h(\omega)\frac{1}{(2\pi)^n}\int_{[0,2\pi]^n} \mathrm{d}^n x\, f\Big(\sum_{i=1}^n \big(a_i(\omega)\sin x_i + b_i(\omega)\cos x_i\big)\Big)\Big]\Big|$$

where Q_j is the (small) hypercube of side $\frac{2\pi}{t}$ constructed starting from the point $(\frac{2\pi}{t}j_1,\dots,\frac{2\pi}{t}j_n)$. Here, we made the change of variables $(x_1,\dots,x_n)=(t\omega_1,\dots,t\omega_n)$ in the last piece.

We majorize the first two pieces by

$$2\pi\|\,\mathrm{d}h\|_\infty\|f\|_\infty\sqrt{n}L^n\frac{1}{t}$$

$$2\pi\|h\|_\infty\|f'\|_\infty\Big[\sum_{i=1}^n \big(\|\,\mathrm{d}a_i\|_\infty + \|\,\mathrm{d}b_i\|_\infty\big)\Big]\sqrt{n}L^n\frac{1}{t}$$

respectively, where $\| \, \|_\infty$ is the supremum norm on suitable compact sets. In order to estimate the last piece, we note that the first part is precisely the Riemann sums of the second part. Analogously, it is majorized by

$$2\pi \| \, \mathrm{d}G \|_\infty \sqrt{n} L^n \frac{1}{t}$$

where G is the smooth function given by

$$G(\omega) := \frac{h(\omega)}{(2\pi)^n} \int_{[0,2\pi]^n} \mathrm{d}^n x \, f\bigg(\sum_{i=1}^{n} \big(a_i(\omega) \sin x_i + b_i(\omega) \cos x_i \big) \bigg) .$$

Collecting together all pieces, we conclude that

$$\bigg| \int_{[0,L]^n} \mathrm{d}^n \omega \, h(\omega) f\bigg(\sum_{i=1}^{n} \big(a_i(\omega) \sin \omega_i t + b_i(\omega) \cos \omega_i t \big) \bigg)$$
$$- \int_{[0,L]^n} \mathrm{d}^n \omega \, h(\omega) \frac{1}{(2\pi)^n} \int_{[0,2\pi]^n} \mathrm{d}^n x \, f\bigg(\sum_{i=1}^{n} \big(a_i(\omega) \sin x_i + b_i(\omega) \cos x_i \big) \bigg) \bigg|$$
$$= \mathrm{O}\left(\frac{1}{t} \right)$$

which leads to the assertion. □

References

[1] H. Araki, *On Quasi-Free States of the Canonical Commutation Relations.* II, Publ. Res. Inst. Math. Sci. **7** (1971–72), 121–152.

[2] H. Araki, M. Shiraishi, *On Quasi-Free States of the Canonical Commutation Relations.* I, Publ. Res. Inst. Math. Sci. **7** (1971–72), 105–120.

[3] O. Bratteli, D.W. Robinson, *Operator Algebras and Quantum Statistical Mechanics.* I, II, Springer, Berlin–Heidelberg–New York, 1981.

[4] P. de Smedt, D. Dürr, J.L. Lebowitz, C. Liverani, *Quantum System in Contact with a Thermal Environment: Rigorous Treatment of a Simple Model.* Commun. Math. Phys. **120** (1988), 195–231.

[5] D. Dürr, V. Naroditsky, N. Zanghi, *On the Hydrodynamic Motion of a Test Particle in a Random Chain.* Ann. Phys. **178** (1987), 74–88.

[6] F. Fidaleo, C. Liverani, *Ergodic Properties for a Quantum Nonlinear Dynamics.* J. Stat. Phys. **97** (1999), 957–1009.

[7] F. Fidaleo, C. Liverani, *Ergodic Properties of a Model Related to Disordered Quantum Anharmonic Crystals.* Commun. Math. Phys. **235** (2003), 169–189.

[8] V. Jakšić, C.-A. Pillet, *Mathematical Theory Of Non-Equilibrium Quantum Statistical Mechanics.* J. Stat. Phys. **108** (2002), 787–829.

[9] D. Kastler, *Equilibrium States of Matter and Operator Algebras.* In Symposia Mathematica Vol. **XX** (1976), 49–107, Academic Press.

[10] H. Kunz, B. Souillard, *Sur le spectre des opérateurs aux différence finies aléatoires.* Commun. Math. Phys. **78** (1986), 201–246.

[11] C.P. Niculescu, A. Ströh, L. Zsidó, *Noncommutative Extension of Classical and Multiple Recurrence Theorems.* J. Operator Theory **50** (2003), 3–52.

[12] M. Spivak, *Calculus on Manifolds.* Benjamin, New York–Amsterdam, 1965.

[13] G. Szegő, *Orthogonal Polynomials.* Coll. Math. Vol. **23**, American Mathematical Society, Providence, 1981.

[14] M. Takesaki, *Theory of Operator Algebras.* I, Springer, Berlin–Heidelberg–New York 1979.

Francesco Fidaleo
Dipartimento di Matematica
Università di Tor Vergata
via della Ricerca Scientifica 1
I-00133 Roma
Italy
e-mail: `fidaleo@mat.uniroma2.it`

Carlangelo Liverani
Dipartimento di Matematica
Università di Tor Vergata
via della Ricerca Scientifica 1
I-00133 Roma
Italy
e-mail: `liverani@mat.uniroma2.it`

Operator Theory:
Advances and Applications, Vol. 153, 143–156

On Operator Periodically Correlated Random Fields

Păstorel Gaşpar

Abstract. In [3] Gladyshev gave an interesting characterization of periodically correlatedness for the second-order one continuous time parameter univariate continuous stochastic processes, in terms of the support of an associated spectral bimeasure. Recently, Makagon [6] extended this result to the case, where continuity is weakened to locally square summability. It is our aim now to give such a characterization of periodically correlatedness for second order n continuous time parameters infinite variate locally square integrable random fields.

Mathematics Subject Classification (2000). 66G12, 47N30.

Keywords. Stochastic process, Random field.

1. Notation and preliminaries

We give in this first section some notation and preliminary notions. First $\mathbb{Z}, \mathbb{R}, \mathbb{C}$ stand for the sets of integers, real and complex numbers respectively, $\mathscr{B}(\mathbb{R}^n)$ denotes the σ-algebra of Borel sets in the euclidian space $\mathbb{R}^n$, $\mathrm{m}_n(\cdot)$ the Lebesgue measure on $\mathbb{R}^n$. The scripts $\mathcal{E}, \mathcal{F}, \mathcal{G}$ are reserved for complex separable Hilbert spaces, $\mathcal{M}(\Omega, \mathcal{E})$ and $L^p(\Omega, \mathcal{E})$, $p \in [0, \infty]$, are the $\mathcal{E}$-valued, strongly (or weakly) measurable, respectively p-integrable functions, or – in probabilistic setting – infinite variate random variables respectively p-order infinite variate random variables on the probability space $(\Omega, \Sigma, \wp)$. A *second-order Hilbert space valued random field* (with n continous time parameters) is a function (see [4])

$$\mathbb{R}^n \ni s \mapsto X(s) \in L^2(\Omega, \mathcal{E}). \tag{1.1}$$

If the function (1.1) is continuous, or locally square integrable on $\mathbb{R}^n$, then the field is called *continuous*, respectively *locally square integrable*. Further we need other notions for a suitable organization of the random variables into functional analytic structures.

So $\mathcal{B}(\mathcal{X},\mathcal{Y})$ means the space (algebra if $\mathcal{X} = \mathcal{Y}$) of all continuous linear operators between the linear Hausdorff topological spaces $\mathcal{X}$ and $\mathcal{Y}$, whereas for the Hilbert spaces $\mathcal{F}$ and $\mathcal{G}$, for $\alpha = 1, 2$, $\mathscr{C}^\alpha(\mathcal{F},\mathcal{G})$ is the trace class ideal, respectively the Hilbert-Schmidt class of operators from $\mathcal{B}(\mathcal{F},\mathcal{G})$. We also mention the well-known duality between $\mathscr{C}^1(\mathcal{E})$ and $\mathcal{B}(\mathcal{E})$, in the sense that $\mathcal{B}(\mathcal{E})$ is the normed dual of $\mathscr{C}^1(\mathcal{E})$, the pairing being given by $\langle C, A\rangle := \operatorname{tr}(CA)$; $A \in \mathcal{B}(\mathcal{E}), C \in \mathscr{C}^1(\mathcal{E})$. It is well known that the space of all second order $\mathcal{E}$-valued random variables $L^2(\Omega,\mathcal{E})$ is a normal Hilbert left $\mathcal{B}(\mathcal{E})$-module and it can be identified with the Hilbert-Schmidt class $\mathscr{C}^2(\mathcal{F},\mathcal{E})$, where $\mathcal{F} = L^2(\Omega,\mathbb{C}) = L^2(\Omega)$ (see [4, Corollary 7, p. 30]). In the following we will use for $L^2(\Omega,\mathcal{E})$ and $\mathscr{C}^2(L^2(\Omega),\mathcal{E})$ the same notation $\mathcal{H}$. We recall that the canonical normal Hilbert left $\mathcal{B}(\mathcal{E})$-module structure of $\mathcal{H}$ is given by the $\mathscr{C}^1(\mathcal{E})$-valued inner product $[T,S] := TS^*$, $T,S \in \mathcal{H}$ which is also called its gramian, the Hilbert space structure being given by the scalar product $\operatorname{tr}([T,S])$, $T,S \in \mathcal{H}$. Then the direct sum $\boxed{+}_{l\in\mathbb{Z}^n} \mathcal{H}_l$, where, for each $l \in \mathbb{Z}^n$, $\mathcal{H}_l$ is a normal Hilbert left $\mathcal{B}(\mathcal{E})$-module, is defined as the set of all $(h_l)_{l\in\mathbb{Z}^n}$ from $\prod_{l\in\mathbb{Z}^n} \mathcal{H}_l$, such that $\sum_{l\in\mathbb{Z}^n} h_l$ is gramian square integrable, i.e., $\sum_{l\in\mathbb{Z}^n}[h_l,h_l]$ is norm-convergent in $\mathscr{C}^1(\mathcal{E})$. It is easy to observe that this direct sum is a normal Hilbert left $\mathcal{B}(\mathcal{E})$-module with the inner product

$$[(h_l)_l,(g_l)_l] := \sum_{l\in\mathbb{Z}^n} [h_l,g_l]_{\mathcal{H}_l}.$$

For the particular case $\mathcal{H}_l = \mathcal{H}$, $l \in \mathbb{Z}^n$ we shall use for this direct sum the notation $\ell^2[\mathbb{Z}^n,\mathcal{H}]$. We shall also use another $\mathcal{B}(\mathcal{E})$-module, which is gramian unitarily equivalent to the last one. Namely, $L^2[\mathbb{T}^n,\mathcal{H}]$ is defined as the set of all functions

$$[0,2\pi) \times \cdots \times [0,2\pi) \ni (u_1,\ldots,u_n) = u \mapsto h(u) \in \mathcal{H},$$

which are (strongly) measurable as $\mathcal{H}$-valued functions and gramian square integrable, i.e., $u \mapsto [h(u),h(u)]$ are Bochner integrable as $\mathscr{C}^1(\mathcal{E})$-valued functions on $\mathbb{T}^n$ w.r.t. m_n. Particularly, from the easy relation $\|T\|^2 = \|TT^*\| \le \operatorname{tr}(TT^*)$, $(T \in \mathcal{B}(\mathcal{F},\mathcal{E}))$, we have that $\|h(\cdot)\|_{\mathcal{B}(\mathcal{F},\mathcal{E})}$ is square integrable on $\mathbb{T}^n$ and, consequently, also integrable on $\mathbb{T}^n$. On the other hand, from a polarization formula it holds that $[h(\cdot),k(\cdot)]_{\mathcal{H}}$ is Bochner integrable as $\mathscr{C}^1(\mathcal{E})$-valued function and so the gramian on the $\mathcal{B}(\mathcal{E})$-module $L^2[\mathbb{T}^n,\mathcal{H}]$ can be defined by

$$[h,k] := \frac{1}{(2\pi)^n}\int_0^{2\pi}\cdots\int_0^{2\pi}[h(u),k(u)]_{\mathcal{H}}\,\mathrm{dm}_n(u); \tag{1.2}$$

$h,\ k \in L^2[\mathbb{T}^n,\mathcal{H}]$. As in the case where $\mathcal{H}$ is a Hilbert space, we have in this setting also the gramian unitarity of the Fourier transform

$$L^2[\mathbb{T}^n,\mathcal{H}] \ni h \mapsto \big(\hat{h}(l)\big)_{l\in\mathbb{Z}^n} \in \ell^2[\mathbb{Z}^n,\mathcal{H}],$$

where, by putting $\langle l, u\rangle = l_1u_1 + \cdots + l_nu_n$,

$$\hat{h}(l) = \frac{1}{(2\pi)^n} \int_0^{2\pi} \cdots \int_0^{2\pi} e^{-\mathrm{i}\langle l,u\rangle} h(u)\ \mathrm{d}m_n(u), \quad l \in \mathbb{Z}^n. \tag{1.3}$$

We also remark that it is possible to work with $u = (u_1, \ldots, u_n)$ from $[0, T_1) \times \cdots \times [0, T_n)$, $(T_j > 0, j = 1, 2, \ldots, n)$ instead of $[0, 2\pi) \times \cdots \times [0, 2\pi)$. In this case the formulas (1.2) and (1.3) become

$$[h, k] := \frac{1}{T_1 \cdots T_n} \int_0^{T_1} \cdots \int_0^{T_n} [h(u), k(u)]_{\mathcal{H}}\ \mathrm{d}m_n(u), \tag{1.2'}$$

respectively

$$\hat{h}(l) = \frac{1}{T_1 \cdots T_n} \int_0^{T_1} \cdots \int_0^{T_n} e^{-2\pi \mathrm{i}\langle l, \frac{u}{T}\rangle} h(u)\ \mathrm{d}m_n(u), \quad l \in \mathbb{Z}^n, \tag{1.3'}$$

where $\frac{u}{T}$ means $\left(\frac{u_1}{T_1}, \ldots, \frac{u_2}{T_2}\right)$, and the corresponding $\mathcal{B}(\mathcal{E})$-module will be denoted by $L^2[[0,T), \mathcal{H}]$.

We shall also use the classical notation for the basic spaces from distribution theory $\mathcal{D}(\mathbb{R}^n) = \mathcal{D}_n$, $\mathcal{S}(\mathbb{R}^n) = \mathcal{S}_n$ (see [1]) and for the $\mathcal{H}$-valued distributions and $\mathcal{H}$-valued tempered distributions $\mathcal{D}'_n(\mathcal{H}) = \mathcal{B}(\mathcal{D}_n, \mathcal{H})$, $\mathcal{S}'_n(\mathcal{H}) = \mathcal{B}(\mathcal{S}_n, \mathcal{H})$ respectively, where $\mathcal{H}$ is a normal Hilbert left $\mathcal{B}(\mathcal{E})$-module of the above type endowed with a Hilbert space structure. A continuous $\mathcal{H}$-valued mapping on $\mathcal{D}_n \times \mathcal{D}_n$ ($\mathcal{S}_n \times \mathcal{S}_n$ respectively) will be called a bi-distribution (tempered bi-distribution) on $\mathbb{R}^n \times \mathbb{R}^n$. We shall work with the usual Fourier transform on $\mathbb{R}^n$

$$\hat{\varphi}(u) = \frac{1}{(2\pi)^n} \int_{\mathbb{R}^n} e^{-\mathrm{i}\langle s,u\rangle} \varphi(s)\ \mathrm{d}m_n(s); \quad \varphi \in \mathcal{S}_n \tag{1.4}$$

and with a non-standard normalization of it on $\mathbb{R}^n \times \mathbb{R}^n$

$$\hat{\varphi}(u, v) = \frac{1}{(2\pi)^{2n}} \int_{\mathbb{R}^n}\int_{\mathbb{R}^n} e^{-\mathrm{i}(\langle s,u\rangle - \langle t,v\rangle)} \varphi(s,t)\ \mathrm{d}m_{2n}(s,t); \quad \varphi \in \mathcal{S}_{2n}, \tag{1.5}$$

where $u = (u_1, \ldots, u_n), v = (v_1, \ldots, v_n) \in \mathbb{R}^n$.

The Fourier transform of D, which is a tempered $\mathcal{H}$-valued distribution on $\mathbb{R}^n$, or a tempered $\mathcal{H}$-valued bi-distribution on $\mathbb{R}^n \times \mathbb{R}^n$, is $\widehat{D}$ defined for $\varphi \in \mathcal{S}_n$, respectively for $\varphi \in \mathcal{S}_{2n}$ by the formula

$$\widehat{D}(\varphi) : = D(\hat{\varphi}). \tag{1.6}$$

2. Operator periodically correlated random fields

We continue in this section to present basic notions regarding infinite variate random fields and their correlation functions, which are used to formulate and prove the main result of the next section. Particularly, we state some results regarding operator periodically correlated random fields.

In the following we shall consider the second-order Hilbert space-valued random field

$$\mathbb{R}^n \ni (s_1, \dots, s_2) = s \mapsto X(s) \in \mathcal{H}, \tag{2.1}$$

which is locally square integrable on $\mathbb{R}^n$. It is well known that if the random field is continuous, it is more convenient to work instead of $\mathcal{H}$, with $\mathcal{H}_X$, the time module of the field, i.e., the closed normal $\mathcal{B}(\mathcal{E})$-submodule in $\mathcal{H}$ generated by $\{X(s),\ s \in \mathbb{R}^n\}$. In our case $\mathcal{H}_X$ will be generated by

$$\left\{ \int_{\mathbb{R}^n} \varphi(s) X(s) \mathrm{d}m_n(s),\ \varphi \in B_c(\mathbb{R}^n) \right\},$$

where $B_c(\mathbb{R}^n)$ is the set of all compactly supported complex measurable functions on $\mathbb{R}^n$.

The $\mathscr{C}^1(\mathcal{E})$-valued function

$$K_X(s,t) = [X(s), X(t)], \quad s, t, \in \mathbb{R}^n \tag{2.2}$$

is the *operator correlation function* of the field X. This function is, by hypothesis, (Bochner) locally integrable. A scalar correlation function can also be defined as $\mathrm{tr} K_X(s,t)$ (see [4, pp. 149]). We shall work here only with the operator correlation function.

The random field X is *strongly operator harmonizable*, if there is a $\mathscr{C}^1(\mathcal{E})$-valued measure $M = M_X$ on $\mathbb{R}^n \times \mathbb{R}^n$, called the *spectral bi-distribution*[1] of X, such that

$$K_X(s,t) = \int\limits_{\mathbb{R}^n \times \mathbb{R}^n} e^{\mathrm{i}\langle s,u \rangle} e^{-\mathrm{i}\langle t,v \rangle} \mathrm{d}M(u,v). \tag{2.3}$$

If in (2.3) M is only a $\mathscr{C}^1(\mathcal{E})$-valued bi-measure on $\mathbb{R}^n \times \mathbb{R}^n$ of bounded operator semivariation (see [4, pp. 155]), then X will be called *weakly operator harmonizable.* If the random field $\{X(s)\}_{s \in \mathbb{R}^n}$ is continuous and its correlation function $K_X(s,t)$ depends only on the difference $s - t$, it will be called *operator stationary.* For an operator stationary random field X, the one variable operator correlation function k_X can be defined as $k_X(s) = K_X(s,0)$. So $K_X(s,t) = k_X(s-t)$. It is well known

[1] In the context of random fields the term "distribution" or "bi-distribution" used together with the word "spectral", should not be understood in the sense of L. Schwartz. These can be measures or bi-measures respectively. Some authors use even the terms of "spectral measures" ("spectral bi-measures" respectively) associated to a random field. We prefer to avoid confusion with the term "spectral measures" from operator theory.

(see [4, chap. IV]) that for such an operator stationary random field X, there exists a $\mathscr{C}^1(\mathcal{E})$-valued non-negative measure F_X on $\mathbb{R}^n$ such that

$$k_X(s) = \int_{\mathbb{R}^n} e^{i\langle s,u\rangle} dF_X(u); \quad s \in \mathbb{R}^n, \tag{2.4}$$

therefore

$$K_X(s,t) = k_X(s-t) = \int_{\mathbb{R}^2} e^{i(\langle s,u\rangle - \langle t,u\rangle)} dF_X(u),$$

wherefrom it is clear that an operator stationary random field is strongly operator harmonizable, (2.3) being satisfied with $M_X(\sigma) := F_X(\sigma \cap \Delta)$, where $\sigma \in \mathscr{B}(\mathbb{R}^{2n})$ and $\Delta = \Delta_{\mathbb{R}^{2n}}$ is the "diagonal" from $\mathbb{R}^n \times \mathbb{R}^n$.

Definition 2.1. The infinite variate locally square integrable random field $\{X(s)\}_{s\in\mathbb{R}^n}$ is called operator periodically correlated (o.p.c.) with period $T = (T_1, \ldots, T_n) \in \mathbb{R}^n$, $T_1, \ldots, T_n > 0$, if for every $s \in \mathbb{R}^n$, the modified operator correlation function, $B_X(\cdot, s)$ defined by

$$\mathbb{R}^n \ni u \mapsto K_X(s+u, u) \in \mathscr{C}^1(\mathcal{E}) \tag{2.5}$$

is periodical (as a class of measurable functions) with period T, i.e.,

$$B_X(u+T, s) = B_X(u, s), \quad \mathrm{m}_n(u) \text{ a.e.}, \tag{2.6}$$

For $\mathcal{E} = \mathbb{C}$ and $n = 1$ such processes were studied by A. Makagon ([5, 6]), and before, under the stronger condition of continuity, by Gladyshev ([3]). These were also studied under the condition of strong operator harmonizability in [4, Section 5.5, pp. 230–234], for an arbitrary Hilbert space $\mathcal{E}$. The Fourier coefficients of the modified operator correlation B_X, of an o.p.c. random field defined by

$$A_l(s) := \frac{1}{T_1 \cdots T_n} \int_0^{T_1} \cdots \int_0^{T_n} e^{-2\pi i \langle l, \frac{u}{T}\rangle} B_X(u,s)\, d\mathrm{m}_n(u), \tag{2.7}$$

where $\frac{u}{T}$ means $\left(\frac{u_1}{T_1}, \cdots, \frac{u_n}{T_n}\right)$ and $l \in \mathbb{Z}^n$, $s \in \mathbb{R}^n$, play an important role in establishing a bijective correspondence between the o.p.c. random fields X and certain operator stationarily cross-correlated families indexed by $\mathbb{Z}^n$ of random fields. Let us now define such a family.

Definition 2.2. Let $\mathcal{K}$ be a normal $\mathcal{B}(\mathcal{E})$-module. An operator stationarily cross-correlated family (o.s.c.c.f.) of random fields on $\mathbb{R}^n$ (as $\mathcal{K}$-valued functions) is a family $\mathcal{X} = \{X^j(\cdot),\ j \in \mathbb{Z}^n\}$, for which each random field is continuous and for any $j, l \in \mathbb{Z}^n$ the cross-correlation function of $X^j(\cdot)$ and $X^l(\cdot)$

$$K_{X^j, X^l}(s,t) = [X^j(s), X^l(t)] \tag{2.8}$$

depends only on the difference $s - t$. It will be denoted by $k^{j,l}(s-t)$. The vector "matrix" with the entries $k^{j,l}_{\mathcal{X}}(\cdot)$, $j, l \in \mathbb{Z}^n$, defined on $\mathbb{R}^n$ will be called the cross-correlation function of $\mathcal{X}$.

Note that, in particular, each field $X^j(\cdot)$ of this family is operator stationary. Let $\mathcal{X} := \{X^j(\cdot), j \in \mathbb{Z}^n\}$ be an o.s.c.c.f. on $\mathbb{Z}^n$ of random fields. Then there exists the shift gramian unitary representation $\{U_s, s \in \mathbb{R}^n\}$ on $\mathbb{R}^n$ (the dual group of the locally compact abelian group $\mathbb{R}^n$) in the time module $\mathcal{K}_{\mathcal{X}}$ of $\mathcal{X}$ (i.e., the normal $\mathcal{B}(\mathcal{E})$-submodule of $\mathcal{K}$ generated by $\{X^j(s) : \ s \in \mathbb{R}^n, \ j \in \mathbb{Z}^n\}$), such that

$$X^j(s) = U_s X^j(0), \quad j \in \mathbb{Z}^n, \, s \in \mathbb{R}^n.$$

This representation is called *the shift group of the o.s.c.c.f.* $\mathcal{X}$. By a natural extension of [4, Proposition 4, p. 49], there exists a regular gramian spectral $\mathcal{B}(\mathcal{K}_{\mathcal{X}})$-valued measure E on $\mathbb{R}^n$ such that

$$U_s = \int_{\mathbb{R}^n} e^{\mathrm{i}\langle s,u\rangle} \mathrm{d}E(u), \quad s \in \mathbb{R}^n. \tag{2.9}$$

It is easy to see that

$$k_{\mathcal{X}}^{j,l}(s) = \int_{\mathbb{R}^n} e^{\mathrm{i}\langle s,u\rangle} \mathrm{d}F_{\mathcal{X}}^{j,l}(u) \quad s \in \mathbb{R}^n, \, j,l \in \mathbb{Z}^n, \tag{2.10}$$

where the $\mathscr{C}^1(\mathcal{E})$-valued measure from above is defined for $\sigma \in \mathscr{B}(\mathbb{R}^n)$ by $F_{\mathcal{X}}^{j,l}(\sigma) := [E(\sigma)X^j(0), X^l(0)]$. This last measure is called the *spectral distribution of the o.s.c.c.f.* $\mathcal{X}$. Let us observe that the correlation function $k_{\mathcal{X}}^{j,l}(s)$ of an o.s.c.c.f. $\mathcal{X}$, defines a $\mathscr{C}^1(\mathcal{E})$-valued positive definite function, i.e., for each $m \in \mathbb{N}$, each $s_p \in \mathbb{R}^n, l_p \in \mathbb{Z}^n, a_p \in \mathcal{B}(\mathcal{E}), \ p = 1,2,\dots,m$, we have

$$\sum_{p=1}^{m}\sum_{q=1}^{m} a_p k_{\mathcal{X}}^{l_p,l_q}(s_p - s_q) a_q^* \geq 0. \tag{2.11}$$

Conversely, as in the casc of a single field, if a $\mathscr{C}^1(\mathcal{E})$-valued function $k^{j,l}(s), \quad j,l \in \mathbb{Z}^n, \, s \in \mathbb{R}^n$, satisfies the positivity condition (2.11), then there exists a unique – up to a unitary equivalence – o.s.c.c.f. $\mathcal{X}$, such that

$$k^{j,l}(s) = k_{\mathcal{X}}^{j,l}(s), \quad j,l \in \mathbb{Z}^n, \ s \in \mathbb{R}^n.$$

Now we can formulate the existence of the correspondence mentioned above.

Theorem 2.3. *For each o.p.c. random field* $X(\cdot)$ *on* $\mathbb{R}^n$, *there is a uniquely determined (modulo a unitary equivalence) o.s.c.c.f.* $\mathcal{X}$ *indexed by* $\mathbb{Z}^n$ *and* $L^2\big[[0,T),\mathcal{H}_X\big]$*-valued, such that the family* $\{A_l(\cdot), \ l \in \mathbb{Z}^n\}$ *of the Fourier coefficients of the modified correlation function* B_X *satisfies the condition that* $e^{-2\pi\mathrm{i}\langle j,\frac{s}{T}\rangle} A_{j-l}(s)$ *is the cross-correlation of* $\mathcal{X}$ *(i.e., is just* $k_{\mathcal{X}}^{j,l}(s)$*).*

Let us remark that the members of the above associated o.s.c.c.f. $\mathcal{X}$, are $L^2\big[[0,T),\mathcal{H}_X\big]$-valued functions defined by

$$(X^j(s))(\cdot) \ := \ e^{-2\pi\mathrm{i}\langle j,\frac{s+\cdot}{T}\rangle} \, X(s+\cdot), \quad j \in \mathbb{Z}^n, \, s \in \mathbb{R}^n. \tag{2.12}$$

By reformulating the condition from Theorem 2.3 the "onto part" of the correspondence $X \mapsto \mathcal{X}$ from above will be stated in

Theorem 2.4. *If $\mathcal{X}$ is an o.s.c.c.f. indexed by $\mathbb{Z}^n$, such that its cross-correlation function satisfies the condition*

$$e^{2\pi i\langle j,s\rangle} k_{\mathcal{X}}^{j,l}(s) \quad \text{depends only on } j-l, \quad (j,l \in \mathbb{Z}^n), \tag{2.13}$$

then there exists a uniquely determined – up to a unitary equivalence – o.p.c. random field X having as Fourier coefficients for $B_X(\cdot, s)$ just the functions defined in (2.13) (i.e., X is the pre-image of $\mathcal{X}$ through the correspondence from Theorem 2.3).

Let us mention that X can be constructed with the aid of the shift group $\{U_s,\ s \in \mathbb{R}^n\}$ of $\mathcal{X}$ as $X(s) = U_s p(s)$, where $p(\cdot)$ is an $L^2\big[[0,T), \mathcal{K}_{\mathcal{X}}\big]$-valued function having the vectors $\frac{1}{T_1 \cdots T_n} \int_0^{T_1} \cdots \int_0^{T_n} X^j(s-T) \mathrm{d}m_n(s)$ as Fourier coefficients. The details of proof for Theorems 2.3 and 2.4 were given in [2].

3. Temperedly correlated processes

In this section we define the class of operator temperedly correlated random fields with their spectral operator bi-distribution and characterize the class of o.p.c. random field in terms of the associated operator spectral bi-distribution as a subclass of the operator temperedly correlated ones. The idea of such a characterization appears in the recent paper [6], where the case $n = 1$ and $\mathcal{E} = \mathbb{C}$ was considered. We shall see that such a result can be obtained also for our setting following essentially the steps from [6]. If $X(\cdot)$ is a locally gramian square integrable random field on $\mathbb{R}^n$, then, as observed, its operator correlation function $K_X(\cdot,\cdot)$ is locally Bochner integrable on $\mathbb{R}^{2n}$ and consequently it defines a $\mathscr{C}^1(\mathcal{E})$-valued distribution on $\mathbb{R}^{2n}$ given by

$$K_X(\varphi) := \iint_{\mathbb{R}^{2n}} K_X(s,t)\varphi(s,t)\ \mathrm{d}m_n(s)\ \mathrm{d}m_n(t),\ \varphi \in \mathcal{D}_{2n}. \tag{3.1}$$

Definition 3.1. A locally gramian square integrable random field is called operator temperedly correlated if its operator correlation function is an operator tempered distribution.

One first easy example of an operator temperedly correlated field is a strongly operator harmonizable one. Another easy example is given by the following proposition.

Proposition 3.2. *Each operator periodically correlated random field is an operator temperedly correlated one.*

Proof. Suppose $X(\cdot)$ is an operator periodically correlated random field. Denote $I_{j,l} := [jT, (j+(1,\ldots,1))T) \times [lT, (l+(1,\ldots,1))T)$ be the product of two "rectangles" in $\mathbb{R}^{2n}$. Put $w = (s,t)$. For $\varphi \in \mathcal{S}_{2n}$, let $C > 0$ be such that

$$|\varphi(w)| \le C(1+|w|^2)^{-n}, \quad \text{for all } w \in \mathbb{R}^{2n}. \tag{3.2}$$

Then

$$\|K_X(w)\varphi(w)\|_{\mathscr{C}^1(\mathcal{E})} \leq CM(w)(1+|w|^2)^{-n},$$

where

$$M(w) = \|X(s)\|_{\mathscr{C}^2(\mathcal{F},\mathcal{E})}\|X(t)\|_{\mathscr{C}^2(\mathcal{F},\mathcal{E})}.$$

From the square integrability of $X(\cdot)$, it results that the function $M(w)$ is square integrable over each rectangle $I_{j,l}$ and from operator periodical correlatedness we have $\iint\limits_{I_{j,l}} M^2(w)\mathrm{dm}_{2n}(w) = B^2 < \infty$, for all $j, l \in \mathbb{Z}^n$. Then it holds

$$\begin{aligned}
&\iint \|K_X(w)\varphi(w)\|_{\mathscr{C}^1(\mathcal{E})}\mathrm{dm}_{2n}(w) \\
&\quad \leq C\sum_j\sum_l \int\limits_{I_{j,l}} M(w)(1+|w|^2)^{-n}\mathrm{dm}_{2n}(w) \\
&\quad \leq C\sum_j\sum_l \Big(\int\limits_{I_{j,l}} M^2(w)\mathrm{dm}_{2n}(w)\Big)^{\frac{1}{2}}\Big(\int\limits_{I_{j,l}} (1+|w|^2)^{-2n}\mathrm{dm}_{2n}(w)\Big)^{\frac{1}{2}} \qquad (3.3)\\
&\quad = BC\sum_j\sum_l \Big(\int\limits_{I_{j,l}} (1+|w|^2)^{-2n}\mathrm{dm}_{2n}(w)\Big)^{\frac{1}{2}} < \infty
\end{aligned}$$

Therefore $K_X(w)\varphi(w)$ is Bochner integrable over $\mathbb{R}^{2n}$ and consequently, regarding K_X as distribution on $\mathbb{R}^{2n}$, $K_X(\varphi)$ is well defined for every $\varphi \in \mathcal{S}_{2n}$.

For the (sequential) continuity of $\varphi \mapsto K_X(\varphi)$, let φ_j be a sequence converging to zero in $\mathcal{S}_{2n}$. Then there are constants $C_j \to 0$, for $j \to \infty$, such that $|\varphi_j(w)| \leq C_j(1+|w|^2)^{-n}$, $j \in \mathbb{Z}_+$. From (3.3) it results $K_X(\varphi_j) \to 0$ as $j \to \infty$. □

Now, since $K_X \in \mathcal{S}'_n(\mathscr{C}^1(\mathcal{E}))$, its Fourier transform in the sense of distributions is a well defined operator tempered distribution from $\mathcal{S}'_n(\mathscr{C}^1(\mathcal{E}))$.

Definition 3.3. The spectral bi-distribution F_X of an operator temperedly correlated random field $X(\cdot)$ on $\mathbb{R}^n$, is the operator tempered distribution on $\mathbb{R}^{2n}$ defined by

$$F_X(\varphi) := K_X(\hat{\varphi}) = \iint\limits_{\mathbb{R}^{2n}} K_X(w)\hat{\varphi}(w)\ \mathrm{dm}_{2n}(w),\ \ \varphi \in \mathcal{S}_{2n}. \qquad (3.4)$$

Remark 3.4. Let F be the $\mathscr{C}^1(\mathcal{E})$-measure on $\mathbb{R}^{2n}$ associated to a strongly operator harmonizable random field $X(\cdot)$. Now, if we regard F as a $\mathscr{C}^1(\mathcal{E})$-valued distribution on $\mathbb{R}^{2n}$, by applying the inverse Fourier transform and (3.4) we have, for $\varphi \in \mathcal{D}_{2n}$

$$\begin{aligned} F(\varphi) &= \int\limits_{\mathbb{R}^{2n}} \varphi(u,v)\mathrm{d}F(u,v) \\ &= \int\limits_{\mathbb{R}^{2n}}\int\limits_{\mathbb{R}^{2n}} e^{\mathrm{i}\langle s,u\rangle} e^{-\mathrm{i}\langle t,v\rangle}\hat{\varphi}(s,t)\ \mathrm{dm}_{2n}(s,t)\mathrm{d}F(u,v) \\ &= \int\limits_{\mathbb{R}^{2n}} K_X(s,t)\hat{\varphi}(s,t)\ \mathrm{dm}_{2n}(s,t) = F_X(\varphi), \end{aligned}$$

consequently *the spectral bi-distribution for a strongly operator harmonizable field is the bi-measure (which is already a measure) assuring the integral representation of the correlation function of the field.*

Proposition 3.5. *Each spectral bi-distribution of an operator temperedly correlated random field $X(\cdot)$ with n continuous time parameters, is positive definite (p.d.), i.e., for each $m \in \mathbb{Z}_+$ and any $\varphi_1, \varphi_2, \ldots, \varphi_m \in \mathcal{D}_n$ and $a_1, a_2, \ldots, a_m \in \mathcal{B}(\mathcal{E})$*

$$\sum_{p,q=1}^{m} a_p F_X(\varphi_p \otimes \bar{\varphi}_q) a_q^* \geq 0,$$

where we used the notation $(\varphi \otimes \bar{\psi})(s,t) = \varphi(s)\bar{\psi}(t),\ \ s,t \in \mathbb{R}^n,\ \ \varphi, \psi \in \mathcal{D}_n$.

Proof. Indeed we have successively

$$\begin{aligned} \sum_{p,q=1}^{m} a_p F_X(\varphi_p \otimes \bar{\varphi}_q) a_q^* &= \sum_{p,q=1}^{m} a_p K_X(\hat{\varphi}_p \otimes \bar{\hat{\varphi}}_q) a_q^* \\ &= \sum_{p,q=1}^{m} a_p \int\limits_{\mathbb{R}^{2n}} \hat{\varphi}_p(s)\bar{\hat{\varphi}}_q(t) K_X(s,t)\mathrm{dm}_{2n}(s,t)\ a_q^* \\ &= \sum_{p,q=1}^{m} a_p \int\limits_{\mathbb{R}^{2n}} [\hat{\varphi}_p(s)X(s), \hat{\varphi}_q(t)X(t)]\mathrm{dm}_{2n}(s,t)\ a_q^* \\ &= \sum_{p,q=1}^{m} a_p \left[\int\limits_{\mathbb{R}^n} \hat{\varphi}_p(s)X(s)\mathrm{dm}_n(s), \int\limits_{\mathbb{R}^n} \hat{\varphi}_q(t)X(t)\mathrm{dm}_n(t) \right] a_q^* \\ &= \left[\sum_{p=1}^{m} a_p \int\limits_{\mathbb{R}^n} \hat{\varphi}_p(s)X(s)\mathrm{d\ m}_n(s), \sum_{q=1}^{m} a_q \int\limits_{\mathbb{R}^n} \hat{\varphi}_q(t)X(t)\mathrm{d\ m}_n(t) \right] \geq 0. \qquad \square \end{aligned}$$

It appears naturally the problem of the characterization of operator temperedly correlated random fields with the aid of the corresponding spectral bi-distribution. For the moment we can answer this question only for the subclass of the o.p.c. random fields. Let us first bring up a special class of $\mathscr{C}^1(\mathcal{E})$-valued tempered distributions. Denote, for $j \in \mathbb{Z}^n$ the linear manifolds

$$S_j := \left\{ \left(s, s + 2\pi \frac{j}{T} \right) \in \mathbb{R}^{2n}, \ s \in \mathbb{R}^n \right\},$$

which are "parallel" to the diagonal manifold $\{(s,s), \ s \in \mathbb{R}^n\}$ in $\mathbb{R}^n \times \mathbb{R}^n = \mathbb{R}^{2n}$. Let G_j be some $\mathscr{C}^1(\mathcal{E})$-valued measures on $\mathbb{R}^{2n}$ supported respectively on the manifolds S_j, $j \in \mathbb{Z}^n$, for which their total variations are uniformly bounded. Since, for each $\varphi \in \mathcal{D}_{2n}$, only a finite number of S_j intersects the support of φ, the sum $\sum_{j\in\mathbb{Z}^n} G_j(\varphi)$ makes sense. Then we can define in the sense of $\mathscr{C}^1(\mathcal{E})$-valued distributions the sum

$$\sum_{j\in\mathbb{Z}^n} G_j = G. \tag{3.5}$$

Proposition 3.6. *Let G_j, $j \in \mathbb{Z}^n$ be as above. Then the sum* (3.5) *defines G as a $\mathscr{C}^1(\mathcal{E})$-valued tempered distribution on $\mathbb{R}^{2n}$.*

Proof. Let $\varphi \in \mathcal{S}_{2n}$ and $C > 0$ such that

$$|\varphi(w)| \leq C(1 + |w|^2)^{-n}, \quad w \in \mathbb{R}^{2n}.$$

Then we have that

$$\|G_j(\varphi)\|_{\mathscr{C}^1(\mathcal{E})} = \left\| \int_{\mathbb{R}^{2n}} \varphi(w) \mathrm{d}G_j(w) \right\|_{\mathscr{C}^1(\mathcal{E})} \leq \sup_{w\in S_j} \frac{C}{(1+|w|^2)^n} \operatorname{var}(G_j)$$
$$\leq \frac{C}{(1 + C_1|j|^2)^n} \operatorname{var}(G_j),$$

which, leads to the fact that in (3.5) the sum $\sum_{j\in\mathbb{Z}^n} G_j(\varphi)$ makes sense. The same inequality gives the (sequential) continuity of G from (3.5). □

Let us observe that G from (3.5) is well defined for every bounded Borel function with compact support. So we can define $G(\sigma) := G(\chi_\sigma)$, where χ_σ is the indicator function of a bounded $\sigma \in \mathscr{B}(\mathbb{R}^{2n})$. It is not hard to see that the distribution G is p.d. iff for every natural number m and each $a_1, \ldots, a_m \in \mathcal{B}(\mathcal{E})$ and any bounded sets $\Delta_1, \ldots, \Delta_m$ in $\mathscr{B}(\mathbb{R}^{2n})$ we have

$$\sum_{p,q=1}^{m} a_p G(\Delta_p \times \Delta_q) a_q^* \geq 0. \tag{3.6}$$

Now we are in a position to state the characterization of an o.p.c. random field in terms of its spectral bi-distribution.

Theorem 3.7. *A tempered $\mathscr{C}^1(\mathcal{E})$-valued distribution F on $\mathbb{R}^{2n}$ is the spectral bi-distribution of an o.p.c. random field X with n continuous time parameters, iff F is p.d. and is of the form $F = \sum_{j\in\mathbb{Z}^n} F_j$, where F_j, $(j \in \mathbb{Z}^n)$ are $\mathscr{C}^1(\mathcal{E})$-valued measures* *on $\mathbb{R}^{2n}$ supported respectively on the manifolds S_j, $j \in \mathbb{Z}^n$, having uniformly bounded total variations. If this is the case, then F_j are given by*

$$F_j(\sigma) := K_j(s_j^{-1}(\sigma)), \qquad \sigma \in \mathscr{B}(\mathbb{R}^{2n}),\ j \in \mathbb{Z}^n, \tag{3.7}$$

where $s_j : \mathbb{R}^n \to \mathbb{R}^{2n}$ is given by $s_j(t) := \left(t, t + 2\pi \frac{j}{T}\right)$ and K_j is a measure defined by the relation

$$\int_{\mathbb{R}^n} e^{\mathrm{i}\langle t,u\rangle}\, \mathrm{d}K_j(u) = \frac{1}{T_1 \cdots T_2} \int_{[0,T)} e^{-2\pi\mathrm{i}\langle j, \frac{u}{T}\rangle} K_X(t+u,u)\, \mathrm{dm}_n(u), \quad j \in \mathbb{Z}^n.$$

Proof. Let $X(\cdot)$ be an operator periodically correlated random field on $\mathbb{R}^n$ and F_X its spectral bi-distribution, which by Proposition 3.2 is a tempered $\mathscr{C}^1(\mathcal{E})$-valued distribution. Since the $\mathscr{C}^1(\mathcal{E})$-valued functions

$$A_j(t) := \frac{1}{T_1 \cdots T_2} \int_{[0,T)} e^{-2\pi i\langle j, \frac{u}{T}\rangle} K_X(t+u,u)\mathrm{dm}_n(u), \quad t \in \mathbb{R}^n,$$

are just the Fourier coefficients of the modified operator correlation function B_X, from Theorem 2.3, there exists a stationarily cross-correlated family

$$\mathcal{X} = \{X^j(\cdot),\ j \in \mathbb{Z}^n\}$$

in the $\mathcal{B}(\mathcal{E})$-module $L^2\left[[0,T), \mathcal{H}_X\right]$ such that

$$[X^j(t), X^l(0)] = k^{j,l}(t) = e^{-2\pi\mathrm{i}\langle j, \frac{t}{T}\rangle} A_{j-l}(t), \quad j,l \in \mathbb{Z}^n,\ t \in \mathbb{R}^n. \tag{3.8}$$

In particular, $A_j(t) = k^{0,-j}(t)$. Applying now the formula (2.10), for each $j \in \mathbb{Z}^n$, it holds

$$A_j(t) = \int e^{\mathrm{i}\langle t,u\rangle} \mathrm{d}K_{-j}(u), \tag{3.9}$$

where $K_j(\cdot) = F_{\mathcal{X}}^{0,j}(\cdot)$.

Now, the measures K_j have uniformly bounded total variations, because of the fact that (3.8) implies $\left[X^l(0), X^l(0)\right] = A_0(0)$, $l \in \mathbb{Z}^n$. Indeed by this last inequality and by [4, Lemma 2 (3), pp. 18], for each $\sigma \in \mathscr{B}(\mathbb{R}^n)$ and $j \in \mathbb{Z}^n$, we have

$$\begin{aligned} K_j(\sigma)K_j(\sigma)^* &= \left[E(\sigma)X^0(0), X^j(0)\right]\left[X^j(0), E(\sigma)X^0(0)\right] \\ &\leq \operatorname{tr}\left[E(\sigma)X^0(0), X^0(0)\right] A_0(0), \end{aligned}$$

from where $\|K_j(\sigma)\|_{\mathscr{C}^1(\mathcal{E})} \leq \|K_0(\sigma)\|_{\mathscr{C}^1(\mathcal{E})}(\mathrm{tr}A_0(0))^{\frac{1}{2}}$, which leads to $|K_j|(\mathbb{R}^n) \leq |K_0|(\mathbb{R}^n)(\mathrm{tr}A_0(0))^{\frac{1}{2}}$, where we used the notation of the total variation of a $\mathscr{C}^1(\mathcal{E})$-valued measure from [4, pp. 54]. The fact that $|K_0|(\mathbb{R}^n)$ is finite can be obtained by the following reasoning. By [4, Theorem 5 (3), pp. 57], $K_0(\cdot) = E(\cdot)X^0(0)$ being gramian orthogonally scattered, is of operator bounded variation, which by the same Theorem leads to the fact that $\left[E(\cdot)X^0(0), X^0(0)\right] = K_0$ has finite total variation. Define now $F_j(\sigma) := K_j(s_j^{-1}(\sigma))$, $\sigma \in \mathscr{B}(\mathbb{R}^n)$, with s_j as above. Then by Proposition 3.6 it results that $F = \sum_{j \in \mathbb{Z}^n} F_j$ is an operator tempered distribution.

It remains to prove that $F = F_X$. Since $\{\varphi, \varphi \in \mathcal{S}_{2n} \text{ s. t. } \hat{\varphi} \in \mathcal{D}_{2n}\}$ is dense in $\mathcal{S}_{2n}$, it is enough to verify $F_X(\varphi) = F(\varphi)$ for such functions. Calculate first

$$F(\varphi) = \sum_{j\in\mathbb{Z}^n} \int_{\mathbb{R}^n}\int_{\mathbb{R}^n} \varphi(u,v)\,\mathrm{d}F_{-j}(u,v) = \sum_{j\in\mathbb{Z}^n}\int_{\mathbb{R}^n} \varphi\left(u, u - 2\pi\frac{j}{T}\right)\mathrm{d}K_{-j}(u)$$
$$= \sum_{j\in\mathbb{Z}^n} \int_{\mathbb{R}^n}\int_{\mathbb{R}^{2n}} e^{i\langle s-t,u\rangle} e^{2\pi i\langle j,\frac{s}{T}\rangle}\hat{\varphi}(s,t)\,\mathrm{d}m_{2n}(s,t)\,\mathrm{d}K_{-j}(u).$$

The calculus of F_X for the same φ runs as follows

$$F_X(\varphi) = \int_{\mathbb{R}^{2n}} K_X(w)\hat{\varphi}(w)\,\mathrm{d}m_{2n}(w)$$
$$= \int_{\mathbb{R}^n}\int_{\mathbb{R}^n} K_X(u+s,s)\hat{\varphi}(u+s,s)\,\mathrm{d}m_n(s)\mathrm{d}m_n(u)$$
$$= \int_{\mathbb{R}^n}\int_{\mathbb{R}^n}\left(\sum_{j\in\mathbb{Z}^n} e^{2\pi i\langle j,\frac{s}{T}\rangle}\int_{\mathbb{R}^n} e^{i\langle u,v\rangle}\,\mathrm{d}K_{-j}(v)\right)\hat{\varphi}(u+s,s)\,\mathrm{d}m_n(s)\,\mathrm{d}m_n(u)$$
$$= \sum_{j\in\mathbb{Z}^n}\int_{\mathbb{R}^n}\int_{\mathbb{R}^{2n}} e^{i\langle s-t,v\rangle}e^{2\pi i\langle j,\frac{s}{T}\rangle}\hat{\varphi}(t,s)\,\mathrm{d}m_{2n}(t,s)\,\mathrm{d}K_{-j}(v) = F(\varphi).$$

We have used the fact that the series $\sum_{j\in\mathbb{Z}^n} e^{2\pi \mathrm{i}\langle j,\frac{s}{T}\rangle}A_j(u)$ converges to $K_X(u+s,s)$ in $L^2[I,\mathcal{H}]$, for each compact $I \subset \mathbb{R}^n$ and $u \in \mathbb{R}^n$, and that $\hat{\varphi}$ has compact support.
For the "only if" part suppose that F is a positive definite tempered distribution of the form $F = \sum F_j$ as in the statement of the theorem. Denoting $K_j(\sigma) := F_j(s_j(\sigma))$, $j \in \mathbb{Z}^n$, $\sigma \in \mathscr{B}(\mathbb{R}^n)$, since for $j \in \mathbb{Z}^n$ and $\sigma, \tau \in \mathscr{B}(\mathbb{R}^n)$, $s_j(t) \in \sigma\times\tau$ iff $t \in \sigma \cap \left(\tau - \frac{2\pi j}{T}\right)$ holds, it follows $F_j(\sigma\times\tau) = K_j\left(\sigma\cap\left(\tau - 2\pi\frac{j}{T}\right)\right)$. Now, defining

$$K^{j,l}(\sigma) := K_{l-j}\left(\sigma + 2\pi\frac{j}{T}\right); \quad j,l\in\mathbb{Z}^n, \quad \sigma\in\mathscr{B}(\mathbb{R}^n), \tag{3.10}$$

it results

$$\begin{aligned} K^{j,l}(\sigma\cap\tau) &= K_{l-j}\left(\sigma\cap\left(\tau+2\pi\frac{j}{T}\right)\right)\\ &= K_{l-j}\left(\left(\sigma+2\pi\frac{j}{T}\right)\cap\left(\tau+2\pi\frac{j}{T}\right)\right)\\ &= K_{l-j}\left(\left(\sigma+2\pi\frac{j}{T}\right)\cap\left(\tau+2\pi\frac{j}{T}-2\pi\frac{(l-j)}{T}\right)\right)\\ &= F_{l-j}\left(\left(\sigma+2\pi\frac{j}{T}\right)\times\left(\tau+2\pi\frac{l}{T}\right)\right). \end{aligned} \tag{3.11}$$

By forming the set function $K = [K^{j,l}]_{j,l\in\mathbb{Z}^n}$, it is positive definite in the sense that, for each $m \in \mathbb{Z}_+$ and any $j_1,\dots,j_m \in \mathbb{Z}^n$, any operators $a_1,\dots,a_m \in \mathcal{B}(\mathcal{E})$

and any bounded sets $D_1, \dots, D_m \in \mathscr{B}(\mathbb{R}^n)$

$$\sum_{p,q=1}^{m} a_p K^{j_p,j_q}(D_p \cap D_q) a_q^* \geq 0. \tag{3.12}$$

We shall suppose that each D_p is included in exactly one of the intervals $\big[2\pi\frac{j}{T}, 2\pi\frac{(j+(1,\dots,1))}{T}\big)$ (if this is not the case, simply split D_p into smaller sets). In this way $D_p \times D_q$ intersects at most one manifold S_j. By using (3.11) we obtain

$$\sum_{p,q=1}^{m} a_p K^{j_p,j_q}(D_p \cap D_q) a_q^*$$
$$= \sum_{p,q=1}^{m} a_p F_{j_p - j_q}\left(\left(D_p + 2\pi\frac{j_p}{T}\right) \times \left(D_q + 2\pi\frac{j_q}{T}\right)\right) a_q^*. \tag{3.13}$$

But, by the hypothesis, F is positive definite and as

$$\left(D_p + 2\pi\frac{j_p}{T}\right) \times \left(D_q + 2\pi\frac{j_q}{T}\right)$$

intersects at most one manifold S_j, using (3.10), we have

$$\sum_{p,q=1}^{m} a_p K^{j_p,j_q}(D_p \cap D_q) a_q^*$$
$$= \sum_{p,q=1}^{m} a_p F\left(\left(D_p + 2\pi\frac{j_p}{T}\right) \times \left(D_q + 2\pi\frac{j_q}{T}\right)\right) a_q^* \geq 0. \tag{3.14}$$

Consequently $[K^{j,l}]$ is the spectral distribution of a stationarily cross-correlated family indexed by $\mathbb{Z}^n$, say $\mathcal{Y} = \{Y^j(\cdot),\ j \in \mathbb{Z}^n\}$. The cross-correlation function of $\mathcal{Y}$ is given by

$$k_{\mathcal{Y}}^{j,l}(t) = [Y^j(t), Y^l(0)] = \int_{\mathbb{R}^n} e^{\mathrm{i}\langle t,u\rangle}\, \mathrm{d}K^{j,l}(u) = \int_{\mathbb{R}^n} e^{\mathrm{i}\langle t,u\rangle}\, \mathrm{d}K_{l-j}\left(u + 2\pi\frac{j}{T}\right)$$
$$= e^{-2\pi\mathrm{i}\langle j,\frac{t}{T}\rangle} \int_{\mathbb{R}^n} e^{\mathrm{i}\langle t,v\rangle} \mathrm{d}K_{l-j}(v) = e^{-2\pi\mathrm{i}\langle j,\frac{t}{T}\rangle} A_{j-l}(t), \quad j,l \in \mathbb{Z}^n,\, t \in \mathbb{R}^n.$$

where the functions $A_l(\cdot)$ are defined by

$$A_l(t) = \int_{\mathbb{R}^n} e^{\mathrm{i}\langle t,u\rangle} \mathrm{d}K_{-l}(u), \quad l \in \mathbb{Z}^n\ ,\ t \in \mathbb{R}^n. \tag{3.15}$$

Now from Theorem 2.4 we infer that there is an o.p.c. random field $X(\cdot)$ on $\mathbb{R}^n$, such that its correlation function $K_X(s,t)$ satisfies

$$A_j(t) = \frac{1}{T_1 \cdots T_n} \int_{[0,T)} e^{-2\pi\mathrm{i}\langle j,\frac{u}{T}\rangle} K_X(t+u,u)\, \mathrm{d}m_n(u), \quad j \in \mathbb{Z}^n,\, t \in \mathbb{R}^n. \tag{3.16}$$

From the "if part", for this random field $X(\cdot)$ we have that its spectral bi-distribution F_X has the form

$$F_X = \sum_{j\in\mathbb{Z}^n} G_j,$$

where G_j are $\mathscr{C}^1(\mathcal{E})$-valued measures on $\mathbb{R}^{2n}$ of uniformly bounded total variations, which are supported on the manifolds S_j, respectively. They are determined by $G_j(\sigma) = \gamma_j(s_j^{-1}(\sigma))$, where γ_j are given by the relation $A_{-j}(t) = \int_{\mathbb{R}^n} e^{i\langle t,u\rangle} d\gamma_j(u)$, $j \in \mathbb{Z}^n$. From (3.15) we have that $\gamma_j = K_j$ and therefore $G_j = F_j$, $j \in \mathbb{Z}^n$. It results then $F = F_X$. □

Acknowledgment

The author is grateful to the referee for most valuable observations and suggestions on this paper.

References

[1] D. Gaşpar, *Analiză funcţională*. Editura Facla, Timişoara, 1984.

[2] P. Gaşpar, *On two Time Continuous Parameters Periodically Correlated Processes*, The National Conf. on Math. An. and Appl. 2001, (eds. M. Megan, N. Suciu), Univ. de Vest Timişoara, 107–117.

[3] E.G. Gladyshev, *Periodically and Almost Periodically Correlated Random Processes with Continuous Time Parameter*. Theory Probab. Appl. **8** (1963), 173–177.

[4] Yu. Kakihara, *Multidimensional Second Order Stochastic Processes*. World Scientific Publ. Comp., River Edge, NI, 1997.

[5] A. Makagon, *Induced Stationary Processes and Structure of Locally Square Integrable Periodically Correlated Processes*. Studia Math. **136** (1999), 71–86.

[6] A. Makagon, *Characterization of the Spectra of Periodically Correlated Processes*. J. Multivar. Anal. **78** (2001), 1–10.

Păstorel Gaşpar
Department of Mathematics
West University of Timişoara
Bul. V. Pârvan nr. 4
300223-Timişoara
Romania

Operator Theory:
Advances and Applications, Vol. 153, 157–167

Weighted Composition Operators on Hardy and Bergman Spaces

Romesh Kumar and Jonathan R. Partington

Abstract. This paper contains an analysis of weighted composition operators between Hardy and Bergman spaces of general simply-connected complex domains. Concepts studied include boundedness, compactness, boundedness below, isometry and invariant subspaces.

Mathematics Subject Classification (2000). Primary 47B33; Secondary 30H05, 47A15.

Keywords. Composition operator, Hardy space, Bergman space, Invariant subspace.

1. Introduction

Let H be a Hilbert space of analytic functions on a domain $\Omega \subseteq \mathbb{C}$ and $\phi : \Omega \to \Omega$ a holomorphic function. Then unweighted composition operators on H have the form $f \mapsto f \circ \phi$. The basic properties of such operators acting on the Hardy space $H^2 = H^2(\mathbb{D})$ and the Bergman space $L^2_a(\mathbb{D})$ of the unit disc are well understood (see, for example, [7, 20, 22]). On arbitrary domains, however, much less is known, and general composition operators induced by mappings between Hardy or Bergman spaces of different domains have not been fully analysed (for example, unlike in the case of the disc, such operators are not necessarily bounded). In fact it turns out to be no more difficult to study weighted composition operators between different spaces, since, by standard unitary equivalences, it is possible to reduce them to weighted composition operators $f \mapsto h.(f \circ \psi)$ on the Hardy and Bergman spaces of the disc.

In this paper we analyse such weighted composition operators, first on the Hardy space and then on the Bergman space. The main properties that we study are boundedness, compactness, boundedness below (closed range), isometry, and invariant subspaces. For the Hardy space, conditions for boundedness have recently been given in [6], but it seems that our results characterizing boundedness below, isometries and certain classes of invariant subspaces are new. For the Bergman space, we are able to use measure-theoretic arguments to characterize boundedness, compactness and boundedness below, apparently for the first time.

2. Hardy spaces and composition operators

2.1. Hardy spaces on general domains

Let Ω be a simply connected proper subdomain of $\mathbb{C}$, and let $\alpha : \mathbb{D} \to \Omega$ be a conformal bijection (Riemann mapping). Then the Hardy–Smirnoff space $E^2(\Omega)$ is the space of all functions $f : \Omega \to \mathbb{C}$ such that the function $Jf : z \mapsto f(\alpha(z))[\alpha'(z)]^{1/2}$ lies in the Hardy class $H^2 = H^2(\mathbb{D})$. It is clear that $E^2(\Omega)$ is a Hilbert space if we define the norm by $\|f\| = \|Jf\|_2$ for $f \in E^2(\Omega)$, so that the mapping $J : E^2(\Omega) \to H^2$, defined above, is an isometry. Note that

$$(J^{-1}g)(w) = g(\alpha^{-1}(w))[(\alpha^{-1})'(w)]^{1/2} = \frac{g(\alpha^{-1}(w))}{[\alpha'(\alpha^{-1}w)]^{1/2}}, \qquad (g \in H^2,\ w \in \Omega).$$

The space $E^2(\Omega)$ coincides with Rudin's space of all analytic functions f such that $|f|^2$ has a harmonic majorant, if and only if there exist constants a, $b > 0$ such that $a \le |\alpha'(z)| \le b$ for all $z \in \mathbb{D}$. We refer to the book of Duren [8] for further background on these spaces.

Let $\phi : \Omega_1 \to \Omega_2$ be an analytic mapping between two such domains and suppose that $v \in \mathrm{Hol}(\Omega_1)$, the space of analytic functions defined on Ω_1. Then a weighted composition mapping $C_{v,\phi} : \mathrm{Hol}(\Omega_2) \to \mathrm{Hol}(\Omega_1)$ can be defined by $(C_{v,\phi}g)(z) = v(z)g(\phi(z))$, for $z \in \Omega_1$. If $v \equiv 1$, then we denote the unweighted composition operator $C_{v,\psi}$ by C_ψ.

Let us write $\psi = \alpha_2^{-1}\phi\alpha_1$. If in fact $C_{v,\phi}$ restricts to a bounded operator from $E^2(\Omega_2)$ to $E^2(\Omega_1)$, then we have the following commutative diagrams showing that $C_{v,\phi}$ is unitarily equivalent to a bounded operator T on H^2. Here, J_1 and J_2 are the isometries induced by Riemann mappings $\alpha_j : \mathbb{D} \to \Omega_j$ for $j = 1, 2$.

$$\begin{array}{ccc} \Omega_1 & \overset{\phi}{\longrightarrow} & \Omega_2 \\ \alpha_1 \uparrow & & \uparrow \alpha_2 \\ \mathbb{D} & \overset{\psi}{\longrightarrow} & \mathbb{D} \end{array} \qquad \begin{array}{ccc} E^2(\Omega_1) & \overset{C_{v,\phi}}{\longleftarrow} & E^2(\Omega_2) \\ J_1 \downarrow & & \downarrow J_2 \\ H^2(\mathbb{D}) & \overset{T}{\longleftarrow} & H^2(\mathbb{D}) \end{array}.$$

We recall that an unweighted composition operator C_ψ is automatically bounded on $H^2(\mathbb{D})$ by Littlewood's subordination theorem (see, for example, [17, 16]).

The following simple result is given (in the unweighted case) in [21] for the case $\Omega_1 = \Omega_2$, and the example $\Omega_1 = \Omega_2 = \mathbb{C}_+$, the right-hand half-plane, is studied in [3], for example.

Proposition 2.1. *With the above notation, the operator* $T = J_1 C_{v,\phi} J_2^{-1}$ *is another weighted composition operator, given by*

$$(Tf)(z) = h(z)f(\psi(z)), \qquad \textit{where } h(z) = v(\alpha_1(z)) \left(\frac{\alpha_1'(z)}{\alpha_2'(\psi(z))} \right)^{1/2}.$$

Proof. For $f \in H^2(\mathbb{D})$, we have

$$\begin{aligned} (T_\psi f)(z) &= (J_1 C_\phi J_2^{-1} f)(z) \\ &= (C_\phi J_2^{-1} f)(\alpha_1(z))[\alpha_1'(z)]^{1/2} \\ &= v(\alpha_1(z))(J_2^{-1} f)(\phi\alpha_1(z))[\alpha_1'(z)]^{1/2} \\ &= v(\alpha_1(z)) f(\alpha_2^{-1}\phi\alpha_1(z))[\alpha_1'(z)]^{1/2}/[\alpha_2'(\alpha_2^{-1}\phi\alpha_1)(z)]^{1/2}, \end{aligned}$$

as required. □

Example 2.1. We give a few instructive examples corresponding to the unweighted case $v \equiv 1$.

1. Let $\Omega_1 = \Omega_2 = \mathbb{C}_+$, and define $\alpha : \mathbb{D} \to \mathbb{C}_+$ by $\alpha(z) = \frac{1-z}{1+z}$, so that $\alpha'(z) = -2/(1+z)^2$. We arrive at $Tf(z) = \frac{1+\psi(z)}{1+z}(C_\psi f)(z)$, where C_ψ is the unweighted composition operator on H^2.
2. Let $\Omega_1 = \mathbb{D}$ and $\Omega_2 = \mathbb{C}_+$, with $\alpha_1(z) = z$ and $\alpha_2(z) = \frac{1-z}{1+z}$ as before. Every composition operator from $E^2(\mathbb{C}_+)$ to $H^2(\mathbb{D})$ is unitarily equivalent to an operator such that $Tf(z) = \frac{1}{2i}(1+\psi(z))(C_\psi f)(z)$ on H^2, and is automatically bounded since C_ψ is bounded and $1 + \psi \in H^\infty$. (Warning: the notation $H^2(\mathbb{C}_+)$ is generally used for what we are here denoting $E^2(\mathbb{C}_+)$ and not for the Rudin space defined earlier.)
3. Reversing the roles of Ω_1 and Ω_2 in the previous example, we see that every composition operator from $H^2(\mathbb{D})$ to $E^2(\mathbb{C}_+)$ is equivalent to an operator satisfying $(Tf)(z) = \frac{2i}{1+z}(C_\psi f)(z)$. In fact, since the constant functions do not lie in $E^2(\mathbb{C}_+)$, there are no such operators, but it is possible instead to consider composition operators defined on proper subspaces of H^2. For example, if $\psi(z) = (1+z)\xi(z)$, and $\xi \in H^\infty$, then T defines a bounded operator on $H_0^2 = zH^2$, since $T(z.g) : z \mapsto 2i\xi(z)(C_\psi g)(z)$.
4. Another example is the left semi-disc $\mathbb{D}_L = \{z \in \mathbb{D} : \operatorname{Re} z < 0\}$. Multiplication operators on $E^2(\mathbb{D}_L)$ were studied in [4] in the context of invariant subspaces. A suitable map $\alpha : \mathbb{D} \to \mathbb{D}_L$ is the inverse mapping to $\beta : \mathbb{D}_L \to \mathbb{D}$ where $\beta(w) = (w^2 - 2w - 1)/(w^2 + 2w - 1)$ for $w \in \mathbb{D}_L$.

2.2. Isometric weighted composition operators on H^2

We are thus led to study the weighted composition operator $T_{h,\psi} : f \mapsto h \cdot (f \circ \psi)$, for holomorphic $\psi : \mathbb{D} \to \mathbb{D}$. Since ψ lies in $H^\infty(\mathbb{D})$ it has a canonical nontangential extension to the unit circle $\mathbb{T}$, and we shall denote this by ψ also. For $T_{h,\psi}$ to be bounded it is clearly necessary that h lie in $H^2(\mathbb{D})$ (since the image of the constant function 1 is h), and it is sufficient that h lie in $H^\infty(\mathbb{D})$, since $\|T_{h,\psi} f\|_2 \le \|h\|_\infty \|C_\psi f\|_2$.

An exact necessary and sufficient condition for boundedness was given in [6], namely that $T_{h,\psi}$ is bounded if and only if the measure $\mu_{h,\psi}$, given for measurable subsets $E \subset \overline{\mathbb{D}}$ by

$$\mu_{h,\psi}(E) = \int_{\psi^{-1}(E)\cap\mathbb{T}} |h|^2\, dm, \tag{2.1}$$

where dm is Lebesgue measure on $\mathbb{T}$, is a Carleson measure.

It is known from [17] that an unweighted composition operator C_ψ on $H^2(\mathbb{D})$ is isometric if and only if ψ is inner and $\psi(0) = 0$; for the half-plane, the necessary and sufficient condition was given in [3], namely that the unitarily equivalent operator T, where $Tf : z \mapsto \frac{1+\psi(z)}{1+z}(C_\psi f)(z)$, is an isometry if and only if ψ is inner and the function h given by $h : z \mapsto \frac{1+\psi(z)}{1+z}$ lies in H^2 and has unit norm.

Theorem 2.1. *The operator $T_{h,\psi}$ is an isometry on H^2 if and only if ψ is inner, $\|h\|_2 = 1$, and $\langle h, h.\psi^n\rangle = 0$ for $n \geq 1$.*

Proof. It is clear that $T_{h,\psi}$ is an isometry if and only if the sequence $(h.\psi^n)_{n\geq 0}$ is orthonormal in H^2. This implies directly that $\|h\|_2 = 1$. Now

$$\|h.\psi\|_2^2 = \int_0^{2\pi} |h(e^{it})|^2|\psi(e^{it})|^2\, dt = \int_0^{2\pi} |h(e^{it})|^2\, dt = \|h\|_2^2, \tag{2.2}$$

and since $|h(e^{it})| > 0$ almost everywhere, this implies that ψ is inner or we would have a strict inequality in (2.2). Now for ψ inner we have $\langle h.\psi^m, h.\psi^n\rangle = \langle h, h.\psi^{n-m}\rangle$ for $n \geq m$, and thus the given conditions are both necessary and sufficient. □

Lance and Stessin ([12]), in their work on multiplication-invariant subspaces of H^p, introduced the notion of a *ψ-2-inner function*. Namely, if ψ is inner and nonconstant, then h is said to be ψ-2-inner if $\|h\|_2 = 1$ and $\langle |h|^2, \psi^n\rangle = 0$ for $n = 1, 2, \ldots$. The same condition arises here.

Corollary 2.1. *The operator $T_{h,\psi}$ is an isometry if and only if ψ is inner and h is ψ-2-inner.*

Proof. This follows immediately from Theorem 2.1, since $\langle h, h.\psi^n\rangle = \langle |h|^2, \psi^n\rangle$ for all $n \geq 1$. □

2.3. Operators bounded below

An operator $T : X \to Y$ is said to be bounded below if there is a constant $\delta > 0$ such that $\|Tx\| \geq \delta\|x\|$ for every $x \in X$. If X and Y are Banach spaces and T is injective, this is the same as saying that T has closed range.

In [5], Cima, Thomson and Wogen gave a necessary and sufficient condition for an unweighted composition operator on H^2 to be bounded below. We now generalize their results to give a necessary and sufficient condition for a weighted composition operator on H^2 (and thus for weighted composition operators between more general Hardy spaces) to be bounded.

For a function $\psi : \mathbb{D} \to \mathbb{D}$ and a function $h \in H^2$, recall that we defined a Borel measure $\mu_{h,\psi}$ on subsets of $\overline{\mathbb{D}}$ by (2.1). In fact $\mu_{h,\psi}$, restricted to subsets of $\mathbb{T}$, is absolutely continuous with respect to Lebesgue measure: to see this, it is sufficient to consider the case $h \equiv 1$, in which case the arguments of [17] and [5] apply. We write $g_{h,\psi}$ for the Radon–Nikodym derivative of $\mu_{h,\psi}$ with respect to Lebesgue measure.

Theorem 2.2. *The weighted composition operator $T_{h,\psi}$ is bounded below if and only if $g_{h,\psi}$ is essentially bounded away from zero.*

Proof. Take $f \in H^2$ and suppose that $g_{h,\psi}$ is essentially bounded below. In particular $\psi(\mathbb{T})$ must include the whole of $\mathbb{T}$ (to within a set of measure 0). Then

$$\begin{aligned} \|T_{h,\psi} f\|^2 &= \int_{\mathbb{T}} |h(e^{i\theta})|^2 |f(\psi(e^{i\theta}))|^2 \, d\theta \\ &\geq \int_{\mathbb{T}} |f(e^{it})|^2 \, d\mu_{h,\psi}(t) \\ &= \int_{\mathbb{T}} |f(e^{it})|^2 g_{h,\psi}(t) \, dt \end{aligned}$$

and hence $T_{h,\psi}$ is bounded below. Note that the inequality above may be strict if $\psi(\mathbb{T})$ includes points of the open disc $\mathbb{D}$.

Conversely, if $g_{h,\psi}$ is not essentially bounded away from zero, then for any $\epsilon > 0$ we can find a set $E \subseteq \mathbb{T}$ such that

$$\int_{\psi^{-1}(E) \cap \mathbb{T}} |h(e^{i\theta})|^2 \, d\theta < \epsilon m(E),$$

where m is normalized Lebesgue measure. Let $f \in H^2$ be an outer function such that

$$|f(e^{it})| = \begin{cases} 1 & \text{if } e^{it} \in E, \\ \frac{1}{2} & \text{otherwise.} \end{cases}$$

Now f^n (i.e., $f \times \cdots \times f$) satisfies $\|f^n\|^2 \geq m(E)^{1/2}$ for all n, but $f(z)^n$ tends pointwise and monotonically to zero on $\overline{\mathbb{D}} \setminus E$; therefore we have

$$\limsup_{n \to \infty} \|T_{h,\psi} f^n\|^2 = \int_{\psi^{-1} E} |h(e^{i\theta})|^2 \, d\theta \leq \epsilon m(E),$$

and since ϵ was arbitrary, this means that $T_{h,\psi}$ is not bounded below. □

2.4. Invariant subspaces

The characterization of all (closed) invariant subspaces of a composition operator is a difficult problem, known to be equivalent to the Invariant Subspace Problem for Hilbert spaces (see [18]). It is therefore of interest to obtain information on the invariant subspaces for weighted composition operators $T_{h,\psi}$, and one case which can be studied is the case of common invariant subspaces with the shift operator S defined by $(Sf)(z) = zf(z)$ (itself a weighted composition operator of a particularly simple kind). By Beurling's theorem ([11]), all invariant subspaces of S have the form $M = \theta H^2$, where θ is an inner function. As in [3], we shall study the (already rather complicated) case when θ is a Blaschke product.

We recall from [1] the Denjoy–Wolff theorem, which states that for a holomorphic mapping $\psi : \mathbb{D} \to \mathbb{D}$ that is not an elliptic automorphism, one of the following cases must occur.

1. ψ fixes a point $p \in \mathbb{D}$, in which case the sequence of iterates $(\psi^{(n)})_{n\geq 1}$ satisfies $\lim_{n\to\infty} \psi^{(n)}(z) = p$ uniformly on compact subsets of $\mathbb{D}$.
2. ψ has no fixed point in $\mathbb{D}$ but there exists a point $p \in \mathbb{T}$ such that $\lim_{n\to\infty} \psi^{(n)}(z) = p$ uniformly on compact subsets of $\mathbb{D}$.

We call such a p the *Denjoy–Wolff point* for Φ.

The following result extends results in [3] on unweighted composition operators on $H^2(\mathbb{D})$ and $H^2(\mathbb{C}_+)$. Note that the case in which h has no zeroes occurs frequently, for example if we consider unweighted composition operators ($v = 1$) in Proposition 2.1.

Theorem 2.3. *Let B be a Blaschke product, factorizing as $B = bB_0$, where b and B_0 are Blaschke products such that b divides h and b is maximal (that is, $b = \mathrm{GCD}(B, h)$). Let $\{z_k\}$ be the zeroes of B_0, with multiplicities $\{m(z_k)\}$. Then BH^2 is an invariant subspace for $T_{h,\psi}$ if and only if for each k, one has that $\psi(z_k)$ is a zero of B of multiplicity at least $m(z_k)$.*

If h has no zeroes in $\mathbb{D}$, then $B = B_0$, and so BH^2 is an invariant subspace for $T_{h,\psi}$ if and only if for each k one has that $\psi(z_k) = z_\ell$ for some ℓ with $m(z_\ell) \geq m(z_k)$. Moreover, if $B = B_0$ and the Denjoy–Wolff point p of ψ lies in $\mathbb{D}$, then for each k there is an n such that $\psi^{(n)}(z_k) = p$, where $\psi^{(n)}$ is the nth iterate of ψ.

Proof. If $f = Bg$, then $T_{h,\psi} f = h(B \circ \psi)(g \circ \psi)$ and so it is clear that BH^2 is an invariant subspace for $T_{h,\psi}$ if and only if B divides $h(B \circ \psi)$. That is, if and only if $B_0 b$ divides $h(B \circ \psi)$. Because of the maximality of b, the necessary and sufficient condition is that B_0 divide $B \circ \psi$.

In the case $B = B_0$, the iterates $\psi^{(n)}(z_k)$ tend to p, and they are zeroes of B. By the principle of isolated zeroes they must eventually equal p. □

3. Bergman spaces and composition operators

3.1. Bergman spaces on general domains

Let Ω be a simply-connected domain, and let $L^2_a(\Omega)$ denote the Bergman space on Ω, consisting of all holomorphic functions $f : \Omega \to \mathbb{C}$ such that

$$\|f\| := \left(\iint_\Omega |f(z)|^2 \, dA(z) \right)^{1/2} < \infty,$$

where $dA = dx\, dy/\pi$ denotes normalized 2-dimensional Lebesgue measure on Ω. A standard reference for Bergman spaces and their properties is [10].

Let $\alpha : \mathbb{D} \to \Omega$ be a conformal bijection. As in the Hardy space case there is an isometry $J : L^2_a(\Omega) \to L^2_a(\mathbb{D})$, and in this case it is easily verified that it is given by $(Jf)(z) = f(\alpha(z))\alpha'(z)$.

Let $\phi : \Omega_1 \to \Omega_2$ be an analytic mapping between two such domains and suppose that $v \in \mathrm{Hol}(\Omega_1)$, the space of analytic functions defined on Ω_1. Then a weighted composition mapping $C_{v,\phi} : \mathrm{Hol}(\Omega_2) \to \mathrm{Hol}(\Omega_1)$ can be defined by $(C_{v,\phi}g)(z) = v(z)g(\phi(z))$, for $z \in \Omega_1$. If $v \equiv 1$, then we denote the unweighted composition operator $C_{v,\psi}$ by C_ψ.

Let us write $\psi = \alpha_2^{-1}\phi\alpha_1$. If in fact $C_{v,\phi}$ restricts to a bounded operator from $L^2_a(\Omega_2)$ to $L^2_a(\Omega_1)$, then we have the following commutative diagrams showing that $C_{v,\phi}$ is unitarily equivalent to a bounded operator T on $L^2_a(\mathbb{D})$. Here, J_1 and J_2 are the isometries induced by Riemann mappings $\alpha_j : \mathbb{D} \to \Omega_j$ for $j = 1, 2$.

$$\begin{array}{ccc} \Omega_1 & \overset{\phi}{\longrightarrow} & \Omega_2 \\ \alpha_1 \uparrow & & \uparrow \alpha_2 \\ \mathbb{D} & \overset{\psi}{\longrightarrow} & \mathbb{D} \end{array} \qquad\qquad \begin{array}{ccc} L^2_a(\Omega_1) & \overset{C_{v,\phi}}{\longleftarrow} & L^2_a(\Omega_2) \\ J_1 \downarrow & & \downarrow J_2 \\ L^2_a(\mathbb{D}) & \overset{T}{\longleftarrow} & L^2_a(\mathbb{D}) \end{array} .$$

The following analogue of Proposition 2.1 is proved in a similar way, and we omit the details.

Proposition 3.1. *With the above notation, the operator $T = J_1 C_{v,\phi} J_2^{-1}$ is another weighted composition operator, given by*

$$(Tf)(z) = h(z)f(\psi(z)), \qquad \text{where} \quad h(z) = v(\alpha_1(z))\left(\frac{\alpha_1'(z)}{\alpha_2'(\psi(z))}\right).$$

Hence, as in the case of the Hardy spaces, every weighted composition operator on a Bergman space $L^2_a(\Omega)$, with Ω simply connected, is unitarily equivalent to a weighted composition operator on $L^2_a(\mathbb{D})$. Thus we are led to study these in more detail.

3.2. Weighted composition operators on $L^2_a(\mathbb{D})$

In this section we give necessary and sufficient conditions for weighted composition operators $T_{h,\psi} : f \mapsto h.(f \circ \psi)$ to be bounded on $L^2_a(\mathbb{D})$, then conditions for boundedness below. Here $\psi : \mathbb{D} \to \mathbb{D}$ is holomorphic and $h \in L^2_a(\mathbb{D})$, an obvious necessary condition for boundedness since $T_{h,\psi}(1) = h$, where 1 denotes the constant function.

A key role is now played by the following measure on $\mathbb{D}$, which is analogous to the one defined in Section 2.3 in the case of the Hardy space. For Borel subsets $E \subseteq \mathbb{D}$ define $\nu_{h,\psi}(E)$ by

$$\nu_{h,\psi}(E) = \int_{\psi^{-1}(E)} |h(z)|^2 \, dA(z).$$

Again we see that $\nu_{h,\psi}$ is absolutely continuous with respect to normalized two-dimensional Lebesgue measure $dx\,dy/\pi$, a fact which this time can be shown by elementary complex analytic arguments, for example by considering the local behavior of ψ. Let $w = w_{h,\psi}$ denote the Radon–Nikodym derivative of $\nu_{h,\psi}$ with respect to normalized Lebesgue measure, and note that $w \geq 0$.

The following result gives an alternative expression for the norm of a function $T_{h,\psi} f$.

Lemma 3.1. *Let $f \in L^2(\mathbb{D}, dA)$. Then*

$$\int_{\mathbb{D}} |h(z)|^2 |f(\psi(z))|^2 \, dA(z) = \int_{\mathbb{D}} |f(z)|^2 \, d\nu_{h,\psi}(z) = \int_{\mathbb{D}} |f(z)|^2 w(z) \, dA(z).$$

Proof. Suppose first that f is a non-negative simple function

$$f(z) = \sum_{j=1}^{n} a_k \chi_{E_j}(z).$$

Then we have

$$\begin{aligned}\int_{\mathbb{D}} |h(z)|^2 |f(\psi(z))|^2 \, dA(z) &= \int_{\mathbb{D}} |h(z)|^2 \Big(\sum_{j=1}^{n} a_j \chi_{\psi^{-1}(E_j)}(z) \Big) \, dA(z) \\ &= \sum_{j=1}^{n} a_j \int_{\psi^{-1}(E_j)} |h(z)|^2 \, dA(z) = \int_{\mathbb{D}} |f(z)|^2 \, d\nu_{h,\psi}(z).\end{aligned}$$

The general result follows by standard monotone convergence arguments for positive functions, and then by linearity for all functions in $L^2(\mathbb{D}, dA)$ (cf. a Hardy space analogue in [6]). The expression in terms of w is now obvious. □

We are now ready to characterize boundedness and boundedness below. In the following result $A(E)$ denotes the (normalized) 2-dimensional Lebesgue measure (area) of a set E.

Theorem 3.1. *The weighted composition operator $T_{h,\psi}$ is bounded on $L^2_a(\mathbb{D})$ if and only if there is a constant $C_1 > 0$ such that the measure $\nu_{h,\psi}$ satisfies*

$$\nu_{h,\psi}(D \cap \mathbb{D}) \leq C_1 A(D \cap \mathbb{D})$$

for all discs D with centers on $\mathbb{T}$.

The operator $T_{h\psi}$ is compact if and only if

$$\lim_{r \to 1} \sup_{D \in \mathcal{D}_r} \frac{\nu_{h,\psi}(D \cap \mathbb{D})}{A(D \cap \mathbb{D})} = 0, \tag{3.1}$$

where $\mathcal{D}_r$ denotes the set of discs with centers on $\mathbb{T}$ and radii at most r.

It is bounded below if and only if there are constants r, $C_2 > 0$ such that the set $E_r = \{z \in \mathbb{D} : w(z) \geq r\}$ satisfies

$$A(D \cap E_r) \geq C_2 A(D \cap \mathbb{D})$$

for all discs D with centers on $\mathbb{T}$.

Proof. We see from Lemma 3.1 that $T_{h,\psi}$ is bounded if and only the natural injection $J : L^2_a(\mathbb{D}, dA) \to L^2(\mathbb{D}, d\nu_{h,\psi})$ is bounded. This is the condition that $\nu_{h,\psi}$ be a Hastings–Carleson measure (cf. [9, 13]), and the first result follows.

Similarly $T_{h,\psi}$ is compact if and only if the injection J is compact. This is the vanishing condition (3.1), as given in [15]. It can be easily proved by testing when the condition that (f_n) tends weakly (and hence locally uniformly) to 0 implies that $\|Jf_n\| \to 0$.

Finally, $T_{h,\psi}$ is bounded below if and only if there is a constant $\epsilon > 0$ such that

$$\int_{\mathbb{D}} |f(z)|^2 w(z)\, dA(z) \geq \epsilon \int_{\mathbb{D}} |f(z)|^2\, dA(z), \qquad (f \in L^2_a(\mathbb{D})),$$

and now Corollary 1 of Luecking [13] gives the result. □

4. Summary and conclusions

The following table summarizes conditions for composition operators to have the various properties studied in this paper. Older results are given as citations, new ones as references in this paper. In each case there are two entries: Hardy space first, then Bergman.

	Unweighted	Weighted
Bounded	[22] – [22]	[6] – Theorem 3.1
Bounded Below	[5] – [23]	Theorem 2.2 – Theorem 3.1
Compact	[19] – [14]	[6] – Theorem 3.1

Several open questions remain. For example, it would be of interest to classify those subspaces of $H^2(\mathbb{D})$ which are simultaneously invariant under multiplication by an arbitrary inner function and under a weighted composition operator. The recent work of Lance and Stessin ([12]) classifies the multiplication-invariant spaces, but it turns out that they are no longer singly generated. Similar questions apply to shift-invariant subspaces of $L^2_a(\mathbb{D})$, for which the recent paper ([2]) is a good reference.

Finally, we do not know which weighted composition operators act as isometries on the Bergman space. It is easy to see using Schwarz's lemma (cf. [22]) that in the unweighted case the only isometries have the form C_ϕ where $\phi(z) = \alpha z$ and $|\alpha| = 1$. In the weighted case there are others: for example the isometries of the Bergman space $L^2_a(\mathbb{C}_+)$ induced by the mappings $\phi_a(z) = z + ia$, with $a \in \mathbb{R}$, are equivalent to weighted composition operators on $L^2_a(\mathbb{D})$, as in Proposition 3.1.

Acknowledgment

This work was undertaken when the first author was visiting Leeds as a Commonwealth Fellow, and the Commonwealth Scholarships Commission is thanked for its financial support. The authors are also grateful to the referee for some helpful comments.

References

[1] P.S. Bourdon, J.H. Shapiro, *Cyclic Phenomena for Composition Operators.* Memoirs of the A.M.S. **596** (1997).

[2] B.J. Carswell, P.L. Duren, M.I. Stessin, *Multiplication Invariant Subspaces of the Bergman Space.* Indiana Univ. Math. J. **51** (2002), no. 4, 931–961.

[3] I. Chalendar, J.R. Partington, *On the Structure of Invariant Subspaces for Isometric Composition Operators on $H^2(\mathbb{D})$ and $H^2(\mathbb{C}_+)$.* Arch. Math. (Basel) **81** (2003), 2, 193–207.

[4] I. Chalendar, J.R. Partington, *Spectral density for multiplication operators with applications to factorization of L^1 functions.* J. Operator Theory **50** (2003), 2, 411–422.

[5] J.A. Cima, J. Thomson, W. Wogen, *On Some Properties of Composition Operators.* Indiana Univ. Math. J. **24** (1974/75), 215–220.

[6] M.D. Contreras, A.G. Hernández-Díaz *Weighted Composition Operators on Hardy Spaces.* J. Math. Anal. Appl. **263** (2001), 224–233.

[7] C.C. Cowen, B.D. MacCluer, *Composition Operators on Spaces of Analytic Functions.* Studies in Advanced Mathematics, CRC Press, Boca Raton, FL, 1995.

[8] P.L. Duren. *Theory of H^p Spaces.* Pure and Applied Mathematics **38** Academic Press, New York, 1970.

[9] W.W. Hastings, *A Carleson Measure Theorem for Bergman Spaces.* Proc. Amer. Math. Soc. **52** (1975), 237–241.

[10] H. Hedenmalm, B. Korenblum, K. Zhu, *Theory of Bergman Spaces.* Springer-Verlag, New York, 2000.

[11] K. Hoffman, *Banach Spaces of Analytic Functions.* Prentice-Hall, Englewood Cliffs, 1965.

[12] T.L. Lance, M.I. Stessin, *Multiplication Invariant Subspaces of Hardy Spaces.* Canad. J. Math. **49** (1997), 100–118.

[13] D.H. Luecking, *Inequalities on Bergman Spaces.* Illinois J. Math. **25** (1981), 1–11.

[14] B.D. MacCluer, J.H. Shapiro, *Angular Derivatives and Compact Composition Operators on the Hardy and Bergman Spaces.* Canad. J. Math. **38** (1986), 878–906.

[15] G. McDonald, C. Sundberg, *Toeplitz Operators on the Disc.* Indiana Univ. Math. J. **28** (1979), 595–611.

[16] N.K. Nikolski, *Operators, Functions, and Systems: An Easy Reading. Vol. 1. Hardy, Hankel, and Toeplitz.* Translated from the French by Andreas Hartmann. Mathematical Surveys and Monographs **92**, American Mathematical Society, Providence, RI, 2002.

[17] E.A. Nordgren, *Composition Operators.* Canad. J. Math. **20** (1968), 442–449.

[18] E. Nordgren, P. Rosenthal, F.S. Wintrobe, *Invertible Composition Operators on H^p.* J. Funct. Anal. **73** (1987), 324–344.

[19] J.H. Shapiro, *The Essential Norm of a Composition Operator.* Ann. Math. **125** (1987), 375–404.

[20] J.H. Shapiro, *Composition Operators and Classical Function theory.* Springer-Verlag, 1993.

[21] J.H. Shapiro, W. Smith, *Hardy Spaces that Support no Compact Composition Operators.* J. Funct. Anal. **205** (2003), 1, 62–89.

[22] K.H. Zhu, *Operator Theory in Function Spaces.* Marcel Dekker, 1990.

[23] N. Zorboska, *Composition Operators with Closed Range.* Trans. Amer. Math. Soc. **344** (1994), 791–801.

Romesh Kumar
Department of Mathematics
University of Jammu
Jammu – 180 004
India
e-mail: romesh_jammu@yahoo.com

Jonathan R. Partington
School of Mathematics
University of Leeds
Leeds LS2 9JT
UK
e-mail: J.R.Partington@leeds.ac.uk

Operator Theory:
Advances and Applications, Vol. 153, 169–184

Integral Transforms Controlled by Maximal Functions

Mircea Martin and Pawel Szeptycki †

Abstract. The paper characterizes the kernel functions on $\mathbb{R}^n$ with the property that the associated convolution operators are controlled by certain maximal operators.

Mathematics Subject Classification (2000). Primary: 44A05, 42B25; Secondary: 43A15.

Keywords. Integral transform, Maximal operator, Hedberg inequality.

1. Introduction

The theme of this article originates at the borderline between the celebrated Hardy-Littlewood-Sobolev fractional integral theorem and the Hardy-Littlewood-Wiener maximal theorem, two results that guided for quite a while the development of classical harmonic analysis on Euclidean spaces. The former theorem is concerned with the Riesz potential operator of order α on $\mathbb{R}^n$ given by

$$I_\alpha u(x) = \int_{\mathbb{R}^n} k_\alpha(y)u(x-y)dy, \qquad x \in \mathbb{R}^n, \tag{1.1}$$

where $0 < \alpha < n$, and

$$k_\alpha(x) = |x|^{\alpha-n}, \qquad x \in \mathbb{R}^n_0 = \mathbb{R}^n \setminus \{0\}, \tag{1.2}$$

and the latter deals with the maximal operator

$$Mu(x) = \sup_{t>0} \frac{1}{\mathrm{vol}(t\mathbb{B}^n)} \int_{t\mathbb{B}^n} |u(x-y)|dy, \qquad x \in \mathbb{R}^n, \tag{1.3}$$

where $\mathbb{B}^n$ stands for the closed unit ball in $\mathbb{R}^n$.

Both operators are well defined on the standard Lebesgue spaces $L^p(\mathbb{R}^n, \mathbb{R})$ of p-integrable real-valued functions on $\mathbb{R}^n$, for some specific values of p. According to the fractional integral theorem, I_α yields a bounded linear operator from $L^p(\mathbb{R}^n, \mathbb{R})$ into $L^q(\mathbb{R}^n, \mathbb{R})$ whenever $1 < p < q < \infty$ and $q^{-1} = p^{-1} - \alpha n^{-1}$. By the maximal theorem, M acts as a continuous operator from $L^p(\mathbb{R}^n, \mathbb{R})$ into $L^p(\mathbb{R}^n, \mathbb{R})$ for every $1 < p \leq \infty$. The interested reader can find a full historical account and complete proofs in the monographs by Stein ([25, 26]) and Hörmander ([14]). Several related issues with a particular emphasis on sharp norm estimates, applications, and some geometric ideas are discussed in the article by Lieb ([17]) and the survey article by Beckner ([4]). As a highlight of [4], we should mention the excellent list of references therein that indicates many important contributions and the extensive work done in this area over the years.

The two classical theorems stated above are linked by an inequality due to Hedberg ([13]), which points out that

$$|I_\alpha u(x)| \leq A_{p,q}(\alpha)[Mu(x)]^{p/q}\|u\|_p^{1-p/q}, \qquad x \in \mathbb{R}^n, \tag{1.4}$$

for every $u \in L^p(\mathbb{R}^n, \mathbb{R})$, provided $1 \leq p < q < \infty$ and $q^{-1} = p^{-1} - \alpha n^{-1}$, where the constant $A_{p,q}(\alpha)$ is independent of u.

The proofs given in [13] and [26] do not provide the optimal value of $A_{p,q}(\alpha)$. This issue is addressed in Martin and Szeptycki ([21]), where Hedberg's inequality has been improved and generalized. The improvement amounts to determining the best value of the constant $A_{p,q}(\alpha)$, and the generalization is achieved by replacing the kernel functions k_α with a broader class of homogeneous scalar-, vector-, or operator-valued kernels, and by producing the sharp forms of the inequalities derived for those kernels.

The present article is a sequel to [21]. Our primary goal is to completely characterize the kernel functions on $\mathbb{R}^n$ with the property that the associated convolution operators are controlled by certain maximal operators, in a way similar to Hedberg's inequality (1.4).

To begin with, suppose that $k : \mathbb{R}^n \to [-\infty, \infty]$ is a Lebesgue measurable, locally integrable function, finite on $\mathbb{R}^n_0 = \mathbb{R}^n \setminus \{0\}$. In all the interesting situations, the kernel k will be continuous and different from zero away from the origin, where k has a singularity.

The convolution operator $\mathfrak{I} = \mathfrak{I}_k$ associated with k is defined as

$$\mathfrak{I}u(x) = \int_{\mathbb{R}^n} k(y)u(x-y)dy, \qquad x \in \mathbb{R}^n. \tag{1.5}$$

We assume that $u : \mathbb{R}^n \to \mathbb{R}$ is a Lebesgue measurable function such that the integral above, sometimes defined and computed as a principal value, exists.

The behavior of the integral transform $\mathfrak{I}u$ depends on the "size" of k, which is described by the distribution function, $\omega = \omega_k$, of k. To define it we set

$$\Omega_k[t] = \{y \in \mathbb{R}^n_0 : |k(y)| \geq t\} \cup \{0\}, \qquad 0 < t < \infty, \tag{1.6}$$

and let $\omega : (0, \infty) \to [0, \infty]$ be the decreasing function given by

$$\omega(t) = \operatorname{vol} \Omega_k[t], \qquad 0 < t < \infty. \tag{1.7}$$

For the sake of convenience, we will assume that each set $\Omega_k[t]$ has a finite nonzero measure. The sets $\Omega_k[t]$, $0 < t < \infty$, are next used to associate a maximal operator

$\mathfrak{M} = \mathfrak{M}_k$ to k by setting

$$\mathfrak{M}u(x) = \sup_{t>0} \frac{1}{\operatorname{vol}\Omega_k[t]} \int_{\Omega_k[t]} |u(x-y)|dy, \qquad x \in \mathbb{R}^n. \tag{1.8}$$

In the case when $k = k_\alpha$, where k_α is the kernel defined by (1.2), the corresponding sets $\Omega_\alpha[t]$ are Euclidean balls, namely,

$$\Omega_\alpha[t] = \mathbb{B}^n[t^{-1/(n-\alpha)}], \qquad 0 < t < \infty, \tag{1.9}$$

where $\mathbb{B}^n[\rho]$ denotes the closed ball in $\mathbb{R}^n$ centered at the origin with radius ρ. Therefore, the maximal operator $\mathfrak{M}$ equals the classical maximal operator M defined by (1.3), and obviously $\mathfrak{I}$ equals I_α. The distribution function ω_α corresponding to k_α is given by

$$\omega_\alpha(t) = \lambda t^{-\kappa}, \qquad 0 < t < \infty \tag{1.10}$$

where $\lambda = \operatorname{vol}\mathbb{B}^n$, with $\mathbb{B}^n = \mathbb{B}^n[1]$, and $\kappa = n/(n-\alpha)$.

We are now in a position to formulate the first main result of our article.

Theorem A. *Suppose $1 < \kappa < \infty$. The following two statements are equivalent.*

(i) *If $1 \le p < q < \infty$ and $q^{-1} = p^{-1} - 1 + \kappa^{-1}$, then there exists a positive constant $A_{p,q}(k)$ such that*

$$|\mathfrak{I}_k u(x)| \le A_{p,q}(k)[\mathfrak{M}_k u(x)]^{p/q} \|u\|_p^{1-p/q}, \qquad x \in \mathbb{R}^n \tag{1.11}$$

for every $u \in L^p(\mathbb{R}^n, \mathbb{R})$.

(ii) *There exists a positive constant λ such that the distribution function $\omega = \omega_k$ of k satisfies the estimate*

$$\omega(t) \le \lambda t^{-\kappa}, \qquad 0 < t < \infty. \tag{1.12}$$

Whenever (i) *or* (ii) *is true, and λ is the smallest constant in* (1.12), *that is,*

$$\lambda = \sup_{t>0} t^\kappa \omega(t), \tag{1.13}$$

a possible value of $A_{p,q}(k)$ in (1.12) *is given by*

$$A_{p,q}(k) = \frac{q}{q-p} \left(\frac{pq - q + p}{p} \cdot \lambda \right)^{1-p^{-1}+q^{-1}}. \tag{1.14}$$

Moreover, if (1.12) *is an equality for all $t \in (0, \infty)$, the value of $A_{p,q}(k)$ in* (1.14) *is the best constant in inequality* (1.11).

The preceding comments clearly point out that Hedberg's inequality is a consequence of Theorem A, and the best constant in (1.4) is given by (1.14) with $\lambda = \operatorname{vol}\mathbb{B}^n$.

Actually, Theorem A works for more general kernels $k : \mathbb{R}^n \to [-\infty, \infty]$ satisfying a homogeneity condition of the form

$$k(tx) = t^{-n/\kappa} k(x), \qquad t \in (0, \infty),\ x \in \mathbb{R}_0^n,\ 1 < \kappa < \infty, \tag{1.15}$$

because the sets $\Omega_k[t]$, $0 < t < \infty$, for such a kernel are related by the following equations,

$$\Omega_k[t] = t^{-\kappa/n}\Omega_k[1], \qquad 0 < t < \infty, \tag{1.16}$$

whence, the distribution function $\omega = \omega_k$ is given by

$$\omega(t) = \lambda t^{-\kappa}, \qquad 0 < t < \infty, \tag{1.17}$$

where $\lambda = \operatorname{vol}\Omega_k[1]$. Theorem A leads, for homogeneous kernels with property (1.15), to the main result in [21].

Theorem A can be extended by replacing $\mathbb{R}^n$ with a locally compact group equipped with a Haar measure. Such an extension will easily follow from Theorem B stated and proved in Section 2. In contrast to Theorem A above, Theorem B uses more general integral operators defined on measure spaces with or without a topological or group structure, and an appropriate substitute for the maximal operators. More details are presented in Section 2 below.

In Section 3 we outline several consequences and applications of Theorems A and B in approximation theory. These will include a unified approach to some results due to Ahlfors and Beurling ([1]), Alexander ([2, 3]), Gustafsson and Khavinson ([11]), Gürlebeck and Sprössig ([12]), Khavinson ([15, 16]), Martin ([18, 19, 20]), Martin and Szeptycki ([21]), Putinar ([23]), Tarkhanov ([27, Chapter 6]), and Weinstock ([28]).

2. Integral operators dominated by maximal operators

This section deals with a version of Theorem A in the setting of measure spaces.

2.1. Throughout this section, we let $K : X \times Y \to [-\infty, \infty]$ be a measurable function on the Cartesian product of two measure spaces X and Y equipped with positive measures, and assume that the set

$$\Sigma = \{(x, y) \in X \times Y : |K(x, y)| = \infty\}$$

has measure zero with respect to the product measure on $X \times Y$. In all the specific situations we will be interested in, the spaces X and Y are metric spaces furnished with positive Borel measures, Σ is a closed set, and K will be a continuous function on $X \times Y \setminus \Sigma$.

The spaces of p-integrable real-valued functions on X or Y, with $1 \le p \le \infty$, are denoted subsequently by $L^p(X, \mathbb{R})$ and $L^p(Y, \mathbb{R})$, respectively, and their standard norms will be denoted by $\|\cdot\|_{X,p}$ and $\|\cdot\|_{Y,p}$, or just $\|\cdot\|_p$.

We begin by associating to K the integral operator $\mathfrak{I} = \mathfrak{I}_K$ given by

$$\mathfrak{I}u(x) = \int_Y K(x, y)u(y)dy, \qquad x \in X, \tag{2.1}$$

and defined for measurable functions $u : Y \to \mathbb{R}$ such that the integral above exists for almost all $x \in X$.

Next, for every x in X and $0 < t < \infty$ we introduce the set $\Omega[x,t] = \Omega_K[x,t]$ defined as

$$\Omega[x,t] = \{y \in Y : |K(x,y)| \geq t\} \tag{2.2}$$

and let $\omega = \omega_K : X \times (0,\infty) \to [0,\infty]$ be the function given by

$$\omega(x,t) = \text{measure}\,\Omega[x,t]. \tag{2.3}$$

In other words, $\omega(x,\cdot) = \omega_K(x,\cdot) : (0,\infty) \to [0,\infty]$ is the distribution function corresponding to the measurable function $K(x,\cdot) : Y \to \mathbb{R}$. We will always assume that $\omega_K(x,\cdot)$ has finite values on $(0,\infty)$.

Finally, to every measurable function $u : Y \to \mathbb{R}$ we associate its maximal function $\mathfrak{M}u = \mathfrak{M}_K u$ on X by setting

$$\mathfrak{M}u(x) = \sup_{t>0} \frac{1}{\omega(x,t)} \int_{\Omega[x,t]} |u(y)|dy, \qquad x \in X. \tag{2.4}$$

It goes without saying that whenever $\omega(x,t) = 0$ we set

$$\frac{1}{\omega(x,t)} \int_{\Omega[x,t]} |u(y)|dy = 0.$$

We are now in a position to state the main result of this section.

Theorem B. *Suppose $1 < \kappa < \infty$. The following two statements are equivalent.*

(i) *If $1 \leq p < q < \infty$ and $q^{-1} = p^{-1} - 1 + \kappa^{-1}$, then for each $x \in X$ there exists a positive constant $A_{p,q}(K,x)$ such that*

$$|\mathfrak{I}_K u(x)| \leq A_{p,q}(K,x)[\mathfrak{M}_K u(x)]^{p/q} \|u\|_p^{1-p/q}, \tag{2.5}$$

for every $u \in L^p(Y,\mathbb{R})$.

(ii) *There exists a measurable function $\lambda : X \to [0,\infty)$ such that*

$$\omega_K(x,t) \leq \lambda(x) t^{-\kappa}, \qquad x \in X,\ 0 < t < \infty. \tag{2.6}$$

Whenever (i) *or* (ii) *is true, and $\lambda(x)$ in* (2.6) *is defined as*

$$\lambda(x) = \sup_{t>0} t^\kappa \omega(x,t), \qquad x \in X, \tag{2.7}$$

a possible value of $A_{p,q}(K,x)$ in (2.5) *is given by*

$$A_{p,q}(K,x) = \frac{q}{q-p} \left[\frac{pq-q+p}{p} \cdot \lambda(x)\right]^{1-p^{-1}+q^{-1}}. \tag{2.8}$$

Moreover, at points $x \in X$ where (2.6) *is an equality for all $t \in (0,\infty)$, the value of $A_{p,q}(K,x)$ in* (2.8) *is the best constant in inequality* (2.5).

2.2. Before proceeding with a proof of Theorem B, we notice that Theorem A follows by applying Theorem B to the kernel $K(x,y) = k(x-y)$, $(x,y) \in \mathbb{R}^n \times \mathbb{R}^n$. Since the Lebesgue measure is invariant to translations, we have $\omega_K(x,t) = \omega_k(t)$ and, moreover, $\mathfrak{I}_K = \mathfrak{I}_k$ and $\mathfrak{M}_K = \mathfrak{M}_k$.

2.3. Proof of Theorem B. Part 1. We first prove that (i) implies (ii). Suppose $x \in X$ and $\tau \in (0,\infty)$ are given such that $\omega(x,\tau) \neq 0$, and let χ_τ be the characteristic functions of $\Omega[x,\tau]$. We are going to apply (2.5) to the function $u(\cdot) = K(x,\cdot)|K(x,\cdot)|^{-1}\chi_\tau(\cdot)$ in the particular case when $p = 1$. On the one hand we get

$$\mathfrak{I}_K u(x) = \int_{\Omega[x,\tau]} |K(x,y)|dy \geq \tau\omega(x,\tau) \tag{2.9}$$

and

$$\|u\|_1 = \omega(x,\tau). \tag{2.10}$$

On the other hand we observe that

$$\mathfrak{M}_K u(x) \leq \|u\|_\infty = 1. \tag{2.11}$$

Substituting (2.9), (2.10), and (2.11) in (2.5) with $p = 1$ and $q = \kappa$ we get

$$\tau\omega(x,\tau) \leq A(x)[\omega(x,\tau)]^{1-1/\kappa}, \tag{2.12}$$

where $A(x) = A_{1,\kappa}(K,x)$.

To conclude the proof, we notice that (2.12) implies

$$\omega(x,\tau) \leq [A(x)]^\kappa \tau^{-\kappa}, \tag{2.13}$$

hence (2.7) defines a measurable function $\lambda : X \to [0,\infty)$ such that (2.6) is true.

As this proof points out, inequality (2.13) follows from the next weak form of (2.5),

$$|\mathfrak{I}_K u(x)| \leq A_{1,K}(K,x)\|u\|_\infty^{1/\kappa}\|u\|_1^{1-1/\kappa}, \tag{2.14}$$

so (2.14) implies (ii). Since in 2.4 below we will prove that (ii) implies (i), we conclude that the weak form (2.14) of (2.5) is actually equivalent to (2.5). An even weaker equivalent form of (2.5) is indicated in Section 2.8.

2.4. Proof of Theorem B. Part 2. We now show that (ii) implies (i). Suppose $x \in X$ is given and let $u : Y \to \mathbb{R}$ be a measurable function. If $\mathfrak{M}u(x) = \infty$, then (2.5) is obvious, so we may assume that $\mathfrak{M}u(x) < \infty$. This assumption makes it possible to associate to u the decreasing function $\mu_{u,x} : (0,\infty) \to [0,\infty)$ given by

$$\mu_{u,x}(t) = \int_{\Omega[x,t]} |u(y)|dy. \tag{2.15}$$

When $u \equiv 1$ we clearly have $\mu_{u,x}(t) = \omega(x,t)$ and from now we will denote the distribution function $\omega(x,\cdot)$ by μ_x. From (2.15) and (2.4) using the previous notations we have

$$\mathfrak{M}u(x) = \sup_{t>0} \frac{\mu_{u,x}(t)}{\mu_x(t)}. \tag{2.16}$$

Condition (2.6) can be rewritten as

$$\mu_x(t) \leq \lambda(x)t^{-\kappa}, \qquad 0 < t < \infty. \tag{2.17}$$

We next select $\tau \in (0, \infty)$ and introduce the inner and outer parts of K defined as

$$K_{\text{inn}}(x, y) = \begin{cases} |K(x, y)| - \tau, & \text{if } y \in \Omega[x, \tau] \\ 0, & \text{if } y \in Y \setminus \Omega[x, \tau] \end{cases} \tag{2.18}$$

and

$$K_{\text{out}}(x, y) = \begin{cases} \tau, & \text{if } y \in \Omega[x, \tau] \\ |K(x, y)|, & \text{if } y \in Y \setminus \Omega[x, \tau]. \end{cases} \tag{2.19}$$

Since $|K(x, y)| = K_{\text{inn}}(x, y) + K_{\text{out}}(x, y)$, we clearly have

$$|\mathfrak{J}u(x)| \leq \mathfrak{J}_{\text{inn}}|u|(x) + \mathfrak{J}_{\text{out}}|u|(x), \qquad x \in X, \tag{2.20}$$

where $\mathfrak{J}_{\text{inn}} = \mathfrak{J}_{K_{\text{inn}}}$ and $\mathfrak{J}_{\text{out}} = \mathfrak{J}_{K_{\text{out}}}$.

We claim that

$$\mathfrak{J}_{\text{inn}}|u|(x) \leq \frac{\lambda(x)\mathfrak{M}u(x)}{\kappa - 1} \cdot \tau^{-\kappa+1} \tag{2.21}$$

and

$$\mathfrak{J}_{\text{out}}|u|(x) \leq \left[\frac{p}{p - \kappa(p-1)} \cdot \lambda(x)\right]^{(p-1)/p} \|u\|_p \tau^{1-\kappa(p-1)/p}. \tag{2.22}$$

To prove inequality (2.21), we express $\mathfrak{J}_{\text{inn}}|u|(x)$ as a Stieltjes integral associated with the decreasing function $\mu_{u,x}$ defined in (2.15), namely,

$$\mathfrak{J}_{\text{inn}}|u|(x) = \int_{\Omega[x,\tau]} [|K(x, y)| - \tau]|u(y)|dy = -\int_{\tau}^{\infty} (t - \tau)d\mu_{u,x}(t).$$

From (2.16) and (2.17) we get

$$\mu_{u,x}(t) \leq \mu_x(t)\mathfrak{M}u(x) \leq \lambda(x)\mathfrak{M}u(x)t^{-\kappa}, \tag{2.23}$$

whence

$$\lim_{t\to\infty} t\mu_{u,x}(t) = 0.$$

An integration by parts for the last integral yields

$$\mathfrak{J}_{\text{inn}}|u|(x) = \int_{\tau}^{\infty} \mu_{u,x}(t)dt,$$

which, in conjunction with (2.23), results in

$$\mathfrak{J}_{\text{inn}}|u|(x) \leq \lambda(x)\mathfrak{M}u(x) \int_{\tau}^{\infty} t^{-\kappa}dt,$$

an inequality equivalent to (2.21).

We next prove inequality (2.22). By Hölder's inequality we have

$$\mathfrak{J}_{\text{out}}|u|(x) \leq \|K_{\text{out}}(x, \cdot)\|_{p'} \|u\|_p \tag{2.24}$$

where $1 < p' \leq \infty$ is the conjugate exponent to p, i.e., $p' = p/(p-1)$ if $p > 1$, and $p' = \infty$ when $p = 1$.

In the case when $p = 1$, from (2.19) we easily get $\|K_{\text{out}}(x, \cdot)\|_\infty = \tau$, and (2.22) clearly follows from (2.24).

If $p > 1$, then we successively have

$$\begin{aligned}\|K_{\text{out}}(x,\cdot)\|_{p'}^{p'} &= \int_{\Omega[x,\tau]} \tau^{p'} dy + \int_{Y\setminus\Omega[x,\tau]} |K(x,y)|^{p'} dy \\ &= \tau^{p'}\mu_x(\tau) - \int_0^\tau t^{p'} d\mu_x(t) \\ &= p' \int_0^\tau t^{p'-1}\mu_x(t)dt.\end{aligned}$$

The last estimate in conjunction with (2.17) yields

$$\|K_{\text{out}}(x,\cdot)\|_{p'}^{p'} \le \frac{p'}{p'-\kappa}\cdot \tau^{p'-\kappa}\lambda(x),$$

and, as before, (2.22) is now derived from (2.24) after some direct computations.

By (2.20), (2.21), and (2.22) we get

$$\begin{aligned}|\mathfrak{J}u(x)| \le{}& \frac{\lambda(x)\mathfrak{M}u(x)}{\kappa-1}\cdot\tau^{-\kappa+1} \\ &+ \left[\frac{p}{p-\kappa(p-1)}\cdot\lambda(x)\right]^{(p-1)/p} \|u\|_p\cdot\tau^{1-\kappa(p-1)/p}.\end{aligned} \tag{2.25}$$

To minimize the right side of (2.25) we choose

$$\tau = \left[\frac{p-\kappa(p-1)}{p}\cdot\lambda(x)\right]^{1/\kappa}\cdot\left[\frac{\mathfrak{M}u(x)}{\|u\|_p}\right]^{p/\kappa}.$$

A direct calculation shows that for this specific value of τ inequality (2.25) takes the form (2.5) with a constant $A_{p,q}(K,x)$ as in (2.8).

2.5. Proof of Theorem B. Part 3. We yet have to show that (2.8) provides the best value of $A_{p,q}(K,x)$ at points $x \in X$ where

$$\mu_x(t) = \lambda(x)t^{-\kappa}, \qquad 0 < t < \infty. \tag{2.26}$$

This can be done by indicating an extremal function $u : Y \to \mathbb{R}$ in the cases when $p = 1$ or $p > 1$.

We suppose that $x \in X$ is given such that (2.26) holds true and observe that

$$\lambda(x) = \mu_x(1) = \text{measure}\,\Omega[x,1]. \tag{2.27}$$

If $p > 1$, we set

$$u(y) = \begin{cases} K(x,y)\cdot|K(x,y)|^{-1}, & \text{if } y\in\Omega[x,1] \\ K(x,y)\cdot|K(x,y)|^{(2-p)/(p-1)}, & \text{if } y\in Y\setminus\Omega[x,1]. \end{cases} \tag{2.28}$$

The left side in inequality (2.5) becomes

$$|\mathfrak{J}_K u(x)| = \int_{\Omega[x,1]} |K(x,y)|dy + \int_{Y\setminus\Omega[x,1]} |K(x,y)|^{p/(p-1)}dy,$$

whence, by using (2.26) we get

$$\begin{aligned}|\mathfrak{I}_K u(x)| &= -\int_1^\infty t d\mu_x(t) - \int_0^1 t^{p/(p-1)} d\mu_x(t) \\ &= \kappa\lambda(x)\left[\int_1^\infty t^{-\kappa}dt + \int_0^1 t^{-\kappa+1/(p-1)}dt\right] \\ &= \kappa\lambda(x)\left[\frac{1}{\kappa-1} + \frac{p-1}{p-\kappa(p-1)}\right].\end{aligned}$$

Since $\kappa = pq/(pq-q+p)$ we end up with

$$|\mathfrak{I}_K u(x)| = \frac{q}{q-p}\cdot\frac{pq-q+p}{p}\cdot\lambda(x). \tag{2.29}$$

A similar computation leads to

$$\|u\|_p^p = \int_{\Omega[x,1]} 1dy + \int_{Y\setminus\Omega[x,1]} |K(x,y)|^{p/(p-1)}dy = \frac{pq-q+p}{p}\cdot\lambda(x),$$

therefore

$$\|u\|_p^{1-p/q} = \left[\frac{pq-q+p}{p}\cdot\lambda(x)\right]^{p^{-1}-q^{-1}}. \tag{2.30}$$

Finally, since $|u(y)| = 1$ on $\Omega[x,1]$ and $|u(y)| < 1$ on $Y\setminus\Omega[x,1]$ we have

$$\mathfrak{M}_K u(x) = 1. \tag{2.31}$$

Substituting now (2.29), (2.30), and (2.31) in (2.5) with $A_{p,q}(K,x)$ as in (2.8) we get an equality, hence inequality (2.5) is sharp.

In the case when $p = 1$, we define an extremal function by

$$u(y) = \begin{cases} K(x,y)\cdot|K(x,y)|^{-1}, & \text{if } y\in\Omega[x,1] \\ 0, & \text{if } y\in Y\setminus\Omega[x,1]. \end{cases} \tag{2.32}$$

Repeating the previous calculations, we recover (2.29), (2.30), and (2.31) with $p = 1$ and $q = \kappa$ so, once again, (2.5) turns out to be an equality.

The proof of Theorem B is complete. □

2.6. Inequality (2.5) in Theorem B and, consequently, inequality (1.11) in Theorem A, can be slightly improved by replacing the maximal operator $\mathfrak{M}$ with the operator $\mathfrak{M}^*$ defined as

$$\mathfrak{M}^* u(x) = \inf_{r\geq 1}[\mathfrak{M}|u|^r(x)]^{1/r}. \tag{2.33}$$

This is done by improving estimate (2.21) in Section 2.4 as follows. Suppose $r > 1$ is given, let r' be the conjugate exponent, and observe that by Hölder's inequality we have

$$\begin{aligned}\mathfrak{J}_{\mathrm{inn}}|u|(x) &= \int_{\Omega[x,\tau]} K_{\mathrm{inn}}^{1/r'}(x,y)K_{\mathrm{inn}}^{1/r}(x,y)|u(y)|dy \\ &\le \left[\int_{\Omega[x,\tau]} K_{\mathrm{inn}}(x,y)dy\right]^{1/r'} \left[\int_{\Omega[x,\tau]} K_{\mathrm{inn}}(x,y)|u(y)|^r dy\right]^{1/r} \\ &= \left[\mathfrak{J}_{\mathrm{inn}}\chi_\tau(x)\right]^{1/r'} \left[\mathfrak{J}_{\mathrm{inn}}|u|^r(x)\right]^{1/r},\end{aligned}$$

where χ_τ is the characteristic function of $\Omega[x,\tau]$.

Since $\mathfrak{M}\chi_\tau(x) = 1$, by (2.21) we get

$$\mathfrak{J}_{\mathrm{inn}}\chi_\tau(x) \le \frac{\lambda(x)}{\kappa - 1} \cdot \tau^{-\kappa+1},$$

as well as

$$\mathfrak{J}_{\mathrm{inn}}|u|^r(x) \le \frac{\lambda(x)\mathfrak{M}|u|^r(x)}{\kappa - 1} \cdot \tau^{-\kappa+1}.$$

Combining the last three estimates we obtain

$$\mathfrak{J}_{\mathrm{inn}}|u|(x) \le \lambda(x)\frac{[\mathfrak{M}|u|^r(x)]^{1/r}}{\kappa - 1} \cdot \tau^{-\kappa+1},$$

whence

$$\mathfrak{J}_{\mathrm{inn}}|u|(x) \le \frac{\lambda(x)\mathfrak{M}^* u(x)}{\kappa - 1} \cdot \tau^{-\kappa+1}. \tag{2.34}$$

2.7. The constants $A_{p,q}(K,x)$ in (2.5) are no longer the best for nonnegative functions. Nevertheless, the existing estimates can be used to find the best constants whenever u is nonnegative and K takes positive as well as negative values. We first split the kernel function K by setting $K = K_+ - K_-$, where

$$K_\pm(x,y) = \max\{\pm K(x,y), 0\}, \qquad (x,y) \in X \times Y.$$

If $u \ge 0$, then obviously

$$|\mathfrak{J}_K u(x)| \le \max\left\{\mathfrak{J}_{K_+}u(x), \mathfrak{J}_{K_-}u(x)\right\}, \qquad x \in X.$$

Therefore, by applying (2.5) to the nonnegative kernels K_+ and K_- we get two estimates for $\mathfrak{J}_{K_+}u(x)$ and $\mathfrak{J}_{K_-}u(x)$, respectively, that can be combined into an estimate for $\mathfrak{J}_K u(x)$. In effect, if u is a nonnegative function, then instead of (2.5) we get

$$|\mathfrak{J}_K u(x)| \le A^*_{p,q}(K,x)[\mathfrak{M}_K u(x)]^{p/q}\|u\|p^{1-p/q}, \tag{2.35}$$

where

$$A^*_{p,q}(K,x) = \max\left\{A_{p,q}(K_+,x), A_{p,q}(K_-,x)\right\}. \tag{2.36}$$

The actual value of $A^*_{p,q}(K,x)$ can be found from (2.8) by replacing the function $\lambda(x)$ with the function $\lambda^*(x)$ given by

$$\lambda^*(x) = \max\{\lambda_+(x), \lambda_-(x)\}, \qquad x \in X, \tag{2.37}$$

where $\lambda_+(x)$ and $\lambda_-(x)$ are associated to K_+ and K_-, respectively, as in (2.7).

2.8. Now, let us apply (2.35) to the function $u = \chi_\Omega$, the characteristic function of a set $\Omega \subseteq Y$ with a bounded measure, in the particular case when $p = 1$ and $q = \kappa$. Since obviously $\mathfrak{M}\chi_\Omega(x) \leq \|\chi_\Omega\|_\infty = 1$, from (2.35), (2.36), and (2.37) we get

$$\left|\int_\Omega K(x,y)dy\right| \leq \frac{\kappa}{\kappa - 1}[\lambda^*(x)]^{1/\kappa}[\text{measure}(\Omega)]^{1-1/\kappa}, \qquad x \in X. \tag{2.38}$$

The last estimate, though weaker than its forbear, inequality (2.5), turns out to be equivalent to (2.5). By Theorem B, all we need is to show that (2.38) implies statement (ii) in that theorem. To this end, we apply (2.38) to the disjoint sets

$$\Omega^\pm[x,\tau] = \{y \in Y : \pm K(x,y) \geq \tau\}, \qquad x \in X,\ \tau \in (0,\infty),$$

and get, after a straightforward calculation,

$$\text{measure}\,\Omega^\pm[x,\tau] \leq \left(\frac{\kappa}{\kappa - 1}\right)^\kappa \lambda^*(x)\tau^{-\kappa}, \qquad x \in X,\ \tau \in (0,\infty).$$

Since $\Omega[x,\tau] = \Omega^-[x,\tau] \cup \Omega^+[x,\tau]$, we conclude that the distribution function $\omega = \omega_K$ has the required property (2.6) in Theorem B.

2.9. We end this section with a brief comment regarding the behavior of the operator $\mathfrak{I} = \mathfrak{I}_K$ in certain special circumstances.

As a first basic assumption, let us suppose that the maximal operator $\mathfrak{M}$ acts continuously from $L^p(Y,\mathbb{R})$ into $L^p(X,\mathbb{R})$ for every $1 < p \leq \infty$. One can formulate sufficient conditions for this property in terms of the kernel K, by merely adapting to this setting the postulates discussed in Stein [26, Chapter 1], that guarantee the use of Vitali- or Whitney-type covering lemmas for the collection of "balls" $\Omega[x,\tau]$. Under such assumptions, from Theorem B we get that whenever the function λ defined by (2.7) is essentially bounded, or a constant, as in Theorem A, then the integral operator $\mathfrak{I}$ yields a bounded linear operator from $L^p(Y,\mathbb{R})$ into $L^q(X,\mathbb{R})$, for $1 < p < q < \infty$ and $q^{-1} = p^{-1} - 1 + \kappa^{-1}$. In the more general case when λ is a function in $L^r(X,\mathbb{R})$, with $\max\{1, p/[\kappa(2p-1) - p\} \leq r \leq \infty$, one gets that $\mathfrak{I}$ is bounded as an operator from $L^p(Y,\mathbb{R})$ into $L^{q_*}(X,\mathbb{R})$, where $q_*^{-1} = p^{-1} - 1 + \kappa^{-1} + (\kappa r)^{-1}$. An interesting situation occurs if $\kappa \leq 2$ and $r = (\kappa - 1)^{-1}$, whence $q_* = p$.

3. Uniform approximations by solutions of elliptic equations

This section points out consequences of Theorems A and B that provide quantitative Hartogs-Rosenthal-type theorems. Several general qualitative results in this area are presented in the monograph by Tarkhanov ([27, Chapter 6]). The classical result, extensions, and specific means of studying rational approximation are presented in the monographs by A. Browder ([5]) and Gamelin ([9]). Early contributions addressing the problem on approximation by solutions of elliptic equations can be found in the articles by F. Browder ([6, 7]). Other additional references related to the main result in this section have been already mentioned at the end of Section 1.

3.1. Throughout this section we let X stand for a smooth Riemannian manifold of dimension n equipped with the volume measure and assume that E is a smooth Hermitean vector bundle over X of complex rank m.

The spaces of all smooth sections or compactly supported smooth sections from X to E will be denoted by $\mathcal{E}(X, E)$ and $\mathcal{D}(X, E)$, respectively.

The fiber of E at a point $x \in X$ is denoted in what follows by E_x. For a later use, we also introduce the complex vector bundle $\mathcal{L}(E, E)$ over $X \times X$ whose fiber at a point $(x, y) \in X \times X$ equals the space $\mathcal{L}(E_y, E_x)$ consisting of linear maps from E_y into E_x. The norm in E_x and the operator norm in $\mathcal{L}(E_y, E_x)$ are denoted by $\|\cdot\|_x$ and $\|\cdot\|_{x,y}$, respectively.

On the space $\mathcal{D}(X, E)$ we introduce the norms $\|\cdot\|_p$, with $1 \leq p < \infty$, by setting

$$\|u\|_p = \left(\int_X \|u(x)\|_x^p dx\right)^{1/p}, \qquad u \in \mathcal{D}(X, E), \tag{3.1}$$

and the uniform norm $\|\cdot\|_\infty$ defined as

$$\|u\|_\infty = \sup_{x \in X} \|u(x)\|_x, \qquad u \in \mathcal{D}(X, E). \tag{3.2}$$

In addition, we will let $L^p(X, E)$, $1 \leq p \leq \infty$, denote the Lebesgue spaces of p-integrable sections from X to E with the norms defined as in (3.1) and (3.2).

3.2. We are going to apply Theorem B to integral operators associated to smooth sections from $X \times X \setminus \Delta$ into $\mathcal{L}(E, E)$, where $\Delta = \{(x, x) : x \in X\}$. If Φ is such a section, then $\Phi(x, y) : E_y \to E_x$ is a linear map for every $(x, y) \in X \times X \setminus \Delta$. We define the operator $\mathfrak{I} = \mathfrak{I}_\Phi$ acting on $\mathcal{D}(X, E)$ by setting

$$\mathfrak{I}u(x) = \lim_{\rho \downarrow 0} \int_{X \setminus B(x,\rho)} \Phi(x, y) u(y) d(y), \qquad u \in \mathcal{D}(X, E),\ x \in X, \tag{3.3}$$

where $B(x, \rho)$ stands for the closed geodesic ball in X centered at $x \in X$, with radius $\rho > 0$.

We assume that the limit in (3.3) exists and $\mathfrak{I}u \in \mathcal{E}(X, E)$ for every $u \in \mathcal{D}(X, E)$. In order to use Theorem B to its full extent, we will also assume that there exists $\kappa \in (1, \infty)$ and $A(\Phi) \in (0, \infty)$ such that for any compact set $\Omega \subseteq X$ we have

$$\int_\Omega \|\Phi(x, y)\|_{x,y} dy \leq A(\Phi)[\text{measure}(\Omega)]^{1-1/\kappa}, \qquad x \in \Omega. \tag{3.4}$$

This estimate works as a substitute for inequality (2.38) in Section 2.8 and makes it possible to apply Theorem B in this setting.

Actually, based on the comments made in 2.8, we know that inequality (3.4) implies a stronger estimate similar to (2.14), namely,

$$\|\mathfrak{I}u(x)\|_x \leq A(\Omega)\|u\|_\infty^{1/\kappa}\|u\|_1^{1-1/\kappa}, \qquad \text{for every } x \in \Omega. \tag{3.5}$$

for every $u \in \mathcal{D}(X, E)$, or $u \in L^\infty(X, E) \cap L^1(X, E)$. According to (2.7) and (2.8) in Theorem B, a possible value of the constant $A(\Omega)$ in (3.5) is given by

$$A(\Omega) = \frac{\kappa}{\kappa - 1} \sup_{x\in\Omega} \sup_{t>0} t[\omega(x,t)]^{1/\kappa}, \tag{3.6}$$

where $\omega : X \times (0, \infty) \to [0, \infty)$ is defined as

$$\omega(x,t) = \text{measure}\{y \in Y : y \neq x, \ \|\Phi(x,y)\|_{x,y} \geq t\}. \tag{3.7}$$

Whenever $\omega(x,t)$ does not depend on $x \in X$, we clearly get that $A(\Omega) = A$, a constant that depends solely on Φ.

Finally, we suppose that $\mathfrak{D} : \mathcal{E}(X, E) \to \mathcal{E}(X, E)$ is an elliptic differential operator of order α, with $1 \leq \alpha \leq n - 1$, that has a fundamental solution, i.e., a smooth section $\Phi : X \times X \setminus \Delta \to \mathcal{L}(E, E)$ such that the associated operator $\mathfrak{I} : \mathcal{D}(X, E) \to \mathcal{E}(X, E)$ has the property

$$\mathfrak{D}\mathfrak{I}u = u, \qquad u \in \mathcal{D}(X, E), \tag{3.8}$$

and, in addition, satisfies the estimates (3.5) with $\kappa = n/(n - \alpha)$. In other words, $\mathfrak{I}$ is a right inverse to $\mathfrak{D}$ for which we can use Theorem B.

3.3. Further, we let $\Omega \subseteq X$ be a compact set and denote by $\mathcal{C}(\Omega, E)$ the Banach space of all continuous sections from Ω to the restriction of the bundle E to Ω, with the uniform norm given by

$$\|u\|_{\Omega,\infty} = \sup_{x\in\Omega} \|u(x)\|_x, \qquad u \in \mathcal{C}(\Omega, E). \tag{3.9}$$

We also introduce the norm $\|\cdot\|_{\Omega,1}$ defined as

$$\|u\|_{\Omega,1} = \int_\Omega \|u(x)\|_x dx, \qquad u \in \mathcal{C}(\Omega, E). \tag{3.10}$$

We next associate to the differential operator $\mathfrak{D} : \mathcal{E}(X, E) \to \mathcal{E}(X, E)$ the solution subspace $\mathcal{S}_\mathfrak{D}(\Omega, E)$ of $\mathcal{C}(\Omega, E)$ defined as the uniform closure in $\mathcal{C}(\Omega, E)$ of the set consisting of restrictions to Ω of sections u from $\mathcal{E}(X, E)$ such that $\mathfrak{D}u = 0$ on certain open neighborhoods of Ω in X that may vary from a section to another.

3.4. The main result of this section estimates the distance in $\mathcal{C}(\Omega, E)$ from a compactly supported section u to the solution subspace $\mathcal{S}_\mathfrak{D}(\Omega, E)$.

Theorem C. *Suppose $\mathfrak{D} : \mathcal{E}(X, E) \to \mathcal{E}(X, E)$ is an elliptic operator of order α, with $1 \leq \alpha \leq n - 1$, and let $u \in \mathcal{D}(X, E)$ be a given section. Then*

$$\text{dist}_{\mathcal{C}(\Omega,E)}[u, \mathcal{S}_\mathfrak{D}(\Omega, E)] \leq A(\Omega)\|\mathfrak{D}u\|_{\Omega,\infty}^{1/\kappa}\|\mathfrak{D}u\|_{\Omega,1}^{1-1/\kappa}, \tag{3.11}$$

where $A(\Omega)$ is as in (3.6).

Proof. We begin by observing that (3.8) implies $\mathfrak{D}\mathfrak{I}\mathfrak{D}u = \mathfrak{D}u$, whence we get that the restriction to Ω of the section $v = u - \mathfrak{I}\mathfrak{D}u$ in $\mathcal{E}(X, E)$ belongs to $\mathcal{S}_{\mathfrak{D}}(\Omega, E)$. Therefore,

$$\text{dist}_{\mathcal{C}(\Omega,E)}[u, \mathcal{S}_{\mathfrak{D}}(\Omega, E)] \leq \|u - v\|_{\Omega,\infty} = \|\mathfrak{I}\mathfrak{D}u\|_{\Omega,\infty}. \tag{3.12}$$

On the other hand, by (3.5) we have

$$\|\mathfrak{I}\mathfrak{D}u(x)\|_x \leq A(\Omega)\|\mathfrak{D}u\|_\infty^{1/\kappa}\|\mathfrak{D}u\|_1^{1-1/\kappa}, \qquad x \in \Omega. \tag{3.13}$$

We next notice that the integral operator $\mathfrak{I}$ in (3.12) and (3.13) can be replaced with an operator $\mathfrak{I}'$ defined as in (3.3), where instead of X we employ a relatively compact open neighborhood X' of Ω. In effect, by combining (3.12) and (3.13) for $\mathfrak{I}'$ we get

$$\text{dist}_{\mathcal{C}(\Omega,E)}[u, \mathcal{S}_{\mathfrak{D}}(\Omega, E)] \leq A(\Omega)\|\mathfrak{D}u\|_{X',\infty}^{1/\kappa}\|\mathfrak{D}u\|_{X',1}^{1-1/\kappa}, \tag{3.14}$$

Inequality (3.11) follows from (3.14) by letting X' approach Ω. □

3.5. Since by the Stone-Weierstrass theorem the space consisting of restrictions to Ω of sections from $\mathcal{D}(X, E)$ is dense in $\mathcal{C}(\Omega, E)$, from Theorem C we get the next Hartogs-Rosenthal-type theorem.

Corollary. *If $\Omega \subseteq X$ is a compact set of measure zero, then $\mathcal{S}_{\mathfrak{D}}(\Omega, E) = \mathcal{C}(\Omega, E)$.*

Proof. It is enough to notice that for any $u \in \mathcal{D}(X, E)$ we have

$$\|\mathfrak{D}u\|_{\Omega,1} \leq \|\mathfrak{D}u\|_{\Omega,\infty}\text{measure}(\Omega) = 0,$$

whence, by (3.11),

$$\text{dist}_{\mathcal{C}(\Omega,E)}[u, \mathcal{S}_{\mathfrak{D}}(\Omega, E)] = 0.$$

3.6. As direct applications of Theorem C or its corollary, we may take the manifold $X = \mathbb{R}^n$, the trivial bundle $E = \mathbb{R}^n \times \mathbb{C}^m$, and an elliptic differential operator $\mathfrak{D} : \mathcal{E}(\mathbb{R}^n, \mathbb{C}^m) \to \mathcal{E}(\mathbb{R}^n, \mathbb{C}^m)$ of order α, with $1 \leq \alpha \leq n - 1$. The fundamental solution for $\mathfrak{D}$ is an operator-valued map $k : \mathbb{R}_0^n \to \mathcal{L}(\mathbb{C}^m, \mathbb{C}^m)$ homogeneous of degree $\alpha - n$, and the corresponding convolution operator $\mathfrak{I} = \mathfrak{I}_k$ defined in a natural way on $\mathcal{D}(\mathbb{R}^n, \mathbb{C}^m)$ satisfies condition (3.8).

Actually, this was the setting developed in Martin and Szeptycki ([21]) and studied there by relying on a simpler form of Theorem A.

Among the basic examples of differential operators that fall within this framework we mention the $\bar{\partial}$ operator on $\mathbb{R}^2 = \mathbb{C}$, the Laplace operator on $\mathbb{R}^n$, $n \geq 3$, or the Euclidean Dirac operator on $\mathbb{R}^n$, $n \geq 2$, which generalizes $\bar{\partial}$ to higher dimensions. More details on the Dirac operator can be found in the monographs by Brackx, Delange, and Sommen ([8]), Gilbert and Murray ([10]), Gürlebeck and Sprössig ([12]), Mitrea ([22]), and the article by Ryan ([24]). Refined forms of Theorem A for Dirac operators and applications in Clifford analysis are discussed in Martin ([18, 19]).

When applied to the $\bar{\partial}$ operator, Theorem C results in an inequality proved by Alexander ([2, 3]) regarding rational approximation. An estimate similar to (3.4) for the Cauchy kernel is due to Ahlfors and Beurling ([1]), and a generalization that results from Theorem A is indicated in Putinar ([23]). For second-order differential operators, Theorem C and its corollary were established by Khavinson ([16]) and estimates like (3.4) for kernels related to such operators have been proved by Khavinson ([15]) and Gustafsson and Khavinson ([11]).

The first Hartogs-Rosenthal-type theorem for general elliptic operators was proved by Weinstock ([28]). For a proof in the setting of Clifford analysis, regarding approximation by monogenic functions, we refer to Brackx, Delanghe, and Sommen ([8, Section 18]). Inequalities for the Teodorescu transform that can be derived from (1.11) in Theorem A are proved in the monograph by Gürlebeck and Sprössig ([12, Chapter 3]).

Note added in proof by Mircea Martin. On January 30, 2004, the mathematical community lost one of its finest: Professor Pawel Szeptycki, PhD. He was a dear friend of mine and a distinguished researcher in the field of mathematical analysis. I would like to dedicate to his memory my contribution in writing this joint paper. He will be missed.

References

[1] L. Ahlfors, A. Beurling, *Conformal Invariants and Function Theoretic Null Sets.* Acta Math. **83** (1950), 101–129.

[2] H. Alexander, *Projections of Polynomial Hulls.* J. Funct. Anal. **13** (1973), 13–19.

[3] H. Alexander, *On the Area of the Spectrum of an Element of a Uniform Algebra.* In *Complex Approximation Proceedings, Quebec, July 3–8, 1978*, Birkhäuser, 1980, 3–12.

[4] W. Beckner, *Geometric Inequalities in Fourier Analysis.* In *Essays on Fourier Analysis in Honor of Elias M. Stein*, Princeton University Press, Princeton, 1995, 36–68.

[5] A. Browder, *Introduction to Function Algebras.* Benjamin, New York, 1969.

[6] F. Browder, *Approximation by Solutions of Partial Differential Equations.* Amer. J. Math. **84** (1962), 134–160.

[7] F. Browder, *Functional Analysis and Partial Differential Equations.* II, Math. Ann. **145** (1961), 81–226.

[8] F. Brackx, R. Delanghe, F. Sommen, *Clifford Analysis.* Pitman Research Notes in Mathematics Series, 76, 1982.

[9] T.W. Gamelin, *Uniform Algebras.* Prentice Hall, 1969.

[10] J.E. Gilbert, and M.A.M. Murray, *Clifford Algebras and Dirac Operators in Harmonic Analysis.* Cambridge Studies in Advanced Mathematics, 26, Cambridge University Press, 1991.

[11] B. Gustafsson, and D. Khavinson, *On Approximation by Harmonic Vector Fields.* Houston J. Math. **20** (1994), 75–92.

[12] K. Gürlebeck, and W. Sprössig, *Quaternionic and Clifford Calculus for Physicists and Engineers.* John Wiley & Sons, New York, 1997.

[13] L. Hedberg, *On Certain Convolution Inequalities.* Proc. Amer. Math. Soc. **36** (1972), 505–510.

[14] L. Hörmander, *The Analysis of Linear Partial Differential Operators, Vol I: Distribution Theory and Fourier Analysis.* Springer-Verlag, Berlin, 1983.
[15] D. Khavinson, *On Uniform Approximation by Harmonic Functions.* Mich. Math. J.**34** (1987), 465–473.
[16] D. Khavinson, *Duality and Uniform Approximation by Solutions of Elliptic Equations.* Operator Theory: Advances and Applications **35** (1988), Birkhäuser Verlag, Basel, 129–141.
[17] E.H. Lieb, *Sharp Constants in the Hardy-Littlewood-Sobolev and Related Inequalities.* Annals of Math. **118** (1983), 349–379.
[18] M. Martin, *Higher-Dimensional Ahlfors-Beurling Inequalities.* Proc. Amer. Math. Soc. **126** (1998), 2863–2871.
[19] M. Martin, *Convolution and Maximal Operator Inequalities in Clifford Analysis.* In *Clifford Algebras and Their Applications in Mathematical Physics, Vol. 2: Clifford Analysis*, Progress in Physics **9**, Birkhäuser Verlag, Basel, 2000, 95–113.
[20] M. Martin, *Uniform Approximation by Closed Forms in Several Complex Variables.* Preprint 2002.
[21] M. Martin, and P. Szeptycki, *Sharp Inequalities for Convolution Operators with Homogeneous Kernels and Applications.* Indiana Univ. Math. **46** (1997), 975–988.
[22] M. Mitrea, *Singular Integrals, Hardy Spaces, and Clifford Wavelets.* Lecture Notes in Mathematics, 1575, Springer-Verlag, Heidelberg, 1994.
[23] M. Putinar, *Extreme Hyponormal Operators.* Operator Theory: Advances and Applications **28** (1988), 249–265.
[24] J. Ryan, *Dirac Operators, Conformal Transformations and Aspects of Classical Harmonic Analysis.* Journal of Lie Theory **8** (1998), 67–82.
[25] E.M. Stein, *Singular Integrals and Differentiability Properties of Functions.* Princeton Univ. Press, Princeton, NJ, 1970.
[26] E.M. Stein, *Harmonic Analysis: Real-Variable Methods, Orthogonality, and Oscillatory Integrals.* Princeton Univ. Press, Princeton, NJ,, 1993.
[27] N.N. Tarkhanov, *The Cauchy Problem for Solutions of Elliptic Equations.* Akademie Verlag, Berlin, 1995.
[28] B.M. Weinstock, *Uniform Approximations by Solutions of Elliptic Equations.* Proc. Amer. Math. Soc. **41** (1973), 513–517.

Mircea Martin
Department of Mathematics
Baker University
Baldwin City, KS 66006
USA
e-mail: mircea.martin@bakeru.edu

Pawel Szeptycki †
Department of Mathematics
University of Kansas
Lawrence, KS 66045
USA
e-mail: szeptyck@math.ukans.edu

Operator Theory:
Advances and Applications, Vol. 153, 185–195

Uniform Exponential Dichotomy and Admissibility for Linear Skew-Product Semiflows

Mihail Megan, Adina Luminiţa Sasu and Bogdan Sasu

Abstract. We give necessary and sufficient conditions for uniform exponential dichotomy of linear skew-product semiflows. We present connections between the uniform exponential dichotomy of a linear skew-product semiflow and the uniform admissibility of the pair $(C_b(\mathbf{R}_+, X), \Delta(\mathbf{R}_+, X))$, where $\Delta(\mathbf{R}_+, X)$ denotes the unit ball in $C_c(\mathbf{R}_+, X)$.

Mathematics Subject Classification (2000). 34D09, 34D05.

Keywords. Linear skew-product semiflow, Exponential dichotomy, Admissibility.

1. Introduction

Exponential dichotomy is one of the most important asymptotic properties of evolution equations. Significant results in this field have been obtained in [1]–[15], [17], [19]–[21].

In the last few years an important progress was made in the study of the asymptotic behavior of linear skew-product semiflows. In [5], Chow and Leiva introduced the concept of pointwise discrete dichotomy for a skew-product sequence $(\Phi_n(\theta), \sigma(\theta, n))_{n\in\mathbf{N}}$, over $X \times \Theta$, extending a theorem due to Henry (see [8]) for the equations of type $x_{n+1} = \Phi_n(\theta)x_n + f_n$. In [21], Sacker and Sell presented characterizations for exponential dichotomy of a weakly hyperbolic linear skew-product semiflow. There, this property has been obtained by imposing the condition of finite dimension for the unstable manifold. Other important generalizations for the dichotomy theorems due to Henry, for the case of linear skew-product semiflows, have been presented by Pliss and Sell in [20].

Exponential dichotomy of linear skew-product semiflows, defined on compact spaces, has been also studied in [6]. The case of linear skew-product semiflows, defined on locally compact spaces has been considered in [9] and in [10], respectively. In [10], dichotomy of linear skew-product flows has been expressed in terms of the hyperbolicity of a family of weighted shift operators acting on $c_0(\mathbf{Z}, X)$. In this paper, Latushkin and Schnaubelt characterized dichotomy in terms of the existence and uniqueness of bounded, continuous mild solutions of the inhomogeneous equation

$$u(\sigma(\theta,t)) = e^{-\lambda t}\Phi(\theta,t)u(\theta) + \int_0^t e^{-\lambda(t-\tau)}\Phi(\sigma(\theta,\tau),t-\tau)g(\sigma(\theta,\tau))\,d\tau$$

on $C_0(\Theta, X)$ and $C_b(\Theta, X)$, respectively.

The starting point of the present paper, is a dichotomy theorem due to Van Minh, Räbiger and Schnaubelt (see [19]), given by

Theorem 1.1. *An evolution family* $\mathcal{U} = \{U(t,s)\}_{t\geq s\geq 0}$ *is uniformly exponentially dichotomic if and only if for every* $u \in C_0(\mathbf{R}_+, X)$ *there is* $f \in C_0(\mathbf{R}_+, X)$ *such that*

$$(E) \qquad f(t) = U(t,s)f(s) + \int_s^t U(t,\tau)u(\tau)\,d\tau, \quad \forall t \geq s \geq 0$$

and the space $X_1 = \{x \in X : \sup_{t\geq 0} \|U(t,0)x\| < \infty\}$ *is closed and complemented in* X.

The above result has been extended in [13] for the case of evolution families with nonuniform exponential growth. In this case the solvability of the equation (E) in $C_0(\mathbf{R}_+, X)$ is a sufficient condition for nonuniform exponential dichotomy. A generalization of Theorem 1.1, using discrete-time methods was given in [14].

In [15] we have characterized the exponential dichotomy of linear skew-product semiflows using discrete-time methods. Our approach was based on the equivalence between discrete dichotomy and exponential dichotomy for linear skew-product semiflows. As a consequence of the discrete characterizations, we expressed the uniform exponential dichotomy of linear skew-product semiflows in terms of the admissibility of the pair $(C_0(\mathbf{R}_+, X), C_{00}(\mathbf{R}_+, X))$.

The aim of this paper is to give necessary and sufficient conditions for uniform exponential dichotomy of linear skew-product semiflows, without using discrete-time techniques. For a Banach space X we shall denote by $C_b(\mathbf{R}_+, X)$ the space of all continuous bounded functions $u : \mathbf{R}_+ \to X$, by $C_c(\mathbf{R}_+, X)$ the space of all continuous functions with compact support contained in $(0,\infty)$ and by $\Delta(\mathbf{R}_+, X)$ the unit ball in $C_c(\mathbf{R}_+, X)$. For a linear skew-product semiflow $\pi = (\Phi, \sigma)$, defined on $\mathcal{E} = X \times \Theta$, we shall consider an integral equation and we shall characterize the uniform exponential dichotomy in terms of uniform admissibility of the pair $(C_b(\mathbf{R}_+, X), \Delta(\mathbf{R}_+, X))$. The methods are direct and the proofs are based on input-output techniques. In this manner we generalize the results presented in [15] and consequently, the dichotomy theorem for evolution families due to van Minh, Räbiger and Schnaubelt.

2. Preliminaries

Let X be a Banach space, let Θ be a metric space and let $\mathcal{E} = X \times \Theta$.

Definition 2.1. A continuous mapping $\sigma : \Theta \times \mathbf{R} \to \Theta$ is said to be a *flow* on Θ, if $\sigma(\theta, 0) = \theta$, for all $\theta \in \Theta$ and $\sigma(\theta, s+t) = \sigma(\sigma(\theta, s), t)$, for all $(\theta, s, t) \in \Theta \times \mathbf{R}^2$.

Definition 2.2. A pair $\pi = (\Phi, \sigma)$ is called *linear skew-product semiflow* on $\mathcal{E} = X \times \Theta$ if σ is a flow on Θ and $\Phi : \Theta \times \mathbf{R}_+ \to \mathcal{B}(X)$ satisfies the following conditions:

(i) $\Phi(\theta, 0) = I$, the identity operator on X, for all $\theta \in \Theta$;

(ii) $\Phi(\theta, t+s) = \Phi(\sigma(\theta, s), t)\Phi(\theta, s)$, for all $(\theta, t, s) \in \Theta \times \mathbf{R}_+^2$ (*the cocycle identity*);

(iii) there are $M \geq 1$ and $\omega > 0$ such that $\|\Phi(\theta, t)\| \leq Me^{\omega t}$, for all $(\theta, t) \in \Theta \times \mathbf{R}_+$;

(iv) for every $x \in X$ the mapping $(\theta, t) \mapsto \Phi(\theta, t)x$ is continuous.

Diverse examples of linear skew-product semiflows can be found in [4]–[6], [9], [10], [15]–[18], [20], [21].

Definition 2.3. A linear skew-product semiflow $\pi = (\Phi, \sigma)$ is said to be *uniformly exponentially dichotomic* if there exist a family of projections $\{P(\theta)\}_{\theta \in \Theta} \subset \mathcal{B}(X)$ and two constants $N \geq 1$ and $\nu > 0$ such that

(i) $\Phi(\theta, t)P(\theta) = P(\sigma(\theta, t))\Phi(\theta, t)$, for all $(\theta, t) \in \Theta \times \mathbf{R}_+$;

(ii) $\|\Phi(\theta, t)x\| \leq Ne^{-\nu t}\|x\|$, for all $(\theta, t) \in \Theta \times \mathbf{R}_+$ and all $x \in ImP(\theta)$;

(iii) $\|\Phi(\theta, t)x\| \geq (1/N)e^{\nu t}\|x\|$, for all $(\theta, t) \in \Theta \times \mathbf{R}_+$ and all $x \in KerP(\theta)$;

(iv) the restriction $\Phi(\theta, t)_| : KerP(\theta) \to KerP(\sigma(\theta, t))$ is an isomorphism, for every $(\theta, t) \in \Theta \times \mathbf{R}_+$.

Proposition 2.4. *Let $\pi = (\Phi, \sigma)$ be a linear skew-product semiflow on $\mathcal{E} = X \times \Theta$. If π is uniformly exponentially dichotomic relative to the family of projections $\{P(\theta)\}_{\theta \in \Theta}$, then*

(i) $\sup\limits_{\theta \in \Theta} \|P(\theta)\| < \infty$;

(ii) *for every $(\theta, t) \in \Theta \times \mathbf{R}_+^*$ and every $x \in KerP(\sigma(\theta, t))$ the mapping $s \mapsto \Phi(\sigma(\theta, s), t-s)_|^{-1}x$ is continuous on $[0, t]$;*

(iii) *for every $(x, \theta) \in \mathcal{E}$ the mapping $t \mapsto P(\sigma(\theta, t))x$ is continuous on $\mathbf{R}_+$.*

Proof. (i) The idea is the same as in [7]. For every $\theta \in \Theta$ we define

$$\delta_\theta := \inf\{\|x_1 + x_2\| : x_1 \in ImP(\theta), x_2 \in KerP(\theta), \|x_1\| = \|x_2\| = 1\}.$$

Let $\theta \in \Theta$ and $x \in X$ with $P(\theta)x \neq 0$ and $(I - P(\theta))x \neq 0$. Then

$$\begin{aligned}\delta_\theta &\leq \left\| \frac{P(\theta)x}{\|P(\theta)x\|} + \frac{(I - P(\theta))x}{\|(I - P(\theta))x\|} \right\| \\ &= \frac{1}{\|P(\theta)x\|} \left\| x + \frac{\|P(\theta)x\| - \|(I - P(\theta))x\|}{\|(I - P(\theta))x\|}(I - P(\theta))x \right\| \leq \frac{2\|x\|}{\|P(\theta)x\|}.\end{aligned}$$

It results that $\|P(\theta)\| \leq 2/\delta_\theta$, for all $\theta \in \Theta$.

If $x_1 \in ImP(\theta)$ and $x_2 \in KerP(\theta)$ such that $\|x_1\| = \|x_2\| = 1$, then, for every $t \geq 0$, we have

$$\|x_1 + x_2\| \geq \frac{1}{M}e^{-\omega t}\|\Phi(\theta,t)x_1 + \Phi(\theta,t)x_2\| \geq \frac{1}{M}e^{-\omega t}\Big(\frac{1}{N}e^{\nu t} - Ne^{-\nu t}\Big),$$

where M, ω are given by Definition 2.2 and N, ν are given by Definition 2.3. It follows that there is $c > 0$ such that $\delta_\theta \geq c$, for all $\theta \in \Theta$, which ends the proof of (i). For (ii) and (iii) see [15]. □

Definition 2.5. Let $\pi = (\Phi, \sigma)$ be a linear skew-product semiflow on $\mathcal{E} = X \times \Theta$ and let $(x, \theta) \in \mathcal{E}$. We say that Φ *has a negative continuation* relative to (x, θ) if there is a function $\varphi : \mathbf{R}_- \to X$ such that $\varphi(0) = x$ and $\varphi(s + t) = \Phi(\sigma(\theta, s), t)\varphi(s)$, for all $(s, t) \in \mathbf{R}_- \times \mathbf{R}_+$ with $s + t \leq 0$.

For a linear skew-product semiflow $\pi = (\Phi, \sigma)$ on $\mathcal{E} = X \times \Theta$, for every $\theta \in \Theta$, we consider the linear subspaces

$$X_1(\theta) = \{x \in X : \sup_{t \geq 0} \|\Phi(\theta, t)x\| < \infty\}$$

$$X_2(\theta) = \{x \in X : \Phi \text{ has a negative continuation } \varphi \text{ relative to } (x, \theta)$$

$$\text{such that } \sup_{s \leq 0} \|\varphi(s)\| < \infty\}.$$

Lemma 2.6. *If $\pi = (\Phi, \sigma)$ is linear skew-product semiflow on $\mathcal{E} = X \times \Theta$ then $\Phi(\theta, t)X_1(\theta) \subset X_1(\sigma(\theta, t))$ and $\Phi(\theta, t)X_2(\theta) \subset X_2(\sigma(0, t))$, for all $(\theta, t) \in \Theta \times \mathbf{R}_+$.*

Proof. Let $(\theta, t) \in \Theta \times \mathbf{R}_+$. The first inclusion obviously holds. To prove the second, let $x \in X_2(\theta)$ and let φ be a negative continuation of Φ relative to (x, θ) with $\sup_{s \leq 0} \|\varphi(s)\| < \infty$. We denote by $y = \Phi(\theta, t)x$ and we define the function $\psi : \mathbf{R}_- \to X, \psi(s) = \Phi(\sigma(\theta, s), t)\varphi(s)$. It is easy to see that ψ is a negative continuation of Φ relative to $(y, \sigma(\theta, t))$ and $\sup_{s \leq 0} \|\psi(s)\| < \infty$, so $y \in X_2(\sigma(\theta, t))$.□

Proposition 2.7. *Let $\pi = (\Phi, \sigma)$ be a linear skew-product semiflow on $\mathcal{E} = X \times \Theta$. If π is uniformly exponentially dichotomic relative to the family of projections $\{P(\theta)\}_{\theta \in \Theta}$, then $ImP(\theta) = X_1(\theta)$ and $KerP(\theta) = X_2(\theta)$, for all $\theta \in \Theta$.*

Proof. It can be done in a similar manner like Proposition 2.2 from [5]. □

Remark 2.8. From the previous theorem it follows that if the linear skew-product semiflow $\pi = (\Phi, \sigma)$ is uniformly exponentially dichotomic, then the family of projections $\{P(\theta)\}_{\theta \in \Theta}$ is uniquely determined by the conditions from Definition 2.3.

3. Uniform admissibility and exponential dichotomy

In what follows, we shall generalize a characterization for uniform exponential dichotomy of linear skew-product semiflows in terms of the solvability of an attached integral equation, obtained in [15]. As in the second section, let X be a Banach space, let Θ be a metric space and let $\mathcal{E} = X \times \Theta$.

Let $C_b(\mathbf{R}_+, X)$ be the space of all bounded continuous functions $u : \mathbf{R}_+ \to X$ which is a Banach space with respect to the norm $|||u||| = \sup_{t \geq 0} \|u(t)\|$.

We denote by $C_c(\mathbf{R}_+, X)$ the space of all continuous functions $u : \mathbf{R}_+ \to X$ with compact support contained in $(0, \infty)$ and let $\Delta(\mathbf{R}_+, X) = \{u \in C_c(\mathbf{R}_+, X) : |||u||| \leq 1\}$.

Let $\pi = (\Phi, \sigma)$ be a linear skew-product semiflow on $\mathcal{E} = X \times \Theta$. For every $\theta \in \Theta$ we consider the integral equation (E_θ) given by

$$f(t) = \Phi(\sigma(\theta, s), t - s)f(s) + \int_s^t \Phi(\sigma(\theta, \tau), t - \tau)u(\tau)\, d\tau, \quad t \geq s \geq 0$$

with $u \in \Delta(\mathbf{R}_+, X)$ and $f \in C_b(\mathbf{R}_+, X)$.

We consider the set $D_\theta = \{f \in C_b(\mathbf{R}_+, X) : f(0) \in X_2(\theta) \text{ and } \exists u \in \Delta(\mathbf{R}_+, X) \text{ such that } (f, u) \text{ verifies } (E_\theta)\}$.

Definition 3.1. The pair $(C_b(\mathbf{R}_+, X), \Delta(\mathbf{R}_+, X))$ is said to be *uniformly admissible* for the linear skew-product semiflow $\pi = (\Phi, \sigma)$ on $\mathcal{E} = X \times \Theta$ if the following conditions are satisfied

(i) for every $(\theta, u) \in \Theta \times \Delta(\mathbf{R}_+, X)$ there is $f_{\theta,u} \in C_b(\mathbf{R}_+, X)$ such that $(f_{\theta,u}, u)$ verifies the equation (E_θ);
(ii) if $\theta \in \Theta$ and $f_\theta \in D_\theta$ such that $(f_\theta, 0)$ verifies (E_θ) then $f_\theta = 0$;
(iii) there is $L > 0$ such that $|||f_\theta||| \leq L$, for all $f_\theta \in D_\theta$ and all $\theta \in \Theta$.

Lemma 3.2. *If the pair $(C_b(\mathbf{R}_+, X), \Delta(\mathbf{R}_+, X))$ is uniformly admissible for the linear skew-product semiflow $\pi = (\Phi, \sigma)$, then $X_1(\theta) \cap X_2(\theta) = \{0\}$, for all $\theta \in \Theta$.*

Proof. Let $\theta \in \Theta$ and $x \in X_1(\theta) \cap X_2(\theta)$. We consider the function $f_\theta : \mathbf{R}_+ \to X, f_\theta(t) = \Phi(\theta, t)x$ and we have that $f_\theta \in C_b(\mathbf{R}_+, X)$. We observe that the pair $(f_\theta, 0)$ verifies the equation (E_θ) and $f_\theta(0) = x \in X_2(\theta)$, so $f_\theta \in D_\theta$. Since the pair $(C_b(\mathbf{R}_+, X), \Delta(\mathbf{R}_+, X))$ is uniformly admissible for π it results that $f_\theta = 0$ and hence $x = f_\theta(0) = 0$. □

Theorem 3.3. *Let $\pi = (\Phi, \sigma)$ be a linear skew-product semiflow on $\mathcal{E} = X \times \Theta$. If the pair $(C_b(\mathbf{R}_+, X), \Delta(\mathbf{R}_+, X))$ is uniformly admissible for π, then there are $N \geq 1, \nu > 0$ such that*

$$\|\Phi(\theta, t)x\| \leq Ne^{-\nu t}\|x\|, \quad \forall x \in X_1(\theta), \forall(\theta, t) \in \Theta \times \mathbf{R}_+.$$

Proof. By hypothesis, there is $\nu > 0$ such that

$$|||f_\theta||| \leq \frac{1}{\nu}, \quad \forall f_\theta \in D_\theta, \forall \theta \in \Theta. \tag{3.1}$$

Let $\theta \in \Theta, x \in X_1(\theta) \setminus \{0\}, t_0 = \sup\{t \geq 0 : \Phi(\theta, t)x \neq 0\}$ and $n_0 \in \mathbf{N}^*$ with $2/n_0 < t_0$. For every $n \in \mathbf{N}, n \geq n_0$ let $\alpha_n : \mathbf{R}_+ \to [0, 1]$ be a continuous function with compact support contained in $(0, t_0)$ and $\alpha_n(t) = 1$, for all $t \in [1/n, \min\{n, t_0 - 1/n\}]$. We define the functions

$$u_n : \mathbf{R}_+ \to X, \quad u_n(t) = \alpha_n(t) \frac{\Phi(\theta, t)x}{\|\Phi(\theta, t)x\|}$$

$$f_n : \mathbf{R}_+ \to X, \quad f_n(t) = \int_0^t \frac{\alpha_n(\tau)}{\|\Phi(\theta, \tau)x\|} d\tau \; \Phi(\theta, t)x.$$

We have that $u_n \in \Delta(\mathbf{R}_+, X)$ and because $x \in X_1(\theta)$, it follows that $f_n \in C_b(\mathbf{R}_+, X)$, for all $n \geq n_0$. Moreover, it is easy to see that $f_n(0) = 0$ and the pair (f_n, u_n) verifies the equation (E_θ), for every $n \geq n_0$. It follows that $f_n \in D_\theta$, for all $n \geq n_0$. From (3.1) we obtain that

$$\int_0^t \frac{\alpha_n(s)}{\|\Phi(\theta, s)x\|} ds \leq \frac{1}{\nu} \frac{1}{\|\Phi(\theta, t)x\|}, \qquad \forall t \in [0, t_0), \forall n \geq n_0.$$

For $n \to \infty$ it results

$$\int_0^t \frac{1}{\|\Phi(\theta, s)x\|} ds \leq \frac{1}{\nu} \frac{1}{\|\Phi(\theta, t)x\|}, \qquad \forall t \in [0, t_0). \tag{3.2}$$

Let M, ω be given by Definition 2.2. We shall prove that

$$\|\Phi(\theta, t)x\| \leq N e^{-\nu t} \|x\|, \quad \forall t \in [0, t_0), \tag{3.3}$$

where $N = M e^{\omega + \nu}/\nu$. Indeed, for $t \in [0, 1]$ the relation (3.3) obviously holds. If $t_0 > 1$ let

$$F : [1, t_0) \to \mathbf{R}_+^*, \quad F(t) = \int_0^t \frac{1}{\|\Phi(\theta, s)x\|} ds.$$

Using (3.2) we obtain

$$F(1)\, e^{\nu(t-1)} \leq F(t) \leq \frac{1}{\nu} \frac{1}{\|\Phi(\theta, t)x\|}, \qquad \forall t \in [1, t_0),$$

so $\|\Phi(\theta, t)x\| \leq (e^\nu/\nu F(1))\, e^{-\nu t}$, for all $t \in [1, t_0)$. Taking into account that $F(1) \geq 1/(M e^\omega \|x\|)$ we deduce that (3.3) holds for $t \in [1, t_0)$. It follows $\|\Phi(\theta, t)x\| \leq N e^{-\nu t} \|x\|$, for all $t \geq 0$. Since ν and N do not depend on θ or x we obtain the conclusion. □

Corollary 3.4. *Let $\pi = (\Phi, \sigma)$ be a linear skew-product semiflow on $\mathcal{E} = X \times \Theta$. If the pair $(C_b(\mathbf{R}_+, X), \Delta(\mathbf{R}_+, X))$ is uniformly admissible for π, then $X_1(\theta)$ is a closed linear subspace, for all $\theta \in \Theta$.*

Proof. Let $\theta \in \Theta$ be fixed and let $(x_p) \subset X_1(\theta)$ converging to $x \in X$. It follows that there is $L > 0$ such that $\|x_p\| \leq L$, for all $p \in \mathbf{N}$. If N, ν are given by Theorem 3.3, we deduce that $\|\Phi(\theta, t)x_p\| \leq N L e^{-\nu t}$ for all $t \geq 0$ and all $p \in \mathbf{N}$. Hence, we obtain $\|\Phi(\theta, t)x\| \leq N L e^{-\nu t}$ for all $t \geq 0$, so $x \in X_1(\theta)$. It follows that $X_1(\theta)$ is closed. □

Theorem 3.5. *Let $\pi = (\Phi, \sigma)$ be a linear skew-product semiflow on $\mathcal{E} = X \times \Theta$. If the pair $(C_b(\mathbf{R}_+, X), \Delta(\mathbf{R}_+, X))$ is uniformly admissible for π, then there exist $N \geq 1$ and $\nu > 0$ such that*

$$\|\Phi(\theta, t)x\| \geq \frac{1}{N}\, e^{\nu t}\, \|x\|, \quad \forall x \in X_2(\theta), \forall(\theta, t) \in \Theta \times \mathbf{R}_+.$$

Proof. Let $\nu \in (0,1)$ such that $|||f_\theta||| \leq 1/\nu$, for all $f_\theta \in D_\theta$ and all $\theta \in \Theta$. Let $\theta \in \Theta$ and $x \in X_2(\theta) \setminus \{0\}$. From Lemma 3.2 it follows that $\Phi(\theta, t)x \neq 0$, for all $t \geq 0$.

For every $n \in \mathbf{N}^*$, let $\alpha_n : \mathbf{R}_+ \to [0,1]$ be a continuous function with compact support contained in $(0, \infty)$ and with $\alpha_n(t) = 1$, for all $t \in [1/n, n]$. For every $n \in \mathbf{N}^*$, we define

$$u_n : \mathbf{R}_+ \to X, \quad u_n(t) = -\alpha_n(t) \frac{\Phi(\theta, t)x}{\|\Phi(\theta, t)x\|}$$

and

$$f_n : \mathbf{R}_+ \to X, \quad f_n(t) = \int_t^\infty \frac{\alpha_n(\tau)}{\|\Phi(\theta, \tau)x\|}\, d\tau\, \Phi(\theta, t)x.$$

Then $u_n \in \Delta(\mathbf{R}_+, X), f_n \in C_b(\mathbf{R}_+, X)$ and (f_n, u_n) verifies the equation (E_θ), for all $n \in \mathbf{N}^*$. Moreover

$$f_n(0) = \int_0^\infty \frac{\alpha_n(\tau)}{\|\Phi(\theta, \tau)x\|}\, d\tau\; x \in X_2(\theta),$$

so $f_n \in D_\theta$, for all $n \in \mathbf{N}^*$. Using an analogous argument as in the proof of Theorem 3.3 it follows

$$\int_t^\infty \frac{1}{\|\Phi(\theta, \tau)x\|}\, d\tau \leq \frac{1}{\nu} \frac{1}{\|\Phi(\theta, t)x\|}, \qquad \forall t \geq 0.$$

Let

$$F : \mathbf{R}_+ \to \mathbf{R}_+, \quad F(t) = \int_t^\infty \frac{1}{\|\Phi(\theta, \tau)x\|}\, d\tau.$$

It results that $\nu F(t) \leq -\dot{F}(t)$, for all $t \geq 0$, so

$$F(t) \leq e^{-\nu t} F(0) \leq \frac{1}{\nu}\, e^{-\nu t}\, \frac{1}{\|x\|}, \qquad \forall t \geq 0. \tag{3.4}$$

If M and ω are given by Definition 2.2, then

$$\frac{1}{\|\Phi(\theta, t)x\|} \leq M e^{\omega} \int_t^{t+1} \frac{1}{\|\Phi(\theta, \tau)x\|}\, d\tau \leq M e^{\omega} F(t), \quad \forall t \geq 0,$$

and from (3.4) we obtain $\|\Phi(\theta, t)x\| \geq \nu e^{\nu t}/M e^{\omega}\|x\|$, for all $(x, \theta, t) \in \mathcal{E} \times \mathbf{R}_+$, which ends the proof. □

Corollary 3.6. *If the pair $(C_b(\mathbf{R}_+, X), \Delta(\mathbf{R}_+, X))$ is uniformly admissible for the linear skew-product semiflow π on $\mathcal{E} = X \times \Theta$, then $X_2(\theta)$ is a closed linear subspace, for all $\theta \in \Theta$.*

Proof. Let $\theta \in \Theta$. If $y \in X_2(\theta)$ and φ is a negative continuation for Φ relative to (y,θ) with $\sup_{s\le 0} \|\varphi(s)\| < \infty$, then it is easy to see that $\varphi(s) \in X_2(\sigma(\theta,s))$ for all $s \le 0$.

Let $(x_p) \subset X_2(\theta)$ converging to $x \in X$. For every $p \in \mathbf{N}$ there is a negative continuation φ_p for Φ relative to (x_p,θ) such that $\sup_{s\le 0} \|\varphi_p(s)\| < \infty$. Since

$$\varphi_p(s+t) = \Phi(\sigma(\theta,s),t)\varphi_p(s), \quad \forall s \le 0, t \ge 0, s+t \le 0, \forall p \in \mathbf{N}, \tag{3.5}$$

for N,ν given by Theorem 3.5, it follows that

$$\begin{aligned} \|x_p - x_k\| &= \|\Phi(\sigma(\theta,-t),t)(\varphi_p(-t) - \varphi_k(-t))\| \\ &\ge \frac{1}{N} e^{\nu t} \|\varphi_p(-t) - \varphi_k(-t)\|, \qquad \forall t \ge 0, \forall p,k \in \mathbf{N}. \end{aligned} \tag{3.6}$$

Using the fact that $(x_p)_{p\in\mathbf{N}}$ is fundamental, from (3.6) it follows that for every $s \le 0$ the sequence $(\varphi_p(s))_p$ is fundamental, so it is convergent. We denote by $\varphi(s) := \lim_{p\to\infty} \varphi_p(s)$, for all $s \le 0$. Hence, $\varphi(0) = x$ and from (3.5) we obtain $\varphi(s+t) = \Phi(\sigma(\theta,s),t)\varphi(s)$, for all $s \le 0, t \ge 0$ with $s+t \le 0$. From (3.6) we deduce

$$\|\varphi(-t)\| \le N e^{-\nu t} \|x_p - x\| + \|\varphi_p(-t)\|, \quad \forall t \ge 0, \forall p \in \mathbf{N}.$$

It results that $\sup_{s\le 0} \|\varphi(s)\| < \infty$, so $x \in X_2(\theta)$, and hence $X_2(\theta)$ is closed. □

Proposition 3.7. *If the pair $(C_b(\mathbf{R}_+,X), \Delta(\mathbf{R}_+,X))$ is uniformly admissible for the linear skew-product semiflow π on $\mathcal{E} = X \times \Theta$ and $X_1(\theta) + X_2(\theta) = X$, for all $\theta \in \Theta$, then $X_2(\sigma(\theta,t)) = \Phi(\theta,t)X_2(\theta)$, for all $(\theta,t) \in \Theta \times \mathbf{R}_+$.*

Proof. Let $M,\omega \in (0,\infty)$ be given by Definition 2.2. Let $(\theta,t) \in \Theta \times \mathbf{R}_+^*$. To prove that $X_2(\sigma(\theta,t)) \subset \Phi(\theta,t)X_2(\theta)$, it is sufficient to show that for every $x \in X_2(\sigma(\theta,t))$ with $\|x\| \le 1/Me^{2\omega}$, we have that $x \in \Phi(\theta,t)X_2(\theta)$.

Let $x \in X_2(\sigma(\theta,t))$ with $\|x\| \le 1/Me^{2\omega}$. Let $\alpha : \mathbf{R}_+ \to [0,1]$ be a continuous function with compact support contained in $(t,t+2)$ and with $\int_t^\infty \alpha(s)\, ds = 1$. We consider the functions

$$u : \mathbf{R}_+ \to X, \quad u(\tau) = -\alpha(\tau)\Phi(\sigma(\theta,t),\tau - t)x$$

$$f : \mathbf{R}_+ \to X, \quad f(\tau) = \begin{cases} x, & \tau \in [0,t] \\ \int_\tau^\infty \alpha(s)\, ds\, \Phi(\sigma(\theta,t),\tau-t)x, & \tau \ge t. \end{cases}$$

We observe that $u \in \Delta(\mathbf{R}_+,X), f \in C_b(\mathbf{R}_+,X)$ and

$$f(\tau) = \Phi(\sigma(\theta,s),\tau-s)f(s) + \int_s^\tau \Phi(\sigma(\theta,\xi),\tau-\xi)u(\xi)\, d\xi, \quad \forall \tau \ge s \ge t.$$

By hypothesis there is $g \in C_b(\mathbf{R}_+,X)$ such that the pair (g,u) verifies the equation (E_θ). It follows that $f(\tau) - g(\tau) = \Phi(\sigma(\theta,t),\tau-t)(f(t)-g(t))$ for all $\tau \ge t$. Since $f,g \in C_b(\mathbf{R}_+,X)$ we have $f(t) - g(t) \in X_1(\sigma(\theta,t))$.

But $g(t) = \Phi(\theta,t)g(0)$ and from hypothesis there are $y_1 \in X_1(\theta)$ and $y_2 \in X_2(\theta)$ such that $g(0) = y_1 + y_2$. It results that $x - \Phi(\theta,t)y_2 = (f(t)-g(t)) +$

$\Phi(\theta,t)y_1$. From Lemma 2.6 and Lemma 3.2 we deduce that $x - \Phi(\theta,t)y_2 = 0$, so $x \in \Phi(\theta,t)X_2(\theta)$. In this manner, we deduce that $X_2(\sigma(\theta,t)) \subset \Phi(\theta,t)X_2(\theta)$. Applying once again Lemma 2.6 we obtain the conclusion. □

Theorem 3.8. *Let $\pi = (\Phi,\sigma)$ be a linear skew-product semiflow on $\mathcal{E} = X \times \Theta$. Then π is uniformly exponentially dichotomic if and only if the pair $(C_b(\mathbf{R}_+, X), \Delta(\mathbf{R}_+, X))$ is uniformly admissible for π and $X_1(\theta) + X_2(\theta) = X$, for all $\theta \in \Theta$.*

Proof. Necessity. Suppose that π is uniformly exponentially dichotomic relative to the family of projections $\{P(\theta)\}_{\theta\in\Theta}$. From Proposition 2.7 we have $X_1(\theta) = ImP(\theta)$ and $X_2(\theta) = KerP(\theta)$, for all $\theta \in \Theta$, so $X_1(\theta)+X_2(\theta) = X$, for all $\theta \in \Theta$.

Let $\theta \in \Theta$ and $u \in \Delta(\mathbf{R}_+, X)$. We define the function $f_{\theta,u} : \mathbf{R}_+ \to X$,

$$\begin{aligned} f_{\theta,u}(t) = &\int_0^t \Phi(\sigma(\theta,s),t-s)P(\sigma(\theta,s))u(s)\,ds \\ &- \int_t^\infty \Phi(\sigma(\theta,t),s-t)_|^{-1}(I-P(\sigma(\theta,s)))u(s)\,ds, \end{aligned} \tag{3.7}$$

where $\Phi(\sigma(\theta,t),s-t)_|^{-1}$ denotes the inverse of the operator $\Phi(\sigma(\theta,t),s-t)_|$: $KerP(\sigma(\theta,t)) \to KerP(\sigma(\theta,s))$. It follows that $f_{\theta,u} \in C_b(\mathbf{R}_+, X)$ and the pair $(f_{\theta,u}, u)$ verifies the equation (E_θ). Moreover

$$f_{\theta,u}(0) = -\int_0^\infty \Phi(\theta,s)_|^{-1}(I-P(\sigma(\theta,s)))u(s)\,ds \in X_2(\theta).$$

Let $\theta \in \Theta$ and $f_\theta \in D_\theta$. Then, $f_\theta(0) \in X_2(\theta)$ and there is $u \in \Delta(\mathbf{R}_+, X)$ such that the pair (f_θ, u) verifies the equation (E_θ). Let $f_{\theta,u}$ be given by (3.7) and let $g = f_\theta - f_{\theta,u}$. Then, $g(0) \in X_2(\theta)$ and

$$g(t) = \Phi(\theta,t)g(0), \quad \forall t \geq 0. \tag{3.8}$$

Since $g \in C_b(\mathbf{R}_+, X)$, by (3.8) we deduce that $g(0) \in X_1(\theta)$. It follows that $g(0) = 0$ and from (3.8), we obtain that $f_\theta = f_{\theta,u}$.

From these arguments, we have that if $\theta \in \Theta$ and $f_\theta \in D_\theta$ such that $(f_\theta, 0)$ verifies the equation (E_θ), then $f_\theta = 0$. Moreover, if $\theta \in \Theta$, $f_\theta \in D_\theta$ and N, ν are given by Definition 2.3, we obtain that $\|f_\theta(t)\| \leq 2N(K+1)/\nu$, for all $t \geq 0$, where $K = \sup\limits_{\theta\in\Theta} \|P(\theta)\|$. Denoting by $L = 2N(K+1)/\nu$, we conclude that $|||f_\theta||| \leq L$, for all $f_\theta \in D_\theta$ and all $\theta \in \Theta$. So, the pair $(C_b(\mathbf{R}_+, X), \Delta(\mathbf{R}_+, X))$ is uniformly admissible for π.

Sufficiency. From hypothesis, Corollary 3.4, Corollary 3.6 and Lemma 3.2 it follows that $X_1(\theta) \oplus X_2(\theta) = X$, for all $\theta \in \Theta$. For every $\theta \in \Theta$ let $P(\theta)$ be the projection corresponding to $X_1(\theta)$ such that $ImP(\theta) = X_1(\theta)$ and $KerP(\theta) = X_2(\theta)$. Using Lemma 2.6 it follows that $\Phi(\theta,t)P(\theta) = P(\sigma(\theta,t))\Phi(\theta,t)$, for all $(\theta,t) \in \Theta \times \mathbf{R}_+$. From Theorem 3.5 and Proposition 3.7 it follows that the restriction $\Phi(\theta,t)_|$: $KerP(\theta) \to KerP(\sigma(\theta,t))$ is an isomorphism. Finally, using Theorem 3.3 and Theorem 3.5 we obtain that π is uniformly exponentially dichotomic. □

Remark 3.9. The above result is a generalization of Theorem 3.2 from [15].

Acknowledgment

The authors would like to thank the referee for important suggestions and comments, that essentially led to the improvement of the paper.

References

[1] A. Ben-Artzi, I. Gohberg, *Dichotomies of Systems and Invertibility of Linear Ordinary Differential Operators.* Oper. Theory Adv. Appl. **56** (1992), 90–119.

[2] A. Ben-Artzi, I. Gohberg, *Dichotomies of Perturbed Time-Varying Systems and the Power Method*, Indiana Univ. Math. J. **42** (1993), 699–720.

[3] A. Ben-Artzi, I. Gohberg, M.A. Kaashoek, *Invertibility and Dichotomy of Differential Operators on the Half-Line.* J. Dynam. Differential Equations **5** (1993), 1–36.

[4] C. Chicone, Y. Latushkin, *Evolution Semigroups in Dynamical Systems and Differential Equations*, Math. Surveys and Monographs, vol. 70, 1999.

[5] S.-N. Chow, H. Leiva, *Existence and Roughness of the Exponential Dichotomy for Linear Skew-Product Semiflows in Banach Spaces.* J. Differential Equations **120** (1995), 429–477.

[6] S.-N. Chow, H. Leiva, *Unbounded Perturbation of the Exponential Dichotomy for Evolution Equations.* J. Differential Equations **129** (1996), 509–531.

[7] J. Daleckii, M. Krein, *Stability of Differential Equations in Banach Space.* Amer. Math. Soc., Providence, RI, 1974.

[8] D. Henry, *Geometric Theory of Semilinear Parabolic Equations.* Springer-Verlag, New York, 1981.

[9] Y. Latushkin, S. Montgomery-Smith, T. Randolph, *Evolutionary Semigroups and Dichotomy of Linear Skew-Product Flows on Locally Compact Spaces with Banach Fibers.* J. Differential Equations **125** (1996), 73–116.

[10] Y. Latushkin, R. Schnaubelt, *Evolution Semigroups, Translation Algebras and Exponential Dichotomy of Cocycles.* J. Differential Equations **159** (1999), 321–369.

[11] B.M. Levitan, V.V. Zhikov, *Almost Periodic Functions and Differential Equations.* Cambridge University Pres, 1982.

[12] J.L. Massera, J.J. Schäffer, *Linear Differential Equations and Function Spaces.* Academic Press, New York, 1966.

[13] M. Megan, B. Sasu, A.L. Sasu, *On Nonuniform Exponential Dichotomy of Evolution Operators in Banach Spaces.* Integral Equations Operator Theory, **44** (2002), 71–78.

[14] M. Megan, A.L. Sasu, B. Sasu, *Discrete Admissibility and Exponential Dichotomy for Evolution Families.* Discrete Contin. Dynam. Systems **9** (2003), 383–397.

[15] M. Megan, A.L. Sasu, B. Sasu, *On Uniform Exponential Dichotomy of Linear Skew-Product Semiflows.* Bull. Belg. Math. Soc. Simon Stevin **10** (2003), 1–21.

[16] M. Megan, A.L. Sasu, B. Sasu, *Perron Conditions for Uniform Exponential Expansiveness of Linear Skew-Product Flows.* Monatsh. Math. **138** (2003), 145–157.

[17] M. Megan, A.L. Sasu, B. Sasu, *Perron Conditions for Pointwise and Global Exponential Dichotomy of Linear Skew-Product Flows.* Integral Equations Operator Theory, to appear.

[18] M. Megan, A.L. Sasu, B. Sasu, *Theorems of Perron Type for Uniform Exponential Stability of Linear Skew-Product Semiflows.* Dynam. Contin. Discrete Impulsive Systems, to appear.

[19] N. Van Minh, F. Räbiger, R. Schnaubelt, *Exponential Stability, Exponential Expansiveness and Exponential Dichotomy of Evolution Equations on the Half-Line.* Integral Equations Operator Theory **32** (1998), 332–353.

[20] V.A. Pliss, G.R. Sell, *Robustness of Exponential Dichotomies in Infinite-Dimensional Dynamical Systems.* J. Dynam. Differential Equations **3** (1999), 471–513.

[21] R.J. Sacker, G.R. Sell, *Dichotomies for Linear Evolutionary Equations in Banach Spaces.* J. Differential Equations **113** (1994), 17–67.

Mihail Megan
Department of Mathematics
West University of Timişoara
Bul. V. Pârvan 4
300223-Timişoara
Romania
e-mail: `megan@math.uvt.ro`

Adina Luminiţa Sasu
Department of Mathematics
West University of Timişoara
Bul. V. Pârvan 4
300223-Timişoara
Romania
e-mail: `sasu@math.uvt.ro`

Bogdan Sasu
Department of Mathematics
West University of Timişoara
Bul. V. Pârvan 4
300223-Timişoara
Romania
e-mail: `lbsasu@yahoo.com`

Operator Theory:
Advances and Applications, Vol. 153, 197–209

On a Class of Stochastic Integral Operators of McShane Type

Romeo Negrea

Abstract. The aim of this note is to give an extension of a result by McShane to a general stochastic integral operator with a non-Lipschitz condition on the coefficient functions.

Mathematics Subject Classification (2000). Primary 60H20, 62L20; Secondary 41A35.

Keywords. McShane integral, Stochastic integral equations, Stochastic integral operators.

1. Introduction

Let $(\Omega, \mathcal{A}, \mathbb{P})$ be a complete probability space and let $\mathcal{F} = \{\mathcal{F}_t,\ t \in [0,T]\}$ ($T \in \mathbb{R}_+$) be a complete filtration of the measurable space $(\Omega, \mathcal{A})$. If $f, g : [0,T] \times \mathbb{R} \to \mathbb{R}$ are continuous functions in (t,x), a well-known class of stochastic equations is given by the stochastic integral equation of Itô-Doob type

$$x(t,\omega) = x_0(\omega) + \int_0^t f(\tau, x(\tau,\omega))d\tau + \int_0^t g(\tau, x(\tau,\omega))dW(\tau) \qquad (*)$$

where $t \in [0,T]$, the first integral is a Lebesgue integral and the second is an Itô-type stochastic integral defined with respect to a scalar Brownian motion process $\{W(t),\ t \in [0,T]\}$.

The operators W_1 and W_2 from $\mathcal{C}([0,T], L_2(\Omega,\mathcal{F},\mathbb{P}))$ into $\mathcal{C}([0,T], L_2(\Omega,\mathcal{F},\mathbb{P}))$ defined by

$$(W_1 x)(t,\omega) = \int_0^t x(\tau,\omega)d\tau \qquad \text{and} \qquad (W_2 x)(t,\omega) = \int_0^t x(t,\omega)dW(\tau)$$

are linear and continuous. If $f(t,\cdot)$ and $g(t,\cdot)$ are linear for every fixed $t \in [0,T]$, then there exists a unique random solution to equation $(*)$, i.e., $x(t,\omega)$ belongs to $\mathcal{C}([0,T], L_2(\Omega,\mathcal{F},\mathbb{P}))$ and satisfies the equation $\mathbb{P}$ a.e.

A generalization of this class of stochastic integral equations has been given by McShane ([5]) with a weaker condition for the integrator process. Moreover, a stochastic calculus and a method of constructing mathematical models which permit a unified theory that applies equally well when the input noise is determinate and smooth and when it is "white" or "coloured", has introduced by E.J. McShane ([5], [6]).

The problem of the existence and uniqueness of the solutions processes in stochastic integral equations of McShane's type has been treated by E.J. McShane himself ([5], [6]) and by K.P. Elworthy (1982), J.M.A. Ibáñez and A.G. Jáimez ([4]), Ph. Protter (1992), A. Constantin ([2], [3]). We consider the problem of the existence and uniqueness of the solution processes in stochastic integral equations of McShane's type:

$$\begin{aligned} X(t) = X(t_0) + \int_{t_0}^{t} f(s, X(s))ds + \sum_{\rho=1}^{r} \int_{t_0}^{t} g_\rho(s, X(s))dz^\rho(s) \\ + \sum_{\rho,\sigma=1}^{r} \int_{t_0}^{t} h_{\rho\sigma}(s, X(s))dz^\rho(s)dz^\sigma(s), \quad t \in [t_0, T], T > t_0 \geq 0 \end{aligned} \tag{1.1}$$

with non-Lipschitz conditions on $f, g_\rho, h_{\rho,\sigma}$; see T. Yamada and S. Watanabe ([8], [9]), T. Yamada ([10]) and T. Taniguchi ([7]).

2. Definitions and previous results

Let $[t_0, T]$ be a closed real interval ($t_0 > 0$). Let $(\Omega, \mathcal{A}, \mathbb{P})$ be a probability space and $\mathcal{F} = \{\mathcal{F}_t, t \in [t_0, T]\}$ a complete filtration of the measurable space $(\Omega, \mathcal{A})$.

We consider the following types of processes: $z(t)$, $t \geq 0$, which satisfy a $\mathbb{K}$ condition (E.J. McShane [6], Definition 5.3, p. 132), i.e., $\|z(t_0)\|$ is finite and there exists a number K such that whenever $t_0 \leq u \leq v \leq T$, then a.s.

$$\begin{aligned} |\mathbb{E}(z(v) - z(u)/\mathcal{F}_u)| \leq K(v-u) \\ \mathbb{E}([z(v) - z(u)]^2/\mathcal{F}_u) \leq K(v-u). \end{aligned} \tag{2.1}$$

It is known (McShane [6], Lemma 5.5, p. 132–133) that if z^ρ satisfy a $\mathbb{K}$ condition and $\mathbb{E}(\|g_\rho(t)\|^2)$, ($\rho = 1, \ldots, r$), are Lebesgue integrable on $[t_0, T]$, then

$$\left\| \int_{t_0}^{T} g_\rho(t)dz^\rho(t) \right\| \leq C\left\{ \int_{t_0}^{T} \|g_\rho(t)\|^2 dt \right\}^{\frac{1}{2}}, \quad \rho = 1, \ldots, r, \tag{2.2}$$

where $C = 2K(T-t_0)^{\frac{1}{2}} + K^{\frac{1}{2}}$.

Moreover, if z^ρ satisfy a $\mathbb{K}$ condition and an additing condition

$$\mathbb{E}([z^\rho(t) - z^\rho(s)]^4/\mathcal{F}_s) \leq K(t-s) \tag{2.3}$$

(McShane [6], Lemma 11.3, p. 153) and $h_{\rho,\sigma}$ are $\mathcal{F}_t$-adapted and $\mathbb{E}(\|h_{\rho,\sigma}(t)\|^2)$ are Lebesgue integrable from t_0 to T, then:

$$\left\| \int_{t_0}^{T} h_{\rho,\sigma}(t)dz^{\rho}(t)dz^{\sigma}(t) \right\| \leq C\left\{ \int_{t_0}^{T} \|h_{\rho,\sigma}(t)\|^2 dt \right\}^{\frac{1}{2}} \tag{2.4}$$

for a same constant C as in (2.2).

3. Existence and uniqueness

In this paragraph we prove the existence and uniqueness of solution processes for the equation of type (1.1). Previously we define the set of hypotheses we will require in relation to the functions that intervene in the integrals of (1.1).

3.1. Hypotheses

We consider the following set of hypotheses relative to the function f, g_ρ and $h_{\rho,\sigma}$ and the integrators z^ρ, $\rho, \sigma = 1, 2, \ldots, r$:

[H-(i)] *$z^1, \ldots, z^r$ satisfy a $\mathbb{K}$ condition and for each s and $t > s$ in $[t_0, T]$ a.s. $\mathbb{E}([z^\rho(t) - z^\rho(s)]^4/\mathcal{F}_s) \leq K(t-s)$;*

[H-(ii)] *functions $f, g_\rho, h_{\rho,\sigma}$ are bounded functions on $[t_0, T] \times \mathbb{R}$, $0 \leq t_0 < T$ and continuous in x for each fixed $t \in [t_0, T]$;*

[H-(iii)] *there exists a function $H(t,u) : [t_0, T] \times \mathbb{R}^+ \longrightarrow \mathbb{R}^+$ such that*

$$\mathbb{E}|f(t,X)|^2 + \sum_{\rho=1}^{r} \mathbb{E}|g_\rho(t,X)|^2 + \sum_{\rho,\sigma=1}^{r} \mathbb{E}|h_{\rho\sigma}(t,X)|^2 \leq H(t, \mathbb{E}|X|^2) \tag{3.1}$$

for all $t \in [t_0, T]$ and all $X \in S(X_0, R) = \{X \in L^2(\Omega, \mathbb{R}) \mid \mathbb{E}\,|X - X_0|^2 \leq R\}$, where $X_0 = X(t_0)$ and $R > 0$;

[H-(iv)] *$H(t,u)$ is locally integrable in t for each fixed $u \in \mathbb{R}^+$ and is continuous, monotone, nondecreasing in u for each fixed $t \in [t_0, \infty)$.*

Remark 3.1. The hypothesis [H-(i)] is similar to that established by McShane ([5], [6]) and the hypotheses [H-(ii)]–[H-(iv)], are analogous to those established by Taniguchi ([7]).

3.2. Existence and uniqueness local theorem

First we prove a similar result as that of Taniguchi ([7], Theorem 1).

Theorem 3.2. *We assume that z^ρ satisfy the hypothesis [H-(i)] and f, g_ρ, $h_{\rho\sigma}$ satisfy the hypotheses [H-(ii)], [H-(iii)] and [H-(iv)], ($\rho, \sigma = 1, 2, \ldots, r$). Let $\{X_n(t)\}$,*

$n \geq 0$, be the sequence of stochastic processes which are defined by the successive approximation:

$$\begin{aligned} X_n(t) = &\ X_0 + \int_{t_0}^{t} f(s, X_{n-1}(s))ds + \sum_{\rho=1}^{r} \int_{t_0}^{t} g_\rho(s, X_{n-1}(s))dz^\rho(s) \\ &+ \sum_{\rho,\sigma=1}^{r} \int_{t_0}^{t} h_{\rho\sigma}(s, X_{n-1}(s))dz^\rho dz^\sigma, \\ X_0 = &\ X(t_0) \end{aligned} \tag{3.2}$$

and let X_0 be independent of $z^\rho(t)$, $t \geq 0$ and $\mathbb{E}|X_0|^2 < \infty$.

Then, there exists a time t_1 such that $t_0 < t_1 < T$ and the sequence of the functions defined on $[t_0, t_1]$, $\{\mathbb{E}|X_n(t)|^2\}$, $t_0 < t < t_1$ is uniformly bounded.

Proof. First we note by condition [H-(iv)] and Caratheodory's Theorem that the differential equation:

$$\frac{du}{dt} = 4(rC^2 + r^2C^2 + T)H(t, u) \tag{3.3}$$

has a local solution with any initial value (t_0, u_0). $u_0 \geq 0$.

Take a $u_0 \in \mathbb{R}^+$ such that $u_0 > 4\mathbb{E}|X_0|^2$ and let $u(t) = u(t; t_0, u_0)$ be the local solution of (3.3) with the initial value (t_0, u_0). Now, by condition [H-(iii)] and the relations (2.2), (2.4), we obtain:

$$\begin{aligned} &\mathbb{E}|X_1(t)|^2 \\ &\leq 4\mathbb{E}\Bigg(|X_0|^2 + \left|\int_{t_0}^{t} f(s, X_0)ds\right|^2 + \left|\sum_{\rho=1}^{r} \int_{t_0}^{t} g_\rho(s, X_0)dz^\rho(s)\right|^2 \\ &\quad + \left|\sum_{\rho,\sigma=1}^{r} \int_{t_0}^{t} h_{\rho\sigma}(s, X_0)dz^\rho(s)dz^\sigma\right|^2\Bigg) \\ &\leq 4\mathbb{E}\Bigg(|X_0|^2 + \left|\int_{t_0}^{t} f(s, X_0)ds\right|^2 + r\sum_{\rho=1}^{r} \left|\int_{t_0}^{t} g_\rho(s, X_0)dz^\rho(s)\right|^2 \\ &\quad + r^2 \sum_{\rho,\sigma=1}^{r} \left|\int_{t_0}^{t} h_{\rho\sigma}(s, X_0)dz^\rho(s)dz^\sigma\right|^2\Bigg) \\ &\leq 4\mathbb{E}(|X_0|^2) + 4T\mathbb{E}\left(\int_{t_0}^{t} |f(s, X_0)|^2 ds\right) + 4rC^2 \sum_{\rho=1}^{r} \mathbb{E}\left(\int_{t_0}^{t} |g_\rho(s, X_0)|^2 ds\right) \\ &\quad + 4r^2C^2 \sum_{\rho,\sigma=1}^{r} \mathbb{E}\left(\int_{t_0}^{t} |h_{\rho\sigma}(s, X_0)|^2 ds\right) \end{aligned}$$

$$\le 4\mathbb{E}(|X_0|^2) + 4(rC^2 + r^2C^2 + T)\mathbb{E}\Big(\int_{t_0}^t |f(s, X_0)|^2 ds\Big)$$

$$+ 4(rC^2 + r^2C^2 + T)\sum_{\rho=1}^r \mathbb{E}\Big(\int_{t_0}^t |g_\rho(s, X_0)|^2 ds\Big)$$

$$+ 4(rC^2 + r^2C^2 + T)\sum_{\rho,\sigma=1}^r \mathbb{E}\Big(\int_{t_0}^t |h_{\rho\sigma}(s, X_0)|^2 ds\Big)$$

$$\le 4\mathbb{E}(|X_0|^2) + 4(rC^2 + r^2C^2 + T)\int_{t_0}^t H(s, \mathbb{E}(|X_0|^2))ds$$

for all $t \in [t_0, T]$, from which by condition [H-(iv)] we obtain a time T_0 such that $t_0 < T_0 < T$ and

$$u(t) - \mathbb{E}|X_1|^2 > 4(rC^2 + r^2C^2 + T)\int_{t_0}^t [H(s, u(s)) - H(s, \mathbb{E}|X_0|^2)]ds \ge 0$$

for all $t \in [t_0, T_0]$, because $u_0 > 4\mathbb{E}|X_0|^2$ and $u(t)$ is the local solution of (3.3) starting with the initial point u_0.

Next, since the function $u(t)$ is continuous on $[t_0, T_0]$, set $p_0 = \max\{u(t)| \, t \in [t_0, T_0]\} < \infty$. Thus, $H(s, u(s)) \le H(s, p_0)$ for each $s \in [t_0, T_0]$ by condition [H-(iv)]. Hence, the function $H(s, u(s))$ is integrable on $[t_0, T_0]$ by condition [H-(iv)]. Therefore, we obtain a time t_1 such that $t_0 < t_1 < T_0$ and from the above relation and the continuity of the integral, we have:

$$\mathbb{E}(|X_1(t) - X_0|^2 \le 3\mathbb{E}\Big(\Big|\int_{t_0}^t f(s, X_0)ds\Big|^2 + r\sum_{\rho=1}^r \Big|\int_{t_0}^t g_\rho(s, X_0)dz^\rho(s)\Big|^2$$

$$+ r^2 \sum_{\rho,\sigma=1}^r \Big|\int_{t_0}^t h_{\rho\sigma}(s, X_0)dz^\rho(s)dz^\sigma(s)\Big|^2\Big)$$

$$\le 3(rC^2 + r^2C^2 + T)\int_{t_0}^t H(s, \mathbb{E}(|X_0|^2))ds$$

$$\le 3(rC^2 + r^2C^2 + T)\int_{t_0}^t H(s, u(s))ds$$

$$\le 3(rC^2 + r^2C^2 + T)\int_{t_0}^t H(s, p_0)ds \le R$$

for all $t \in [t_0, t_1]$.

Now, let us continue the proof of the theorem by mathematical induction. Hence, let $n = k$ and suppose that the inequalities hold, i.e.,

$$\mathbb{E}|X_k(t)|^2 < u(t), \quad \mathbb{E}|X_k(t) - X_0|^2 \le R$$

for all $t \in [t_0, t_1]$.

Then,

$$\begin{aligned}\mathbb{E}|X_{k+1}(t)|^2 \leq 4\mathbb{E}(|X_0|^2) + 4T\mathbb{E}\Big(\int_{t_0}^t |f(s,X_k(s))|^2 ds\Big)\\ + 4rC^2\mathbb{E}\Big(\sum_{\rho=1}^r \int_{t_0}^t |g_\rho(s,X_k(s))|^2 ds\Big)\\ + 4r^2C^2\mathbb{E}\Big(\sum_{\rho,\sigma=1}^r \int_{t_0}^t |h_{\rho\sigma}(s,X_k(s))|^2 ds\Big)\\ \leq 4\mathbb{E}(|X_0|^2) + 4(rC^2 + r^2C^2 + T)\int_{t_0}^t H(s,\mathbb{E}(|X_k(s)|^2))ds\end{aligned}$$

for all $t \in [t_0, T]$, from which since $u_0 > 4\mathbb{E}|X_0|^2$, we obtain that

$$\begin{aligned}\mathbb{E}(|X_{k+1}(t)|^2) - 4(rC^2 + r^2C^2 + T)\int_{t_0}^t H(s,\mathbb{E}(|X_k(s)|^2))ds\\ < u(t) - 4(rC^2 + r^2C^2 + T)\int_{t_0}^t H(s,\mathbb{E}(|X_k(s)|^2))ds.\end{aligned}$$

Hence, by the assumption of mathematical induction,

$$u(t) - \mathbb{E}|X_{k+1}(t)|^2 > 4(rC^2 + r^2c^2 + T)\int_{t_0}^t [H(s,u(s)) - H(s,\mathbb{E}|X_k(s)|^2)]ds \geq 0$$

for all $t \in [t_0, t_1]$. Next,

$$\begin{aligned}\mathbb{E}(|X_{k+1}(t) - X_0|^2) \leq 3\mathbb{E}\Big(\Big|\int_{t_0}^t f(s,X_k(s))ds\Big|^2 + r\sum_{\rho=1}^r \Big|\int_{t_0}^t g_\rho(s,X_k(s))dz^\rho(s)\Big|^2\\ + r^2\sum_{\rho,\sigma=1}^r \Big|\int_{t_0}^t h_{\rho\sigma}(s,X_k(s))dz^\rho(s)dz^\sigma\Big|^2\Big)\\ \leq 3(rC^2 + r^2C^2 + T)\int_{t_0}^t H(s,\mathbb{E}(|X_k(s)|^2))ds\\ \leq 3(rC^2 + r^2C^2 + T)\int_{t_0}^t H(s,u(s))ds \leq R\end{aligned}$$

for all $t \in [t_0, t_1]$. Consequently, we obtain that

$$\mathbb{E}|X_{k+1}(t)|^2 < u(t), \quad \mathbb{E}|X_{k+1}(t) - X_0|^2 \leq R$$

for all $t \in [t_0, t_1]$. Since $u(t)$ is continuous on $[t_0, t_1]$, there exists a real number $M > 0$ such that $\mathbb{E}|X_n(t)|^2 < M$ for all $t \in [t_0, t_1]$ and every integer $n \geq 0$. □

Theorem 3.3. *For the stochastic differential equation* (1.1), *suppose that the following conditions are satisfied:*

(i) *z^ρ satisfy the hypotheses of previously theorem, $\rho = 1, 2, \ldots, r$;*

(ii) *the functions* $f(t,x), g_\rho(t,x), h_{\rho\sigma}(t,x)$ *satisfy hypotheses [H-(ii)], [H-(iii)] and [H-(iv)],* $\rho,\sigma = 1,2,\dots,r$;

(iii) *there exists a nonnegative real valued function* $G(t,u)$ *defined on* $[t_0,T]\times[0,4R]$ *which is monotone nondecreasing, continuous in* $u\in[0,4R]$ *for each fixed* $t\in[t_0,T]$ *and is locally integrable in* $t\in[t_0,T]$ *for each fixed* $u\in[0,4R]$ *such that*

$$G(t,0)=0,$$

$$\mathbb{E}|f(t,X)-f(t,Y)|^2+\sum_{\rho=1}^{r}\mathbb{E}|g_\rho(t,X)-g_\rho(t,Y)|^2 + \sum_{\rho,\sigma=1}^{r}\mathbb{E}|h_{\rho\sigma}(t,X)-h_{\rho\sigma}(t,Y)|^2 \le G(t,\mathbb{E}|X-Y|^2)$$

for all $t\in[t_0,T]$ *and all* $X,Y\in S(X_0,R)$;

(iv) *the function* $G(t,u)$ *satisfies a sufficient condition under which if a nonnegative, continuous function* $z(t)$ *satisfies that*

$$z(t)\le A\int_{t_0}^{t}G(s,z(s))ds \quad \textit{for all } t\in[t_0,T_1]$$

where $A=3(rC^2+r^2C^2+T)$, *and* T_1 *is a time with* $t_0<T_1\le T$, *and if* $z(t_0)=0$, *then* $z(t)\equiv 0$ *for all* $t\in[t_0,T_1]$.

Let X_0 *be a random variable independent of* $z^\rho(s)$, $s\ge 0$, $\rho=1,\dots,r$ *and* $\mathbb{E}|X_0|^2<\infty$. *Then, the sequence* $\{X_n(t)\}$ *defined by the successive approximations* (3.2) *converges uniformly on some interval* $[t_0,t_1]$ *to a unique local solution of* (1.1) *with an initial variable* (t_0,X_0), $t_0\ge 0$.

Proof. Let the t_1 be the time which was obtained in Theorem 3.2, with $t_0<t_1<T$. Now, we define the functions $a_{mn}(t)$ and $a_n(t)$ on $[t_0,t_1]$ for all integer $m\ge n\ge 0$ as (Taniguchi [7]):

$$a_{mn}=\mathbb{E}|X_m(t)-X_n(t)|^2,\quad a_n(t)=\sup\{a_{pq}(t);p\ge q\ge n\}. \tag{3.4}$$

First, we shall show that the sequence $\{a_n(t)\}$ has a subsequence $\{a_{n_k}(t)\}$ which converge uniformly on $[t_0,t_1]$ to some continuous function $a(t)$ defined on $[t_0,t_1]$.

Since the sequence$\{\mathbb{E}|X_n(t)|^2\}$, $t_0\le t\le t_1$ is uniform bounded by Theorem 3.2, we obtain a positive and real number M such that

$$a_{mn}(t)\le 2\mathbb{E}(|X_n(t)|^2+|X_m(t)|^2)<M \tag{3.5}$$

for all $t\in[t_0,t_1]$. From [7], we have that:

$$\begin{aligned}|a_{mn}(t)-a_{mn}(s)|^2 &= |\mathbb{E}(|X_m(t)-X_n(t)|^2)-\mathbb{E}(|X_m(s)-X_n(s)|^2)|\\ &\le \{\mathbb{E}(|X_m(t)-X_n(t)|+|X_m(s)-X_n(s)|)^2\}^{\frac12}\\ &\quad \{\mathbb{E}(2|X_m(t)-X_m(s)|^2+2|X_n(t)-X_n(s)|^2)\}^{\frac12}.\end{aligned}$$

On the other side,

$$\begin{aligned}\mathbb{E}(|X_n(t)-X_n(s)|^2 &\le 3\Big(T\int_s^t \mathbb{E}(|f(\tau,X_{n-1}(\tau))|^2)d\tau\\ &\quad + rC^2\sum_{\rho=1}^{r}\int_s^t \mathbb{E}(|g_\rho(\tau,X_{n-1}(\tau))|^2)d\tau\\ &\quad + r^2C^2\sum_{\rho,\sigma=1}^{r}\int_s^t \mathbb{E}(|h_{\rho\sigma}(\tau,X_{n-1}(\tau))|^2)d\tau\Big)\\ &\le 3(rC^2+r^2C^2+T)\int_s^t H(\tau,\mathbb{E}(|X_{n-1}(\tau)|^2))d\tau\\ &\le 3(rC^2+r^2C^2+T)|M(t)-M(s)|,\end{aligned}$$

for all $t,s\in[t_0,t_1]$, and $n\ge 1$ integer, where $M(t)=\int_{t_0}^t H(\tau,u(\tau))d\tau$ and $u(t)$ is a local solution of (3.3). Thus, we obtain a positive and real number Q such that

$$|a_{mn}(t)-a_{mn}(s)|\le Q|M(t)-M(s)|^{\frac{1}{2}} \tag{3.6}$$

for all $m\ge n\ge 0$ and $t,s\in[t_0,t_1]$.

Therefore, by (3.5) and (3.6), we have that $0\le a_n(t)<M$ and $|a_n(t)-a_n(s)|\le Q|M(t)-M(s)|^{\frac{1}{2}}$, $\forall\, n\ge 0$ and $t,s\in[t_0,t_1]$, which implies that by Ascoli-Arzela Theorem there exists a subsequence $\{a_{n_k}(t)\}$ which converges uniformly to some continuous function $a(t)$ defined on $[t_0,t_1]$.

Now, for $m\ge n\ge n_{k+1}$, since $m-1\ge n=1\ge n_k$, by condition (iii),

$$\begin{aligned}a_{mn}(t)=\mathbb{E}(|X_m(t)-X_n(t)|^2)&\le 3\mathbb{E}\Big(\Big|\int_{t_0}^t[f(\tau,X_{m-1}(\tau))-f(\tau,X_{n-1}(\tau))]d\tau\Big|^2\\ &\quad+\Big|\sum_{\rho=1}^{r}\int_{t_0}^t[g_\rho(\tau,X_{m-1}(\tau))-g_\rho(\tau,X_{n-1}(\tau))]dz^\rho(\tau)\Big|^2\\ &\quad+\Big|\sum_{\rho,\sigma=1}^{r}\int_{t_0}^t[h_{\rho\sigma}(\tau,X_{m-1}(\tau))-h_{\rho\sigma}(\tau,X_{n-1}(\tau))]dz^\rho(\tau)dz^\sigma(\tau)\Big|^2\Big)\\ &\le A\int_{t_0}^t G(\tau,\mathbb{E}|X_{m-1}(\tau)-X_{n-1}(\tau)|^2)d\tau\\ &\le A\int_{t_0}^t G(\tau,a_{n(k)}(\tau))d\tau \qquad t\in[t_0,t_1].\end{aligned}$$

Thus, since $\{a_{n_k}(t)\}$ converges uniformly to $a(t)$ as $k\to\infty$, by the continuity of $G(t,u)$ in u for each fixed $t\in[t_0,t_1]$ and the Lebesgue's Dominated Convergence Theorem:

$$a(t)\le A\int_{t_0}^t G(\tau,a(\tau))d\tau,\qquad t\in[t_0,t_1].$$

Therefore, by condition (iv), we have that $a(t) \equiv 0$ on $[t_0, t_1]$ (see Taniguchi, Lemma 3, p. 156 [7]).

Next, for $m \geq n \geq n_{k+1}$

$$\begin{aligned}
&\mathbb{E}(\sup_{t_0 \leq t \leq t_1} |X_m(t) - X_n(t)|^2) \\
&\quad \leq 3\mathbb{E}\Bigg(\sup_{t_0 \leq t \leq t_1} \left| \int_{t_0}^{t} [f(\tau, X_{m-1}(\tau)) - f(\tau, X_{n-1}(\tau))] d\tau \right|^2 \\
&\qquad + \sup_{t_0 \leq t \leq t_1} \left| \sum_{\rho=1}^{r} \int_{t_0}^{t} [g_\rho(\tau, X_{m-1}(\tau)) - g_\rho(\tau, X_{n-1}(\tau))] dz^\rho(\tau) \right|^2 \\
&\qquad + \sup_{t_0 \leq t \leq t_1} \left| \sum_{\rho,\sigma=1}^{r} \int_{t_0}^{t} [h_{\rho\sigma}(\tau, X_{m-1}(\tau)) - h_{\rho\sigma}(\tau, X_{n-1}(\tau))] dz^\rho(\tau) dz^\sigma(\tau) \right|^2 \Bigg) \\
&\quad \leq 3T \int_{t_0}^{t_1} \mathbb{E}|f(\tau, X_{m-1}(\tau)) - f(\tau, X_{n-1}(\tau))|^2 d\tau \\
&\qquad + rC^2 \sum_{\rho=1}^{r} \int_{t_0}^{t_1} \mathbb{E}|g_\rho(\tau, X_{m-1}(\tau)) - g_\rho(\tau, X_{n-1}(\tau))|^2 d\tau \\
&\qquad + r^2 C^2 \sum_{\rho,\sigma=1}^{r} \int_{t_0}^{t} t_1 \mathbb{E}|h_{\rho\sigma}(\tau, X_{m-1}(\tau)) - h_{\rho\sigma}(\tau, X_{n-1}(\tau))|^2 d\tau \\
&\quad \leq A \int_{t_0}^{t_1} G(\tau, \mathbb{E}(|X_{m-1}(\tau) - X_{n-1}(\tau)|^2)) d\tau \\
&\quad \leq A \int_{t_0}^{t_1} G(\tau, a_{n_k}(\tau)) d\tau \longrightarrow 0, \qquad k \to \infty
\end{aligned}$$

which implies that the sequence $\{X_n(t)\}$ is a Cauchy sequence in Banach space $B(t_0, t_1)$[1] (see Taniguchi [7], Lemma 1., p. 155). Therefore, there exists a stochastic process $X(t)$ such that:

$$\mathbb{E}(\sup_{t_0 \leq t \leq t_1} |X_n(t) - X(t)|^2) \longrightarrow 0, \qquad n \to \infty \ .$$

Furthermore, we shall show that the stochastic process $X(t)$ is a local solution of the stochastic differential equation (1.1),

$$\begin{aligned}
&\mathbb{E}\Bigg(\sup_{t_0 \leq t \leq t_1} \Bigg| X_n(t) - [X_0 + \int_{t_0}^{t} f(\tau, X(\tau)) d\tau \\
&\qquad + \sum_{\rho=1}^{r} \int_{t_0}^{t} g_\rho(\tau, X(\tau)) dz^\rho(\tau) + \sum_{\rho,\sigma=1}^{r} \int_{t_0}^{t} h_{\rho\sigma}(\tau, X(\tau)) dz^\rho(\tau) dz^\sigma(\tau)] \Bigg|^2 \Bigg)
\end{aligned}$$

[1] $B(t_0, t_1)$ is the space of all function $\xi(t, \omega) : [t_0, t_1] \times \Omega \longrightarrow \mathbb{R}$ with $\xi(t, \omega)$ measurable in ω for each fixed $t \in [t_0, t_1]$ and it is continuous in t for a.e. fixed $\omega \in \Omega$; $\|\xi(t, \omega)\|_{B(t,t_0)} = \{\sup |\xi(t, \omega)^2|\}^{\frac{1}{2}}$.

$$\leq 3(rC^2 + r^2C^2 + T) \int_{t_0}^{t_1} G(\tau, \mathbb{E}(|X_{n-1}(\tau) - X(\tau)|^2))d\tau$$

$$\leq A \int_{t_0}^{t_1} G(\tau, \mathbb{E} \sup_{t_0 \leq t \leq t_1} |X_{n-1}(\tau) - X(\tau)|^2)d\tau \longrightarrow 0, \qquad k \to \infty$$

from which we obtain that

$$X(t) = X(t_0) + \int_{t_0}^{t} f(\tau, X(\tau))d\tau + \sum_{\rho=1}^{r} \int_{t_0}^{t} g_\rho(\tau, X(\tau))dz^\rho(\tau)$$

$$+ \sum_{\rho,\sigma=1}^{r} \int_{t_0}^{t} h_{\rho\sigma}(\tau, X(\tau))dz^\rho(\tau)dz^\sigma(\tau)$$

for all $t \in [t_0, t_1]$ and a.s. $\omega \in \Omega$. Moreover, we can prove the uniqueness of the local solution. Let $X(t)$ and $Y(t)$ be two local solutions existing on $[t_0, t']$ with $X(t_0) = Y(t_0) = X_0$ Then, we obtain:

$$\mathbb{E}|X(t) - Y(t)|^2 \leq 3(T + rC^2 + r^2C^2) \int_{t_0}^{t} G(\tau, \mathbb{E}|X(\tau) - Y(\tau)|^2)d\tau$$

for all $t \in [t_0, t']$, from which using condition (iv), we have that $\mathbb{E}|X(t) - Y(t)|^2 = 0$ for all $t \in [t_0, t']$. This show the uniqueness of the local solution. Therefore, the proof of the theorem is complete. □

3.3. Global existence and uniqueness theorem

Theorem 3.4. *For the stochastic differential equation* (1.1), *suppose that the following conditions are satisfied:*

(i) *the coefficients $f(t,x), g_\rho(t,x)$ and $h_{\rho\sigma}(t,x)$ satisfy conditions* (ii), (iii) *and* (iv) *from Theorem* 3.3 *and z^ρ satisfies the condition* (i) *from Theorem* 3.3, *for $R = \infty$ and $T = \infty$, $(\rho, \sigma = 1, 2, \ldots, r)$;*

(ii) *for any fixed $T > 0$, the differential equation*

$$\frac{du}{dt} = 3(T + rC^2 + r^2C^2)H(t, u) \tag{3.7}$$

has a global solution for any initial value (t_0, u_0), $t_0 \geq 0$, $u_0 \geq 0$;

(iii) *for any fixed $T > 0$ condition* (iv) *from Theorem* 3.3 *holds.*

Let X_0 be independent of the $z^\rho(t)$, $t \geq 0$, $\rho = 1, \ldots, r$ and let $\mathbb{E}|X_0|^2 < \infty$. Then the sequence $\{X_n(t)\}$ defined by the successive approximations (3.2) *converges uniformly on any finite subinterval $[t_0, T]$ of $[t_0, \infty)$, to a unique solution of* (1.1).

Proof. Let S be the set of times s such that the sequence $\{X_n(t)\}$ converges uniformly on the interval $[t_0, s]$. Let $s_1 = \sup\{s \in S\}$. By Theorem 3.3, we have that $s_1 > 0$. Now, we suppose that $s_1 < \infty$. Then, we can choose a time T_0 such that $s_1 < T_0 < \infty$. Then, by (3.2) and condition (i),

$$\mathbb{E}|X_{n+1}(t)|^2 \leq 4\mathbb{E}|X_0|^2 + 4(rC^2 + r^2C^2 + T) \int_{t_0}^{t} H(\tau, \mathbb{E}|X_n(\tau)|^2)d\tau$$

for all $t \in [t_0, T_0]$. Take a $u_0 \in \mathbb{R}^+$ such that $u_0 > 4\mathbb{E}|X_0|^2$. Then, by condition (ii), we have a solution of $u(t) = u(t, t_0, u_0)$ of (3.7) with $T = T_0$ and with the initial value (t_0, u_0), and the remainder of the proof follows as in Theorem 3.3, replacing t by T_0, which completes the proof of the theorem . □

Also, we can prove a similar result as Corollary 1 from [7] (here we suppose $t_0 = 0$).

Corollary 3.5. *For the stochastic differential equation* (1.1) *suppose that the following conditions are satisfied:*

(i)
$$|f(t,x) - f(t,y)|^2 + \left|\sum_{\rho=1}^{r} g_\rho(t,x) - g_\rho(t,y)\right|^2 + \left|\sum_{\rho,\sigma=1}^{r} h_{\rho\sigma}(t,x) - h_{\rho\sigma}(t,y)\right|^2 \le \lambda(t)\alpha(|x-y|^2).$$

(ii) $|f(t,0)|$, $|g_\rho(t,0)|$, $|h_{\rho\sigma}(t,0)| \in L^2_{loc}([0,\infty), \mathbb{R}^+)$, $\rho, \sigma = 1, \dots, r$, *for all* $t \in [0,\infty)$ *and all* $x, y \in \mathbb{R}$, *where* $\lambda(t) : [0,\infty) \to \mathbb{R}^+$ *is a locally integrable function and* $\alpha(u) : \mathbb{R}^+ \to \mathbb{R}^+$ *is a continuous, monotone nondecreasing and concave function with* $\alpha(0) = 0$ *such that* $\int_{0+} (1/\alpha(u))du = \infty$.

Let $\mathbb{E}|X_0|^2 < \infty$. *Then, on any finite interval* $[0, T]$, *the sequence* $\{X_n(t)\}$, $0 \le t \le T$, *defined by the successive approximations* (3.2), *converges uniformly to a unique solution of* (1.1).

Proof. Since $\alpha(u)$ is concave on $[0, \infty)$, there exists positive real numbers $a > 0, b > 0$ such that $\alpha(u) \le au + b$. Set $\gamma(t) = 3a\lambda(t)$ and $\beta(t) = 3b\lambda(t) + |f(t,0)|^2 + r|\sum_\rho g_\rho(t,0)|^2 + r^2|\sum_{\rho\sigma}(t,0)|^2$ (these are defined in an analogous way to that of Taniguchi [7]).

From condition (i), we have:

$$\left(|f(t,x) - f(t,0)| + \left|\sum_{\rho=1}^{r} g_\rho(t,x) - g_\rho(t,0)\right| + \left|\sum_{\rho,\sigma=1}^{r} h_{\rho\sigma}(t,x) - h_{\rho\sigma}(t,0)\right|\right)^2$$
$$\le 3\left\{|f(t,x) - f(t,0)|^2 + \left|\sum_{\rho=1}^{r} g_\rho(t,x) - g_\rho(t,0)\right|^2 + \left|\sum_{\rho,\sigma=1}^{r} h_{\rho\sigma}(t,x) - h_{\rho\sigma}(t,0)|^2\right\}$$
$$\le 3\lambda(t)\alpha(|x|^2).$$

But,

$$\begin{aligned}|f(t,x) - f(t,0)|^2 &\ge \big|\,|f(t,x)| - |f(t,0)|\,\big|^2\\ &= |f(t,x)|^2 + |f(t,0)|^2 - 2|f(t,x)|\,|f(t,0)|\\ &\ge \big|\,|f(t,x)|^2 - |f(t,0)|^2\,\big|\\ &\ge |f(t,x)|^2 - |f(t,0)|^2.\end{aligned}$$

Analogous, we obtain

$$\left|\sum_{\rho=1}^{r} g_\rho(t,x) - g_\rho(t,0)\right|^2 \geq r\left(\sum_{\rho=1}^{r} |g_\rho(t,x)|^2 - \sum_{\rho=1}^{r} |g_\rho(t,0)|^2\right)$$

and

$$\left|\sum_{\rho,\sigma=1}^{r} h_{\rho\sigma}(t,x) - h_{\rho\sigma}(t,0)\right|^2 \geq r^2\left(\sum_{\rho,\sigma=1}^{r} |h_{\rho\sigma}(t,x)|^2 - \sum_{\rho,\sigma=1}^{r} |h_{\rho\sigma}(t,0)|^2\right).$$

Therefore, we have

$$\begin{aligned}
&|f(t,x)|^2 + \left|\sum_{\rho=1}^{r} g_\rho(t,x)\right|^2 + \left|\sum_{\rho,\sigma=1}^{r} h_{\rho\sigma}(t,x)\right|^2 \\
&\leq 3\lambda(t)\alpha(|x|^2) + \left[|f(t,0)|^2 + r\left|\sum_{\rho=1}^{r} g_\rho(t,0)\right|^2 + r^2\left|\sum_{\rho,\sigma=1}^{r} h_{\rho\sigma}(t,0)\right|^2\right] \\
&\leq \gamma(t)|x|^2 + \beta(t).
\end{aligned}$$

Next, put $H(t,u) = \beta(t) + \gamma(t)u$, $u \geq 0$, for all $t \in [0,\infty)$. Then, since H is linear in u, condition (ii) from Theorem 3.4 holds. Conditions (i) and (iii) from Theorem 3.4 are also satisfied. Therefore, by Theorem 3.4, we obtain the desirable conclusion. □

Acknowledgement

I express my sincere gratitude to an anonymous referee for his/her helpful comments.

References

[1] A.T. Bharucha-Reid, *Random Integral Equation.* Academic Press, NY, 1972.

[2] A. Constantin, *On the Existence, Uniqueness and Parametric Dependence on the Coefficients of the Solution Processes in McShane's Stochastic Integral Equations.* Publ. Mathématiques **38** (1994), 11–24.

[3] A. Constantin, *Global existence of solutions for perturbed differential equations.* Annali di Matematica Pura ed Applicada Serie IV, **CLXVIII** (1995), 237–299.

[4] J.M.A. Ibáñez, R.G. Jáimez, *On the Existence and Uniqueness of the Solution Processes in McShane's Stochastic Integral Equation Systems.* Annales Scientifique de l'Université Blaise Pascal, Probabilités et Applications 7e fascicule, **92** (1988), 1–9.

[5] E.J. McShane, *Stochastic Calculus and Stochastic Models.* Academic Press, NY, 1974.

[6] E.J. McShane, *Stochastic Differential Equations* Journal of Multivariate Analysis **5** (1975), 121–177.

[7] T. Taniguchi, *Successive Approximations to Solutions of Stochastic Differential Equations.* Journal of Differential Equations **96** (1992), 152–169.

[8] S. Watanabe, T. Yamada, *On the Uniqueness of Solutions of Stochastic Differential Equations.* J. Math. Kyoto Univ. **11** (1971), 155–167.

[9] T. Yamada, S. Watanabe, *On the Uniqueness of Solutions of Stochastic Differential Equations II.* J. Math. Kyoto Univ.**11** (1971), 553–563.

[10] T. Yamada, *On the Successive Approximation of Solutions of Stochastic Differential Equations.* J. Math. Kyoto Univ. **21** (1981), 501–515.

Romeo Negrea
Faculty of Mathematics
West University of Timişoara
Bvd. V.Pârvan No. 4, 1900
Timişoara
Romania
e-mail: negrea@hilbert.math.uvt.ro

Operator Theory:
Advances and Applications, Vol. 153, 211–227

Regularized Traces of Differential Operators

A.S. Pechentsov

Abstract. The paper discusses properties of the characteristic determinant and the regularized trace associated to certain differential operators.

Mathematics Subject Classification (2000). 47E05, 34L20.

Keywords. Differential operator, Characteristic determinant, Regularized trace, Asymptotic representation.

1. Introduction

In 1953 I.M. Gel'fand and B.M. Levitan ([1]) calculated the regularized trace for the Sturm–Liouville problem:

$$-y'' + q(x)y = \lambda y, \quad y(0) = y(\pi) = 0, \quad q(x) \in C^1[0,\pi], \quad q : [0,\pi] \to \mathbb{R}.$$

The eigenvalues λ_n of this problem have the asymptotic representation

$$\lambda_n = n^2 + c_0 + O\left(\frac{1}{n^2}\right), \quad n \to \infty, \quad c_0 = \frac{1}{\pi}\int_0^\pi q(x)dx,$$

and

$$\sum_{n=1}^{\infty} \left[\lambda_n - n^2 - c_0\right] = \frac{1}{2}c_0 - \frac{q(0)+q(\pi)}{4}.$$

This is the well-known Gel'fand–Levitan formula. The quantity is called the regularized trace of Sturm–Liouville's operator.

Let μ_k, $k = 1, 2, \ldots$ be the eigenvalues of some operator A. The sum

$$\sum_{k=1}^{\infty} [\mu_k^m - A_m(k)] = S_m, \quad m \in \mathbb{N},$$

where $A_m(k)$ are fixed numbers, providing convergence of the series, is called the regularized m-order trace of the operator A.

The eigenvalues for wide classes of differential operators are zeros of the characteristic determinant $\Delta(\lambda)$ which is an entire function.

Consider an entire function $\Delta(\lambda)$ having an asymptotic representation as $\lambda \to \infty$

$$\Delta(\lambda) \sim \sum_{k=1}^{H} \exp\left(\sum_{\ell=0}^{N-1} \theta_{k\ell}\lambda^{\frac{N-\ell}{h}}\right) \cdot \lambda^{\frac{n_k}{h}} \sum_{\nu=0}^{\infty} \beta_\nu^{(k)} \lambda^{-\frac{\nu}{h}}, \tag{1.1}$$

where $H,\ h,\ N \in \mathbb{N}$, $H > 1$, $n_k \in \mathbb{Z}$, $\theta_{k\ell},\ \beta_\nu^{(k)} \in \mathbb{C}$, $\beta_0^{(k)} \neq 0$.
The asymptotic representation (1.1) of the entire function $\Delta(\lambda)$ means that $\forall r \in \mathbb{N}$ $\exists R \in \mathbb{R}$ such that for $|\lambda| > R$ the following equality holds

$$\Delta(\lambda) = \sum_{k=1}^{H} \exp\left(\sum_{\ell=0}^{N-1} \theta_{k\ell}\lambda^{\frac{N-\ell}{h}}\right) \cdot \lambda^{\frac{n_k}{h}} \left(\sum_{\nu=0}^{r} \beta_\nu^{(k)} \lambda^{-\frac{\nu}{h}} + O\left(\lambda^{-\frac{r+1}{h}}\right)\right).$$

We assume here that a cut has been done in the complex plane λ and some branch of $\sqrt[h]{\lambda}$ is fixed, for example, by its principal value. In representation (1.1) it is supposed that

$$\theta_{k0} = \theta_{k1} = \cdots = \theta_{k,\ell_k-1} = 0, \quad \theta_{k,\ell_k} \neq 0,\ 0 \leq \ell_k \leq N-1, \text{ for } k,\ k = 1, \ldots, H.$$

We do not exclude the case when

$$\theta_{\tilde{k}0} = \theta_{\tilde{k}1} = \cdots = \theta_{\tilde{k},N-1} = 0$$

for some $k = \tilde{k}$. The functions having the representation (1.1) with $h = N = 1$ are well known in the literature. They are called entire functions of class K ([2]). The eigenvalues of some general boundary problems on a segment are zeros of entire functions of class K. Namely, let us consider a boundary problem on the segment $[0,\ 1]$, generated by the differential equation

$$\frac{d^n y}{dx^n} + P_1(x,\lambda)\frac{d^{n-1}y}{dx^{n-1}} + P_2(x,\lambda)\frac{d^{n-2}y}{dx^{n-2}} + \cdots + P_n(x,\lambda)y = 0 \tag{1.2}$$

and the boundary conditions

$$U_i(y,\lambda) = \sum_{j=0}^{n-1}[a_{ij}(\lambda)y^{(j)}(0) + b_{ij}(\lambda)y^{(j)}(1)] = 0, \qquad i = \overline{1,n}. \tag{1.3}$$

Here $\lambda \in \mathbb{C}$ is the spectral parameter, entering equation (1.2) and the boundary conditions (1.3) polynomially:

$$P_\alpha(x,\lambda) = \sum_{\beta=0}^{\alpha} P_{\alpha\beta}(x)\lambda^{\alpha-\beta}, \quad \alpha = \overline{1,n} \tag{1.4}$$

and $a_{ij}(\lambda)$, $b_{ij}(\lambda)$ are arbitrary polynomials on λ. It is supposed here that for each fixed λ the boundary conditions (1.3) are linearly independent and $P_{\alpha\beta}(x)$ are smooth functions on the segment $[0,\ 1]$.

We associate with the equation (1.2) the characteristic polynomial

$$\pi(\omega) = \omega^n + P_{10}\omega^{n-1} + \cdots + P_{\alpha 0}\omega^{n-\alpha} + \cdots + P_{n0} \tag{1.5}$$

where the coefficients $P_{\alpha 0}$, $\alpha = \overline{1,n}$ are the leading coefficients in (1.4) (we consider only the case when the coefficients $P_{\alpha 0}$, $\alpha = \overline{1,n}$ are constant on the segment

[0, 1]). The characteristic determinant $f(\lambda)$ of the problem (1.2)–(1.3) is an entire function of class K, provided that the polynomial $\pi(\omega)$ has only simple roots.

In the general case of multiple roots of the polynomial $\pi(\omega)$ the characteristic determinant $\Delta(\lambda)$ of the problem (1.2)–(1.3) does not belong to the class K. The asymptotic representation (1.1) has fractional degrees of λ in exponents ([6]). If the spectral parameter λ enters the coefficients $P_\alpha(x,\lambda)$ with arbitrary degree m:

$$P_\alpha(x,\lambda) = \lambda^{m\alpha} \sum_{\beta=0}^{m\alpha} P_{\alpha\beta}(x)\lambda^{-\beta}, \quad \alpha = \overline{1,n},$$

then the characteristic determinant $\Delta(\lambda)$ of the problem (1.2)–(1.3) has the representation (1.1) with $N = mh$.

The entire functions with representation (1.1) can also arise from studying the spectrum of singular differential operators. For example, consider the differential operator in $L_2[0,\infty)$ generated by the expression

$$l(y) \equiv (-1)^n \frac{d^{2n}y}{dx^{2n}} + xy, \quad n \in \mathbb{N}$$

and by general boundary conditions at the point $x = 0$, fixing a self-adjoint extension:

$$U_m(y) \equiv \sum_{j=0}^{k_m} a_{mj} y^{(k_m-j)}(0) = 0,$$

$$a_{mj} \in \mathbb{R}, \quad m = \overline{1,n}, \quad a_{m0} = 1, \quad k_n < k_{n-1} < \cdots < k_1 < 2n.$$

Eigenvalues of such operators are the zeros of the entire function $\Delta(\lambda)$, having the representation (1.1) with $h = 2n$, $N = 2n + 1$ (see [7]).

Special cases of entire functions $\Delta(\lambda)$ with representation (1.1) were considered in [3], [6]. However, a mistake was found in [3]. In what follows an example will be constructed showing that the basic relation (1.4) given on the page 466 of that paper is not true.

2. Zeros of the function $\Delta(\lambda)$

Denoting

$$\mu_k(\lambda) = \sum_{\ell=0}^{N-1} \theta_{k\ell}\lambda^{\frac{N-\ell}{h}}, \quad F_k(\lambda) \sim \sum_{\nu=0}^{\infty} \beta_\nu^{(k)}\lambda^{-\frac{\nu}{h}}, \quad \lambda \to \infty, \quad k = \overline{1,H},$$

the representation (1.1) can be rewritten in the form

$$\Delta(\lambda) = \sum_{k=1}^{H} \exp\left(\mu_k(\lambda)\right)\lambda^{\frac{n_k}{h}} F_k(\lambda). \tag{2.1}$$

Define the set

$$\Gamma_{pq} = \{\lambda \in \mathbb{C} : \operatorname{Re}\mu_p(\lambda) = \operatorname{Re}\mu_q(\lambda)\} \tag{2.2}$$

of points which satisfy the equation

$$\operatorname{Re}\left(\sum_{\ell=0}^{N-1}(\theta_{p\ell}-\theta_{q\ell})\,\lambda^{\frac{N-\ell}{h}}\right)=0\,. \tag{2.3}$$

We shall be interested in the part of the set Γ_{pq}, located outside the disk of sufficiently large radius with the center at the point $\lambda = 0$.

Let

$$\theta_{p\ell}-\theta_{q\ell}=r_{pq\ell}e^{i\varphi_{pq\ell}}\,,\quad \ell=\overline{0,N-1}\,,\quad \lambda=|\lambda|e^{i\varphi}\,,\quad r=|\lambda|^{\frac{1}{h}}\,. \tag{2.4}$$

In this case it is possible that

$$r_{pq0}=\cdots=r_{pq\ell_{pq}-1}=0\,,\quad r_{pq\ell_{pq}}\neq 0\,,\quad 0\le \ell_{pq}\le N-1\,. \tag{2.5}$$

Then the equation (2.3) can be rewritten in the form

$$\sum_{\ell=\ell_{pq}}^{N-1} r_{pq\ell}\cos\left(\varphi_{pq\ell}+\frac{N-\ell}{h}\varphi\right)r^{N-\ell}=0\,, \tag{2.6}$$

where $r_{pq\ell_{pq}} \neq 0$.

If $\cos\left(\varphi_{pq\ell_{pq}}+\frac{N-\ell_{pq}}{h}\varphi\right)$ is not equal to zero on the segment $[0,2\pi]$, then all roots $r_1(\varphi),\dots,r_{N-\ell_{pq}}(\varphi)$, $0\le\varphi\le 2\pi$, of the equation (2.6) lie on the disk of radius $\tilde{R}$ ($\tilde{R}$ being a sufficiently large positive number) with the center at the origin. Therefore infinitely large positive roots $r(\varphi)$ of the equation (2.6) (i.e., the roots $r(\varphi)\to+\infty$ as $\varphi\to\varphi^*$, φ^* some number in the interval $[0,\ 2\pi)$) can arise only in arbitrarily small neighborhoods of the roots of the function $\cos\left(\varphi_{pq\ell_{pq}}+\frac{N-\ell_{pq}}{h}\varphi\right)$ lying in the interval $[0,\ 2\pi)$. Let $0\le\varphi_1^{(pq)}<\varphi_2^{(pq)}<\cdots<\varphi_{M_{pq}}^{(pq)}$ be the roots of the function $\cos\left(\varphi_{pq\ell_{pq}}+\frac{N-\ell_{pq}}{h}\varphi\right)$ lying in the interval $[0,\ 2\pi)$. Take $\varepsilon>0$ sufficiently small and set

$$I^{(pq)}(\varepsilon)=\bigcup_{k=1}^{M_{pq}} I_k^{(pq)}\,,\quad I_k^{(pq)}=\left(\varphi_k^{(pq)}-\varepsilon,\ \varphi_k^{(pq)}+\varepsilon\right)\,.$$

If $\varphi_1^{(pq)}=0$ then we put

$$I_1^{(pq)}=[0,\ \varepsilon)\cup(2\pi-\varepsilon,\ 2\pi]\,.$$

For sufficiently small $\varepsilon>0$

$$I_k^{(pq)}\cap I_\ell^{(pq)}=\emptyset\quad \forall k\neq\ell\ \text{ and } I_{M_{pq}}^{(pq)}\subset[0,\ 2\pi)\,.$$

Then the function $\cos\left(\varphi_{pq\ell_{pq}}+\frac{N-\ell_{pq}}{h}\varphi\right)$ has no zeros on the compact set $K^{(pq)}(\varepsilon) = [0,2\pi]\setminus I^{(pq)}(\varepsilon)$.

Theorem 2.1. *There exist positive constants* $\mu_0^{(pq)}$, $R_0^{(pq)}$ *such that*

$$|\operatorname{Re}(\mu_p(\lambda) - \mu_q(\lambda))| \geq \mu_0^{(pq)} |\lambda|^{\frac{N-\ell_{pq}}{h}}$$

$\forall \lambda = |\lambda| e^{i\varphi}$, *provided that* $\varphi \in K^{(pq)}(\varepsilon)$ *and* $|\lambda| \geq R_0^{(pq)}$.

Proof. Let $r = |\lambda|^{\frac{1}{h}}$; then

$$\operatorname{Re}(\mu_p(\lambda) - \mu_q(\lambda)) = \operatorname{Re}\left(\sum_{\ell=0}^{N-1} (\theta_{p\ell} - \theta_{q\ell})\, \lambda^{\frac{N-\ell}{h}} \right)$$
$$= \sum_{\ell=\ell_{pq}}^{N-1} r_{pq\ell} \cos\left(\varphi_{pq\ell} + \frac{N-\ell}{h}\varphi \right) r^{N-\ell}.$$

Suppose

$$\mu_0^{(pq)} = \frac{1}{2} \min_{\varphi \in K^{(pq)}(\varepsilon)} \left| r_{pq\ell_{pq}} \cos\left(\varphi_{pq\ell_{pq}} + \frac{N-\ell_{pq}}{h}\varphi \right) \right|.$$

If $\ell_{pq} < N-1$, then we put

$$\mu_\ell^{(pq)} = \max_{\varphi \in K^{(pq)}(\varepsilon)} \left| r_{pq\ell} \cos\left(\varphi_{pq\ell} + \frac{N-\ell}{h}\varphi \right) \right|, \quad \ell = \overline{\ell_{pq}+1, N-1}.$$

Since the function $\left| r_{pq\ell_{pq}} \cos\left(\varphi_{pq\ell_{pq}} + \frac{N-\ell_{pq}}{h}\varphi \right) \right|$ is continuous on the compact set $K^{(pq)}(\varepsilon)$ and has no zeros on $K^{(pq)}(\varepsilon)$ we have $\mu_0^{(pq)} > 0$. Then for $r > 1$ we get the inequality

$$|\operatorname{Re}(\mu_p(\lambda) - \mu_q(\lambda))| = r^{N-\ell_{pq}} \left| \sum_{\ell=\ell_{pq}}^{N-1} r_{pq\ell} \cos\left(\varphi_{pq\ell} + \frac{N-\ell}{h}\varphi \right) \frac{1}{r^{\ell-\ell_{pq}}} \right|$$
$$\geq r^{N-\ell_{pq}} \left(2\mu_0^{(pq)} - \sum_{\ell=\ell_{pq}}^{N-1} \mu_\ell^{(pq)} \cdot \frac{1}{r} \right).$$

If $r \geq \sum_{\ell=\ell_{pq}}^{N-1} \mu_\ell^{(pq)} \cdot \frac{1}{\mu_0^{(pq)}} + 1$, then

$$|\operatorname{Re}(\mu_p(\lambda) - \mu_q(\lambda))| \geq \mu_0^{(pq)} \cdot r^{N-\ell_{pq}}.$$

Taking $R_0^{pq} = \left(\sum_{\ell=\ell_{pq}}^{N-1} \mu_\ell^{(pq)} \cdot \frac{1}{\mu_0^{(pq)}} + 1 \right)^h$ we get the statement of Theorem 2.1.

The ray $\ell_k^{(pq)}$ in the λ-plane outgoing from the origin and having the angle $\varphi_k^{(pq)}$ with real axis, $k = \overline{1, M_{pq}}$, is called the critical direction of the set Γ_{pq}. We shall construct M_{pq} sectors $E_1^{(pq)}$, $E_2^{(pq)}, \ldots,$ $E_{M_{pq}}^{(pq)}$ in the λ-plane with vertices at the origin. The critical direction $\ell_k^{(pq)}$ is the bisectrix of the sector $E_k^{(pq)}$, $k = \overline{1, M_{pq}}$ of angle 2ε. Remove all these sectors $E_1^{(pq)}, \ldots,$ $E_{M_{pq}}^{(pq)}$ from the λ-plane. The

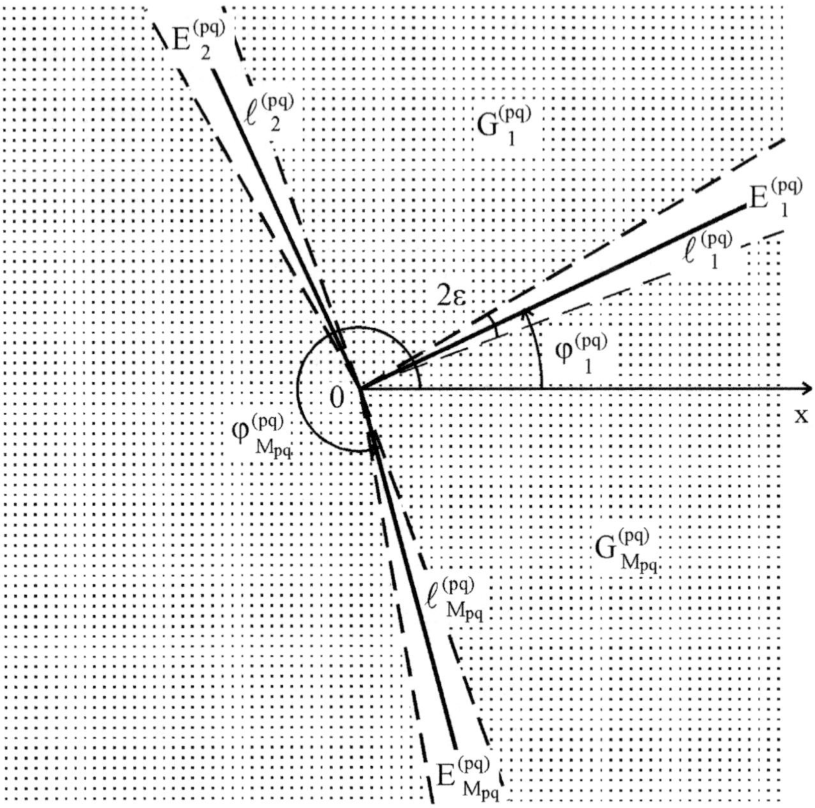

Fig. 1

remaining part of the complex plane decomposes into M_{pq} closed sectors $G_k^{(pq)}$, $k = \overline{1, M_{pq}}$ (see Fig. 1).

Corollary 2.2. *If* $\lambda \in G_k^{(pq)}$, $k = \overline{1, M_{pq}}$ *and* $|\lambda| \geq R_0^{(pq)}$, *then*

$$\text{either} \quad \operatorname{Re}\mu_p(\lambda) - \operatorname{Re}\mu_q(\lambda) \geq \mu_0^{(pq)}|\lambda|^{\frac{N-\ell_{pq}}{h}},$$
$$\text{or} \quad \operatorname{Re}\mu_q(\lambda) - \operatorname{Re}\mu_p(\lambda) \geq \mu_0^{(pq)}|\lambda|^{\frac{N-\ell_{pq}}{h}}.$$

Take the union of all sets Γ_{pq} with $p < q$, $p = 1, \ldots, H-1$, $q = 2, \ldots, H$ and denote it by Γ:

$$\Gamma = \bigcup_{1 \leq p < q \leq H} \Gamma_{pq}.$$

Suppose the critical directions of the set Γ are obtained as the union ℓ of all critical directions $\ell_k^{(pq)}$, $k = \overline{1, M_{pq}}$ with $p < q$, $p = \overline{1, H-1}$, $q = \overline{2, H}$:

$$\ell = \bigcup_{p<q} \bigcup_{k=1}^{M_{pq}} \ell_k^{(pq)} .$$

We enumerate the angles of the critical directions of the set Γ in ascending order: $0 \leq \varphi_1 < \varphi_2 < \cdots < \varphi_L < 2\pi$. Take $\varepsilon > 0$ sufficiently small such that the intervals $I_k = (\varphi_k - \varepsilon,\ \varphi_k + \varepsilon)$, $k = \overline{1, L}$ be disjoint and $I_L \subset [0,\ 2\pi)$ (in the case of $\varphi_1 = 0$ we put $I_1 = [0,\ \varepsilon) \cup (2\pi - \varepsilon,\ 2\pi])$. Then there will be no zeros of the function $\cos\left(\varphi_{pq\ell_{pq}} + \frac{N-\ell_{pq}}{h}\varphi\right)$ on the compact set $K(\varepsilon) = [0,\ 2\pi] \setminus \bigcup_{k=1}^{L} I_k$ for all $p \neq q$, $p = \overline{1, H}$, $q = \overline{1, H}$.

Let

$$\mu_0 = \min_{p<q} \mu_0^{(pq)} = \frac{1}{2} \min_{p<q} \min_{\varphi \in K(\varepsilon)} \left| r_{pq\ell_{pq}} \cos\left(\varphi_{pq\ell_{pq}} + \frac{N - \ell_{pq}}{h}\varphi\right)\right| ,$$

$$\mu_\ell = \max_{p<q} \max_{\varphi \in K(\varepsilon)} \left| r_{pq\ell} \cos\left(\varphi_{pq\ell} + \frac{N-\ell}{h}\varphi\right)\right| , \quad \ell = 1,\ 2, \dots, N-1 ,$$

$$R_0 = \left(\frac{1}{\mu_0} \sum_{\ell=1}^{N-1} \mu_\ell + 1\right)^h .$$

Taking into account the inequality $\ell_{pq} \leq N - 1\ \forall p \neq q$ we get the following theorem:

Theorem 2.3. $\forall p \neq q$, $p = \overline{1, H}$, $q = \overline{1, H}$ $\exists \mu_0,\ R_0 > 0$ *such that*

$$|\mathrm{Re}\,(\mu_p(\lambda) - \mu_q(\lambda))| \geq \mu_0 |\lambda|^{\frac{1}{h}}$$

$\forall \lambda = |\lambda| e^{i\varphi}$, *provided that* $\varphi \in K(\varepsilon)$ *and* $|\lambda| > R_0$.

Construct L sectors $E_1,\ E_2, \dots, E_L$ in the λ-plane of angle 2ε with vertexes at the origin. The bisectrix of the sector E_k, $k = \overline{1, L}$ is the critical direction ℓ_k of the set Γ. Remove these sectors E_k, $k = \overline{1, L}$ from λ-plane. Then the remaining part G of the complex plane decomposes into L closed sectors G_k, $k = \overline{1, L}$.

Denote $D_k = G_k \cap \{|\lambda| \geq R_0\}$, $k = \overline{1, L}$. From Theorem 2.3 and directly from the construction the connected sets D_k we have:

Corollary 2.4. *If* $\lambda \in D_k$, $k = \overline{1, L}$, *then there exists a permutation* $k_1,\ k_2, \dots, k_H$ *of the numbers* $1,\ 2, \dots, H$ *such that*

$$\mathrm{Re}\,\mu_{k_H}(\lambda) - \mathrm{Re}\,\mu_{k_j}(\lambda) \geq \mu_0 |\lambda|^{\frac{1}{h}} \quad \forall k_j \neq k_H .$$

Theorem 2.5. *There are no zeros of the function* $\Delta(\lambda)$ *in the intersection* $G \cap \{|\lambda| \geq R\}$, *where* R *is sufficiently large positive number.*

Proof. In each sector G_k for $|\lambda| \geq R_0$ according to Corollary 2.5 the function $\Delta(\lambda)$ has as $\lambda \to \infty$ a representation of the form

$$\Delta(\lambda) = \exp\left(\mu_{k_H}(\lambda)\right)\left(\lambda^{\frac{n_{k_H}}{h}} F_k(\lambda) + O\left(\exp\left(-\mu_0|\lambda|^{\frac{1}{h}}\right)\right)\right),$$

$$F_k(\lambda) \sim \sum_{\nu=0}^{\infty} \beta_\nu^{(k_H)} \lambda^{-\frac{\nu}{h}}, \quad \beta_0^{(k_H)} \neq 0.$$

From this representation for $\Delta(\lambda)$ we have that for $|\lambda| \geq R_k$, where R_k is some positive number, $R_k \geq R_0$, the factor

$$\left(\lambda^{\frac{n_{k_H}}{h}} F_k(\lambda) + O\left(\exp\left(-\mu_0|\lambda|^{\frac{1}{h}}\right)\right)\right)$$

has no zeros in the sector G_k. Hence the function $\Delta(\lambda)$ has no zeros in the set $G_k \cap \{|\lambda| \geq R_k\}$. Setting $R = \max_{k=1,\ldots,L} R_k$, we get the conclusion of the theorem.

Corollary 2.6. *The zeros λ_n, $n = 1, 2, 3, \ldots$ of the entire function $\Delta(\lambda)$, except for a finite number of them, lie in the sectors E_k, $k = \overline{1, L}$ of an arbitrary small opening.*

Proof. From Theorem 2.5 we have that all zeros λ_k of the function $\Delta(\lambda)$ lying outside the disk $K_R(0)$, centered at the origin and of radius R, are located in sectors $E_1, E_2, \ldots, E_L$. The number φ_k, $0 \leq \varphi_k < 2\pi$, $k = \overline{1, L}$, which is the opening of the angle formed by the bisectrix of the sector E_k with the real axis, is a zero of the function $\cos\left(\varphi_{pq\ell_{pq}} + \frac{N-\ell_{pq}}{h}\varphi\right)$ for some fixed pair of numbers p and q, $p = \overline{1, H-1}$, $q = \overline{2, H}$. The number ℓ_{pq} and the angle $\varphi_{pq\ell_{pq}}$ are determined by the parameters of asymptotic of the function $\Delta(\lambda)$, using the simple relations (2.5) and (2.4) respectively. But the entire function $\Delta(\lambda)$ can only have a finite number of zeros inside the disk $K_R(0)$.

Example. Consider the entire function

$$\begin{aligned} f(z) &= \frac{1}{2}\left(e^{z^2+iz} + e^{z^2-iz}\right) + \frac{1}{2i}\left(e^{-z^2+iz} - e^{-z^2-iz}\right) \\ &= \frac{e^{iz}+e^{-iz}}{2} e^{z^2} + \frac{e^{iz}-e^{-iz}}{2i} e^{-z^2} = \cos z \cdot e^{z^2} + \sin z \cdot e^{-z^2}. \end{aligned}$$

With the notation of [3] the function $f(z)$ is an entire function of order $m = 2$ from class K (see Section 2 p. 464). Keeping this notation we have the following representation (see equality (3), p. 465)

$$f(z) = f_1(z)e^{z^2} + f_2(z)e^{-z^2},$$

where $f_1(z) = \cos z$, $f_2(z) = \sin z$. Here $N_1 = 2$, $\alpha_{12} = 1$, $\alpha_{22} = -1$, $R_1 = [-1, 1]$. The normals to R_1 are the rays outgoing from the origin and having the angles $\varphi_1^{(1)} = \frac{\pi}{2}$, $\varphi_2^{(1)} = \frac{3\pi}{2}$ (see Fig. 2). According to [3], using $\varphi_1^{(1)}$ we determine the rays outgoing from the origin and having the angles $\varphi_{10}^{(1)} = \frac{\pi}{4}$, $\varphi_{11}^{(1)} = \frac{\pi}{4} + \pi$, and using $\varphi_2^{(1)}$ – the rays with the angles $\varphi_{20}^{(1)} = \frac{3\pi}{4}$, $\varphi_{21}^{(1)} = \frac{3\pi}{4} + \pi$. Remove from z-plane

4 sectors of arbitrary small angle. The bisectrices of these sectors are the rays of angles $\varphi_{10}^{(1)}$, $\varphi_{11}^{(1)}$, $\varphi_{20}^{(1)}$, $\varphi_{21}^{(1)}$. The remaining part of the complex plane decomposes into 4 sectors $\Omega_{10}^{(1)}$, $\Omega_{11}^{(1)}$, $\Omega_{20}^{(1)}$, $\Omega_{21}^{(1)}$ (see Fig. 2).

The representation (4) in [3], p. 466 is not correct. Indeed,

$$f(z) = f_1(z)e^{z^2}\left(1 + \operatorname{tg} z \cdot e^{-2z^2}\right), \tag{2.7}$$

but $\operatorname{tg} z \cdot e^{-2z^2} \neq O\left(e^{-\delta z^2}\right)$ $\forall \delta > 0$ for $z \in \Omega_{10}^{(1)}$, $z \to \infty$.

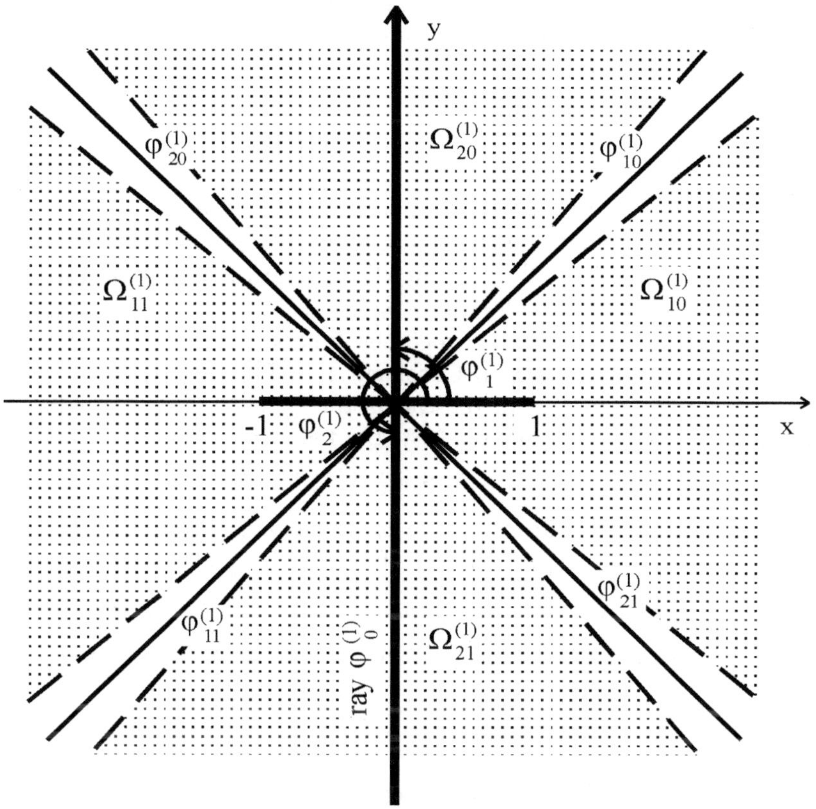

Fig. 2

Indeed, suppose we restrict ourselves to positive z (the positive semi-axis belongs to the sector $\Omega_{10}^{(1)}$). Since $\forall n \in \mathbb{N}$ and $\forall \gamma \in \mathbb{R}$

$$\lim_{x \to \pi n + \frac{\pi}{2} - 0} \operatorname{tg} x \cdot e^{\gamma x^2} = +\infty, \qquad \lim_{x \to \pi n - \frac{\pi}{2} + 0} \operatorname{tg} x \cdot e^{\gamma x^2} = -\infty,$$

then $\operatorname{tg} x \cdot e^{-2x^2} \neq O\left(e^{-\delta x^2}\right)$ as $x \to +\infty$. Further, from that erroneous relation an incorrect conclusion was made that the zeros of the function $f(z)$ in the sector

$\Omega_{10}^{(1)}$ coincide with the zeros of the function $f_1(z)$ (see second indentation on p. 466, [3]). Indeed, the function

$$\varphi(z) = 1 + \operatorname{tg} z e^{-2z^2}$$

on each interval

$$\left(\pi n - \frac{\pi}{2},\ \pi n + \frac{\pi}{2}\right), \quad n \in \mathbb{N}$$

has a zero. Therefore from representation (2.7) we obtain that in the domain $\Omega_{10}^{(1)}$ the set of zeros of the function $f(z)$ consists of the zeros of the function $f_1(z)$ and of the infinite number of zeros of the function $\varphi(z) = 1 + \operatorname{tg} z e^{-2z^2}$. Hence, the inductive passage to the study of the zeros of the function $f_1(z)$ from class K of the order 1 given in [3] is incorrect.

3. Zeta-function associated with the function $\Delta(\lambda)$

Without loss of generality, we can suppose that $\Delta(0) \neq 0$ (if $\Delta(0) = 0$, then, dividing $\Delta(\lambda)$ by λ^q, where q is the multiplicity of zero of function $\Delta(\lambda)$, we get the entire function $\tilde{\Delta}(\lambda) = \frac{\Delta(\lambda)}{\lambda^q}$, having the asymptotic representation (1.1)).

For definiteness of consideration, in the sector G_1 we choose the sector S with the vertex at the origin in which the function $\Delta(\lambda)$ has no zeros (such a sector always exists according to Corollary 2.6). Let the ray $s = te^{i\varphi_0}$, $t \geq 0$, $\varphi_0 \in [0,\ 2\pi)$ be the bisectrix of the the sector S. Then with $\lambda \in s$ for the zeros λ_n, $n = 1,\ 2,\ 3, \dots$ of the function $\Delta(\lambda)$ the following estimate is valid

$$|\lambda_n - \lambda| > \delta |\lambda_n|, \quad \delta > 0 \tag{3.1}$$

(δ may be chosen equal to $\sin\varphi^*$, where $2\varphi^*$ is the angle of the sector S). Suppose that on the set $D_1 = G_1 \cap \{|\lambda| \geq R_0\}$ the following inequality holds (Corollary 2.4)

$$\operatorname{Re}\mu_1(\lambda) - \operatorname{Re}\mu_k(\lambda) \geq \mu_0 |\lambda|^{\frac{1}{h}} \quad \forall k = 2,\ 3, \dots,\ H\,.$$

Then the function $\Delta(\lambda)$ has the asymptotic expansion as $\lambda \to \infty$ in the sector S

$$\Delta(\lambda) \sim \exp\left(\mu_1(\lambda)\right) \lambda^{\frac{n_1}{h}} \sum_{\nu=0}^{\infty} \beta_\nu^{(1)} \lambda^{-\frac{\nu}{h}}\,. \tag{3.2}$$

Differentiating termwise the asymptotic expansion (3.2) of the entire function $\Delta(\lambda)$ ([5]) we obtain the asymptotic expansion of the logarithmic derivative of the function $\Delta(\lambda)$ as $\lambda \in S$, $\lambda \to \infty$

$$\frac{\Delta'(\lambda)}{\Delta(\lambda)} \sim \lambda^{\frac{N-h-\ell_1}{h}} \sum_{\nu=0}^{\infty} \omega_\nu^{(1)} \lambda^{-\frac{\nu}{h}}, \tag{3.3}$$

where $\omega_m^{(1)} = \theta_{1,(\ell_1+m)}\left(\frac{N-(\ell_1+m)}{h}\right)$, $m = 0,\ 1, \dots,\ N-1-\ell_1$, $\omega_{N-\ell_1} = \frac{n_1}{h}$, $\omega_{N-\ell_1+1} = \frac{\beta_1^{(1)}}{\beta_0^{(1)}}$ and so on. Construct the Riemann surface for $\sqrt[h]{\lambda}$ with the cut along the ray s.

Take the contour Γ consisting of the ray s lying on the first sheet of the Riemann surface of $\sqrt[h]{\lambda}$ and going through from ∞ to the point $s_1 = t_0 e^{i\varphi_0}$, $t_0 > 0$, of the curve γ, consisting of the h-fold clockwise loop around zero with radius t_0 and of the ray s lying on the hth sheet and going through from the point $s_h = t_0 e^{i(-2\pi h + \varphi_0)}$ to ∞. Choose some sufficiently small $t_0 > 0$ such that $\Delta(\lambda)$ has no zeros in the circle of radius t_0 with center at the origin. The contour Γ on the Riemann surface of $\sqrt{\lambda}$ is shown on Fig. 3.

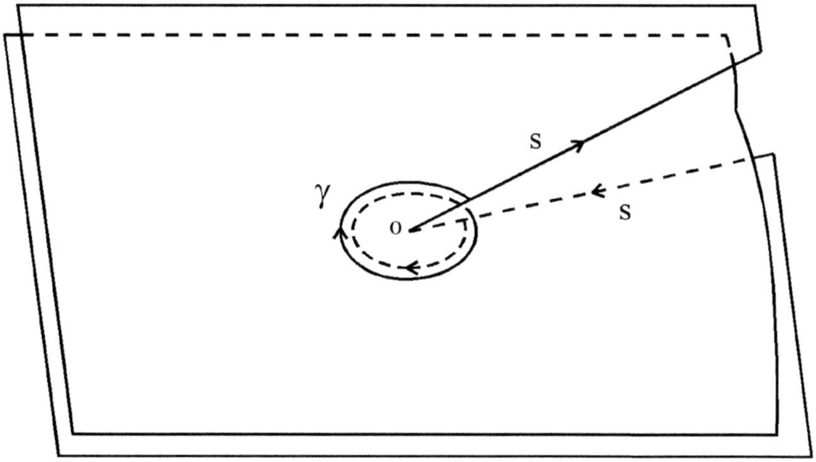

Fig. 3

Lidsky–Sadovnichy's method [2] of introducing the zeta-function associated with the class K of entire functions can also be extended to the entire functions $\Delta(\lambda)$ admitting the asymptotic representation (1.1). The zeta-function associated with the function $\Delta(\lambda)$ is introduced via the integral

$$Z(\sigma) = \frac{1}{2\pi i} \int\limits_{\Gamma} \lambda^{-\sigma} \frac{\Delta'(\lambda)}{\Delta(\lambda)} d\lambda \,, \tag{3.4}$$

where $\lambda^{-\sigma} = e^{-\sigma \operatorname{Ln} \lambda}$ and the principal value of $\operatorname{Ln} \lambda = \ln \lambda$ is fixed on the first sheet of the Riemann surface of $\sqrt[h]{\lambda}$. By virtue of the asymptotic representation (3.3) the integral (3.4) converges on the half-plane $\operatorname{Re} \sigma > 1 + \frac{N-h-\ell_1}{h}$.

Lemma 3.1. *The zeta-function $Z(\sigma)$ can be extended analytically into the whole σ-plane as an entire function; moreover*

$$Z\left(-\frac{p}{h}\right) = h\omega^{(1)}_{p+N-\ell_1}\,, \quad p = 0,\ 1,\ 2,\ \ldots\,.$$

Proof. Let us fix an arbitrary real number x_0 and choose an integer number ν_0 such that $\nu_0 > N - hx_0 - \ell_1 - 1$. Write $Z(\sigma)$ in the following form

$$Z(\sigma) = \frac{1}{2\pi i}\int\limits_{\gamma} \lambda^{-\sigma}\frac{\Delta'(\lambda)}{\Delta(\lambda)}d\lambda + \frac{1}{2\pi i}\int\limits_{\Gamma^*} \lambda^{-\sigma}\left(\frac{\Delta'(\lambda)}{\Delta(\lambda)} - \lambda^{\frac{N-h-\ell_1}{h}}\sum_{\nu=0}^{\nu_0}\frac{\omega_\nu^{(1)}}{\lambda^{\frac{\nu}{h}}}\right)d\lambda$$
$$+ \frac{1}{2\pi i}\int\limits_{\Gamma} \lambda^{-\sigma+\frac{N-h-\ell_1}{h}}\sum_{\nu=0}^{\nu_0}\frac{\omega_\nu^{(1)}}{\lambda^{\frac{\nu}{h}}}d\lambda - \frac{1}{2\pi i}\int\limits_{\gamma} \lambda^{-\sigma+\frac{N-h-\ell_1}{h}}\sum_{\nu=0}^{\nu_0}\frac{\omega_\nu^{(1)}}{\lambda^{\frac{\nu}{h}}}d\lambda,$$

where Γ^* denotes the path consisting of the ray s passing from ∞ to the point $s_1 = t_0e^{i\varphi_0}$ on the first sheet of the Riemann surface of $\sqrt[h]{\lambda}$ and from the point $s_h = t_0e^{i(-2\pi h+\varphi_0)}$ to ∞ on the hth sheet.

For each $\lambda \in \gamma$ the function $\lambda^{-\sigma}$ is an entire function with respect to σ and therefore the integrals

$$I_1(\sigma) = \frac{1}{2\pi i}\int\limits_{\gamma} \lambda^{-\sigma+\frac{N-h-\ell_1}{h}}\frac{\Delta'(\lambda)}{\Delta(\lambda)}d\lambda$$

and

$$I_4(\sigma) = \frac{1}{2\pi i}\int\limits_{\gamma} \lambda^{-\sigma+\frac{N-h-\ell_1}{h}}\sum_{\nu=0}^{\nu_0}\frac{\omega_\nu^{(1)}}{\lambda^{\frac{\nu}{h}}}d\lambda$$

admit differentiation under the integral and hence are entire functions with respect to σ.

In the integral

$$I_2(\sigma) = \frac{1}{2\pi i}\int\limits_{\Gamma^*} \lambda^{-\sigma}\left(\frac{\Delta'(\lambda)}{\Delta(\lambda)} - \lambda^{\frac{N-h-\ell_1}{h}}\sum_{\nu=0}^{\nu_0}\frac{\omega_\nu^{(1)}}{\lambda^{\frac{\nu}{h}}}\right)d\lambda$$

the difference

$$\frac{\Delta'(\lambda)}{\Delta(\lambda)} - \lambda^{\frac{N-h-\ell_1}{h}}\sum_{\nu=0}^{\nu_0}\frac{\omega_\nu^{(1)}}{\lambda^{\frac{\nu}{h}}} \quad \text{is } O\left(\frac{1}{\lambda^{\frac{\nu_0-N+h+\ell_1+1}{h}}}\right),$$

therefore on the domain $\operatorname{Re}\sigma > x_0$ we can differentiate under the integral. Hence $I_2(\sigma)$ is analytic in $\operatorname{Re}\sigma > x_0$. Since x_0 is an arbitrary real number, $I_2(\sigma)$ can be extended as an entire function.

The integral

$$I_3(\sigma) = \frac{1}{2\pi i}\int\limits_{\Gamma} \lambda^{-\sigma+\frac{N-h-\ell_1}{h}}\sum_{\nu=0}^{\nu_0}\frac{\omega_\nu^{(1)}}{\lambda^{\frac{\nu}{h}}}d\lambda$$

for $\operatorname{Re}\sigma > 1 + \frac{N-h-\ell_1}{h}$ is equal to zero, and hence it can be continued analytically as zero into the whole plane.

We shall find the values of $Z(\sigma)$ at the points $\sigma = -\frac{p}{h}$, $p = 0, 1, 2, \dots$. We make the change $\lambda = w^h$; then

$$I_1\left(-\frac{p}{h}\right) = \frac{1}{2\pi i}\int\limits_{\gamma} \lambda^{\frac{p}{h}} \frac{\Delta'(\lambda)}{\Delta(\lambda)} d\lambda = h \int\limits_{\gamma'} w^{p+h-1} \frac{\Delta'(w^h)}{\Delta(w^h)} dw\,.$$

There are no zeros of $\Delta(w^h)$ inside γ', therefore the integrand is an analytic function inside the circle γ' and hence $I_1\left(-\frac{p}{h}\right) = 0$.

Furthermore

$$I_2\left(-\frac{p}{h}\right) = \frac{1}{2\pi i}\int\limits_{\Gamma^*} \lambda^{\frac{p}{h}} \left(\frac{\Delta'(\lambda)}{\Delta(\lambda)} - \lambda^{\frac{N-h-\ell_1}{h}} \sum_{\nu=0}^{\nu_0} \frac{\omega_\nu^{(1)}}{\lambda^{\frac{\nu}{h}}}\right) d\lambda = 0\,,$$

since the integration is done in opposite directions along the lower (in the first sheet of the Riemann surface) and upper (in the hth sheet) sides of the cut s, on which the integrand takes the equal values. From the previous arguments $I_3\left(-\frac{p}{h}\right) = 0$.

Calculate

$$I_4\left(-\frac{p}{h}\right) = \frac{1}{2\pi i}\int\limits_{\gamma} \lambda^{\frac{p+N-h-\ell_1}{h}} \sum_{\nu=0}^{\nu_0} \frac{\omega_\nu^{(1)}}{\lambda^{\frac{\nu}{h}}} d\lambda$$

$$= \frac{1}{2\pi i}\sum_{\nu=0}^{\nu_0} \omega_\nu^{(1)} \int\limits_{\gamma} \lambda^{\frac{p+N-h-\ell_1-\nu}{h}} d\lambda = -h\omega_{p+N-\ell_1}^{(1)}.$$

We get

$$Z\left(-\frac{p}{h}\right) = h\omega_{p+N-\ell_1}^{(1)}\,.$$

Lemma 3.2. *For* $\operatorname{Re}\sigma > N/h$

$$Z(\sigma) = \Lambda(\sigma)\sum_{\ell} \lambda_\ell^{-\sigma}\,,$$

where λ_ℓ *are the zeros of the function* $\Delta(\lambda)$, *taking account of multiplicity and*

$$\Lambda(\sigma) = \begin{cases} h, & \sigma = 0,\ \pm 1,\ \pm 2,\ \dots \\ \frac{e^{-2\pi h\sigma i}-1}{e^{-2\pi\sigma i}-1}, & \textit{otherwise}\,. \end{cases}$$

Proof. For the entire function $\Delta(\lambda)$ of order N/h, by Hadamard's Theorem ([8]), the following expansion holds

$$\Delta(\lambda) = e^{Q_{N_1}(\lambda)} \prod_{k=1}^{\infty} G\left(\frac{\lambda}{\lambda_k},\ N_1\right), \quad N_1 = \left[\frac{N}{h}\right],$$

where

$$G\left(\frac{\lambda}{\lambda_k},\ N_1\right) = \left(1 - \frac{\lambda}{\lambda_k}\right)\exp\left(\frac{\lambda}{\lambda_k} + \frac{1}{2}\left(\frac{\lambda}{\lambda_k}\right)^2 + \cdots + \frac{1}{N_1}\left(\frac{\lambda}{\lambda_k}\right)^{N_1}\right),$$

and $Q_{N_1}(\lambda)$ is a polynomial of degree not greater then N_1.

For the logarithmic derivative of the function $\Delta(\lambda)$ we get the following representation

$$\begin{aligned}\frac{\Delta'(\lambda)}{\Delta(\lambda)} &= Q'_{N_1}(\lambda) + \sum_{k=1}^{\infty}\left[\frac{1}{\lambda-\lambda_k} + \frac{1}{\lambda_k}\left(1+\dots+\left(\frac{\lambda}{\lambda_k}\right)^{N_1-1}\right)\right] \\ &= Q'_{N_1}(\lambda) + \sum_{k=1}^{\infty}\frac{\lambda^{N_1}}{\lambda_k^{N_1}(\lambda-\lambda_k)}.\end{aligned}$$

For $\lambda \in \Gamma$ using the estimate (3.1) we have the inequality

$$\left|\frac{\lambda^{N_1}}{\lambda_k^{N_1}(\lambda-\lambda_k)}\right| \le \delta' \frac{|\lambda|^{N_1}}{|\lambda_k^{N_1+1}|}, \qquad \delta' > 0\,.$$

The numerical series $\sum\limits_{k=1}^{\infty} \frac{1}{|\lambda_k|^{N_1+1}}$ converges, therefore the functional series

$$\sum_{k=1}^{\infty}\frac{\lambda^{N_1}}{\lambda_k^{N_1}(\lambda-\lambda_k)}$$

converges absolutely and uniformly on $\Gamma \cap \{|\lambda| \le R\}$, where R is an arbitrary positive number. Hence this series admits termwise integration along the contour Γ and for $\operatorname{Re}\sigma > N/h$ we obtain

$$\begin{aligned}Z(\sigma) &= \frac{1}{2\pi i}\int\limits_{\Gamma}\lambda^{-\sigma}\frac{\Delta'(\lambda)}{\Delta(\lambda)}d\lambda \\ &= \frac{1}{2\pi i}\int\limits_{\Gamma}\lambda^{-\sigma}Q'_{N_1}(\lambda)d\lambda + \frac{1}{2\pi i}\int\limits_{\Gamma}\lambda^{-\sigma}\sum_{k=1}^{\infty}\frac{\lambda^{N_1}}{\lambda_k^{N_1}(\lambda-\lambda_k)}d\lambda \\ &= \sum_{k=1}^{\infty}\frac{1}{2\pi i}\int\limits_{\Gamma}\frac{\lambda^{N_1-\sigma}}{\lambda_k^{N_1}(\lambda-\lambda_k)}d\lambda\,.\end{aligned}$$

Make the substitution $\lambda = w^h$ and denote by Γ' the contour in which the contour Γ is transformed by the map $w = \sqrt[h]{\lambda}$. Then for $\operatorname{Re}\sigma > N/h$ we have

$$\begin{aligned}\frac{1}{2\pi i}\int\limits_{\Gamma}\frac{\lambda^{N_1-\sigma}}{\lambda_k^{N_1}(\lambda-\lambda_k)}d\lambda &= \frac{1}{2\pi i}\int\limits_{\Gamma'}\frac{w^{(N_1-\sigma)h}\cdot hw^{h-1}}{\lambda_k^{N_1}(w^h-\lambda_k)}dw \\ &= \sum_{\ell=0}^{h-1} w_{k\ell}^{-\sigma h} = \lambda_k^{-\sigma}\cdot\frac{1-e^{-2\pi i\sigma h}}{1-e^{-2\pi i\sigma}},\end{aligned}$$

where $w_{k\ell} = {}_{\ell}\sqrt[h]{\lambda_k} = \sqrt[h]{|\lambda_k|}\left(e^{i\left(\frac{\arg\lambda_k}{h}+\frac{2\pi\ell}{h}\right)}\right)$.

Therefore

$$Z(\sigma) = \Lambda(\sigma)\sum_{\ell}\lambda_\ell^{-\sigma}, \qquad \text{where} \qquad \Lambda(\sigma) = \begin{cases} h, & \sigma = 0,\ \pm1,\ \pm2,\ \dots \\ \frac{e^{-2\pi h\sigma i}-1}{e^{-2\pi\sigma i}-1}, & \text{otherwise}.\end{cases}$$

4. Regularized sums of roots of the function $\Delta(\lambda)$

In order to obtain the regularized sums of roots of the function $\Delta(\lambda)$ it is required to have a detailed information about the asymptotic of the zeros λ_ℓ, namely, it is necessary to have an asymptotic expansion of λ_ℓ in powers of ℓ. Sometimes the zeros z_ℓ of functions from class K have not such asymptotic expansions in powers of ℓ ([4]). This fact leads to the necessity of applying perturbation theory. Therefore, extending Lidskii–Sadovnichii's method ([2]) to a larger class of entire functions than class K, we shall suppose that the following assumption is valid.

Assumption. *Assume that the roots λ_ℓ, $\ell = 1, 2, 3, \ldots$ of the function $\Delta(\lambda)$ lying in the sectors E_k, $k = \overline{1, L}$, decompose into M_k series, and, in addition, for each series the following asymptotic representation is valid*

$$\lambda_{nkr} \sim a_{kr} n^{\frac{h}{m_{kr}}} \left(1 + \sum_{s=1}^{\infty} \frac{R_{skr}(\ln n)}{n^{\frac{s}{m_{kr}}}}\right), \qquad r = 1, \ldots, M_k\,, \tag{4.1}$$

where $a_{kr} \in \mathbb{C}$, $a_{kr} \neq 0$, $m_{kr} \in \mathbb{N}$, $1 \leq m_{kr} \leq N$, $R_{skr}(\ln n)$ are polynomials in $\ln n$ of degree not greater than $s - h + 1$ for $s \geq h$ and of degree zero for $1 \leq s \leq h - 1$.

Taking the asymptotic relation (4.1) to the power $-\sigma$, by Taylor formula we obtain

$$\lambda_{nkr}^{-\sigma} \sim a_{kr}^{-\sigma} n^{-\frac{h}{m_{kr}}\sigma} \left(1 + \sum_{s=1}^{\infty} \frac{Q_{skr}(\sigma, \ln n)}{n^{\frac{s}{m_{kr}}}}\right), \tag{4.2}$$

where

$$Q_{skr}(\sigma, \ln n) = \sum_{\nu=0}^{s-h+1} d_{s\nu kr}(\sigma) \ln^\nu n$$

and $d_{s\nu kr}(\sigma)$ are polynomials in σ of degree s.

Fix some sufficiently large natural number τ. Consider the function

$$\Psi_\tau(\sigma) = {\sum_{n=1}^{\infty}}' \sum_{k=1}^{L} \sum_{r=1}^{M_k} \left[\lambda_{nkr}^{-\sigma} - a_{kr}^{-\sigma} n^{-\frac{h}{m_{kr}}\sigma} \left(1 + \sum_{s=1}^{\tau} \frac{Q_{skr}(\sigma, \ln n)}{n^{\frac{s}{m_{kr}}}}\right)\right]. \tag{4.3}$$

A finite number of zeros of the function $\Delta(\lambda)$ may be non-accounted for. These zeros are included in the sum (4.3) and as a result the prime is attached to the sum sign in (4.3).

By the asymptotic expansion (4.2) we conclude that the function $\Psi_\tau(\sigma)$ can be extended analytically into the half-plane

$$\operatorname{Re}\sigma > -\frac{\tau + 1 - N}{h}.$$

Consider the function

$$\Phi_\tau(\sigma) = \sum_{n=1}^{\infty} \sum_{k=1}^{L} \sum_{r=1}^{M_k} a_{kr}^{-\sigma} n^{-\frac{h}{m_{kr}}\sigma} \left(1 + \sum_{s=1}^{\tau} \frac{Q_{skr}(\sigma, \ln n)}{n^{\frac{s}{m_{kr}}}}\right), \tag{4.4}$$

that is regular for $\operatorname{Re}\sigma > N/h$.

By the relation (4.3) and Lemma 3.2, for $\operatorname{Re}\sigma > N/h$ we have

$$\Psi_\tau(\sigma) = \frac{Z(\sigma)}{\Lambda(\sigma)} - \Phi_\tau(\sigma)\,. \tag{4.5}$$

From here it follows (since $Z(\sigma)$ and $\Lambda(\sigma)$ are entire functions, and $\Psi_\tau(\sigma)$ can be analytically extended into the half-plane $\operatorname{Re}\sigma > -\frac{\tau+1-N}{h}$) that the function $\Phi_\tau(\sigma)$ can be analytically extended as a meromorphic function into the half-plane $\operatorname{Re}\sigma > -\frac{\tau+1-N}{h}$. We remark that $\Phi_\tau(\sigma)$ can be expressed in terms of the Riemann ζ-function and its derivatives. By virtue of (4.4) we have

$$\begin{aligned}
\Phi_\tau(\sigma) &= \sum_{k=1}^{L}\sum_{r=1}^{M_k} a_{kr}^{-\sigma}\zeta\left(\frac{h}{m_{kr}}\sigma\right) + \sum_{k=1}^{L}\sum_{r=1}^{M_k}\sum_{s=1}^{\tau}\sum_{n=1}^{\infty}\frac{a_{kr}^{-\sigma}Q_{skr}(\sigma,\ln n)}{n^{\frac{s+h\sigma}{m_{kr}}}}\\
&= \sum_{k=1}^{L}\sum_{r=1}^{M_k} a_{kr}^{-\sigma}\zeta\left(\frac{h}{m_{kr}}\sigma\right) + \sum_{k=1}^{L}\sum_{r=1}^{M_k}\sum_{s=1}^{\tau}\sum_{\nu=0}^{s-h+1} a_{kr}^{-\sigma}d_{s\nu kr}(\sigma)\sum_{n=1}^{\infty}\frac{\ln^\nu n}{n^{\frac{s+h\sigma}{m_{kr}}}}\\
&= \sum_{k=1}^{L}\sum_{r=1}^{M_k} a_{kr}^{-\sigma}\zeta\left(\frac{h}{m_{kr}}\sigma\right)\\
&\quad + \sum_{k=1}^{L}\sum_{r=1}^{M_k}\sum_{s=1}^{\tau}\sum_{\nu=0}^{s-h+1} a_{kr}^{-\sigma}d_{s\nu kr}(\sigma)(-1)^\nu\zeta^{(\nu)}\left(\frac{s+h\sigma}{m_{kr}}\right).
\end{aligned} \tag{4.6}$$

Evaluating the values $\Psi_\tau(\sigma)$ by formula (4.5) for $\sigma = 0,\ -1,\ -2,\ldots$, we obtain the following theorem

Theorem 4.1. *For any integer $m < \frac{\tau+1-N}{h}$ the following equalities hold*

$$\sum_{n=1}^{\infty}{}'\sum_{k=1}^{L}\sum_{r=1}^{M_k}\left[\lambda_{nkr}^m - a_{kr}^m n^{\frac{hm}{m_s}} - a_{kr}^m\sum_{s=1}^{\tau}\frac{Q_{skr}(-m,\ln n)}{n^{\frac{s-hm}{m_{kr}}}}\right] = \omega^{(1)}_{mh+N-\ell_1} - \Phi_\tau(-m)\,,$$

where $\omega^{(1)}_{mh+N-\ell_1}$ are expansion coefficients in (3.3), $\Phi_\tau(-m)$ are determined in terms of the Riemann ζ-function and its derivatives by formula (4.6), a_{kr} are coefficients in the principal term in the asymptotic (4.2), $Q_{skr}(-m,\ln n)$ is the polynomial in $\ln n$ in the representation (4.2).

References

[1] I.M. Gelfand, B.M. Levitan, *On a Simple Identity for the Eigenvalues of a Differential Equation.* Dokl. Akad. Nauk SSSR, **88** (1953) N. 4, 593–596, (in Russian).

[2] V.B. Lidskii, V.A. Sadovnichii, *Regularized Sums of Roots of One Class of Entire Functions.* Func. anal. and its apps., **1** (1967), N. 2, 52–59, (in Russian); English translation in Functional Anal. Appl. **1** (1967).

[3] V.A. Ljubishkin, *Regularized Sums of Roots of Generalized Quasipolynomials.* Matematicheskii Sbornik v. **135** (1988), (177), N. 4, 463–471, 1988 (in Russian).

[4] V.A. Ljubishkin, V.A. Sadovnicii, *Regularized Sums of Roots of One Class of Entire Functions of Exponential Type*. Dokl. Akad. Nauk SSSR **256** (1981), N. 4, 794–798, (in Russian).

[5] F.W. Olver, *Asymptotics and Special functions*. N.Y., Academic Press, 1974.

[6] A.S. Pechentsov, *Regularized Traces of Boundary Problems in Case of Multiple Roots of Characteristic Polynomial*. Fundamentalnaya i prikladnaya matematica **4** (1998), N. 2, 567–583, (in Russian).

[7] A.S. Pechentsov, *Traces of One Class of Singular Differential Operators: The Lidskii-Sadovnichii Method*. Vestnik Mosk. Univer., Series Math. and Mech. N. 5, 35–42, 1999.

[8] E.C. Titchmarsh, *Theory of Functions*. (2nd edition) London: Oxford University Press, 1939.

A.S. Pechentsov
Moscow Lomonosov State University
Faculty of Mechanics and Mathematics

and

Institute for Mathematical Study of Complex Systems

Russia

Operator Theory:
Advances and Applications, Vol. 153, 229–247

Irreducible Subfactors Derived from Popa's Construction for Non-Tracial States

Florin Rădulescu

Abstract. For an inclusion of the form $\mathbb{C} \subseteq M_n(\mathbb{C})$, where $M_n(\mathbb{C})$ is endowed with a state with diagonal weights $\lambda = (\lambda_1, \ldots, \lambda_n)$, we use Popa's construction, for non-tracial states, to obtain an irreducible inclusion of II_1 factors, $N^\lambda(Q) \subseteq M^\lambda(Q)$ of index $\sum \frac{1}{\lambda_i}$. $M^\lambda(Q)$ is identified with a subfactor inside the centralizer algebra of the canonical free product state on $Q \star M_N(\mathbb{C})$. Its structure is described by "infinite" semicircular elements as in [32].

The irreducible subfactor inclusions obtained by this method are similar to the first irreducible subfactor inclusions, of index in $[4, \infty)$ constructed in [24], starting with the Jones' subfactors inclusion $R^s \subseteq R$, $s > 4$. In the present paper, since the inclusion we start with has a simpler structure, it is easier to control the algebra structure of the subfactor inclusions.

If the weights correspond to a unitary, finite-dimensional representation of a Woronowicz's compact quantum group G, then the factor $M^\lambda(Q)$ is contained in the fixed point algebra of an action of the quantum group on $Q \star M_N(\mathbb{C})$, with equality if G is $SU_q(N)$, (or $SO_q(3)$ when $N = 2$). By Takesaki duality, the factor $M^\lambda(L(F_N))$ is Morita equivalent to $\mathcal{L}(F_\infty)$.

This method gives also another approach to find, as also recently proved in [36], irreducible subfactors of $\mathcal{L}(F_\infty)$ for index values bigger than 4.

Mathematics Subject Classification (2000). 46L37, 46L54.

Keywords. Subfactor, Non-tracial state, Semicircular element.

0. Introduction and definitions

In this paper we consider the structure of subfactors obtained from Popa's construction, for non-tracial states, for the inclusion $M_N(\mathbb{C}) \subseteq M_N(\mathbb{C}) \otimes M_N(\mathbb{C})$. We fix a diagonal matrix with non-zero weights $\lambda_1, \ldots, \lambda_N$. The states on the two algebras are respectively $tr(D\cdot)$ and $tr(D \otimes D^{-1}\cdot)$. This is then the Jones' iterated basic construction ([11], [12]) for $\mathbb{C} \subseteq M_N(\mathbb{C})$, where $M_N(\mathbb{C})$ is endowed with the state $tr(D\cdot)$. The algebra $M_N(\mathbb{C}) \otimes 1$ is invariated by the modular group of the state $tr(D \otimes D^{-1}\cdot)$ on $M_N(\mathbb{C}) \otimes M_N(\mathbb{C})$. Hence by [19], there exists a conditional expectation and a corresponding Jones projection, which is also invariated by the modular group. The value of the state on the Jones projection is $\sum \frac{1}{\lambda_i}$. Let Q be a diffuse finite von Neumann algebra with faithful trace τ.

We apply Popa's construction for the inclusion $M_N(\mathbb{C}) \subseteq M_N(\mathbb{C}) \otimes M_N(\mathbb{C})$. This yields an irreducible inclusion $N^\lambda(Q) \subseteq M^\lambda(Q)$ of type II_1 factors, inside the type III factor $(Q \otimes M_N(\mathbb{C})) *_{M_N(\mathbb{C})} (M_N(\mathbb{C}) \otimes M_N(\mathbb{C}))$, considered with the amalgamated free product state.

By applying the reduction method described in [29], [33], we can reduce this procedure to the case when the algebra over which we amalgamate is abelian. Thus, we end up with a concrete description of the factor and the subfactor inside $Q\star M_N(\mathbb{C})$, where the later algebra is endowed with the free product state $\tau\star tr(D\cdot)$. This factor for $N = 2$, $Q = L^\infty([0,1])$ can be very explicitly described in terms of the "infinite" Voiculescu's type ([40]) free semicircular element used to identify the core of $L^\infty([0,1]) \star M_2(\mathbb{C})$ in [32].

Moreover, by analogy with the case of a trace on $M_N(\mathbb{C})$, it turns out that $M^\lambda(Q)$ is contained in the fixed point algebra for a free product coaction (of the type considered in [37]) on $Q \star M_N(\mathbb{C})$.

Let α be any finite-dimensional unitary selfadjoint coaction of a Woronowicz quantum group G on the finite-dimensional algebra $M_N(\mathbb{C})$. We prove that the algebra $M^\lambda(Q)$ described above is contained in the fixed point algebra of coaction of the quantum group. If G is the quantum group $SU_q(N)$ then the algebra $M^\lambda(Q)$ is exactly the fixed point algebra. This is analogous to the result in [14], [20], where the fixed point algebra of the infinite product action on $M_N(\mathbb{C})$ of $SU_q(N)$ coincides with the hyperfinite factor.

Assume (by analogy with [41]) that $\alpha^{\otimes^n}$ contains any other finite-dimensional unitary representation of G. Then we prove that the fixed point algebra $(Q \star M_N(\mathbb{C}))^G$ is Morita equivalent to the cross product by α. Such a cross product is naturally described ([37]) by a free product with amalgamation. Since the fixed point algebra is a II_1 factor it follows that the fixed point algebra is Morita equivalent with an amalgamated free product of the form $(Q \otimes D) *_D C$, where C, D are (infinite) direct sums of matrix algebras. By the techniques in [29] it follows, for $Q = \mathcal{L}(F_N)$, that the fixed point algebra is the von Neumann algebra of a free group with infinitely many generators. The II_1 factors is invariated by the action of an automorphism scaling the trace on a larger II_∞ factor.

We also reobtain in this way, by a different approach, the result recently proved in the remarkable paper by D. Shlyahtenko and Y. Ueda ([36]) that $L(F_\infty)$ has irreducible subfactors for any index value bigger than 4. One theme in their paper is that one can determine the isomorphism class of a fixed point algebra by looking at the crossed product algebra (via Takesaki's duality). We owe to their paper the use of this philosophy for coactions.

We do not know if the subfactors in this paper coincide the ones constructed in [36]. In both cases the higher relative commutants invariant is A_∞ (though the factors in [36] seem to generate other higher relative commutants invariants) when used for other compact quantum groups). Both algebra of the factors are obtained by taking the the fixed point algebra of a coaction of a quantum group. In one case the fixed point algebra is a II_1 factor and in the other case the fixed point algebra is of type III.

The construction in this paper proves that the non-tracial version of Popa's construction of subfactors, naturally yields irreducible subfactors, In addition the corresponding algebras are fixed algebras of coactions of quantum groups.

These algebras, although they are very similar to the subfactors in the breakthrough construction in [24], are not yet proven to be isomorphic to the algebras in [24], [26]. Though, our result is strong evidence to the conjecture that the subfactors in [24], [25] are free group factors when Q is $\mathcal{L}(F_\infty)$ (the only case when the subfactors in [24], [25] are known to be free group factors is for index values less than 4, and the higher relative commutants are the Temperly-Lieb algebra, see [29]).

The method in this papers also gives an explicit model of the subfactors in terms of (infinite) semicircular random matrices ([40], [30], [32]).

1. Subfactors derived from Popa's construction for non-tracial states

In this section we describe the structure of the algebra of the subfactors that are obtained, by using Popa's construction, from a finite-dimensional inclusion $B \subseteq A$ with Markov state. We apply this construction to the case of the inclusion $\mathbb{C} \subseteq M_n(\mathbb{C})$, where $M_n(\mathbb{C})$ is endowed with a state. This will give irreducible subfactor of indices from 4 to ∞.

We start with the following lemma which proves that the subfactor associated to $B \subseteq A$ only depends on the inclusion matrix. To do this we use a reduction style procedure that was used in [29], to the case when B is abelian. The following lemma is proved in [33]

Lemma 1.1. *Let $C \subseteq B$ be a finite-dimensional inclusion of matrices, with trivial centers intersection and assume that B is endowed with a λ Markov state (that is there exists a normal conditional expectation from B onto C which preserves the state, there exists a state on the corresponding basic construction extending the given one and the corresponding Jones' projection has expectation λ times the scalars).*

Let Q be a type II_1 factor, let $A = \langle B, e_1\rangle$ be the first step in the Jones' basic construction of $C \subseteq B$, where e_1 is the Jones projection. Assume that A is endowed with the canonical Markov state. Let $m_i \in B$ be a Pimsner-Popa orthonormal basis for the inclusion $C \subseteq B$. Let $\mathcal{A}$ be the algebra $(Q \otimes B) \star_B A$ and let Φ be the Popa's map associated to the inclusion defined by

$$\Phi(x) = \sum m_i e_1 x e_1 m_i^*, \quad x \in \mathcal{A}.$$

Then as proved in [24], [25], [26], Φ maps B' into A'.

Let f_i be a family of representatives of minimal projections in B and let F_0 be the sum of this projections. Let $\mathcal{A}_0, A_0, B_0$ be the reductions of respectively the algebras $\mathcal{A}, A, B$ by the projection F_0. In particular B_0 is abelian and, as proved in ([29]), $\mathcal{A}_0$ is identified with $(Q \otimes B_0) \star_{B_0} A_0$. Since F_0 has central support 1, Φ induces a map on $\mathcal{A}_0$ which maps B_0' into A_0'. Let $N(Q) \subseteq M(Q) \subseteq B'$ be the minimal algebra (introduced in [24]) in $\mathcal{A}$ that is closed under Φ and contains Q. Correspondingly let $N_0(Q) \subseteq M_0(Q) \subseteq B_0'$ be the minimal algebra in $\mathcal{A}_0$ that is closed under Φ_0 and contains Q. Then

$$M_0(Q) = (M(Q))_{F_0}; \quad N_0(Q) = (N(Q))_{F_0}.$$

Moreover if m_i' is any orthonormal basis for $B_0 \subseteq A_0$, then

$$\Phi_0(x) = \sum m_i' x (m_i')^*, \quad x \in \mathcal{A}_0.$$

Proof. This lemma will be proved in full generality in [33]. We give here in this section an ad-hoc proof for the case $C = \mathbb{C} \subseteq B = M_N(\mathbb{C})$, where $M_N(\mathbb{C})$ has a state $tr(D\cdot)$ (as in the next lemma). Indeed the next step in the basic construction (as proved in the next lemma) is $A = M_N(\mathbb{C}) \otimes M_N(\mathbb{C})$. Here the algebra $\mathcal{A}$ is $(Q \otimes M_N(\mathbb{C})) *_{M_N(\mathbb{C})} (M_N(\mathbb{C}) \otimes M_N(\mathbb{C}))$ and consequently $\mathcal{A}_0$ is $Q * M_N(\mathbb{C})$. By construction Φ_0 is of the form $\Phi_0(x) = \sum_\alpha n_\alpha x (n_\alpha)^*$, for some fixed n_α in $M_N(\mathbb{C})$. Let e_{ij} be a matrix unit diagonalizing D.

By cutting by a minimal projection, it follows that $\Phi_0(x)$ takes values into $(M_N(\mathbb{C}))' \cap \mathcal{A}_0$. Thus necessary, there exists real $\theta_1, \dots, \theta_N$ numbers such that Φ_0 is of the form

$$\Phi_0(x) = \sum \theta_j e_{ij} x e_{ji}, \quad x \in \mathcal{A}_0.$$

But Φ_0 has also the property that the conditional expectation from $\mathcal{A}_0$ onto $M(Q)$ maps $M_N(\mathbb{C})$ into the scalars. This is only possible if $\theta_j e_{ij}$ is an orthonormal basis for $M_N(\mathbb{C})$, i.e., if θ_i are the inverses of the eigenvalues of D. □

As will see below, even in the case when the inclusion $B \subseteq A$ is $\mathbb{C} \subseteq M_N(\mathbb{C})$, with a trace on $M_N(\mathbb{C})$ has to be handled by a rather a complicated machinery. Indeed in this case the above subfactor is the fixed algebra under the action of $SU(N)$ on $Q * M_N(\mathbb{C})$ (here $SU(N)$ acts trivially on Q and by conjugation on $M_N(\mathbb{C})$). This is because the element $\sum e_{ij} \otimes e_{ji}$ is invariant under the product action of the group $SU(N)$ and all the others are obtained by intercalation or concatenation from this element ([44]).

We generalize below the above construction to the case when the trace on $M_N(\mathbb{C})$ is replaced by a state. First we record the following well known folklore lemma. It deals with the Jones basic construction for states (see, e.g., [27], [12])

Lemma 1.2. *Let $\phi = tr(D\cdot)$ be the state on $M_n(\mathbb{C})$ where D has diagonal eigenvalues $\lambda_1, \dots, \lambda_n$. Then the next step in the Jones basic construction with Markov state is $M_n(\mathbb{C}) \otimes M_n(\mathbb{C})$ with state*

$$\frac{1}{trD^{-1}} tr((D \otimes D^{-1} \cdot).$$

This state is so that the Jones projection is invariated by the modular group and projects onto a scalar multiple of the identity. The value of the scalar is $(\sum \frac{1}{\lambda_i})^{-1}$.

Proof. Indeed if (e_{ij}) is a matrix unit, which also diagonalizes D, then the Jones' projection is

$$\sum_{ij} \lambda_i^{-1/2} \lambda_j^{-1/2} e_{ij} \otimes e_{ij}. \qquad \square$$

In the next lemma we describe what Popa's construction ([24]) is in the non-tracial case in a very particular case (for the more general construction see [33]). This construction allows to obtain irreducible subfactors, in factors with non-trivial fundamental group, of index values in $[4, \infty)$ starting from the inclusion $\mathbb{C} \subseteq M_N(\mathbb{C})$. We will describe this factors explicitly.

Lemma 1.3. *We use the notations from the previous lemma and its proof. Let* e_1 *be the Jones projection for the basic construction in the previous lemma. Let* $(M_\alpha)e_1$, *in the centralizer algebra of the state* ϕ, *be a Pimsner-Popa basis for the inclusion* $M_N(\mathbb{C}) \subseteq M_N(\mathbb{C}) \otimes M_N(\mathbb{C})$. *Note that this is always possible since the centralizer algebra (which contains* $e_{ij} \otimes e_{ij}$*), is isomorphic to* $M_N(\mathbb{C})$ *and contains* e_1.

Let Q *be a diffuse abelian von Neumann algebra or a type* II_1 *factor with faithful trace* τ *and let* $\mathcal{A} = (Q \otimes M_N(\mathbb{C})) *_{M_N(\mathbb{C})} (M_N(\mathbb{C}) \otimes M_N(\mathbb{C}))$ *be the amalgamated free product factor obtained by taking the GNS-construction corresponding to the given state on* $M_N(\mathbb{C}) \otimes M_N(\mathbb{C})$ *and the trace* τ. *This is obviously a type* III *factor if* ϕ *is not a trace* ([8], [35], [5], [39]).

Let as in [24], [25], Φ *be the map defined on* $\mathcal{A}$ *by*

$$\Phi(x) = \sum_\alpha M_\alpha e_1 x e_1 M_\alpha^*, \quad x \in \mathcal{A}.$$

Let $M^\lambda(Q)$ *be the minimal algebra in* $\mathcal{A}$ *containing* Q *that is invariant under the map* Φ. *Let* $N^\lambda(Q)$ *be the image of* $M^\lambda(Q)$ *through* Φ.

Then, (as in [24]*)* $N^\lambda(Q) \subseteq M^\lambda(Q)$ *is an irreducible inclusion of type* II_1 *factors of index* $\sum \frac{1}{\lambda_i}$.

Proof. This is basically proved in [24], [25] (see also [33]). The only additional care comes from the fact that we are dealing with a state instead of a trace. But since $M_\alpha e_1$ are chosen in the centralizer of the state it follows that $M^\lambda(Q)$ is a type II_1 factor. The main part of the argument (the Markovianity of the trace) follows from the fact that the conditional expectation of the algebra $M_N(\mathbb{C}) \otimes M_N(\mathbb{C})$ onto $M^\lambda(Q)$ is the algebra of scalars, which follows from the properties of the Pimsner-Popa basis. $\square$

It is not clear if the above factor is the same as the one constructed in [24] starting from the inclusion derived from the Temperly-Lieb algebra for index values bigger than 4. In that case the algebras stay in a much larger amalgamated free product, since the Pimsner-Popa basis sits in hyperfinite factor (and cannot be chosen to belong to a finite-dimensional algebra).

Remark 1.4. To obtain any additional step in the Jones' basic construction of the above inclusion one has to proceed as follows. One starts with the iterated Jones's basic construction of $\mathbb{C} \subseteq M_N(\mathbb{C})$, where $M_N(\mathbb{C})$ is endowed as above with the state $tr(D\cdot)$. Then the kth, $(k+1)$th steps of this basic construction are $B = M_N(\mathbb{C})^{\otimes^k}$, $A = M_N(\mathbb{C})^{\otimes^{k+1}}$. The states on the iterated steps in the basic construction are states with weights involving consecutive products of D and D^{-1}. Let $e_{k-1}, \dots, e_1$be the corresponding Jones' projections, indexed so that $e_1 \in A$ is exactly the Jones projection for $M_N(\mathbb{C})^{\otimes^{k-1}} \subseteq B$.

Then we perform the same construction as above for $B \subseteq A$ (instead of $M_N(\mathbb{C}) \subseteq M_N(\mathbb{C}) \otimes M_N(\mathbb{C})$). We get a subfactor inclusion $N^\lambda(Q) \subseteq M^\lambda(Q)$. The iterated Jones' basic construction steps (up to $k-1$) are then obtained by adding consecutively the projections $e_1, e_2, \dots, e_{k-1}$. As we will see below by reducing by a minimal projection in B it follows that these subfactors are isomorphic to the original ones. In particular it follows that the factor and subfactor are always isomorphic to the algebra corresponding to D and D^{-1}.

Lemma 1.5. *Let ϕ be a state on $M_N(\mathbb{C})$ with $\phi = tr(D\cdot)$, where D is diagonal with eigenvalues list $\lambda_1, \dots, \lambda_N$. Let (e_{ij}) be a matrix unit for $M_N(\mathbb{C})$, such that e_{ii} are the projections onto the eigenvectors of D. Let Q be a type II_1 factor and consider the type III factor $Q * M_N(\mathbb{C})$ (see, e.g., [4], [8], [32], [35]) where the free product is with respect to the state ϕ on $M_N(\mathbb{C})$ and the trace on Q. Let*

$$\Phi(\alpha) = \sum_{ij} \lambda_j^{-1} e_{ij} \alpha e_{ji}, \qquad \alpha \in Q * M_N(\mathbb{C}),$$

and let $\mathcal{M}^\lambda(Q)$ be the minimal subalgebra of $Q * M_N(\mathbb{C})$ that contains Q and is closed under Φ. Let $\mathcal{N}^\lambda(Q)$ be the image of M through Φ. Then

$$[\mathcal{M}^\lambda(Q) : \mathcal{N}^\lambda(Q)] = \sum_i \frac{1}{\lambda_i}; \quad \mathcal{M}^\lambda(Q) \cap \mathcal{N}^\lambda(Q)' = \mathbb{C}1.$$

Moreover $\mathcal{M}^\lambda(Q)$ has non-trivial fundamental group and the type of the subfactor inclusion is A_∞. The algebra of the subfactor is isomorphic with the factor $M^{(\lambda_i^{-1})}(Q)$.

Proof. We apply Popa's construction for the inclusion $M_N(\mathbb{C}) \subseteq M_N(\mathbb{C}) \otimes M_N(\mathbb{C})$ described in Lemma 1.2. Since this is the basic construction for $\mathbb{C} \subseteq M_N(\mathbb{C})$ we can apply Lemma 1.1 and one obtains an equivalent description of the subfactor $\mathcal{M}^\lambda(Q)$ sitting in $Q * M_N(\mathbb{C})$. Since $\lambda_j^{-1} e_{ij}$ is a Pimsner-Popa basis for the inclusion $\mathbb{C} \subseteq M_N(\mathbb{C})$, the result follows. That the subfactor inclusion is of type A_∞ is proved in a more general context in [33]. □

By analogy with the case of traces, we have the following result which shows that the algebra obtained by Popa's construction from the inclusion of finite-dimensional algebras (with states), $\mathbb{C} \subseteq M_N(\mathbb{C})$ is a minimal fixed point algebra.

Theorem 1.6. *Let A be the function algebra of Woronowicz quantum group G and let V_π be finite-dimensional unitary representation of dimension $N \geqslant 2$. Assume that the operator Q_π associated to the representation as in [43], [2] has eigenvalues $\lambda_1^{-1}, \ldots, \lambda_N^{-1}$. Let $Q * B(V_\pi)$ and consider the Ueda's ([37], [38]) type of free product coaction on $Q * B(V_\pi)$, that acts trivially on Q and by conjugation on $B(V_\pi)$. Let $\mathcal{M}$ be the Popa's type factor constructed in Lemma 1.5.*

Then

$$\mathcal{M} \subseteq (Q * B(V_\pi))^G.$$

If A is the function algebra of $SU_q(2)$, and π is the fundamental representation then we have equality.

Proof. Let (e_{st}) be a matrix unit for $B(V_\pi)$ that diagonalizes Q_π. Assume that the unitary implementing the representation is represented as

$$U = \sum_{s,t} e_{st} \otimes u_{st}.$$

Note that we use another indexing for the entries of the unitary than the one used in [43], Section 4. Since as proved in [2],

$$Q_\pi^{1/2} \overline{U} (Q_\pi)^{-1/2}$$

is the unitary corresponding to the conjugate representation it follows that the matrix

$$\alpha_{ij} = \lambda_i^{-1/2} \lambda_j^{1/2} u_{ji}$$

is a unitary too. Thus we have the following unitarity conditions:

$$\sum_j u_{sj} u_{tj}^* = \delta_{st}; \qquad \sum_j u_{js}^* u_{jt} = \delta_{st}, \qquad s,t = 1, \ldots, N, \tag{1.1}$$

$$\sum_j \lambda_j^{-1} u_{sj}^* u_{tj} = \lambda_s^{-1} \delta_{st}; \qquad \sum_j \lambda_j u_{js} u_{jt}^* = \lambda_s \delta_{st}; \qquad s,t = 1, \ldots, N. \tag{1.2}$$

The corepresentation for the algebra $B(V_\pi)$ obtained by conjugation by the unitary U is given by

$$U(e_{ij} \otimes 1)U^* = \sum_{rs} e_{rs} \otimes u_{ri} u_{sj}^*. \tag{1.3}$$

The elements in the algebra $\mathcal{M}$ have an open and closing parenthesis structure ([24], [7]). This proves that these elements are fixed points under the coaction, by recursively using (first equality in each of) the relations (1.1), (1.2). For example the coaction on an element of the form

$$x = \sum_{\alpha,\beta} \lambda_\beta^{-1} q_1 e_{\alpha,\beta} q_2 e_{\beta,\alpha} q_3$$

gives

$$x = \sum_{r_1,r_2,s_1,s_2} q_1 e_{r_1,s_1} q_2 e_{r_2,s_2} q_3 \left(\sum_{\alpha,\beta} \lambda_\alpha^{-1} u_{r_1\alpha} u^*_{s_1\beta} u_{r_2\beta} u^*_{s_2\alpha} \right). \tag{1.4}$$

By applying first equation 1.1 and then 1.2 it follows that

$$\sum_{\alpha,\beta} \lambda_\alpha^{-1} u_{r_1\alpha} u^*_{s_1\beta} u_{r_2\beta} u^*_{s_2\alpha} = \lambda_{s_1}^{-1} \delta_{r_1 s_2} \delta_{s_1 r_2}. \tag{1.5}$$

This proves that the coaction on x in (1.4) gives $x \otimes 1$. The only other fixed point are obtained by recurrence. Note that the algebra Q is in the fixed point algebra of the coaction.

More precisely if x in the fixed point algebra and h is the Haar measure, we need to show that

$$(Id \otimes h)\left(\sum_{\alpha\beta} \lambda_\beta^{-1} (U(e_{\alpha,\beta} \otimes 1)U^*)(x \otimes 1)(U(e_{\beta,\alpha} \otimes 1)U^*) \right) = \sum_{\alpha\beta} \lambda_\beta^{-1} e_{\alpha,\beta} x e_{\beta,\alpha}.$$

This follows by using the modular properties of the Haar measure and using the relations (1.4). Once we have shown this and since Q is obviously in the fixed point algebra the result follows from the definition of the algebra $M^\lambda(Q)$ as the minimal algebra containing Q closed to the operation in the statement of Lemma 1.5.

Let Q_0 be the subspace of Q consisting of elements of zero trace. The fixed point algebra and Popa's algebra admit a filtration given by the subspaces $B(V_\pi)Q_0B(V_\pi)\cdots Q_0B(V_\pi)$ corresponding to the number of occurrences of copies of $B(V_\pi)$. Moreover the averaging argument (with respect to the Haar measure) used in [18] or [14] to establish that the fixed point algebra in the Powers factor is the algebra generated by the Jones projections, allows to reduce the determination of the the fixed point algebra, to the determination of the intersection of the fixed point algebra with the subspaces in the filtration.

Therefore, to show the equality, when G is $SU_q(2)$, of the algebra in Popa's construction with the fixed point algebra, it is therefore sufficient to check that the intersections of these two algebras with the space

$$B(V_\pi)Q_0B(V_\pi)\cdots Q_0B(V_\pi)$$

coincide. (This comes also to to the determine the fixed point algebra in spaces of the form $B(V_\pi)^{\otimes^n}$, by formally replacing elements in Q_0 with tensor product sign. Counting dimensions of fixed point algebra could probably give an alternative for the argument below.)

Since we already have shown the reverse equality, it is sufficient to check that any element in the fixed point algebra intersected with one of the above finite-dimensional spaces $B(V_\pi)Q_0B(V_\pi)\cdots Q_0B(V_\pi)$ is one of the elements that respects the open and closing paranthesis structure of the recursive construction of the Popa's algebra.

So assume that for given scalars $\mu_{\alpha_0,\ldots,\beta_n}$ we have an fixed point element of the form

$$\begin{aligned}\Big(\sum_{\alpha_0,\ldots,\beta_n} \mu_{\alpha_0,\ldots,\beta_n} e_{\alpha_0\beta_0} q_1 e_{\alpha_1\beta_1} q_2 \cdots q_n e_{\alpha_n\beta_n}\Big) \otimes 1 \\ = \sum_{r_0,\ldots,s_n} e_{r_0 s_0} q_1 e_{r_1 s_1} \cdots q_n e_{r_n s_n} \\ \otimes \Big(\sum_{\alpha_0,\ldots,\beta_n} \mu_{\alpha_0,\ldots,\beta_n} u_{r_0\alpha_0} u^*_{s_0\beta_0} u_{r_1\alpha_1} u^*_{s_1\beta_1} \cdots u_{r_n\alpha_n} u^*_{s_n\beta_n}\Big).\end{aligned} \tag{1.6}$$

Then the fixed point condition comes to the following condition that has to hold for all $r_0,\ldots,s_n$:

$$\sum_{\alpha_0,\ldots,\beta_n} \mu_{\alpha_0,\ldots,\beta_n} u_{r_0\alpha_0} u^*_{s_0\beta_0} u_{r_1\alpha_1} u^*_{s_1\beta_1} \cdots u_{r_n\alpha_n} u^*_{s_n\beta_n} = \mu_{r_0,\ldots,s_n}. \tag{1.7}$$

By denoting Θ the matrix $(\mu_{\alpha_0,ldots,\beta_n})_{\alpha_0,\ldots,\beta_n}$, and by U the matrix $(u_{ij})_{ij}$ and by $U^\#$ the matrix $(u_{ji})_{ij}$, the relation (1.7) comes to the following equation:

$$(U \otimes U^\# \cdots \otimes U \otimes U^\#)\Theta = \Theta. \tag{1.8}$$

Let $I_{(ik)(rs)}$ and $E_{(ik)(rs)}$ be the matrices defined by $I_{(ik)(rs)} = \delta_{rs}$ and $E_{(ik)(rs)} = \lambda_s^{-1}\delta_r s$. Then the equations (1.1), (1.2) correspond to

$$(U \otimes U^\#)I = I; \qquad I(U^\# \otimes U) = I, \tag{1.9}$$

$$(U \otimes U^\#)E = E; \qquad E(U^\# \otimes U) = E, \tag{1.10}$$

where $(U \otimes U^\#)$ and $(U^\# \otimes U)$ are considered as the matrices indexed as follows: $(U \otimes U^\#)_{((st)(ij)} = u_{si}u^*_{tj}$ and $(U^\# \otimes U)_{((st)(ij)} = u_{si}u^*_{tj}$. As in [22], [44], the relations (1.9), (1.10) are the only relations defining $SU_q(2)$. Thus if a matrix Θ verifies the relation (1.8), then Θ is obtained by recurrence (and linear span) from consecutive applications of the relations (1.9), (1.10).

To have a reduction we therefore have to use the relation $\sum_j \lambda_j^{-1} u^*_{sj} u_{tj} = \lambda_s^{-1}\delta_{st}$, or $\sum_j u_{sj}u^*_{tj} = \delta_{st}$ (the other two equations in (1.1), (1.2) involve a summation index which is not appropriate for the corresponding sum). This means that the only possibility to have a simplification in (1.7), by using the defining relations of $SU_q(2)$, is if somewhere in the sum of (1.7), we have that $\mu_{\alpha_0\cdots\beta_n}$ splits as a sum of elements the form

$$\lambda_{\beta_i}^{-1}\delta_{\beta_i a_{i+1}} \times \mu'_{\alpha_0,\ldots,\hat{\beta}_i,\hat{\alpha}_{i+1},\ldots,\beta_n}, \tag{1.11}$$

or in the form

$$\delta_{\alpha_i b_i} \times \mu''_{\alpha_0,\ldots,\hat{\alpha}_i,\hat{\beta}_i,\ldots,\beta_n}. \tag{1.12}$$

The symbol ˆ corresponding to omission of the corresponding symbol. Moreover after applying such a simplification as in (1.11) and respectively (1.12) in (1.6) we obtain a sum in $r_0,\ldots,s_n$ involving a corresponding $\lambda_{s_i}\delta_{s_i r_{i+1}}$ or respectively $\delta_{r_i s_i}$.

Repeated application of the above two procedures outlined in the equations (1.11) and (1.12) to the sum in (1.6), will give, once we have finished the simplification procedure to the sum reducing (1.7) to scalars, it follows that the fixed point element has exactly the open and closing paranthesis structure in the indices $\alpha_0, \dots, \beta_n$ that corresponds to an element in the algebra derived from Popa's construction. □

We can also use the model described in the paper ([32]), where the structure of the free product algebra $Q * M_2(\mathbb{C})$ is described in terms of the "infinite" free semicircular element in [30] (here $M_2(\mathbb{C})$ is endowed with a state). We obtain an explicit random matrix ([40]) model for the structure of the subfactor $\mathcal{N}^\lambda(Q) \subseteq \mathcal{M}^\lambda(Q)$, when Q is a free group factor or $Q = L^\infty([0, 1]$.

We describe this model below:

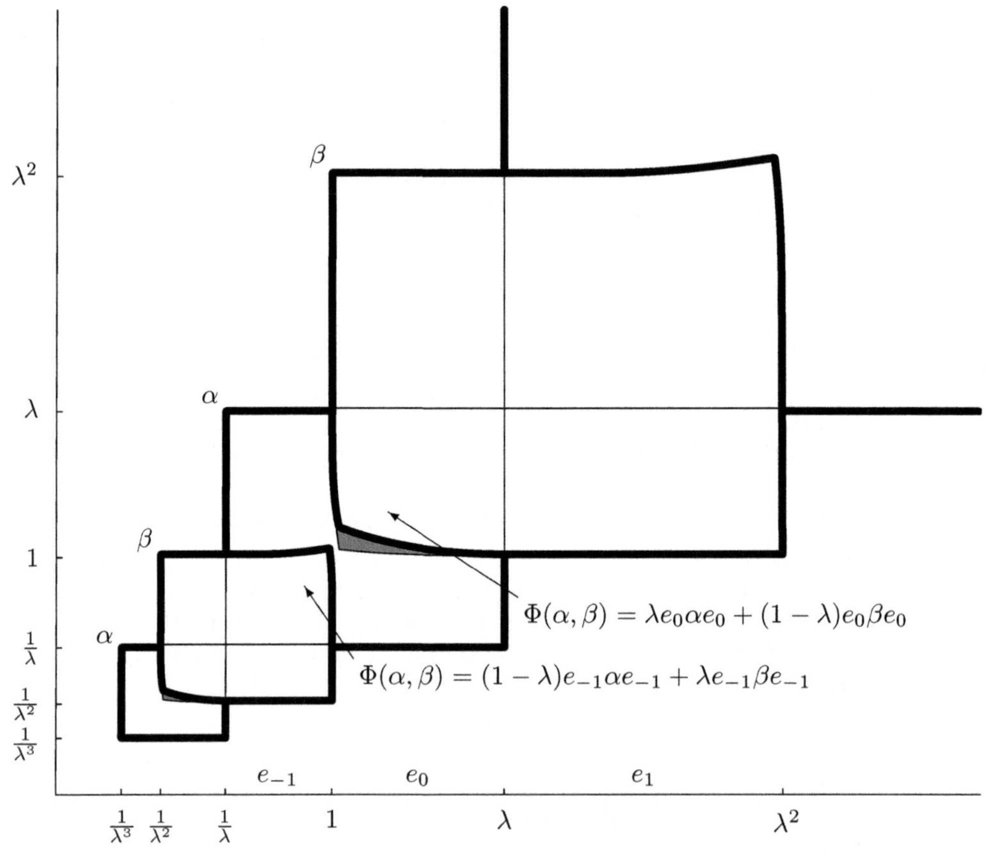

FIGURE 1. Action of Φ

Theorem 1.7. *Let λ be a number in $(0,1)$ and let D be the algebra generated by the characteristic functions of the intervals $e_n = [\lambda^n, \lambda^{n+1}]$, $n \in \mathbb{Z}$. Let X, Y be two "infinite" free semicircular elements as considered in [30] (see also [9]). Let f_i be the projections $e_{2i} + e_{2i+1}$ and let $g_i = e_{2i+1} + e_{2i+2}$. Let F, G be the algebras respectively generated by this projections. Let X_0, Y_0 be the (bounded) free semicircular family obtained by diagonally carving X and respectively Y by the projections f_i and respectively g_i. In [32] it was proved that the algebra $\mathcal{A}$ generated by X_0, Y_0 and D is isomorphic to $M_2(\mathbb{C}) * L^\infty([0,1]$, where on $M_2(\mathbb{C})$ we consider a state with weights λ and $1-\lambda$.*

Note that the automorphism of homothety by λ (which is a restriction of the one parameter group introduced in [30]), scales trace by λ.

Let $\mathcal{B}_0, \mathcal{B}_1$ be the algebras $F' \cap \mathcal{A}$ and respectively $G' \cap \mathcal{A}$. Let E_0, E_1 be the sum of the even, respectively odd, indexed projections in (e_i). On the algebra $\mathcal{B}_0 \oplus \mathcal{B}_1$ we consider the following linear map

$$\Phi(\alpha, \beta) = \lambda E_0 \alpha E_0 + (1-\lambda) E_0 \beta E_0 + (1-\lambda) E_1 \alpha E_1 + \lambda E_1 \beta E_1, \quad (\alpha, \beta) \in \mathcal{B}_0 \oplus \mathcal{B}_1.$$

Let $\mathcal{M}$ be the minimal subalgebra of $\mathcal{B}_0 \oplus \mathcal{B}_1$ containing (X_0, Y_0) and closed under Ψ, and let $\mathcal{N}$ be the image of Φ.

Then $\mathcal{N} \subseteq \mathcal{M}$ is an irreducible subfactor of index $\frac{1}{\lambda} + \frac{1}{1-\lambda}$.

Proof. Indeed, reducing $\mathcal{M}$ by the projection $e_0 + e$, we get the model described in 1.5 (for $N = 2$). □

2. Determination of the cross product algebra

In this section we determine the structure of the crossed product algebra $(Q * B(V_\pi)) \rtimes G$. This, by Takesaki Duality, is in turn used to determine, the structure of the fixed point algebra.

The fixed point algebra could be a type III factor or a type II_1 factor. In the case of the factor in the previous section the fixed point algebra is a type II_1 factor which makes it easier to determine the structure of this algebra. This is similar to the results in [20], [14], [15], where the fixed point algebra of an infinite tensor product of $B(V_\pi)$ under a natural action of $SU_q(N)$ is determined to be the Temperly-Lieb algebra.

To establish that the cross product is stably isomorphic to the fixed point algebra we will need to use instead a quotient group "$SU_q(2)/\mathbb{Z}_2 = SO_q(3)$" (see [21]). Indeed, in the statement of Theorem 1.6 the action of $SU_q(N)$ on $B(V_\pi)$ is not faithful, in the sense that its tensor products do not contain all other representations of the group.

We will use an idea from Wasserman's paper ([41]) that in the case of an infinite tensor product action of a compact group, by a faithful, selfadjoint, unitary irreducible action of a compact group, the Takesaki's duality gives a stable isomorphism between the fixed point algebra and the cross product. We are also indebted to Y. Ueda for pointing out to us that part of the statement in Theorem 5.6 in [20] does not hold, as also mentioned in the abstract of the paper by M. Izumi [10].

Proposition 2.1. *Let V be the Hilbert space of the fundamental representation of $SU_q(2)$. The adjoint coaction of $SU_q(2)$ on $B(V)$ corresponds to a coaction α of $SO_q(3)$ on $B(V)$. Moreover the fixed point algebras for any of the two (free product) coactions on $Q * B(V)$ coincide. Thus*

$$(Q * B(V))^{SU_q(2)} = (Q * B(V))^{SO_q(3)}$$

Proof. Let as in [21] d_0, $d_{1/2}$, d_1, ... be the representations of $SU_q(2)$, with $d_{1/2}$ the fundamental representation. Then the fundamental unitary for $SO_q(3)$ is given by the unitary representing d_1. As $V \oplus \overline{V}$ is $d_0 \oplus d_1$ it follows that the coaction of $SU_q(2)$ gives the coaction α of $SO_q(3)$ on $B(V)$, (for classical groups that means that the representation comes from a representation of the quotient). Since the unitary implementing both coactions is the same, it also follows that we have the same fixed point algebra, (in terms of groups representations if we have a representation that factors through the quotient, then we have the same fixed point algebra). □

In the next proposition we prove that for any finite-dimensional Hilbert space V, if the tensor product of a selfadjoint coaction α of a Woronowicz quantum group G on $B(V)$ contains any other finite-dimensional unitary representation of the quantum group, then the coaction on $Q * B(V)$ is semidual in the sense of [19] and [41]. Thus the fixed point algebra is Morita equivalent to the cross product by the action. We will use Takesaki Duality, in the sense described in [20].

Proposition 2.2. *Let A be the function algebra of a Woronowicz compact quantum group G, with faithful Haar state. Let $L^2(A)$ be the Hilbert space associated to the Haar measure on A. Assume that α is a faithful selfadjoint corepresentation of G (i.e., a corepresentation such that its tensor product contains any other unitary, finite-dimensional irreducible representation of G). Let Q be a II_1 factor (or a diffuse abelian algebra). Then*

$$(Q \star B(V)) \rtimes A \cong (Q \star B(V))^G \otimes B(L^2(A)).$$

Proof. Let $\mathcal{A} = Q \star B(V)$. By the argument in the proof of Theorem 5.6 in [20] (and using also Lemma 20 in [6], as pointed out in [36]) it is sufficient to prove that for any irreducible unitary coaction $\hat{\alpha}$ of G, the corresponding spectral subspace ([6], [41], [13]) $\mathcal{A}_{\hat{\alpha}}$ is non-trivial. But, since $B(V)$ appears in a tensor product situation in $Q \star B(V)$ this follows from the fact that the tensor product of the representation α with itself contains any other representation (see [41], [19]). □

The following was proved in [37]. It expresses the natural fact that the cross product distributes when we have an action on a free product of two algebras, such that the coaction is trivial on one of the two factors in the free product.

Note that one has to specify a state on the algebra over which we amalgamate in order to get a von Neumann algebra.

Proposition 2.3. *([37]) Let A be the function algebra of a Woronowicz quantum group G, let α be a corepresentation of G on the bounded linear operators $B(V)$ acting on a finite-dimensional Hilbert space V. Let ϕ_α be an α invariant faithful state on $B(V)$, that is such that $(\phi_\alpha \otimes Id)(\alpha(x)) = \phi_\alpha(x)1$, for all x in $B(V)$. Let $\hat{A}$ be the dual algebra. Then we have the following isomorphism*

$$(Q * B(V)) \rtimes A \cong (Q \otimes \hat{A}) *_{\hat{A}} (B(V) \rtimes A).$$

The amalgamated free product is with respect to the canonical conditional expectation of $B(V) \rtimes A$ onto $\hat{A}$ (i.e., the restriction of $\phi_\alpha \otimes Id$). Moreover the free amalgamated von Neumann algebra on the right-hand side of the above equality is determined by endowing the algebra $\hat{A}$ with a state (or weight) that is the restriction of a faithful normal state on $B(L^2(A))$.

We note in the case of $SO_q(3)$ the hypothesis are satisfied. Indeed we have the following:

Lemma 2.4. *We use the notations from the above lemma. For the action α in 2.1, if we consider the state $\phi_\alpha = \tau = tr(Q_\pi \cdot)$, where Q_π is the operator associated in [2] to the fundamental representation π of $SU_q(2)$, then ϕ_α is $SO_q(3)$ invariant.*

Proof. Indeed the unitary implementing the adjoint corepresentation $Ad\ \pi$ is the same as the unitary implementing α. Since τ has this property, the same will hold for ϕ_α. □

The algebras $\hat{A} \subseteq B(V) \rtimes A$ are discrete and the structure of the corresponding inclusion matrix can be easily described in the case when the representation ring of A is the same as the representation ring of a classical compact group. We are indebted to V. Toledano, D. Bisch, M. Pimsner and G. Nagy for pointing us the more precise description of this inclusion matrix, which essentially appears in [41]. We will not use the explicit description of this matrix.

Lemma 2.5. *Let A be the function algebra of a Woronowicz quantum group G. Let $\hat{A}$ be the dual algebra and let π be a finite-dimensional unitary corepresentation of A on a Hilbert space V_π. Then $B(V_\pi) \rtimes A$ is isomorphic to $B(V_\pi) \otimes \hat{A}$. Moreover $\hat{A} \subseteq B(V_\pi) \rtimes A$ has a Bratelli inclusion matrix with finite multiplicities.*

Proof. By definition $B(V_\pi) \rtimes A$ is isomorphic to the von Neumann algebra generated in $B(V_\pi) \otimes B(L^2(A)$ by $u^*(B(V(\pi) \otimes 1)u$ and $\hat{A}$. But $u(1 \otimes \hat{A})u^*$ is contained in $B(V_\pi) \otimes \hat{A}$, since $Ad\ u$ acting on $B(L^2(A))$ maps the matrix coefficients (viewed as elements of $\hat{A}$) corresponding to a finite-dimensional unitary representation u^α of A into a linear combination of the matrix coefficients (viewed as elements in $\hat{A}$) of the representation $u \otimes u^\alpha$.

We describe this map in classical setting and then explain the modifications needed for a quantum compact group. Indeed in a classical setting, if G is a compact group, let V be the fundamental unitary viewed as an element in $L^\infty(G) \otimes B(L^2(G))$ defined by $V(g) = \lambda_g$, $g \in G$, where $\lambda_g, g \in G$ is the left regular representation of G into the unitary group on $L^2(G)$.

Assume $\pi(g) = (u_g) \in \mathcal{U}(B(V_\pi))$, $g \in G$, is a finite-dimensional unitary representation of G on a Hilbert space V_π. Let U be the unitary in $L^\infty(G) \otimes B(V_\pi)$ given by u. Then in $B(L^2(G)) \otimes B(V_\pi)$ the following holds (the absorption principle for the left regular representation):

$$U^*(\lambda_g \otimes 1)U = \lambda_g \otimes u_g.$$

By using the fundamental unitary $V = V_{12}$ defined above, in $L^\infty(G) \otimes B(L^2(G)) \otimes B(V_\pi)$, the last equality gives (by using Woronowicz's notations)

$$U^*_{23} V_{12} U_{23} = V_{12} U_{13}. \tag{2.1}$$

Let e_{ij} be a matrix unit for $B(V_\pi)$ and let $u_{ij}(g)$ be the matrix coefficients for u_g in this matrix unit. For convolution operators in $L(G)$ the equality (2.1) corresponds, for f in $L^1(G)$ to the following equality in $B(L^2(G)) \otimes B(V_\pi)$.

$$U^* \left(\int f(g) \lambda_g dg \right) U = \int \sum_{ij} (f(g) u_{ij}(g))(1 \otimes e_{ij}) \lambda_g dg. \tag{2.2}$$

This gives, using the decomposition of tensor products of irreducible unitary representations of G, a complete description of the inclusion $L(G) \subseteq B(V_\pi) \rtimes G$ (since we have described the inclusion $U(L(G) \otimes 1)U^*$ into $B(V_\pi) \otimes L(G)$ in terms of fusion rules of tensor by the representation π (on matrix coefficients).

If we want to generalize the above statement for arbitrary quantum groups, we will use the corresponding absorption principle for arbitrary quantum groups described in formula (5.11) in [45] which replaces (2.1). This formula now holds in $A \otimes B(L^2(A) \otimes B(V_\pi)$ (where π is the representation in our statement), V is now one of the fundamental unitary in $A \otimes B(L^2(A)$ replacing the V used for the classical case. In this context a convolution operator $\int f(g) \lambda_g dg$ is replaced (in $A \otimes B(L^2(A))$ by

$$(h(f \cdot) \otimes Id)(V) \in \hat{A}, \quad f \in A.$$

This is a generic element in a weakly dense subspace of $\hat{A}$ (here we use V_{12} which is the flip of what is usually the fundamental unitary). The formula (2.2) now reads in $A \otimes B(L^2(A))$ as follows

$$U_{23}[(h(f \cdot) \otimes Id)(V_{12})]U^*_{23} \otimes Id_{B(V_\pi)} = (h(f \cdot) \otimes Id \otimes Id_{B(V_\pi)}(U_{23} V_{12} U^*_{23}).$$

By using the formula (5.11) in the paper [45] we get that this is further equal to (and using a matrix unit e_{ij} in $B(V_\pi)$ with respect to which $u \in A \otimes B(V_\pi)$ has components u_{ij})

$$(h(f \cdot) \otimes Id \otimes Id_{B(V_\pi)})(V_{12} U_{13})$$

which is thus the following element in $B(V_\pi) \otimes \hat{A}$,

$$\sum_{ij} (h(f u_{ij} \cdot) \otimes Id)(V) \otimes e_{ij}. \tag{2.3}$$

This shows that in $A \otimes B(L^2(A)$, the transformation $Ad\ u$ maps $B(V_\pi) \rtimes A$ onto $B(V_\pi) \otimes \hat{A}$. □

Remark 2.6. If $G = SU_q(2)$, and π is the fundamental representation, since the representation ring of G is the same as the classical one, it follows that the inclusion matrix of $\hat{A}$ into $B(V_\pi) \rtimes A$, is the same as in the classical one, which is a matrix of type A_∞.

We need to apply the previous lemma to the case of the coaction α of $SO_q(3)$ on $B(V)$ which was described in 2.1.

Lemma 2.7. *Let $\hat{B}$ be the dual Woronowicz algebra for $SO_q(3)$. With the notation in Proposition* 2.1 *we have that $B(V) \rtimes SO_q(3)$ is isomorphic to $B(V) \otimes \hat{B}$. Moreover the inclusion matrix $\hat{B} \subseteq B(V) \rtimes SO_q(3)$ has a Bratelli inclusion matrix with finite multiplicities.*

Proof. Denote $B = SO_q(3)$ and $A = SU_q(2)$ and let $\hat{B}, \hat{A}$ be the dual algebras. The representation $\alpha : B(V) \to B(V) \otimes B$ is the restriction of the adjoint corepresentation induced by π. This holds because $\alpha(x) = u^*(x \otimes 1)u$ belongs to $B(V) \otimes B$, since B is generated by the matrix coefficients of d_1. But then if P is the projection (in $B(L^2(A))$) onto $L^2(B)$, then

$$B(V) \rtimes B = P(B(V) \rtimes A)P,$$

since $P\hat{A}P = \hat{B}$ and since B acts on $L^2(A)$ as the Haar measure on B is the restriction of the Haar measure on A ([21]). □

The following lemma is a direct consequence of the method used in [29]. There it was proved that amalgamated free products of the type $(L(F_N) \otimes D) *_D C$, where $D \subseteq C$ is an inclusion of discrete von Neumann algebras, C with a faithful trace, are isomorphic to a free group factors.

Lemma 2.8. *Let $D \subseteq C$ be von Neumann algebras that are infinite sums of algebras of finite-dimensional matrices, and such that the Bratelli inclusion matrix has finite multiplicities and $\mathcal{Z}(C) \cap \mathcal{Z}(D) = \mathbb{C}1$. Let Q be a free group factor or a diffuse abelian von Neumann algebra. Assume that D is endowed with a (semi)finite faithful trace tr and that the amalgamation is performed* ([40], [24]) *with respect to a normal faithful conditional expectation from C onto D, which then gives a trace state on the amalgamated free product, (which by GNS construction gives the amalgamated free product von Neumann algebra).*

Assume that the amalgamated free product von Neumann algebra is finite or semifinite.

Then $tr \circ E$ is a faithful trace on C and $\mathcal{L}(F_N) \otimes D \star_D C$ is Morita equivalent to a free group factor.

Proof. Indeed if $\tau \circ E$ is not a trace then the modular group of $\tau \circ E$ will be nontrivial on off diagonal elements. If the trace is finite this is exactly the content of Theorem 5.1 in [29] (see also [34] for a different, more recent proof). If the trace is infinite, the arguments in the proof of the theorem mentioned above can obviously be modified to handle this case (by changing the principle of counting the blocks as in the construction of a one parameter group of automorphisms in [30]). □

Remark 2.9. With the notations in Theorem 1.6, one can prove directly that factor $(Q * B(V_\pi)) \rtimes SU_q(2)$ is a finite type II von Neumann algebra. Indeed because of Lemma 2.3, it is sufficient to that the inclusion $\hat{A} \subseteq B(V_\pi) \rtimes A$, and the corresponding conditional expectation $tr(\phi_\alpha \cdot) \otimes Id$ (see also Lemma 2.4) from $B(V_\pi) \rtimes A$ onto A, has the property that the composition of the trace on $\hat{A}$ with the conditional expectation is again a trace. The trace on $\hat{A}$ that we are considering here is the restriction to $\hat{A} \subseteq B(L^2(A)$ of the canonical trace on $B(L^2(A))$.

Let p_π be the projection from $\hat{A}$ onto $B(V_\pi)$ (by viewing $\hat{A}$ as the direct sum over all bounded linear operators on the Hilbert spaces of an enumeration of the irreducible finite-dimensional unitary representations of A, ([16], [23])).

Let $\hat{\phi} : \hat{A} \to \hat{A} \otimes \hat{A}$ be dual comultiplication map. Then the inclusion $\hat{A} \subseteq B(V_\pi) \rtimes A$, (in the identification $B(V_\pi) \rtimes A \cong B(V_\pi) \otimes \hat{A}$, described in Proposition 2.5) is, (because of the description in [23]) exactly the map

$$(p_\pi \otimes Id)\hat{\phi} : \hat{A} \to B(V_\pi) \otimes \hat{A}.$$

Moreover if $\hat{\gamma}$ is the density matrix in $\hat{A}$ of the dual Haar measure $\hat{h}$ on $\hat{A}$, then $\hat{\phi}(\hat{\gamma}) = \hat{\gamma} \otimes \hat{\gamma}$ ([46]). But this amounts exactly to the fact that the composition of the conditional expectation $tr(\phi_\alpha \cdot) \otimes Id$ with $\hat{h}$ gives the state $tr(\phi_\alpha \cdot) \otimes \hat{h}$ on $B(V_\pi) \rtimes A \cong B(V_\pi) \otimes \hat{A}$. Hence, the composition of the trace on $\hat{A}$ (coming from $B(L^2(A))$ with the conditional expectation is a trace.

Corollary 2.10. *The factor $\mathcal{M}^\lambda(\mathcal{L}(F_k))$ in Theorem* 1.6 *(for $N = 2$) is isomorphic to $\mathcal{L}(F_\infty)$.*

Proof. From 2.8, 2.3 it follows that $(Q * B(V_\pi) \rtimes SO_q(3)$ is isomorphic to $\mathcal{L}(F_\infty) \otimes B(H)$, where H is an infinite-dimensional space. The result follows now from 1.6, 2.1, since d_1 is a faithful representation of $SO_q(3)$ (no adjoint is required here). □

In this way we reobtain, by a different method, the result that was recently proved by D. Shlyakhtenko and Y. Ueda in [36]. We do not know if the subfactors obtained by using the method in this paper, which are derived from Popa's construction of irreducible subfactors from the Temperly-Lieb algebra coincide with those constructed in the paper [36]. Both of the two subfactors have A_∞ invariants and both are the fixed point algebra of a coaction of a quantum group, although in one case the fixed point algebra is a type III factor while in the present case this is a type II_1 factor.

Corollary 2.11. *In particular $\mathcal{L}(F_\infty)$ has irreducible subfactors of index λ (and type A_∞) for all index values bigger than* 4.

The previous result, since we are using a non-tracial version of [24] is strong evidence to Popa's conjecture that the subfactors constructed in the breakthrough papers [24] (or [25], [26]) are isomorphic to free group factors. The only case in which Popa's subfactors ([24], [25]) are known to be free group factors is for index values less than 4 ([29]).

Acknowledgement

This work was partially by NSF grant *DMS*. The author is grateful to T. Bănică, D. Bisch, F. Boca, F. Goodman, G. Nagy, M. Pimsner, D. Shlyakhtenko, V. Toledano, Y. Ueda for disscutions on this subject. This work was initiated during the author's stay at the University of Geneva, and was partially supported by the Swiss Science National Foundation. The author is very thankful to P. de la Harpe for the very warm welcome at the University of Geneva. The author is also thankful to the Department of Mathematics of the University of Cardiff, who provided partial support by an EPSRC grant.

Note added in proof: After this preprint has been submitted to publication, Popa and Shlyakhtenko, in *Universal properties of* $L(\mathbb{F}_\infty)$ *in subfactor theory*, generalized the result in [36] to arbitary λ-lattices.

References

[1] S. Baaj, G. Skandalis, *Unitaires multiplicatifs et dualité pour les produits croisés de* C^**-algèbres.* Ann. Sci. Ec. Norm. Supér., IV. Sér. **26** (1993), no. 4, 452–488.

[2] Teodor Bănică, *Representations of Compact Quantum Groups and Subfactors.* J. Reine Angew. Math. **509**, 167–198 (1999).

[3] T. Bănică, *Théorie des représentations du groupe quantique compact libre* $O(n)$. C. R. Acad. Sci., Paris, Sér. I **322** (1996), no. 3, 241–244.

[4] L. Barnett, *Free Product von Neumann Algebras of Type III.* Proc. Amer. Math. Soc. **123** (1995), no. 2, 543–553.

[5] E. Blanchard, K. Dykema, *Embeddings of Reduced Free Products of Operator Algebras.* OA/9911012, to appear in Pacific J. Math. 15 pages.

[6] F.P. Boca, *Ergodic Actions of Compact Matrix Pseudogroups on* C^**-Algebras.* in Connes, A. (ed.), Recent Advances in Operator Algebras, Collection of talks given in the conference on operator algebras held in Orléans, France in July 1992. Paris: Société Mathématique de France, Astérisque. **232** (1995), 93–109.

[7] F.P. Boca, Thesis, UCLA, 1992.

[8] K.J. Dykema, *Amalgamated Free Products of Multi-Matrix Algebras and a Construction of Subfactors of a Free Group Factor*, Amer. J. Math. **117** (1995), no. 6, 1555–1602.

[9] K.J. Dykema, F. Rădulescu, *Compressions of Free Products of von Neumann Algebras.* Math. Ann. **316** (2000), no. 1, 61–82.

[10] M. Izumi, *Actions of Compact Quantum Groups on Operator Algebras.* RIMS Kokyuroku **1024** (1998), 55–60.

[11] V.F.R. Jones, *Index for Subfactors.* Invent. Math. **72** (1983), no. 1, 1–25.

[12] V.F.R. Jones, *Subfactors and Knots.* Expository lectures from the CBMS regional conference, held at the US Naval Academy, Annapolis, USA, June 5–11, 1988, Regional Conference Series in Mathematics **80**, Providence, RI, American Mathematical Society. viii, 113 p. (1991).

[13] R. Hoegh-Krohn, M.B. Landstad, E. Stormer, *Compact Ergodic Groups of Automorphisms.* Ann. Math., II. Ser. **114** (1981), 75–86.

[14] Y. Konishi, *A Note on Actions of Compact Matrix Groups on von Neumann Algebra.* Hihonkai Math. J. **3** (992), 23–29.

[15] Y. Konishi, M. Nagisa, Y. Watatani, *Some Remarks on Actions of Compact Matrix Quantum Groups on C^*-Algebras.* Pac. J. Math. **153** (1992), no.1, 119–128.

[16] H. Kurose, Y. Nakagami, *Compact Hopf *-Algebras, Quantum Enveloping Algebras and Dual Woronowicz Algebras for Quantum Lorentz Groups.* Int. J. Math. **8** (1997), no.7, 959–997.

[17] T. Masuda, Y. Nakagami, *A von Neumann Algebra Framework for the Duality of the Quantum Groups.* Publ. Res. Inst. Math. Sci. **30** (1994), no. 5, 799–850.

[18] G. Nagy, *On the Haar Measure of the Quantum $SU(N)$ Group.* [English] Commun. Math. Phys. **153** (1993), no. 2, 217–228.

[19] Y. Nakagami, M. Takesaki, *Duality for Crossed Products of von Neumann Algebras.* Lecture Notes in Mathematics. **731**, Berlin–Heidelberg–New York, Springer-Verlag. IX, (1979).

[20] Y. Nakagami, *Takesaki Duality for the Crossed product by Quantum Groups.* in Araki, Huzihiro (ed.) et al., Quantum and non-commutative analysis. Past, present and future perspectives. Proceedings of the international Oji seminar on quantum analysis, Kyoto, Japan, June 25–29, 1992 and the symposium on non-commutative analysis, Kyoto, Japan, June 29–July 2, 1992 dedicated to Prof. Huzihiro Araki on the occasion of his 60th birthday. Dordrecht, Kluwer Academic Publishers. Math. Phys. Stud. **16** (1993), 263–281.

[21] P. Podles, *Symmetries of Quantum Spaces. Subgroups and Quotient Spaces of Quantum $SU(2)$ and $SO(3)$ Groups.* Commun. Math. Phys. **170** (1995), no.1, 1–20.

[22] P. Podles, E. Mueller, *Introduction to Quantum Groups.* Rev. Math. Phys. **10** (1998), no. 4, 511–551.

[23] P. Podles, S. L. Woronowicz, *Quantum Deformation of Lorentz Group.* Commun. Math. Phys. **130** (1990), no. 2, 381–431.

[24] S. Popa, *Markov Traces on Universal Jones Algebras and Subfactors of Finite Index.* Invent. Math. **111** (1993), no. 2, 375–405.

[25] S. Popa, *An Axiomatization of the Lattice of Higher Relative Commutants of a Subfactor.* Invent. Math. **120** (1995), no. 3, 427–445.

[26] S. Popa, *The Universal Algebra of a Subfactor.* preprint.

[27] S. Popa, *Classification of Amenable Subfactors of Type II.* Acta Math. **172** (1994), no. 2, 163–255.

[28] S. Popa, A. Wassermann, *Actions of Compact Lie Groups on von Neumann Algebras.* C. R. Acad. Sci. Paris, Ser. I **315** (1992), no. 4, 421–426.

[29] F. Rădulescu, *Random Matrices, Amalgamated Free Products and Subfactors of the von Neumann Algebra of a Free Group, of Noninteger Index.* Invent. Math. **115** (1994), no. 2, 347–389.

[30] F. Rădulescu, *A One-Parameter Group of Automorphisms of $\mathcal{L}(F_\infty) \otimes B(H)$ Scaling the Trace.* C. R. Acad. Sci. Paris Sér. I **314** (1992), no. 13, 1027–1032.

[31] F. Rădulescu, *An Invariant for Subfactors in the von Neumann Algebra of a Free Group.* C. R. Acad. Sci. Paris Sér. I Math. **316** (1993), no. 10, 983–988.

[32] F. Rădulescu, *A Type III_λ Factor with Core Isomorphic to the von Neumann Algebra of a Free Group, Tensor $B(H)$.* in Connes, A. (ed.), Recent advances in operator algebras. Collection of talks given in the conference on operator algebras held in Orléans, France in July 1992. Paris, Société Mathématique de France, Astérisque **232** (1995), 203–209.

[33] F. Rădulescu, *Subfactors Inside Free Group Factors.* In preparation.

[34] D. Shlyakhtenko, *Some Applications of Freeness with Amalgamation.* J. Reine Angew. Math. **500** (1998), 191–212.

[35] D. Shlyakhtenko, *Free Quasi-Free States.* Pac. J. Math. **177** (1997), no. 2, 329–368.

[36] D. Shlyakhtenko, Y. Ueda *Irreducible Subfactors of $L(\mathbb{F}_\infty)$ of Index $\lambda > 4$* . MSRI Preprint, Report number MSRI **2000–030** (2000).

[37] Y. Ueda, *On the Fixed-Point Algebra under a Minimal Free Product-Type Action of the Quantum Group $SU_q(2)$.* Int. Math. Res. Not. **1** (2000), 35–56.

[38] Y. Ueda, *A Minimal Action of the Compact Quantum Group $SU_q(n)$ on a Full Factor.* J. Math. Soc. Japan **51** (1999), no. 2, 449–461.

[39] Y. Ueda, *Amalgamated Free Product over Cartan Subalgebra.* Pac. J. Math. **191** (1999), no. 2, 359–392.

[40] D. Voiculescu, *Limit Laws for Random Matrices and Free Products.* Invent. Math. **104** (1991), no. 1, 201–220.

[41] A. Wassermann, *Ergodic Actions of Compact Groups on Operator Algebras. I: General theory.* Ann. Math. II. Ser. **130** (1989), no. 2, 273–319.

[42] H. Wenzl, *Hecke Algebras of Type A_n and Subfactors.* Invent. Math. **92** (1988), no. 2, 349–383.

[43] S.L. Woronowicz, *Compact Matrix Pseudogroups.* Commun. Math. Phys. **111** (1987), 613–665.

[44] S.L. Woronowicz, *Tannaka-Krein Duality for Compact Matrix Pseudogroups. Twisted SU(N) Groups.* Invent. Math. **93** (1988), no.1, 35–76.

[45] S.L. Woronowicz, *Compact Quantum Groups.* in Connes, A. (ed.) et al., Quantum symmetries. Symétries quantiques. Proceedings of the Les Houches summer school, Session LXIV, Les Houches, France, August 1–September 8, 1995. Amsterdam: North-Holland, (1998), 845–884.

[46] S. Yamagami, *On Unitary Representation Theories of Compact Quantum Groups.* Commun. Math. Phys. **167** (1995), no. 3, 509–529.

Florin Rădulescu
Department of Mathematics
The University of Iowa
Iowa City
IA 52242–1419
USA
e-mail: `radulesc@math.uiowa.edu`
URL: `http://www.math.uiowa.edu/~radulesc/`

Operator Theory:
Advances and Applications, Vol. 153, 249–254

The Structure of some C^*-Algebras Generated by N Idempotents

Mikhail Shchukin and Elena Vatkina

Abstract. We study the structure of some n-homogeneous C^*-algebras generated by flips. The algebra is generated by the flips $u, v, s_1, \dots, s_m$ with the relations between the generators: $us_i = \alpha_i s_i u, vs_i = \beta_i s_i v, s_i s_j = \epsilon_{ij} s_j s_i, \alpha_i = \pm 1, \beta_i = \pm 1, \epsilon_{ij} = \pm 1, 1 \leq i, j \leq m$. The structure of such algebras generated by flips with the relations between generators was studied by Popovich, Samoilenko and Turowska. In the paper we prove that if such an algebra A is n-homogeneous then it is trivial. Such an n-homogeneous C^*-algebra A is isomorphic to the algebra of all continuous matrix-functions of dimension n over some compact subspace of the plane $\mathbb{C}$.

Mathematics Subject Classification (2000). Primary 16G30.

Keywords. C^*-Algebra, Algebraic bundle, Irreducible representations.

1. Introduction

Let A denote an n-homogeneous C^*-algebra. An C^*-algebra A is called *n-homogeneous* if and only if all its irreducible representations are n-dimensional.

Fell ([3]), Tomiyama and Takesaki ([9]) have shown that every n-homogeneous C^*-algebra A is isometrically isomorphic to the algebra of all continuous sections of an appropriate algebraic bundle.

An *algebraic bundle* is a locally trivial bundle with a model fibre homeomorphic to the C^*-algebra $Mat(n) = \mathbb{C}^{n \times n}$ of all $n \times n$ matrices and with the structure group G isomorphic to the group $Aut(n)$ of all automorphisms of this algebra. A n-homogeneous C^*-algebra is called *trivial* if it is isomorphic to the algebra of all continuous matrix-functions $C(X, \mathbb{C}^{n \times n})$ over some topological compact space X.

An element a is called *idempotent* if and only if $a^2 = a$. If $b^2 = e$ then we say that b is a *flip*. There is a linear correspondence between idempotents and flips. If a is an idempotent then the element $b = 2a - e$ is a flip. Therefore, any Banach algebra generated by N flips can be considered as the algebra generated by N idempotents and the identity e. (We say that the Banach algebra A is *generated* by the elements $a_1, \ldots, a_N$ if the smallest Banach algebra that contains the elements $a_1, \ldots, a_N$ coincides with A.)

In the work we study the structure of some n-homogeneous C^*-algebras generated by N flips with some relations between them. The structure of such algebras was studied by Popovich, Samoilenko, Turowska in [7]. The class of such algebras was denoted by A_m. More precisely, let A_m denote the class of Banach algebras that can be generated by the flips $u, v, s_1, \ldots, s_m$ with the relations between generators: $us_i = \alpha_i s_i u, vs_i = \beta_i s_i v, s_i s_j = \epsilon_{ij} s_j s_i, \alpha_i = \pm 1, \beta_i = \pm 1, \epsilon_{ij} = \pm 1, 1 \leq i, j \leq m.$

In this work we prove that if an algebra A from the class A_m is n-homogeneous then the algebra A is trivial. This means that the algebra A is isomorphic to the algebra of all continuous matrix-functions $C(P(A), \mathbb{C}^{n \times n})$, where $P(A)$ denotes the space of primitive ideals of the algebra A in the appropriate topology.

Also we describe precisely the structure of the space $P(A)$. The space $P(A)$ is homeomorphic to some compact subspace X of the plane $\mathbb{C}$ such that

1. X has no interior points.
2. The set $\mathbb{C} \setminus X$ is a connected set.

2. The main results

We start the section with describing the structure of the space of irreducible representations of an n-homogeneous C^*-algebra A generated by two idempotents.

Suppose the algebra A is n-homogeneous. Let A contain two idempotents p, q such that the minimal Banach algebra that contains p, q and e coincides with A. The next theorem describes the structure of the algebra A.

Statement 2.1. ([1]) Let A denotes an n-homogeneous C^*-algebra generated by two idempotents p and q. Then the algebra A is trivial. If A is non-commutative then it is isomorphic to the algebra $C(M, \mathbb{C}^{2\times 2})$, where M denotes the compact space of maximal ideals of the algebra $Alg((p-r)^2)$.

(The symbol $Alg(Y)$ denotes the minimal Banach algebra that contains the element Y.)

This theorem will help us to describe the structure of the space M.

The next statement describes the structure of the commutative algebra $C(M)$.

Statement 2.2. ([4]) Let B denote a Banach algebra with finite number of generators. Suppose $n \geq 2$ is a positive integer

Then the algebra $M_n(B)$ can be generated by two idempotents if and only if $n = 2$ and the algebra B is singly generated.

So, from this theorem it follows that the algebra $C(M)$ is singly generated. The next theorem allows the geometric description of the space M.

Statement 2.3. ([5]) Suppose the algebra $C(M)$ can be generated by one element. Then the space M satisfies the next conditions:

1. M is homeomorphic to some compact subset M_1 of the plane $\mathbb{C}$ such that the set M_1 has not inner points.
2. The set $\mathbb{C}\backslash M_1$ is connected.

On the other hand, if the space M satisfies these conditions then the algebra $C(M)$ is singly generated.

These results allow us to prove the next theorem. This theorem describes the structure of the space of irreducible representations of an n-homogeneous C^*-algebra generated by two idempotents.

Theorem 2.4. *Suppose the C^*-algebra A is n-homogeneous and can be generated by two idempotents p, q and the identity e. If the algebra A is non-commutative then the algebra A is isomorphic to the algebra $C(M, \mathbb{C}^{2\times 2})$, where M denotes the space of maximal ideals of the algebra $Alg(p-r)^2$. The space M satisfies the next conditions:*

1. *The set M is homeomorphic to some compact subspace M_1 of the plane $\mathbb{C}$. The set M_1 has no interior points.*
2. *The set $\mathbb{C}\backslash M_1$ is connected.*

If the algebra A is commutative then the algebra A is isomorphic to the algebra $C(M)$, where the space M is a discrete space that can contain up to four points.

The theorem provides the sufficient and necessary conditions for the space of primitive ideals of an n-homogeneous C^*-algebra A to be generated by two idempotents.

During the paper we need the next fact.

Statement 2.5. ([1]) Let M denote an arbitrary compact subset of the plane $\mathbb{C}$. Then each algebraic bundle E over M is trivial.

Popovich, Samoilenko and Turowska proved in [7] that every algebra A from the class A_m is isomorphic to an algebra of matrices with the elements in some special C^*-algebra.

Let the symbol $Q_{2,m}$ denote the class of Banach algebras that can be generated by the flips $u, v, s_1, \ldots, s_m$, with the next relations between the generators: $u^2 = v^2 = s_i^2 = e; us_i = s_i u; vs_i = s_i v; s_j s_i = s_i s_j, i = \overline{1,m}, j = \overline{1,m}$. The class $Q_{2,m}$ is a sub-class of the class A_m.

Statement 2.6. ([7]) Every algebra A from the class A_k is isomorphic to the algebra $Mat_{2^n}(B)$ or to the algebra $Mat_{2^n}(Z(A))$, where $Z(A)$ denotes the center of the algebra A, while the symbol B denotes some algebra from the class $Q_{2,m}$.

So, to describe the structure of some algebra A from A_k we need to study the structure of the algebra B from $Q_{2,m}$. If the algebra A is n-homogeneous then the algebra B is also l-homogeneous for some positive integer $l \leq n$.

The next theorem describes the structure of a k-homogeneous C^*-algebra B from the class $Q_{2,m}$.

Theorem 2.7. *Let B denote an n-homogeneous C^*-algebra from the class $Q_{2,m}$. Then the algebra B is a trivial C^*-algebra.*

Proof. Suppose the algebraic bundle ξ corresponds to the algebra B.

One can consider the elements $s_i \in Z(B)$ as continuous functions on the space $P(B)$ of all primitive ideals in the appropriate topology.

Define the set $X_i = \{x \in P(B) | s_i(x) = -1\}$ and the set $Y_i = \{x \in P(B) \mid s_i(x) = 1\}$. The sets X_i and Y_i are closed subsets of $P(B)$ and $X_i \bigcup Y_i = P(B)$, $X_i \bigcap Y_i = \emptyset$.

Let $Z_{i_1,\dots,i_m}$ denote the intersection of the sets X_i and Y_i such that X_j corresponds to the positions $i_j = 0$ and Y_j corresponds to the positions $i_j = 1$. For example $Z_{010} = X_1 \bigcap Y_2 \bigcap X_3$.

The functions $s_1, \dots, s_m$ are equal to the constant functions -1 or 1 over each set $Z_{i_1,\dots,i_m}$.

Any two sets $Z_{i_1,\dots,i_m}$ and $Z_{h_1,\dots,h_m}$ do not intersect for the different collections $i_1, \dots, i_m$ and $h_1, \dots, h_m$. The union of all sets $Z_{i_1,\dots,i_m}$ for all collections $i_1, \dots, i_m$ coincides with the space $P(B)$.

Thus, the restriction $\xi_{Z_{i_1,\dots,i_m}}$ of the algebraic bundle ξ to the set $Z_{i_1,\dots,i_m}$ corresponds to the restriction of the algebra B to the set $Z_{i_1,\dots,i_m}$. The algebra of all continuous sections of the algebraic bundle $\xi_{Z_{i_1,\dots,i_m}}$ can be generated by the elements $u_{Z_{i_1,\dots,i_m}}$, $v_{Z_{i_1,\dots,i_m}}$ and the identity $e_{Z_{i_1,\dots,i_m}}$.

So, the algebra of all continuous sections $\Gamma(\xi_{Z_{i_1,\dots,i_m}})$ can be generated by two idempotents and the identity. Therefore, there is a continuous injection $\tau_{i_1,\dots,i_m} : Z_{i_1,\dots,i_m} \to C_{i_1,\dots,i_m}$ where $C_{i_1,\dots,i_m}$ is a closed subset of the plane $\mathbb{C}$.

There is a mapping $\tau : P(B) \to C$ such that $\tau|_{i_1,\dots,i_m} = \tau_{i_1,\dots,i_m}$, because the sets $Z_{i_1,\dots,i_m}$ are closed. The mapping τ is a homeomorphism onto a compact subset of the plane $\mathbb{C}$. Hence the bundle ξ on $P(B)$ is trivial by Statement 2.5.

The proof is finished. □

By using Theorem 2.7 we can describe the structure of an n-homogeneous C^*-algebra A from the class A_m.

Theorem 2.8. *Let the algebra A from A_m be an n-homogeneous C^*-algebra. Then the algebra A is trivial.*

Proof. By Theorem 2.7 and Statement 2.6. □

The next theorem describes the structure of the space of primitive ideals of an algebra $B \in Q_{2,m}$, which is an n-homogeneous C^*-algebra.

Theorem 2.9. *Let B be an n-homogeneous C^*-algebra. Let B belong to the class $Q_{2,m}$.*

Then the space of primitive ideals $P(B)$ has the following properties.

1. *The space $P(B)$ is homeomorphic to some compact subset X of the plane $\mathbb{C}$.*
2. *The space X has no interior points.*
3. *The set $\mathbb{C} \backslash X$ is a connected subset of the plane $\mathbb{C}$.*

Proof. By Theorem 2.4 we have that the set $Z_{1,\dots,i_m}$ from the proof of the previous theorem has the properties 1–3. Now our next goal is to prove that the union of the disjoint sets $Z_{i_1,\dots,i_m}$ has the properties.

We will use induction by the number of sets.

Let us consider two sets Z_1 and Z_2 that satisfy the conditions 1–3. Let the sets Z_1 and Z_2 have empty intersection. We will prove that the space $Z = Z_1 \cup Z_2$ has the properties.

Let us select the homeomorphisms $\tau_1 : Z_1 \to C_1$ and $\tau_2 : Z_2 \to C_2$ such that $C_1 \subset \mathbb{C}'$ and $C_2 \subset \mathbb{C}''$, where $\mathbb{C}'$ and $\mathbb{C}''$ denote two half-planes of the plane $\mathbb{C}$. In this case the space $\mathbb{C}'\backslash C_1$ is connected. The space $\mathbb{C}''\backslash C_2$ is also connected. Let b denotes the line between $\mathbb{C}'$ and $\mathbb{C}''$: $b := \mathbb{C}\backslash(\mathbb{C}' \cup \mathbb{C}'')$. Therefore, the union of two connected sets $(\mathbb{C}' \cup b)\backslash C_1$ and $(\mathbb{C}'' \cup b)\backslash C_2$ is also a connected set. The union is equal to the set $\mathbb{C}\backslash(C_1 \cup C_2)$.

The space $C_1 \cup C_2$ is also compact space without interior points. So, the space Z satisfies the conditions 1–3.

By induction, we obtain that the space $P(B)$ satisfies the conditions 1–3.

The proof is finished. □

Corollary 2.10. *Let the algebra B belong to the class $Q_{2,m}$. Let B be an n-homogeneous C^*-algebra. Then B can be generated by two idempotents and the identity.*

The proof is based on Theorem 2.4 and the previous Theorem 2.9. □

Let us note that in the paper [2] was studied the structure of some Banach algebras generated by N idempotents with some relations between generators.

Antonevich and Krupnuk have shown in [1] that if a Banach algebra generated by N idempotents with the relations between generators is an n-homogeneous C^*-algebra then it is trivial.

Statement 2.11. ([1]) Let A denote the n-homogeneous C^*-algebra generated by the idempotents $p_1, p_2, \dots, p_{2N}, P$ and the identity I with the next relations between them:

(a) $p_i \cdot p_j = \delta_{ij} p_i$ for any $i, j \in \overline{1,\dots,N}$ where δ_{ij} is a Kronecker's delta.
(b) $p_1 + p_2 + \dots + p_{2N} = I$.
(c) $P(p_{2i-1} + p_{2i})P = (p_{2i-1} + p_{2i})P$ and $Q(p_{2i} + p_{2i+1})Q = (p_{2i} + p_{2i+1})Q$, where $Q := I - P$ and $p_{2N+1} := p_1$.
Then the algebra A is trivial.

The authors of the paper [2] described the set of all irreducible representations of such a Banach algebra generated by N idempotents with the relations between generators. If such a Banach algebra A is a n-homogeneous C^*-algebra then the structure of the algebra can be properly described by the space $P(A)$ of primitive ideals of the algebra in the appropriate topology. In this case the algebra A is trivial. But the first author proved in the paper [8] that all non-trivial n-homogeneous ($n \geq 2$) C^*-algebras over the sphere S^2 can be generated by three idempotents. It follows from here that, in general, the structure of an n-homogeneous C^*-algebra generated by idempotents cannot be described properly by the space of primitive ideals of the algebra. We need to know some global topological invariants of such an C^*-algebra generated by idempotents in order to describe the structure of the algebra properly.

References

[1] A. Antonevich, N. Krupnik, *On Trivial and Non-Trivial n-Homogeneous C^*-Algebras.* Integ. Equat. and Oper. Theory **38** (2000), 172–189.

[2] A. Böttcher, I. Gohberg, Yu. Karlovich, N. Krupnik, S. Roch, B. Silbermann, I. Spitkovsky, *Banach Algebras Generated by n Idempotents and Applications.* In Operator Theory. Advances and Applications **90** (1996), pp. 19–54.

[3] J.M.G. Fell, *The Structure of Algebras of Operator Fields.* Acta Math. **106** (1961), 233–280.

[4] N. Krupnik, S. Roch, B. Silbermann, *On C^*-Algebras Generated by Idempotents.* J. Funct. Anal. **137** (1996), 303–319.

[5] M. Lavrentev, *Sur les fonctions d'une variable complexe représentables par des séries de polynômes.* 1936.

[6] V. Ostrovsky, Yu. Samoilenko, *Introduction to the Theory of Representations of Finitely Presented $*$-Algebras.* Rev. Math. and Math. Phys. **11** (1999), 261.

[7] V. Popovich, Yu. Samoilenko, L. Turowska, *Representations of a Cubic Deformation of $SU(2)$ and Parasupersymmetric Commutation Relations.* Symmetry in Nonlinear Math. Physics **2** (1997), 372–383.

[8] M. Shchukin, *Non-Trivial C^*-Algebras Generated by Idempotents.* In International Conference on Nonlinear Operators, Differential Equations and Applications (Cluj-Napoca, 2001), Semin. Fixed Point Theory Cluj-Napoca **3** (2002), 353–359.

[9] J. Tomiyama, M. Takesaki, *Application of Fibre Bundle to Certain Class of C^*-Algebras.* Tohoku Math. Journ. **13** (1961), 498–522.

Mikhail Shchukin
Department of Functional Analysis
Belarusian State University
Scoriny av. 4
Minsk
Belarus
e-mail: `ms@tut.by`

Elena Vatkina
Department of Functional Analysis
Belarusian State University
Scoriny av. 4
Minsk
Belarus
e-mail: `pugach@bsu.by`

Operator Theory:
Advances and Applications, Vol. 153, 255–264

Transfer Functions for "Curved" Conservative Systems

Alexey Tikhonov

Abstract. The aim of this paper is to study relations between "curved" conservative systems (systems for which the main operator is a function of contraction) and their transfer functions. We derive the boundedness property for such transfer functions and apply it to some problems of spectral analysis.

Mathematics Subject Classification (2000). Primary 47A48; Secondary 47A45.

Keywords. Transfer function, Conservative system, Functional model, Hilbert space.

0. Introduction

It is well known (see, e.g., [1]) that there is the one-to-one correspondence between linear conservative discrete-time systems and collections of operators (and Hilbert spaces) that can be organized as unitary block matrices

$$\mathfrak{A}_0 = \begin{pmatrix} T_0 & N_0 \\ M_0 & L_0 \end{pmatrix} \in [H_0 \oplus \mathfrak{N}, H_0 \oplus \mathfrak{M}], \quad \mathfrak{A}_0^* \mathfrak{A}_0 = I,\ \mathfrak{A}_0 \mathfrak{A}_0^* = I,$$

which are called unitary nodes in the other terminology [2]. Here $[H_1, H_2]$ is the set of all bounded operators acting from a separable Hilbert space H_1 into a separable Hilbert space H_2. (In a more standard notation $[H_1, H_2]$ is $\mathcal{L}(H_1, H_2)$). We shall consider pure nodes (i.e., nodes for which $\operatorname{Ker} N_0 = \{0\}$). Since any unitary node $\mathfrak{A}_0$ can be decomposed into the direct sum $\mathfrak{A}_0 = \operatorname{diag}(\mathfrak{A}_0^{pur}, \mathfrak{A}_0^{u})$, where $\mathfrak{A}_0^{pur}$ is a pure node and the state space of $\mathfrak{A}_0^u$ is trivial (i.e., $H_0^u = \{0\}$), our restriction is unessential. By the unitary property of the node, we get the identity $L_0 N_0^* = -M_0 T_0^*$, which uniquely determines the operator L_0 whenever the operators T_0, M_0, N_0 are given. Thus, without loss of essential information, we can consider short conservative systems (T_0, M_0, N_0). To each such system we assign the short transfer function

$$\Upsilon_0(z_0) = M_0 (T_0 - z_0)^{-1} N_0, \qquad z_0 \in \rho(T_0).$$

Obviously, $\Upsilon_0(z_0) = W_0(0) - W_0(1/z_0)$, where $W_0(\zeta) = L_0 + \zeta M_0(I - \zeta T_0)^{-1} N_0$ is the standard transfer function for the system $\mathfrak{A}_0$. The standard transfer function $W_0(\zeta)$ can be restored from $\Upsilon_0(\zeta)$ whenever $\rho(T_0) \cap \mathbb{D} \neq \emptyset$. Without loss of generality we can assume that $0 \in \rho(T_0)$. It suffices to find $W_0(0) = L_0$. We have $L_0 = \Upsilon_0(0)(I - (L_0^* L_0)^{-1})^{-1}$, where $L_0^* L_0 = \psi(\Upsilon_0(0)^* \Upsilon_0(0))$, $\psi(z) = 1/2\,(2 + z - \sqrt{z^2 + 4z})$. Then we can restore (up to unitary equivalence) the initial system $\mathfrak{A}_0$ provided that the node $\mathfrak{A}_0$ is simple ([2]). Recall that a unitary node is simple iff the operator T_0 is a completely non-unitary contraction ([2]).

We shall consider "curved" conservative systems of the form

$$(T, M, N) \;=\; (Z\varphi(T_0)Z^{-1},\, M_0\chi_+(T_0)Z^{-1},\, Z\chi_-(T_0)N_0)\,,$$

where $\mathfrak{A}_0 \in [H_0 \oplus \mathfrak{N}]$ is a pure unitary node, $Z \in [H_0, H]$, $Z^{-1} \in [H, H_0]$, $\varphi \in CM(\mathbb{D}, G_+)$ is a conformal map of $\mathbb{D}$ onto the domain $G_+ = \operatorname{Int} C$, C is a simple closed $C^{1+\varepsilon}$-smooth curve, $G_- = \operatorname{Ext} C$, $\chi_+ = \sqrt{\varphi'}/(\psi_+ \circ \varphi)$, $\chi_- = \sqrt{\varphi'}\,(\psi_- \circ \varphi)$, and $\psi_\pm \in (B : B)(G_+)$. We use $B(G_+)$ for the set of all bounded and analytic functions on the domain G_+ (in a more standard notation $B(G_+)$ is $H^\infty(G_+)$) and $(B : B)(G_+) := \{\psi \in B(G_+) : 1/\psi \in B(G_+)\}$. A "curved" conservative system is called simple if the corresponding unitary node $\mathfrak{A}_0$ is simple. Note that "curved" conservative systems arise in a natural way for Nikolski-Vasyunin functional ([3]) model with changed variables in it ([4]). In particular, linear-fractional transformations of unitary nodes (including the standard Cayley transformation from the unit disk onto the upper half plane) are examples of "curved" conservative systems.

We define the transfer function for a "curved" conservative system (T, M, N) by formula

$$\Upsilon(z) = M(T - z)^{-1} N\,, \qquad z_0 \in \rho(T)\,.$$

In this paper we study relations between "curved" systems and their transfer functions. We find some interesting properties of transfer functions and apply them to problems of spectral analysis and perturbation theory.

The paper is organized as follows. In Section 1, we find the form of transformations between $\Upsilon(z)$ and $\Upsilon_0(z_0)$. Using these transformations, it is shown that a simple "curved" conservative system is determined (up to linearly similarity) by its transfer function. Besides, we give characterization of transfer functions for "curved" systems. Our main tool is the generalized Sz.-Nagy–Foiaş functional model for "curved" systems ([4]). In Section 2, we show that a transfer function $\Upsilon(z)$ is bounded in the domain G_- whenever the curve C and the functions $\psi_\pm$ are sufficiently smooth. We apply this property to prove that the spectral component $N_-(\varphi(T_0))$ ([5, 4]) coincides with the whole Hilbert space. We would also note that transfer functions for "curved" systems play important role in the construction of the functional model for trace class perturbations of normal operators with spectrum on a curve ([4]).

1. Transformations of transfer functions

We attempt to find transformations between transfer functions $\Upsilon(z)$ and $\Upsilon_0(z_0)$. It will be convenient to make this by means of the functional model for "curved" conservative systems ([4]), which is a generalization of the well-known Sz.-Nagy–Foiaş functional model ([6, 3]). (See also [7] for an different approach to a model.) Here we give a brief exposition of the model for "curved" systems from [4]. We assume that a curve C is simple, closed, $C^{1+\varepsilon}$-smooth and the mappings $\pi_\pm \in [L^2(C,\mathfrak{N}),\mathcal{H}]$ satisfy conditions:

(i)$_1$ $\forall\,\psi \in L^\infty(C,[\mathfrak{N}])$ $(\pi_\pm^*\pi_\pm)\psi = \psi(\pi_\pm^*\pi_\pm)$; (i)$_2$ $(\pi_\pm^*\pi_\pm)^{-1} \in [L^2(C,\mathfrak{N})]$;
(ii)$_1$ $(\pi_-^\dagger\pi_+)z = z(\pi_-^\dagger\pi_+)$; (ii)$_2$ $P_-(\pi_-^\dagger\pi_+)P_+ = 0$;
(iii) $\operatorname{Ran}\pi_+ \vee \operatorname{Ran}\pi_- = \mathcal{H}$.

Here $\mathfrak{N},\mathcal{H}$ are separable Hilbert spaces, $\pi_-^\dagger$ is the Moore-Penrose inverse operator for π_-, $P_\pm$ are (non-orthogonal) projections onto the Hardy-Smirnov spaces $E^2(G_\pm,\mathfrak{N})$ [8]: $\operatorname{Ran}P_\pm = E^2(G_\pm,\mathfrak{N})$, $\operatorname{Ker}P_\pm = E^2(G_\mp,\mathfrak{N})$. Let $\Theta^\pm = \pi_\mp^\dagger\pi_\pm$. It is easy to see that $\Theta^\pm \in L^\infty(C,[\mathfrak{N}])$. Moreover, $\Theta^+ \in B(G_+,[\mathfrak{N}])$, that is, Θ^+ admits analytic continuation to the domain G_+ and is a bounded operator-valued function therein. The conditions $U\pi_\pm = \pi_\pm z$ uniquely determine the normal operator U with absolutely continuous spectrum on the curve C. For it, we have $\pi_\pm^\dagger U = z\pi_\pm^\dagger$. We put

$$\begin{aligned} \mathcal{K}_\Theta &= \{\, f \in \mathcal{H} : \pi_+^\dagger f \in E^2(G_-,\mathfrak{N}),\ \pi_-^\dagger f \in E^2(G_+,\mathfrak{N})\,\},\\ P_\Theta &= (I-q_+)(I-q_-), \quad \text{where} \quad q_\pm = \pi_\pm P_\pm \pi_\pm^\dagger . \end{aligned}$$

It can be shown that $P_\Theta^2 = P_\Theta$, $\mathcal{K}_\Theta = \operatorname{Ran}P_\Theta$. The model operators are defined by the formulas:

$$\begin{aligned} \widehat{T}f_0 &= Uf_0 - \pi_+\widehat{M}f_0, \quad \widehat{M}f_0 = \frac{1}{2\pi i}\int_C (\pi_+^\dagger f_0)(z)\,dz = (\pi_+^\dagger U f_0)(\infty),\\ \widehat{N}n &= P_\Theta\pi_- n = (I - \pi_+P_+\pi_+^\dagger)\pi_- n, \quad \text{where } f_0 \in \mathcal{K}_\Theta,\ n \in \mathfrak{N}. \end{aligned}$$

We need the following theorem from the paper [4].

Theorem. *Let (T,M,N) be a simple "curved" conservative system. Then there exist pair $\Pi = (\pi_+,\pi_-)$ satisfying conditions (i), (ii), (iii) and an invertible operator $Z \in [H,\mathcal{K}_\Theta]$ such that $\widehat{T}Z = ZT$, $\widehat{M}Z = M$, $\widehat{N} = ZN$.*

It is easy to show that $\widehat{\Upsilon}(z) = \Upsilon(z)$. In view of this identity, throughout the remaining part of the section we shall work mainly with the model operators. We need the following elementary lemma.

Lemma 1.1. *Let $u \in L^2(C,\mathfrak{N})$, $u_+ = P_+u$, $u_- = P_-u$. Then*

$$P_+\frac{u}{z-a} = \frac{u_+ \mp u_\pm(a)}{z-a}, \qquad P_-\frac{u}{z-a} = \frac{u_- \pm u_\pm(a)}{z-a}, \qquad \textit{where } a \in G_\pm .$$

Proposition 1.2. *Let* $f_0 \in \mathcal{K}_\Theta$. *Then*

1) $(\widehat{T}-a)^{-1}f_0 = (U-a)^{-1}(f_0 - \pi_+ n(a))$;

2) $\widehat{M}(\widehat{T}-a)^{-1}f_0 = -n(a)$, *where* $n(a) = \begin{cases} \Theta^+(a)^{-1}(\pi_-^\dagger f_0)(a), & a \in G_+ \\ (\pi_+^\dagger f_0)(a), & a \in G_- . \end{cases}$

Proof. 1) We put $R_a f_0 = (U-a)^{-1}(f_0 - \pi_+ n(a))$. By Lemma 1.1, it is easy to check that $R_a f_0 \in \mathcal{K}_\Theta$. Further, we have

$$(\widehat{T}-a)R_a f_0 = (U-a)R_a f_0 - \pi_+ m = f_0 - \pi_+ n(a) - \pi_+ m\,,$$

where $m = (\pi_+^\dagger U R_a f_0)(\infty) = (\frac{z}{z-a}(\pi_+^\dagger f_0 - n(a)))(\infty) = -n(a)$. Whence we have $(\widehat{T}-a)R_a = I$ and $(\widehat{T}-a)^{-1} = R_a$.

2) By Cauchy's theorem and formula, we have

$$\begin{aligned}\widehat{M}(\widehat{T}-a)^{-1}f_0 &= \frac{1}{2\pi i}\int_C (\pi_+^\dagger (U-a)^{-1}(f_0 - \pi_+ n(a)))(z)\,dz \\ &= \frac{1}{2\pi i}\int_C \frac{(\pi_+^\dagger f_0)(z) - n(a)}{z-a}\,dz = -n(a)\,. \qquad \square\end{aligned}$$

Denote by $\Theta_\pm^-$ the analytic in $G_\pm$ operator-valued functions such that $\Theta_\pm^-(z)n = (P_\pm \Theta^- n)(z)$ for $n \in \mathfrak{N}$ and $z \in G_\pm$ (see the definition of H^2-strong operator-valued functions in [9]). It is possible that boundary values of the operator-valued functions $\Theta_\mp^-(z)$ do not exist. But we have the existence of boundary values for vector-valued functions $\Theta_\pm^-(z)n \in E^2(G_\pm, \mathfrak{N})$ a.e. on C.

Proposition 1.3. $\Upsilon(z) = \begin{cases} \Theta_+^-(z) - \Theta^+(z)^{-1}, & z \in G_+ \\ -\Theta_-^-(z), & z \in G_- . \end{cases}$

Proof. We have $\widehat{N}n = (I - \pi_+ P_+ \pi_+^\dagger)\pi_- n = \pi_- n - \pi_+ \Theta_+^- n$, $n \in \mathfrak{N}$. By Proposition 1.2(2), we get $\Upsilon(z)n = \widehat{M}(\widehat{T}-z)^{-1}\widehat{N}n = -m(z)$, where

$$\begin{aligned} m(z) &= \Theta^+(z)^{-1}(\pi_-^\dagger(\pi_- n - \pi_+ \Theta_+^- n))(z) = \Theta^+(z)^{-1}(n - \Theta^+(\Theta_+^- n))(z) \\ &= \Theta^+(z)^{-1}(n - \Theta^+(z)(\Theta_+^- n)(z)) = \Theta^+(z)^{-1}n - \Theta_+^-(z)n, \quad z \in G_+\,, \\ m(z) &= (\pi_+^\dagger(\pi_- n - \pi_+ \Theta_+^- n))(z) = (\Theta^- n - \Theta_+^- n)(z) = \Theta_-^-(z)n, \quad z \in G_-\,. \qquad \square \end{aligned}$$

We shall write $\Theta^+ \in (B:F)(G_+, [\mathfrak{N}])$ if $\Theta^+ \in B(G_+, [\mathfrak{N}])$ and $(\Theta^+)^{-1} \in F(G_+, [\mathfrak{N}])$, where $F(G_+, [\mathfrak{N}])$ is the class of operator-valued functions such that their boundary values exist a.e. on C.

Corollary 1.4. *Let* $\Theta^+ \in (B:F)(G_+, [\mathfrak{N}])$. *Then* $\forall n \in \mathfrak{N}$

$$(\Upsilon n)_+(z) - (\Upsilon n)_-(z) = (\Theta^-(z) - \Theta^+(z)^{-1})n, \quad a.e.\, z \in C\,,$$

where $(\Upsilon n)_\pm$ *are boundary values for* $(\Upsilon n)(z)$ *from* $G_\pm$.

Proposition 1.5. *Let* $\Theta^+ \in (B:F)(G_+,[\mathfrak{N}])$. *Then* $\forall n \in \mathfrak{N}$ *and for a.e.* $|z_0| = 1$

$$(\Upsilon n)_+(\varphi(z_0)) - (\Upsilon n)_-(\varphi(z_0)) = \frac{\chi_+(z_0)\chi_-(z_0)}{\varphi'(z_0)}((\Upsilon_0 n)_+(z_0) - (\Upsilon_0 n)_-(z_0)).$$

Proof. Define the unitary operator $C_\varphi \in [L^2(C,\mathfrak{N}), L^2(\mathbb{T},\mathfrak{N})]$ by the formula $C_\varphi f = (f\circ\varphi)\sqrt{\varphi'}$. It is known [4] that $\pi_\pm = \pi_{0\pm} C_\varphi \psi_\pm$. We then have

$$\Theta^\pm = \pi_\mp^\dagger \pi_\pm = \frac{1}{\psi_\mp} C_\varphi^{-1} \pi_{0\mp}^\dagger \pi_{0\pm} C_\varphi \psi_\pm = \frac{\psi_\pm}{\psi_\mp}(\Theta_0^\pm \circ \varphi^{-1}).$$

Whence,

$$\Theta^- \circ\varphi - (\Theta^+)^{-1}\circ\varphi = \frac{\psi_-\circ\varphi}{\psi_+\circ\varphi}(\Theta_0^- - (\Theta_0^+)^{-1}) = \frac{\chi_+\chi_-}{\varphi'}(\Theta_0^- - (\Theta_0^+)^{-1}).$$

It remains to make use of Corollary of Proposition 1.3. □

Define the operator $F_{\varphi,\psi} \in [E^2(G_{1-},\mathfrak{N}), E^2(G_{2-},\mathfrak{N})]$ by the formula $F_{\varphi,\psi} u = P_{2-}((u\circ\varphi^{-1})\psi)$, for $u \in E^2(G_{1-},\mathfrak{N})$, where $\varphi \in CM(G_{1+},G_{2+})$ and $\psi \in (B:B)(G_{2+})$. Consider compositions of such operators.

Lemma 1.6. $F_{\varphi_{32},\psi_3} F_{\varphi_{21},\psi_2} = F_{\varphi_{31},\psi}$, *where* $\varphi_{31} = \varphi_{32}\circ\varphi_{21}$, $\psi = (\psi_2\circ\varphi_{32}^{-1})\psi_3$.

Proof. Let $u \in E^2(G_{1-},\mathfrak{N})$. Put $v = (u\circ\varphi_{21}^{-1})\psi_2$. Then

$$\begin{aligned} F_{\varphi_{32},\psi_3} F_{\varphi_{21},\psi_2} u &= P_{3-}(((P_{2-}v)\circ\varphi_{32}^{-1})\psi_3) \\ &= P_{3-}(((P_{2+}v)\circ\varphi_{32}^{-1} + (P_{2-}v)\circ\varphi_{32}^{-1})\psi_3) \\ &= P_{3-}((v\circ\varphi_{32}^{-1})\psi_3) = P_{3-}((((u\circ\varphi_{21}^{-1})\psi_2)\circ\varphi_{32}^{-1})\psi_3) \\ &= P_{3-}((u\circ\varphi_{31}^{-1})(\psi_2\circ\varphi_{32}^{-1})\psi_3) = F_{\varphi_{31},\psi} u, \end{aligned}$$

because $(P_{2+}v)\circ\varphi_{32}^{-1} \in E^2(G_{3+},\mathfrak{N})$. □

Remarks 1.7. 1) (*algebraic*) This lemma has a simple algebraic interpretation: the product of triangular matrices again is a triangular matrix.

2) (*analytical*) On the other hand, the operator $F_{\varphi,\psi}$ is an analog of the well-known Faber transform.

Proposition 1.8. 1) $\forall n \in \mathfrak{N}$ $(\Upsilon_0 n)_- = F_{\varphi^{-1},(\psi_+/\psi_-)\circ\varphi^{-1}}((\Upsilon n)_-)$;

2) $\Theta_0^+ \in (B:F)(\mathbb{D},[\mathfrak{N}]) \implies (\Upsilon_0 n)_+ = (\Upsilon_0 n)_- + (((\Upsilon n)_+ - (\Upsilon n)_-)\circ\varphi)\frac{\chi_+\chi_-}{\varphi'}$.

Proof. 1) We have

$$\begin{aligned} \Theta_-^- n &= P_-\Theta^- n = P_-\left(\tfrac{\psi_-}{\psi_+}(\Theta_0^-\circ\varphi^{-1})n\right) \\ &= P_-\left(\tfrac{\psi_-}{\psi_+}(((P_+ + P_-)\Theta_0^-)\circ\varphi^{-1})n\right) \\ &= P_-\left(\tfrac{\psi_-}{\psi_+}((P_-\Theta_0^-)\circ\varphi^{-1})n\right) = F_{\varphi,\psi_+/\psi_-}(\Theta_{0-}^- n) \end{aligned}$$

or equivalently, $(\Upsilon n)_- = F_{\varphi,\psi_+/\psi_-}((\Upsilon_0 n)_-)$. Using Lemma 1.6, it is easy to show that $(\Upsilon_0 n)_- = F_{\varphi^{-1},(\psi_+/\psi_-)\circ\varphi^{-1}}((\Upsilon n)_-)$.

2) This follows from Proposition 1.5. □

Thus we have found the procedure that allows us to restore Υ_0 whenever Υ is given. In the introduction we discussed how to compute the standard transfer function W_0 from a short transfer function Υ_0. So we can construct the transformation $W_0 = \mathcal{F}_{\varphi,\psi_+,\psi_-}(\Upsilon)$ ($W_0 = \mathcal{F}(\Upsilon)$ in a brief form) that takes each transfer function for a "curved" system to the corresponding standard transfer function. It is easy to see that this transformation is invertible. Since $W_0(\zeta) = \Theta_0^+(\zeta)^\sim = \Theta_0^+(\bar{\zeta})^*$, we can describe the class of transfer function for "curved" systems with $W_0^{-1} \in F(\mathbb{D}, [\mathfrak{N}])$ as the following:

$$\mathrm{Ob}(Tfn_F) \;=\; \{\Upsilon :\ \mathcal{F}(\Upsilon) \in (B:F)(\mathbb{D}, [\mathfrak{N}]),\ \sup_{|\zeta|<1} ||\mathcal{F}(\Upsilon)(\zeta)|| \leq 1\}$$

or, anticipating (see Corollary of Theorem 2.1), for more smooth C and $\psi_\pm$, we may also write down

$$\begin{aligned}\mathrm{Ob}(Tfn_F) \;=\; \{\Upsilon :\ &\forall n \in \mathfrak{N}\ \ \Upsilon n \in E^2(G_-, \mathfrak{N}),\ \Upsilon \in F(G_+, [\mathfrak{N}]),\\ &\sup_{|\zeta|<1} ||\mathcal{F}(\Upsilon)(\zeta)|| \leq 1\}.\end{aligned}$$

Theorem 1.9. *Let (T, M, N) be a "curved" conservative system, $W_0^{-1} \in F(\mathbb{D}, [\mathfrak{N}])$. Then $\Upsilon(z) = M(T-z)^{-1}N \in \mathrm{Ob}(Tfn_F)$. Conversely, let $\Upsilon \in \mathrm{Ob}(Tfn_F)$. Then there exists (up to linear equivalence) a simple "curved" conservative system (T, M, N) such that $W_0^{-1} \in F(\mathbb{D}, [\mathfrak{N}])$ and $\Upsilon(z) = M(T-z)^{-1}N$.*

Proof. We need only to prove the inverse statement. Let $\Upsilon \in \mathrm{Ob}(Tfn_F)$. For contractive-valued analytic function $W_0 = \mathcal{F}(\Upsilon)$, there exists a realization $W_0(\zeta) = L_0 + \zeta M_0(I - \zeta T_0)^{-1}N_0$, where $\mathfrak{A}_0$ is a simple unitary node [2]. Without loss of generality we may take $\mathfrak{A}_0$ to be a model node, which is constructed from a pair $\Pi_0 = (\pi_{0+}, \pi_{0-})$. We take $\pi_\pm = \pi_{0\pm} C_\varphi \psi_\pm$ and consider the corresponding simple "curved" system $(\widehat{T}, \widehat{M}, \widehat{N})$. We then have $W_0 = \mathcal{F}(\widehat{\Upsilon})$, where $\widehat{\Upsilon}(z) = \widehat{M}(\widehat{T} - z)^{-1}\widehat{N}$. Since $\mathcal{F}$ is invertible, we get $\widehat{\Upsilon} = \mathcal{F}^{-1}(W_0) = \Upsilon$. Thus Υ has the realization $\Upsilon(z) = \widehat{M}(\widehat{T} - z)^{-1}\widehat{N}$. Suppose Υ has another realization $\Upsilon(z) = M'(T' - z)^{-1}N'$, where (T', M', N') is a simple "curved" system. For the system (T', M', N'), there exists (see the beginning of this section) a model system $(\widehat{T}', \widehat{M}', \widehat{N}')$ such that $\widehat{T}'Z' = Z'T'$, $\widehat{M}'Z' = M'$, , $\widehat{N}' = Z'N'$. Since $W_0 = \mathcal{F}(\widehat{\Upsilon}) = \mathcal{F}(\widehat{\Upsilon}') = W_0'$, the corresponding simple unitary nodes $\mathfrak{A}_0$ and $\mathfrak{A}_0'$ are unitarily equivalent. Whence the simple "curved" systems $(\widehat{T}, \widehat{M}, \widehat{N})$ and (T', M', N') are linearly similar. □

Remarks 1.10. 1) If C and $\psi_\pm$ are more smooth, we may assume $\Upsilon \in F(\mathbb{D}, [\mathfrak{N}])$ instead of $W_0^{-1} \in F(\mathbb{D}, [\mathfrak{N}])$ (see Corollary of Theorem 2.1).

2) It may be proved that $\rho(T) \cap G_+ \neq \emptyset$, $M, N \in \mathfrak{S}_2 \implies W_0^{-1} \in F(\mathbb{D}, [\mathfrak{N}])$, where $\mathfrak{S}_2$ is the Hilbert-Schmidt class. For such systems, the assertion of the theorem remains true if we replace $\mathrm{Ob}(Tfn_F)$ by

$$\begin{aligned}\mathrm{Ob}(Tfn_1) \;=\; \{\Upsilon :\ &\mathcal{F}(\Upsilon) \in B(\mathbb{D}, [\mathfrak{N}]),\ \sup_{|\zeta|<1} ||\mathcal{F}(\Upsilon)(\zeta)|| \leq 1,\ \exists \zeta_0 \in \mathbb{D}\\ &\mathcal{F}(\Upsilon)(\zeta_0)^{-1} \in [\mathfrak{N}],\ I - \mathcal{F}(\Upsilon)(\zeta_0)\mathcal{F}(\Upsilon)(\zeta_0)^* \in \mathfrak{S}_2\}.\end{aligned}$$

2. The boundedness of transfer functions

Since $W_0(\zeta)$ is bounded in the unit disk, it is easy to show that $\Upsilon_0(z_0)$ is bounded and analytic in $\mathbb{D}_-$. Transfer functions $\Upsilon(z)$ for "curved" systems possess the analogous property. Namely, we have the following

Theorem 2.1. *Let C be a simple closed $C^{2+\varepsilon}$-smooth curve, and let $\psi_\pm \in C^{1+\varepsilon}(C)$. Then $\Upsilon \in B(G_-,[\mathfrak{N}])$.*

Proof. First we note that $\varphi \in C^{2+\varepsilon}(\mathbb{D})$, $\psi_\pm \in C^{1+\varepsilon}(G_+)$ ([10]). We shall additionally assume that $W_0^{-1} \in F(\mathbb{D},[\mathfrak{N}])$. By Proposition 1.5, $\forall n \in \mathfrak{N}$ and for a.e. $|z_0| = 1$, we have

$$\begin{aligned}(\Upsilon n)_-(\varphi(z_0)) &= \frac{\chi_+(z_0)\chi_-(z_0)}{\varphi'(z_0)}(\Upsilon_0 n)_-(z_0)\\ &+ \left((\Upsilon n)_+(\varphi(z_0)) - \frac{\chi_+(z_0)\chi_-(z_0)}{\varphi'(z_0)}(\Upsilon_0 n)_+(z_0)\right).\end{aligned}$$

Evidently, $||(\Upsilon_0 n)_-||_{L^\infty(\mathbb{T})} \le K_-||n||$. To prove an analogous estimate for the second term, we consider operator-valued function

$$\Omega(z_0) = M(T-\varphi(z_0))^{-1}N - \frac{\chi_+(z_0)\chi_-(z_0)}{\varphi'(z_0)}M_0(T_0 - z_0)^{-1}N_0, \quad |z_0| < 1.$$

We wish to show that $\Omega \in B(\mathbb{D},[\mathfrak{N}])$. Since $\chi_\pm \in C^{1+\varepsilon}(G_+)$, we have

$$(\chi_\pm(T_0) - \chi_\pm(z_0))(T_0 - z_0)^{-1} \in B(\mathbb{D},[H]).$$

So it suffices to check the boundedness of

$$M(\varphi(T_0) - \varphi(z_0))^{-1}N - \frac{1}{\varphi'(z_0)}M(T_0 - z_0)^{-1}N = M\chi(T_0, z_0)N,$$

where

$$\chi(\zeta, z) = \frac{1}{\varphi(\zeta)-\varphi(z)} - \frac{1}{\varphi'(z)}\frac{1}{\zeta - z}.$$

Since $\varphi \in C^{2+\varepsilon}(\mathbb{D})$, it follows that $\sup_{\zeta,z\in\mathbb{D}}|\chi(\zeta,z)| < \infty$. Whence, $\chi(T_0, z_0) \in B(\mathbb{D},[H])$. Therefore we get the second estimate $||(\Omega n)_+||_{L^\infty(\mathbb{T})} \le K_+||n||$. This together with the first estimate imply $||(\Upsilon n)_-||_{L^\infty(\mathbb{T})} \le K||n||$. By the Smirnov maximum principle [11], we get $(\Upsilon n)_- \in B(G_-,\mathfrak{N})$ and $||(\Upsilon n)_-||_{B(G_-)} \le K||n||$. Thus, $\Upsilon \in B(G_-,[\mathfrak{N}])$.

We now turn to the general case (without the restriction $W_0^{-1} \in F(\mathbb{D},[\mathfrak{N}])$). We take $\widetilde{W_0}(\zeta) = 1/2(I - 1/(2c)\Upsilon_0(1/\zeta))$, where $c = \sup_{|\zeta|<1}||\Upsilon_0(1/\zeta)||$. Obviously, we have $\sup_{|\zeta|<1}||\widetilde{W_0}(\zeta)|| \le 1$ and $\widetilde{W_0}^{-1} \in B(\mathbb{D},[\mathfrak{N}]) \subset F(\mathbb{D},[\mathfrak{N}])$. By the above, $\widetilde{\Upsilon} \in B(G_-,[\mathfrak{N}])$. Since $\Upsilon_0 = 4/c\widetilde{\Upsilon}_0$, we obtain

$$(\Upsilon n)_- = F_{\varphi,\psi_+/\psi_-}((\Upsilon_0 n)_-) = 4/c\,F_{\varphi,\psi_+/\psi_-}((\widetilde{\Upsilon}_0 n)_-) = 4/c\,(\widetilde{\Upsilon} n)_-.$$

Whence, $\Upsilon \in B(G_-,[\mathfrak{N}])$. □

Remarks 2.2. In fact, here we have established the property of Faber's transformation to preserve boundedness of a function if the curve is smooth enough. Conversely, it is possible to derive the theorem from this property.

Corollary 2.3. $\Upsilon \in F(G_+, [\mathfrak{N}]) \iff \Upsilon_0 \in F(\mathbb{D}, [\mathfrak{N}]) \iff W_0^{-1} \in F(\mathbb{D}, [\mathfrak{N}])$.

Proof. By Proposition 1.3, we have

$$\Theta^-_- = -\Upsilon \in B(G_-, [\mathfrak{N}]) \iff \Theta^-_+ = \Theta^- - \Theta^-_- \in B(G_+, [\mathfrak{N}])$$

and

$$\Upsilon \in F(G_+, [\mathfrak{N}]) \iff (\Theta^+)^{-1} \in F(G_+, [\mathfrak{N}]) \iff W_0^{-1} \in F(\mathbb{D}, [\mathfrak{N}]). \qquad \square$$

We are going to present some application of the boundedness property to spectral analysis. But beforehand we prove the following auxiliary assertion.

Proposition 2.4. *Let* $f \in \mathcal{H}$, $f_0 = P_\Theta f$, $u_+ = \pi_+^\dagger q_+(I - q_-)f$, $u_- = \pi_-^\dagger q_- f$. *Then*

$$P_\Theta(U-a)^{-1} f = (U-a)^{-1}(f_0 + \pi_- n + \pi_+(m - (\Theta^-_+ n \mp \Theta^-_\pm(a)n))), \; a \in G_\pm,$$

where

$$n = -(\pi_-^\dagger f_0)(a) - \Theta^+(a)u_+(a), \qquad m = u_+(a), \qquad a \in G_+;$$
$$n = u_-(a), \qquad m = -(\pi_+^\dagger f_0)(a), \qquad a \in G_-.$$

Proof. Clearly, $f = \pi_+ u_+ + f_0 + \pi_- u_-$, $f_0 \in \mathcal{K}_\Theta$, $u_\pm \in E^2(G_\pm, \mathfrak{N})$. Then, by Lemma 1.1,

$$q_-(U-a)^{-1} f = \pi_- P_- \frac{u_- + \pi_-^\dagger f_0 + \Theta^+ u_+}{z - a} = (U-a)^{-1}\pi_-(u_- - n).$$

Whence, $(I - q_-)(U-a)^{-1} f = (U-a)^{-1}(f_0 + \pi_+ u_+ + \pi_- n)$. By Lemma 1.1, we have $q_+(U-a)^{-1}(f_0 + \pi_+ u_+) = (U-a)^{-1}\pi_+(u_+ - m)$ and

$$\begin{aligned} q_+(U-a)^{-1}\pi_- n &= \pi_+ P_+ \Theta^- \frac{n}{z-a} = \pi_+ P_+ \frac{\Theta^-_+ n + \Theta^-_- n}{z-a} \\ &= (U-a)^{-1}\pi_+(\Theta^-_+ n \mp \Theta^-_\pm(a)n), \qquad a \in G_\pm. \end{aligned}$$

Gathering these identities, we obtain the desired formula. $\square$

Recall the definition of weak spectral components [4] :

$$N_\pm(T) = \operatorname{clos}\{\, f \in H : \forall g \in H \quad ((T - z)^{-1} f, g) \in E^2(G_\pm)\,\}.$$

For completely non-unitary contraction T_0, we have $N_-(T_0) = H$, $N_+(T_0) = H_{\cdot 1}$ (see [6] for the definition of the subspace $H_{\cdot 1}$). We extend the first of these relations to the case when T is a function of T_0.

Proposition 2.5. *Let* $\Theta^-_\pm \in B(G_\pm, [\mathfrak{N}])$. *Then* $N_-(\widehat{T}) = \mathcal{K}_\Theta$.

Proof. Let $n \in \mathfrak{N}$. Obviously, $P_\Theta \pi_\mp \dfrac{n}{z-a} = 0$, $a \in G_\pm$.
Next, we put $f_\pm = P_\Theta \pi_\pm \dfrac{n}{z-a}$, $a \in G_\pm$. Using Proposition 2.4, we have

$$f_+ = (U-a)^{-1}(\pi_+(I + (\Theta_+^- - \Theta_+^-(a))\Theta^+(a))n - \pi_-\Theta^+(a)n),$$
$$f_- = (U-a)^{-1}(\pi_- n - \pi_+(\Theta_+^- + \Theta_-^-(a))n),$$

Whence,

$$\pi_+^\dagger f_+ = \frac{n - (\Theta_-^- + \Theta_+^-(a))\Theta^+(a)n}{z-a}, \qquad \tau_+^\dagger f_+ = \frac{-\Delta^+\Theta^+(a)n}{z-a};$$
$$\pi_+^\dagger f_- = \frac{(\Theta_-^- - \Theta_-^-(a))n}{z-a}, \qquad \tau_+^\dagger f_- = \frac{\Delta^+ n}{z-a},$$

where $\tau_+^\dagger = (\Delta^+)^{-1}(\pi_-^\dagger - \Theta^+\pi_+^\dagger)$, $\tau_+^\dagger\pi_- = \Delta^+ = (I - \Theta^+\Theta^-)^{1/2} \in L^\infty(C, [\mathfrak{N}])$ (see the paper [4]). So that we obviously have $\pi_+^\dagger f_\pm, \tau_+^\dagger f_\pm \in L^\infty(C, [\mathfrak{N}])$. Further, using the identity $\pi_+\pi_+^\dagger + \tau_+\tau_+^\dagger = I$, we have

$$\begin{aligned}
((\widehat{T} - z)^{-1} f_\pm, g)_\mathcal{H} &= (P_\Theta(U-z)^{-1} f_\pm, g)_\mathcal{H} \\
&= ((\pi_+\pi_+^\dagger + \tau_+\tau_+^\dagger)(U-z)^{-1} f_\pm, P_\Theta^* g)_\mathcal{H} \\
&= (\frac{1}{z-\zeta}(\pi_+^\dagger f_\pm)(\zeta), (\pi_+^* P_\Theta^* g)(\zeta))_{L^2} + (\frac{1}{z-\zeta}(\tau_+^\dagger f_\pm)(\zeta), (\tau_+^* P_\Theta^* g)(\zeta))_{L^2} \\
&= \int_C \frac{h_\pm(\zeta)}{\zeta - z}\,|d\zeta| \in E^2(G_-),
\end{aligned}$$

where $h_\pm(\zeta) = ((\pi_+^\dagger f_\pm)(\zeta), (\pi_+^* P_\Theta^* g)(\zeta))_\mathfrak{N} + ((\tau_+^\dagger f_\pm)(\zeta), (\tau_+^* P_\Theta^* g)(\zeta))_\mathfrak{N} \in L^2(C)$.
Since $\mathcal{K}_\Theta = P_\Theta(\bigvee_{a \in G_+ \cup G_-, n \in \mathfrak{N}} \pi_\pm \frac{n}{z-a})$, we obtain $N_-(\widehat{T}) = \mathcal{K}_\Theta$. □

Corollary 2.6. *Let C be a simple closed $C^{2+\varepsilon}$-smooth curve, $\pi_\pm^*\pi_\pm \in C^{1+\varepsilon}(C)$. Then $N_-(\widehat{T}) = \mathcal{K}_\Theta$.*

Proof. As is known (see [12, 4]), $\pi_\pm^*\pi_\pm \in C^{1+\varepsilon}(C) \iff \psi_\pm \in C^{1+\varepsilon}(C)$. It remains to make use of the proposition and Theorem 2.1. □

Writing Proposition 2.5 in terms of "curved" conservative systems, we obtain the following theorem.

Theorem 2.7. *Let (T, M, N) be a simple "curved" conservative system, C be a simple closed $C^{2+\varepsilon}$-smooth curve, $\psi \in C^{1+\varepsilon}(C)$. Then $N_-(T) = H$.*

Corollary 2.8. *Let T_0 be a completely non-unitary contraction, C be a simple closed $C^{2+\varepsilon}$-smooth curve, and φ be conformal map of $\mathbb{D}$ onto the domain* Int C. *Then $N_-(\varphi(T_0)) = H$.*

Remarks 2.9. 1) Formally the identity $N_-(T) = H$ is a particular case of the duality relations from [4]. But, in fact, we made use of this identity in the proof of the duality theorem (Theorem C in [4]).

2) If we consider the latter result in the context of smooth scalar Smirnov domains, we obtain that bounded functions are dense in the subspace $E_\Theta = \{ u \in E^2(G_+) : \Theta u \in E^2(G_-) \}$ for any $\Theta \in B(G_-)$. These subspaces are analogous (for Smirnov domains) of the subspaces $H^2 \ominus \theta H^2$ for the unit disk, where the function θ is inner. The latter, as is well known [3, 6], describe all co-invariant subspaces of the shift operator in H^2.

References

[1] D.Z. Arov, *Passive Linear Dynamic Systems.* Sibirsk.Math.Zh. **20** (1979), no.2, 211–228.

[2] M.S. Brodskiy, *Unitary Operator Nodes and their Characteristic Functions.* Uspehi mat. nauk **33** (1978), no.4, 141–168.

[3] N.K. Nikolski, V.I. Vasyunin, *Elements of Spectral Theory in Terms of the Free Functional Model. Part I: Basic Constructions.* Holomorphic spaces (eds. Sh. Axler, J. McCarthy, D. Sarason), MSRI Publications **33** (1998), 211–302.

[4] A.S. Tikhonov, *Functional Model and Duality of Spectral Components for Operators with Continuous Spectrum on a Curve.* Algebra i Analiz, **14** (2002), no.4, 158–195.

[5] S.N. Naboko, *Functional Model for Perturbation Theory and its Applications to Scattering Theory.* Trudy Mat. Inst. Steklov **147** (1980), 86–114.

[6] B. Szökefalvi-Nagy, C. Foiaş, *Harmonic Analysis of Operators on Hilbert Space.* North-Holland, Amsterdam-London, 1970.

[7] D.V. Yakubovich, *Linearly Similar Model of Sz.-Nagy–Foias Type in a Domain.* Algebra i Analiz, **15** (2003), no. 2, 180–227.

[8] P.L. Duren, *Theory of H^p Spaces.* Pure Appl. Math., vol. 38, Academic Press, New York–London, 1970.

[9] N.K. Nikolski, *Treatise of the Shift Operator.* Springer-Verlag, Heidelberg, 1986.

[10] C. Pommerenke, *Univalent Functions.* Vandenhoek and Ruprecht, Göttingen, 1975.

[11] J.B. Garnett, *Bounded Analytic Functions.* Pure Appl. Math., vol. 96, Academic Press, New York-London, 1981.

[12] N.A. Shirokov, *Properties of Modulus for Analytic Functions Smooth up to the Boundary.* Dokl. Akad. Nauk SSSR, **269** (1966), no.6, 1320–1323.

Alexey Tikhonov
Mathematical analysis division
Taurida National University
Yaltinskaya str., 4
Simferopol 95007 Crimea
Ukraine
e-mail: `tikhonov@club.cris.net`

Operator Theory:
Advances and Applications, Vol. 153, 265–270

On the Distance between an Operator and an Ideal

Nicolae Ţiţa

Abstract. Let $L(X)$ be the normed algebra of all bounded linear operators $T : X \to X$, where X is a normed space, and let I be an operator ideal, so that $I(X)$ is a two-sided ideal in $L(X)$. If X is a tensor product of normed spaces, endowed with a tensor norm, an estimation is given for the distance of a tensor product operator $T \in L(X)$ to $I(X)$ and it is applied to the study of the quasi-nilpotency of tensor product operators modulo $I(X)$.

Mathematics Subject Classification (2000). 47L20.

Keywords. Operator ideal, Tensor product.

1. Main result

Let $L(X,Y)$ denote, for any normed spaces X and Y, the normed space of all bounded linear operators $T : X \to Y$. For $X = Y$, $L(X) = L(X,X)$ is a normed algebra with respect to the composition as product operation.

Recall (see [10], [7], [15]) that an *operator ideal* I is a way to associate to any normed spaces X, Y some linear subspace $I(X,Y) \subset L(X,Y)$ containing all finite rank operators and such that

$$S_1 T S_2 \in I(X',Y') \quad \text{for } T \in L(X,Y), S_1 \in L(Y,Y'), S_2 \in L(X',X).$$

Furthermore, an operator ideal I is called *quasi-normed* if it is endowed with functions $n : I(X,Y) \to \mathbf{R}_+$ satisfying, for some $c \geq 1$, the conditions:

1. $n(T) = \|T\|$ if $T \in L(X,Y)$, $rank\, T = 1$,
2. $n(T_1 + T_2) \leq c\,(n(T_1) + n(T_2))$ if $T_1, T_2 \in L(X,Y)$,
3. $n(S_1 T S_2) \leq \|S_1\|\, n(T) \|S_2\|$ if $T \in L(X,Y), S_1 \in L(Y,Y'), S_2 \in L(X',X)$.

If $c = 1$ then n is a norm on every $L(X,Y)$ and in this case the pair (I, n) is called a *normed operator ideal.*

Let us recall also the next examples of operator ideals:

(a) Denoting by $K(X,Y)$ the set of all precompact operators in $L(X,Y)$, we get the operator ideal K.

(b) An operator $T \in L(X,Y)$ is called approximable if there exists a sequence $(T_n)_{n\geq 1}$ in $L(X,Y)$, $rank\, T_n < n$, such that $\lim\limits_n \|T - T_n\| = 0$. Denoting by $A(X,Y)$ the set of all approximable operators in $L(X,Y)$, it is known that $A(X,Y) \subseteq K(X,Y)$ and A is again an operator ideal.

(c) An operator $T \in L(X,Y)$ is called nuclear if it can be written as $T = \sum\limits_{i=1}^{\infty} x_i^* \cdot y_i$ with $x_i^* \in X^*$, $y_i \in Y$ and $\sum\limits_{i=1}^{\infty} \|x_i^*\| \cdot \|y_i\| < +\infty$. Putting

$$n(T) = \inf \left\{ \sum_{i=1}^{\infty} \|x_i^*\| \cdot \|y_i\| \ : \ T = \sum_{i=1}^{\infty} x_i^* \cdot y_i \right\}$$

and denoting by $N(X,Y)$ the set of all nuclear operators in $L(X,Y)$, $(N\,,\,n)$ is a normed operator ideal.

If I is an operator ideal and X is a normed space, then the linear subspace $I(X) = I(X,X) \subseteq L(X)$ is a two-sided ideal. We consider the distance

$$d(T, I(X)) = \inf \{ \|T - A\| \ : \ A \in I(X) \}, \quad T \in L(X)$$

and, for fixed I, we shall denote in short $\alpha(T) = d(T, I(X))$. We notice that $L(X) \ni T \mapsto \alpha(T)$ factorizes to a submultiplicative seminorm on the factor algebra $L(X)/I(X)$, which is a norm whenever $I(X)$ is closed.

Let $X \otimes Y$ denote the algebraic tensor product and let σ be a tensor norm on $X \otimes Y$, i.e., a cross norm such that $\|T_1 \otimes T_2\|_\sigma \leq \|T_1\| \cdot \|T_2\|$ for all $T_1 \in L(X)$ and $T_2 \in L(Y)$:

If $z = \sum\limits_{i=1}^{n} x_i \otimes y_i \in X \otimes Y$ then $(T_1 \otimes T_2)(z) = \sum\limits_{i=1}^{n} T_1 x_i \otimes T_2 y_i$ and we have

$$\sigma((T_1 \otimes T_2)(z)) \leq \sum_{i=1}^{n} \sigma(T_1 x_i \otimes T_2 y_i) \leq \|T_1\| \cdot \|T_2\| \cdot \sum_{i=1}^{n} \|x_i\| \, \|y_i\|$$

for any cross norm σ on $X \otimes Y$. The additional condition $\|T_1 \otimes T_2\|_\sigma \leq \|T_1\| \cdot \|T_2\|$ means that we actually have $\sigma((T_1 \otimes T_2)(z)) \leq \|T_1\| \cdot \|T_2\| \cdot \sigma(z)$.

All known relevant norms on $X \otimes Y$ are tensor norms (for example the ϵ and π norms of the injective and projective tensor products [3], [15].)

Let σ be a tensor norm on $X \otimes Y$. Then $X \otimes_\sigma Y$ will denote the space $X \otimes Y$ endowed with the norm σ. An operator ideal I is called *tensor product stable* with respect to σ if $I(X) \otimes I(Y) \subseteq I(X \otimes_\sigma Y)$.

Lemma 1.1. *If the operator ideal I is tensor product stable with respect to the tensor norm σ on $X \otimes Y$, then*

$$\alpha(T_1 \otimes T_2) \leq \alpha(T_1)\, \|T_2\| + \|T_1\|\, \alpha(T_2) + \alpha(T_1) \cdot \alpha(T_2)$$

for all $T_1 \in L(X)$ and $T_2 \in L(Y)$.

Proof. From the definition of $\alpha(T)$ it follows that for given $\epsilon > 0$ there are $A_i \in I(X)$, $i = 1, 2$, such that

$$\|T_i - A_i\| \leq \alpha(T_i) + \epsilon.$$

Since $A_i \in I(X)$, by the tensor stability of I we have $A_1 \otimes A_2 \in I(X \otimes_\sigma X)$ and hence we can write:

$$\begin{aligned}\alpha(T_1 \otimes T_2) &\leq \|T_1 \otimes T_2 - A_1 \otimes A_2\| = \|(T_1 - A_1) \otimes T_2 + A_1 \otimes (T_2 - A_2)\| \\ &\leq \|(T_1 - A_1) \otimes T_2\| + \|A_1 \otimes (T_2 - A_2)\| \\ &\leq (\alpha(T_1) + \epsilon)\,\|T_2\| + \|A_1\| \cdot (\alpha(T_2) + \epsilon) \\ &\leq (\alpha(T_1) + \epsilon)\,\|T_2\| + \|A_1 - T_1 + T_1\|\,(\alpha(T_2) + \epsilon) \\ &\leq (\alpha(T_1) + \epsilon)\,\|T_2\| + (\|T_1\| + \alpha(T_1) + \epsilon)(\alpha(T_2) + \epsilon)\,.\end{aligned}$$

Since ϵ is arbitrary, we obtain the inequality.

Definition 1.2. Given an operator ideal I, $T \in L(X)$ is called asymptotically of quasi-type I if

$$\lim_{n\to\infty} \alpha(T^n)^{\frac{1}{n}} = 0.$$

In this case the canonical image of T in the factor algebra $L(X)/\overline{I(X)}$ is quasi-nilpotent whenever X is complete (hence the spectral radius formula holds in $L(X)/\overline{I(X)}$).

For example, if $I(X) = K(X)$, the closed ideal of the compact operators, we obtain the asymptotically quasi-compact operators of A.F. Ruston ([8]).

From Lemma 1.1 it results in a simple way that:

Proposition 1.3. *If the operator ideal I is tensor product stable with respect to the tensor norm σ on $X \otimes Y$ and the operators $T_1 \in L(X)$, $T_2 \in L(Y)$ are asymptotically of quasi-type I, then also the tensor product operator $T_1 \otimes T_2 \in L(X \otimes_\sigma Y)$ is asymptotically of quasi-type I.*

Proof. By Lemma 1.1 we have for every $n \geq 1$:

$$\begin{aligned}0 \leq \alpha((T_1 \otimes T_2)^n)^{\frac{1}{n}} &= \alpha(T_1^n \otimes T_2^n)^{\frac{1}{n}} \\ &\leq (\alpha(T_1^n)\,\|T_2^n\| + \alpha(T_2^n)\,\|T_1^n\| + \alpha(T_1^n)\,\alpha(T_2^n))^{\frac{1}{n}} \\ &\leq \alpha(T_1^n)^{\frac{1}{n}} \cdot \|T_2\| + \alpha(T_2^n)^{\frac{1}{n}} \cdot \|T_1\| + (\alpha(T_1^n) \cdot \alpha(T_2^n))^{\frac{1}{n}}\,.\end{aligned}$$

Since $\lim_n \alpha(T_i^n)^{\frac{1}{n}} = 0$, it follows that $\lim_n \alpha((T_1 \otimes T_2)^n)^{\frac{1}{n}} = 0$, i.e. $T_1 \otimes T_2$ is asymptotically of quasi-type I.

2. Applications

First we recall the following results:

(a) If $T_1 \in K(X)$ and $T_2 \in K(Y)$ then $T_1 \otimes T_2 \in K(X \otimes_\epsilon Y)$, where ϵ is the injective tensor norm ([6], [12]).

(**b**) If $T_1 \in A(X)$ and $T_2 \in A(Y)$ then $T_1 \otimes T_2 \in A(X \otimes_\sigma Y)$ for any tensor norm σ on $X \otimes Y$ ([6], [11], [12], [15]).

(**c**) If $T_1 \in N(X)$ and $T_2 \in N(Y)$ then $T_1 \otimes T_2 \in N(X \otimes_\epsilon Y)$ ([6], [15]).

As a corollary of Proposition 1.2 we get:

Proposition 2.1. *If $T_1 \in L(X)$ and $T_2 \in L(Y)$ are asymptotically quasi-compact (quasi-nuclear) then the same holds also for $T_1 \otimes T_2 \in L(X \otimes_\epsilon Y)$. Further, if T_1, T_2 are asymptotically quasi-approximable then $T_1 \otimes T_2 \in L(X \otimes_\sigma Y)$ is asymptotically quasi-approximable for any tensor norms σ on $X \otimes Y$.*

Now we denote, for any normed spaces X and Y,

$$L_{c,p}(X,Y) = \left\{ T \in L(X,Y) \,:\, \|T\|_{c,p} = \Big(\sum_n c_n\, a_n(T)^p \Big)^{\frac{1}{p}} < +\infty \right\},$$

where c is a sequence $1 = c_1 \geq c_2 \geq \cdots \geq 0$, $1 \leq p < \infty$ and

$$a_n(T) = \inf \{ \|T - A\| \,:\, rank\, A < n \}.$$

It is known that $L_{c,p}$, endowed with the quasi-norms $T \mapsto \|T\|_{c,p}$, is a "symmetric" quasi-normed operator ideal ([13], [14], [15]).

In the sequel we assume that c is a sequence (c_n) such that

$$c_{n^2} \leq \frac{k}{n} c_n\,, \ \forall n \in \mathbf{N}, \tag{2.1}$$

where k is a constant.

Lemma 2.2. *If the sequence c satisfies* (2.1) *then the operator ideal $L_{c,p}$ is tensor product stable with respect to any tensor norm.*

Proof. Let be $T_1 \in L(X)$, $T_2 \in L(Y)$ and σ a tensor norm on $X \otimes Y$. First we recall the inequality

$$a_{n^2}(T_1 \otimes T_2) \leq 2 \Big(a_n(T_1)\, \|T_2\| + \|T_1\|\, a_n(T_2) \Big)$$

([5], [11], [12], [13], [14]) which is obtained in a similar way as the inequality from Lemma 1.1, because if $A_1 \in L(X)$, $A_2 \in L(Y)$ are such that $rank\, A_i < n$ then $rank\, A_1 \otimes A_2 < n^2$.

Now, since $\|T\| \leq \|T\|_{c,p}$, we obtain: inequality

$$\begin{aligned} \|T_1 \otimes T_2\|_{c,p} &= \Big(\sum_1^\infty c_n\, a_n(T_1 \otimes_\sigma T_2)^p \Big)^{\frac{1}{p}} = \Big(\sum_{n=1}^\infty \sum_{i=n^2}^{(n+1)^2-1} c_i\, a_i(T_1 \otimes T_2)^p \Big)^{\frac{1}{p}} \\ &\leq \Big(\sum_{n=1}^\infty (2n+1)\, c_{n^2}\, a_{n^2}(T_1 \otimes T_2)^p \Big)^{\frac{1}{p}} \\ &\leq \Big(6\,k \sum_{n=1}^\infty c_n \Big(a_n(T_1)\, \|T_2\| + a_n(T_2)\, \|T_1\| \Big)^p \Big)^{\frac{1}{p}} \\ &\leq c(p) \Big(\|T_1\|_{c,p} \cdot \|T_2\| + \|T_2\|_{c,p} \cdot \|T_1\| \Big) \leq 2\, c(p)\, \|T_1\|_{c,p} \cdot \|T_2\|_{c,p}\,. \end{aligned}$$

Remarks.

1. For $c_n = \frac{1}{n}$ we obtain the limit class of Lorentz ideals, which is tensor product stable ([13]). For the particular case of Hilbert spaces, this result is obtained in [2] in a different way.
2. In [1] the quasi-norms $\|\cdot\|_{p,c}$ (firstly considered in [14]) are used in the case of operator ideals on Hilbert spaces.
3. We can consider the general case of the quasinorms

$$\|T\|_{c,\phi,p} = \Phi\Big(\{c_n\, a_n(T)^p\}\Big)^{\frac{1}{p}}\,,$$

where ϕ is a symmetric norming function ([9], [10], [15]). If ϕ is the maximal function then we have the above quasinorms.

As a corollary of Lemma 2.2 and Proposition 1.2 we obtain:

Proposition 2.3. *If the sequence c satisfies* (2.1) *and $T_1 \in L(X)$, $T_2 \in L(Y)$ are asymptotically of quasi-type $L_{c,p}$ then $T_1 \otimes T_2 \in L(X \otimes_\sigma Y$ is asymptotically of quasitype $L_{c,p}$.*

Acknowledgement

The author thanks the referee for his remarks and suggestions.

References

[1] J.-T. Chan, C.W. Li, C.N. Tu, *A Class of Unitarily Invariant Norms on $B(H)$.* Proc. Amer. Soc. **129** (2001), 1065–1076.

[2] F. Cobos, L.M. Fernandez-Cabrera, *On the Tensor Stability of Some Operator Ideals.* Bull. Polish Acad. Sci. Math. **37** (1989), 459–465.

[3] J. Diestel, J. Uhl Jr., *Vector Measures.* Math. Surveys A.M.S., vol. 15, 1978.

[4] G. Khöte, *Topological Vector Spaces.* Springer Verlag, New York (1969), (1979).

[5] G. Khöte, *Letter to N. Ţiţa.* 20.07.1973

[6] J.R. Holub, *Tensor Product Mappings. II.* Proc. Amer. Math. Soc. **42** (1974), 437–441.

[7] A. Pietsch: *Operator Ideals.* North Holland Publ.Co., Amsterdam, 1980.

[8] A. Ruston, *Operators with Fredholm Theory.* J. London Math. Soc. **29** (1954), 318–324.

[9] R. Schatten, *A Theory of Cross-Spaces,* Princeton Univ., 1950.

[10] R. Schatten, *Norm Ideals of Completely Continuous Operators.* Springer Verlag, 1960.

[11] N. Tiţa, *Sur les espaces ϕ-nucléaires.* Univ. din Braşov, 1972.

[12] N. Tiţa, *Remarks on Some Classes of Operators (Romanian).* Noutăţi în ştiinţele naturii **18** (1976), 59–62.

[13] N. Tiţa, *On a Class of $l_{\phi,\varphi}$ Operators*. Collect. Math. **32** (1981), 275–279.

[14] N. Tiţa, *On the Approximation Numbers of Tensor Product Operator*. Ann. St. Univ. "Al. I. Cuza" Iaşi **40** (1994), 329–331.

[15] N. Tiţa, *Ideale de operatori generate de s-numere*. Ed. Univ. "Transilvania", Braşov, 1998.

Nicolae Tiţa
University Transilvania of Braşov
Faculty of Mathematics and Computer Science
Braşov
România
e-mail: n.tita@info.unitbv.ro

Operator Theory:
Advances and Applications, Vol. 153, 271–286

The Gamma Element for Groups which Admit a Uniform Embedding into Hilbert Space

Jean-Louis Tu

Abstract. In [17], it was shown that for every group Γ with a left-invariant metric d such that (Γ, d) has bounded geometry, and which admits a uniform embedding into Hilbert space, the Baum–Connes assembly map with coefficients is split injective. In this paper, we strengthen this result by showing that Γ has a gamma element. 46L80; 19K56, 19L47

Mathematics Subject Classification (2000). 46L80; 19K56, 19L47.

Keywords. Group with bounded geometry, Assembly map, Gamma element.

1. Introduction

Let X be a metric space. Recall that a map f from X to an infinite-dimensional Hilbert space H is said to be a uniform embedding if there exist two non-decreasing functions ρ_1 and ρ_2 on $[0, +\infty)$ such that

(1) $\lim_{r\to+\infty} \rho_i(r) = +\infty$;
(2) $\rho_1(d(x,y)) \le \|f(x) - f(y)\| \le \rho_2(d(x,y))$ for all $x, y \in X$.

Let Γ be a countable group with a left-invariant metric d such that (Γ, d) has bounded geometry (for instance, a finitely generated group with word metric). Suppose that Γ admits a uniform embedding into Hilbert space (that property is independent of the choice of the bounded geometry invariant metric). G. Yu showed that the coarse assembly map for Γ is an isomorphism, which implies that the Novikov conjecture is true for Γ if in addition its classifying space $B\Gamma$ is a finite CW-complex ([21]). Yu introduced a property on metric spaces that he called property A, which is stronger than uniform embeddability into Hilbert space. Higson and Roe ([8]) observed that a discrete group satisfies property A if and only if it acts amenably on a compact space; recall also that a discrete group is exact if and only if it satisfies property A (see, e.g., [1, Theorem 4.6]). Higson then obtained a partial generalization of Yu's result ([5]): if Γ is a discrete group acting amenably on a compact space, then the Baum–Connes assembly map for Γ with coefficients

$$\mu_r\colon K_*^{\mathrm{top}}(\Gamma; B) \to K_*(B \rtimes_r \Gamma)$$

is split injective for any separable Γ-C^*-algebra B. (This is known to imply the Novikov conjecture for Γ without any finiteness assumption on $B\Gamma$.)

Using Higson's descent technique, Skandalis, Tu and Yu showed in [17] that in fact for every discrete group Γ which admits a uniform embedding into Hilbert space, the Baum–Connes assembly map with coefficients is split injective.

The goal of this paper is to prove that Γ in fact has a γ element, in the sense we will explain below.

Let now G be a locally compact group and A a C^*-algebra. The group G is said to act properly on A if there exists a (locally compact Hausdorff) space Z such that A is a G-$C_0(Z)$-algebra in the sense of Kasparov ([12]), that is, a $Z \rtimes G$-algebra in the sense of Le Gall ([16]). Let $\underline{E}G$ the classifying space for proper actions of G ([2]).

Remarks 1.1.

(a) From [4], one may use the classical notion of proper action instead of the stronger one introduced in [2].
(b) By [18, Proposition 6.15], any locally compact group admits a locally compact classifying space for proper actions.

By the universal property of $\underline{E}G$, the group G acts properly on A if and only if A is a $\underline{E}G \rtimes G$-algebra.

The group G is said to have a γ element if there exist a C^*-algebra on which G acts properly, elements

$$\eta \in KK_G(\mathbb{C}, A) \quad \text{and} \quad d \in KK_G(A, \mathbb{C}),$$

called dual-Dirac and Dirac elements respectively, such that $\gamma{=}\eta\otimes_A d{\in}KK_G(\mathbb{C},\mathbb{C})$ satisfies $p^*\gamma = 1 \in KK_{\underline{E}G\rtimes G}(C_0(\underline{E}G), C_0(\underline{E}G))$, where $p\colon \underline{E}G \rtimes G \to G$ is the groupoid homomorphism defined by $p(z, g) = g$ ([18]). With these specifications, the element γ is unique.

The gamma element was first constructed by Kasparov for groups acting properly on a simply connected manifold with non-positive sectional curvature X (in that case, $A = C_\tau(X)$) ([12]). This applies in particular for semisimple Lie groups G acting on the symmetric space $X = G/K$. Then, using an inductive process, Kasparov constructed the gamma element for any connected Lie group ([12]).

The idea of using actions on spaces with geometric properties analogue to non-positive curvature was used by Julg and Valette ([11]) who constructed the gamma element for groups acting properly on a tree, and by Kasparov and Skandalis ([13]) who constructed it for groups acting on Euclidean buildings.

It can be shown that if G has a γ element, then its Baum–Connes assembly map with coefficients is split injective ([20]). More precisely, let $\eta_B = j_{G,r}(\sigma_B(\eta))$, $d_B = j_{G,r}(\sigma_B(d))$, $\gamma_B = \eta_B \otimes_{(A\otimes B)\rtimes_r G} d_B \in KK(B\rtimes_r G, B\rtimes_r G)$, $e_B = d_B \otimes_{B\rtimes_r G}$

$\eta_B \in KK((A \otimes B) \rtimes_r G, (A \otimes B) \rtimes_r G)$. Then e_B and γ_B are projections, the image of μ_r is $K_*(B \rtimes_r G)\gamma_B$ and μ_r is the composition of

$$K_*^{\mathrm{top}}(G;B) \xrightarrow{\simeq} K_*((A \otimes B) \rtimes_r G)e_B \xrightarrow{\cdot \otimes \eta_B} K_*(B \rtimes_r G)\gamma_B \hookrightarrow K_*(B \rtimes_r G).$$

Throughout, we will assume that the reader is familiar with equivariant KK-theory with respect to groupoids ([16]), which is a generalization of Kasparov's theory ([12]). We refer to [2] for an introduction to the Baum–Connes conjecture and to [20] for general notations.

2. The Dirac element for discrete groups

The objective of this section is to give a proof of Theorem 2.1 below; ideas are from Kasparov and Skandalis ([14]).

Theorem 2.1. *Let Γ be a discrete group and let $\underline{E}\Gamma$ its classifying space for proper actions ([2]). There exist a C^*-algebra A on which Γ acts properly, elements*

$$\theta \in KK_{\underline{E}\Gamma \rtimes \Gamma}(C_0(\underline{E}\Gamma), C_0(\underline{E}\Gamma) \otimes A) \quad \text{and} \quad D \in KK_\Gamma(A, \mathbb{C})$$

such that $\theta \otimes_A D = 1 \in KK_{\underline{E}\Gamma \rtimes \Gamma}(C_0(\underline{E}\Gamma), C_0(\underline{E}\Gamma))$.

In the theorem above, the action of the group Γ on $C_0(\underline{E}\Gamma) \otimes A$ is the diagonal one, and the structure of $C_0(\underline{E}\Gamma)$-algebra on $C_0(\underline{E}\Gamma) \otimes A$ is induced from the one on the first factor.

One consequence of Theorem 2.1 is that the topological K-theory of Γ may be seen as a K-theory group, and the assembly map is the product by a KK-element. More precisely,

Theorem 2.2. *With the same notations as in Theorem 2.1, let $E = \sigma_{\underline{E}\Gamma, A}(\theta) \otimes_A D \in KK_\Gamma(A, A)$. Then $E^2 = E$, and for every Γ-algebra B, $K_*^{\mathrm{top}}(\Gamma; B)$ is isomorphic to $K_*((A \otimes B) \rtimes \Gamma)E'$ where $E' = j_\Gamma(\sigma_B(E))$, and the assembly map $K_*((A \otimes B) \rtimes \Gamma)E' \to K_*(B \rtimes_r \Gamma)$ is the product by $j_{\Gamma,r}(\sigma_B(D)) \in KK((A \otimes B) \rtimes_r \Gamma, B \rtimes_r \Gamma)$.*

Proof. Let us prove successively

(a) $E \otimes_A D = D$;

(b) $\theta \otimes_A \sigma_{\underline{E}\Gamma, A}(\theta) = \theta \otimes_{C_0(\underline{E}\Gamma)} \theta$;

(c) $\theta \otimes_A E = \theta$;

(d) $E^2 = E$.

Let us show (a):

$$\begin{aligned} E \otimes_A D &= \sigma_{\underline{E}\Gamma, A}(\theta) \otimes_{A \otimes A} (D \otimes_{\mathbb{C}} D) \\ &= \sigma_{\underline{E}\Gamma, A}(\theta \otimes_A D) \otimes_A D \\ &= 1_A \otimes_A D = D. \end{aligned}$$

Let us show (b): let $pr_i \colon \underline{E}\Gamma \times \underline{E}\Gamma \to \underline{E}\Gamma$ be the two projections. By the universal property of $\underline{E}\Gamma$, they are equivariantly homotopic. Therefore, denoting by pr_i^* $(i = 1, 2)$ the induced morphisms $KK_{\underline{E}\Gamma \rtimes \Gamma}(C_0(\underline{E}\Gamma), C_0(\underline{E}\Gamma) \otimes A) \to$

$KK_{(\underline{E}\Gamma\times\underline{E}\Gamma)\rtimes\Gamma}(C_0(\underline{E}\Gamma\times\underline{E}\Gamma), C_0(\underline{E}\Gamma\times\underline{E}\Gamma)\otimes A)$,

$$\begin{aligned}\theta\otimes_A\sigma_{\underline{E}\Gamma,A}(\theta) &= \theta\otimes_{C_0(\underline{E}\Gamma)\otimes A}\sigma_{\underline{E}\Gamma\times\underline{E}\Gamma,C_0(\underline{E}\Gamma)\otimes A}(pr_2^*(\theta))\\ &= \theta\otimes_{C_0(\underline{E}\Gamma)\otimes A}\sigma_{\underline{E}\Gamma\times\underline{E}\Gamma,C_0(\underline{E}\Gamma)\otimes A}(pr_1^*(\theta))\\ &= \theta\otimes_{C_0(\underline{E}\Gamma)\otimes A}\sigma_A(\theta)\\ &= \theta\otimes_{C_0(\underline{E}\Gamma)}\theta.\end{aligned}$$

Let us show (c): by (b), $\theta\otimes_A E = (\theta\otimes_{C_0(\underline{E}\Gamma)}\theta)\otimes_A D = \theta\otimes_{C_0(\underline{E}\Gamma)}(\theta\otimes_A D) = \theta\otimes_{C_0(\underline{E}\Gamma)} 1 = \theta$.

Let us show (d):

$$\begin{aligned}E^2 &= \sigma_{\underline{E}\Gamma,A}(\theta)\otimes_{A\otimes A}(\sigma_{\underline{E}\Gamma,A}(\theta)\otimes_{\mathbb{C}} D)\otimes_A D\\ &= \sigma_{\underline{E}\Gamma,A}(\theta\otimes_A\sigma_{\underline{E}\Gamma,A}(\theta))\otimes_{A\otimes A}(D\otimes_{\mathbb{C}} D).\end{aligned}$$

By (b),

$$\begin{aligned}E^2 &= \sigma_{\underline{E}\Gamma,A}(\theta\otimes_{C_0(\underline{E}\Gamma)}\theta)\otimes_{A\otimes A}(D\otimes_{\mathbb{C}} D)\\ &= \sigma_{\underline{E}\Gamma,A}(\theta)\otimes_A(\sigma_{\underline{E}\Gamma,A}(\theta\otimes_A D))\otimes_A D\\ &= \sigma_{\underline{E}\Gamma,A}(\theta)\otimes_A 1_A\otimes_A D = E.\end{aligned}$$

The result follows from [18, Théorème 5.19]. □

2.1. Idea of the proof

We note first that Theorem 2.1 is true if Γ has a proper action on a finite-dimensional simplicial complex X, where the space $\underline{E}\Gamma$ is replaced by X. Let us sketch the argument.

Suppose that G is a locally compact group and that M is a manifold on which G acts properly. Then Kasparov ([12, Definitions 4.2 and 4.4]) constructed elements $\theta\in KK_{M\rtimes G}(C_0(M), C_0(M)\otimes C_0(T^*M))$ and $D\in KK_G(C_0(T^*M),\mathbb{C})$ such that $\theta\otimes_{C_0(T^*M)} D = 1\in KK_{M\rtimes G}(C_0(M), C_0(M))$.

Kasparov and Skandalis generalized the construction to non-Hausdorff manifolds ([13]). Let V be a proper non-Hausdorff G-manifold and let $(V_i)_{i\in I}$ be a G-atlas for V ([13, p. 317]). The C^*-algebra of V is defined as the completion of the algebra $C^*(V) = \oplus_{i,j} C_c(U_i\cap U_j)$ acting on $\oplus_{x\in V}\ell^2(\{i|\ x\in U_i\})$. Recall that the product in $C^*(V)$ is defined by $(fg)_{i,j} = \sum_k f_{i,k}g_{k,j}$, if $f = \oplus_{i,j}f_{i,j}$ and $g = \oplus_{i,j}g_{i,j}$.

But if X is a (finite-dimensional) typed simplicial complex on which G acts properly, and if the action of G permutes (or preserves) the type, then X is G-equivariantly homotopically equivalent to a non-Hausdorff manifold M ([13, Lemma 4.8]). Thus [13, Proposition 6.3], one gets elements $D\in KK_G(A,\mathbb{C})$ and $\theta\in KK_{X\rtimes G}(C_0(X), C_0(X)\otimes A)$ such that $\theta\otimes_A D = 1$, where $A = C^*(T^*M)$. (A is, in some sense, a "Poincaré dual" of $C_0(X)$.)

It follows from the above that Theorem 2.1 is true if Γ is a discrete group such that $\underline{E}\Gamma$ is a finite-dimensional simplicial complex. This is not the case for a general discrete group Γ, however the space $\underline{E}\Gamma$ can be approximated by a sequence of finite-dimensional complexes as follows: let $P_n(\Gamma)$ be the Rips' complex whose simplices are the subsets F of Γ of diameter $\leq n$ (for a given left-invariant metric with bounded geometry on Γ). Then Γ acts properly on $P_n(\Gamma)$, and $\underline{E}\Gamma$ is the telescope $\cup_{n\in\mathbb{N}}[n, n+1] \times P_n(\Gamma)$. Let A_n be the C^*-algebra associated as above to $P_n(\Gamma)$ (A_n is a "Poincaré dual" of $C_0(P_n(\Gamma))$). One would like to define the C^*-algebra A in Theorem 2.1 as the inductive limit of the A_n's. However, as there is no obvious way of defining an inductive system we will define directly a C^*-algebra A combining the construction of [13] with the algebra of an infinite-dimensional Hilbert space defined by Higson and Kasparov ([7]).

2.2. Construction of A

The construction of A relies on the C^*-algebra $\mathcal{A}(H)$ that Higson and Kasparov associate to a real affine Euclidean space H (see [6, Definition 4.1], [7, Definition 4.6], [9]). Let us recall its definition:

Let $\mathcal{S}$ be the C^*-algebra $C_0(\mathbb{R})$, graded according to even and odd functions. For every real, finite-dimensional affine space V, denote by V_0 its underlying vector space and by $\mathcal{C}(V)$ the algebra $C_0(V \times V_0, \mathcal{L}(\Lambda^*(V_0) \otimes \mathbb{C}))$.

Suppose that $V' \subset V$ is an inclusion of finite-dimensional affine subspaces of H, then a map

$$\mathcal{S}\hat{\otimes}\mathcal{C}(V') \to \mathcal{S}\hat{\otimes}\mathcal{C}(V)$$

is constructed as follows: let W be the orthogonal complement of V' in V, and define the unbounded multiplier B_W of $\mathcal{C}(W)$ by $B_W(w_1, w_2) = i\bar{c}(w_1) + c(w_2)$ where $c(w)$ is the Clifford multiplication by w. Let X be the identity function from $\mathbb{R}$ to $\mathbb{R}$, and consider X as an unbounded multiplier of $\mathcal{S}$. Then the maps

$$\mathcal{S}\hat{\otimes}\mathcal{C}(V') \to \mathcal{S}\hat{\otimes}\mathcal{C}(W)\hat{\otimes}\mathcal{C}(V') \simeq \mathcal{S}\hat{\otimes}\mathcal{C}(V)$$
$$f\hat{\otimes}h \mapsto f(X\hat{\otimes}1 + 1\hat{\otimes}B_W)h$$

form an inductive system of graded C^*-algebras. The inductive limit is denoted by $\mathcal{A}(H)$.

Denote by $\mathcal{F}_\Gamma$ the set of nonempty finite subsets of Γ. For every $\sigma \in \mathcal{F}_\Gamma$, denote by H_σ the affine subspace of $\ell^2_{\mathbb{R}}(\Gamma)$ spanned by the elements of σ. For every $x \in \ell^2_{\mathbb{R}}(\Gamma)$, let $q_\sigma(x)$ be the point on the simplex $|\sigma|$ spanned by σ such that $d(x, q_\sigma(x))$ is minimal.

For all σ, $\tau \in \mathcal{F}_\Gamma$, let $X_{\sigma,\tau} = \{x \in \ell^2_{\mathbb{R}}(\Gamma) |\ q_\sigma(x) = q_\tau(x)\}$. The following lemma is a trivial consequence of the definitions.

Lemma 2.3. *For all α, σ, $\tau \in \mathcal{F}_\Gamma$, we have $X_{\sigma,\alpha} \cap X_{\alpha,\tau} \subset X_{\sigma,\tau}$.*

Now, let $Y_{\sigma,\tau}$ be the interior of $X_{\sigma,\tau}$ with respect to the weak topology (or equivalently to the norm-topology) and $\Omega_{\sigma,\tau} = Y_{\sigma,\tau} \cap H_{\sigma\cup\tau}$. Note that $Y_{\sigma,\tau} \simeq \Omega_{\sigma,\tau} \times (\ell^2_{\mathbb{R}}(\Gamma) \ominus H_{\sigma\cup\tau})$.

Lemma 2.4. *For all σ, $\tau \in \mathcal{F}_\Gamma$, $\Omega_{\sigma,\tau}$ is the interior of*

$$\{x \in H_{\sigma\cup\tau} |\ q_{\sigma\cup\tau}(x) \in |\sigma \cap \tau|\}.$$

Proof. It suffices to show that for all $x \in H$, $q_\sigma(x) = q_\tau(x) \iff q_{\sigma\cup\tau}(x) \in |\sigma\cap\tau|$.

Let us show "$\Longrightarrow$": let $y = q_\sigma(x) = q_\tau(x) \in |\sigma \cap \tau|$. Then every element of $|\sigma|$ and every element of $|\tau|$ lies in the half-space $\{z \in H|\ \langle y-x, y-z\rangle \leq 0\}$. This is also true for every z in the convex hull of $|\sigma| \cup |\tau|$ which is $|\sigma \cup \tau|$. It follows that $y = q_{\sigma\cup\tau}(x)$.

To show the converse, note that $q_{\sigma\cup\tau}(x) \in |\sigma \cap \tau|$ implies

$$q_{\sigma\cup\tau}(x) = q_\sigma(x) = q_\tau(x) = q_{\sigma\cap\tau}(x).$$

□

For every $\sigma \in \mathcal{F}_\Gamma$, choose a labeling $\sigma = \{g_1, \dots, g_n\}$ and define

$$e_\sigma = e_{g_1} \wedge \cdots \wedge e_{g_n} \in \Lambda(\ell^2(\Gamma)).$$

This vector is well defined, up to a sign. Let $\mathcal{H}_\Gamma = e_\emptyset^\perp \subset \Lambda(\ell^2(\Gamma))$. We define the algebra A as follows: let $A_1 = \mathcal{A}(\ell^2_{\mathbb{R}}(\Gamma))\hat{\otimes}\mathcal{K}(\mathcal{H}_\Gamma)$ and

$$A = \{T \in A_1|\ T_{\sigma,\tau}(x) = 0\ \forall x \notin \Omega_{\sigma,\tau}\},$$

where $T_{\sigma,\tau} = \langle e_\sigma, Te_\tau\rangle \in \mathcal{A}(\ell^2_{\mathbb{R}}(\Gamma))$. The algebra $C_0(H_{\sigma\cup\tau})$ lies in the center of the multiplier algebra of $\mathcal{A}(\ell^2_{\mathbb{R}}(\Gamma))$, hence the notation $T_{\sigma,\tau}(x)$. Note the similarity with the algebra of a finite-dimensional simplicial complex defined by Kasparov and Skandalis ([13, Definition 1.2]). The algebra A can be thought of as the C^*-algebra associated to the infinite-dimensional simplicial complex whose simplices consist of the finite, nonempty subsets of Γ. The fact that A is an algebra results from Lemma 2.3.

(A more canonical way to define A is as follows: $A = \{T \in A_1|\ \langle e', Te\rangle(x) = 0\ \forall x \neq \Omega_{\sigma,\tau}, e' \in L_\tau, e \in L_\sigma\}$, where $L_\sigma = \mathbb{C}e_\sigma$.)

Let M be the set of positive measures μ on Γ such that $1/2 < |\mu| \leq 1$, endowed with the topology of the dual of $C_0(\Gamma)$. Kasparov and Skandalis showed that the space M is a model for the classifying space for proper actions of Γ (see, e.g., [18], Proposition 6.15 for a proof). We show that A is endowed with a structure of $M \rtimes \Gamma$-algebra, which will imply that Γ acts properly on A. Let $f \in C_0(M)$. To f we associate a multiplier m_f of A as follows:

$$m_f = \oplus_{\sigma\in\mathcal{F}_\Gamma}(f \circ q_\sigma \otimes \mathrm{Id}_{L_\sigma}) \in \mathcal{L}(\oplus_{\sigma\in\mathcal{F}_\Gamma}\mathcal{A}(\ell^2_{\mathbb{R}}(\Gamma)) \otimes L_\sigma) = M(A_1),$$

i.e., $(fT)_{\sigma,\tau} = (f \circ q_\sigma) \cdot T_{\sigma,\tau}$. Here, q_σ is a continuous function from H_σ to $|\sigma|$, and we identify $|\sigma|$ with the subspace of M consisting of measures supported in σ with total weight 1; $f \circ q_\sigma$ is thus an element of $C_b(H_\sigma)$, and acts naturally on $\mathcal{A}(\ell^2_{\mathbb{R}}(\Gamma))$. Note that, from the definition of A, one also has $(fT)_{\sigma,\tau} = (f\circ q_\tau)\cdot T_{\sigma,\tau}$, hence the multiplier defined by f is central. Since the action of $C_0(M)$ on A is evidently compatible with the action of Γ, we obtain a structure of $M \rtimes \Gamma$-algebra as claimed. In other words, Γ acts properly on A.

The following lemma (compare with [13], Lemma 1.4) will be needed later.

Lemma 2.5. *$M(A) = \{T \in M(A_1)|\ T_{\sigma,\tau}(x) = 0\ \forall x \notin \Omega_{\sigma,\tau}\}$. In particular, $A = M(A) \cap A_1$.*

Proof. Since the representation of A in A_1 is nondegenerate and faithful, $M(A)$ is contained in $M(A_1)$. Let $A_2 = \{T \in M(A_1) | \, T_{\sigma,\tau}(x) = 0 \; \forall x \notin \Omega_{\sigma,\tau}\}$. It is easy to check, using Lemma 2.3 that for every $T \in A_2$ and $a \in A$, Ta belongs to A, hence $A_2 \subset M(A)$. Conversely, suppose $T \in M(A)$ and let $x \notin \Omega_{\sigma,\tau}$. Choose a strictly positive element a of $\mathcal{A}(\ell^2_{\mathbb{R}}(\Gamma))$. Since $a \otimes \mathrm{Id}_{L_\tau} \in A$, we have $(T(a \otimes \mathrm{Id}_{L_\tau}))_{\sigma,\tau}(x) = 0$. On the other hand, $(T(a \otimes \mathrm{Id}_{L_\tau}))_{\sigma,\tau}(x) = T_{\sigma,\tau}(x)a(x)$, hence $T_{\sigma,\tau}(x) = 0$. □

2.3. A model for $\underline{E}\Gamma$

Recall that Γ may be endowed with a left-invariant bounded geometry metric d, i.e., $d(g,h) = \ell(g^{-1}h)$ where ℓ is a proper length function. For instance, choose a countable set of generators $\{\gamma_1, \gamma_2, \ldots\}$ and let

$$\ell(\gamma) = \inf\Big\{ \sum_{k=1}^{r} i_k \mid \gamma = \gamma_{i_1}^{\pm 1} \cdots \gamma_{i_r}^{\pm 1} \Big\}.$$

Let $\mathrm{Prob}(\Gamma)$ be the space of probability measures on Γ, and let $Z = \{(\mu, t) \in \mathrm{Prob}(\Gamma) \times \mathbb{R}_+ \mid \mathrm{diam}\,(\mathrm{supp}(\mu)) \leq t\}$. Then it is easy to show, using [2], Proposition1.8, that Z is a model for $\underline{E}\Gamma$.

2.4. A few lemmas

Let Q be the orthogonal projection of $\Lambda(\ell^2(\Gamma))$ onto $\mathcal{H}_\Gamma$. For every $(\mu, t) \in Z$, denote by ξ_μ the unit vector $\sqrt{\mu} \in \ell^2_{\mathbb{R}}(\Gamma)$. Let $c(\xi_\mu) \in \mathcal{L}(\Lambda(\ell^2(\Gamma)))$ be the left Clifford multiplication by ξ_μ $(c(\xi) = e(\xi) + e(\xi)^*$ where $e(\xi)v = \xi \wedge v)$, and

$$F_\mu \quad = Qc(\xi_\mu)Q \in \mathcal{L}(\mathcal{H}_\Gamma). \tag{2.1}$$

For all $g \in \Gamma$, define

$$D(g, \mu) = \sum_{h \in \Gamma} \mu(h) d(g, h).$$

The function D is continuous in the pair (g, μ). Since $\mathrm{diam}\,(\mathrm{supp}(\mu)) \leq t$, the function D satisfies $d(g, \mathrm{supp}(\mu)) \leq D(g, \mu) \leq d(g, \mathrm{supp}(\mu)) + t$. Let $c_{g,\mu,t} = \inf(D(g, \mu) - t, 1)_+$. Then

$$0 \leq c_{g,\mu,t} \leq 1 \tag{2.2}$$
$$d(g, \mathrm{supp}(\mu)) \geq t + 1 \implies c_{g,\mu,t} = 1 \tag{2.3}$$
$$g \in \mathrm{supp}(\mu) \implies c_{g,\mu,t} = 0 \tag{2.4}$$

For all $\sigma \in \mathcal{F}_\Gamma$, let

$$c_{\sigma,\mu,t} = \sup\{c_{g,\mu,t} \mid g \in \sigma\} \tag{2.5}$$
$$c'_{\sigma,\mu,t,g} = 6 \inf(2c_{\sigma,\mu,t}, 1) c_{g,\mu,t} \tag{2.6}$$
$$\eta_{\sigma,\mu,t} = \Big(1 - \sum_{g \in \sigma} c'_{\sigma,\mu,t,g}\Big)\mu + \sum_{g \in \sigma} c'_{\sigma,\mu,t,g}\, g \tag{2.7}$$

The idea is that the vector $\eta_{\sigma,\mu,t} \in H_{\sigma \cup \mathrm{supp}(\mu)}$ is equal to μ if $\sigma \subset \mathrm{supp}(\mu)$, and is far away enough from $\mathrm{supp}(\mu)$ if there is $g \in \sigma$ such that $d(g, \mathrm{supp}(\mu))$ is large.

Let f_0 be the element of $\mathcal{S} = C_0(\mathbb{R})$ be defined by $f_0(x) = \sup(-1, \inf(x, 1))$. For every $\eta \in \ell^2_{\mathbb{R}}(\Gamma)$, let $i_\eta \colon \mathcal{S} \xrightarrow{\simeq} \mathcal{A}(\{\eta\}) \to \mathcal{A}(\ell^2_{\mathbb{R}}(\Gamma))$ where the second arrow is induced by the inclusion of affine spaces $\{\eta\} \to \ell^2_{\mathbb{R}}(\Gamma)$. Define $\beta_{\mu,t} \in M(A_1) \simeq \mathcal{L}(\oplus_{\sigma\in\mathcal{F}_\Gamma} \mathcal{A}(\ell^2_{\mathbb{R}}(\Gamma)) \otimes L_\sigma)$ by

$$\beta_{\mu,t} = \oplus_{\sigma\in\mathcal{F}_\Gamma} i_{\eta_{\sigma,\mu,t}}(f_0) \otimes \mathrm{Id}_{L_\sigma}. \tag{2.8}$$

Let

$$\chi_{\mu,t}(\sigma) = (2c_{\sigma,\mu,t} - 1)_+. \tag{2.9}$$

Lemma 2.6. *For every $(\mu, t) \in Z$, $\{\sigma \in \mathcal{F}_\Gamma | \ \chi_{\mu,t}(\sigma) < 1\}$ is finite.*

Proof. If $c_{\sigma,\mu,t} < 1$, then $\sigma \subset B(\mathrm{supp}(\mu), t + 1)$. □

Lemma 2.7. *Let $(e_i)_{i\in\tau}$ be a finite orthonormal family of vectors in a real Euclidean space, $|\tau|$ be the simplex spanned by $(e_i)_{i\in\tau}$ and H be the affine space spanned by $(e_i)_{i\in\tau}$. Denote by $q_\tau \colon H \to |\tau|$ the map which associates to every point of H its closest point on the simplex $|\tau|$.*

Let $\eta \in H$ and $\lambda = q_\tau(\eta)$. Denote by η_i and λ_i the coordinates of η and λ. Then there is $a \in \mathbb{R}$ unique such that $\sum_{i\in\tau}(\eta_i - a)_+ = 1$. Moreover,

(a) $a \geq 0$;
(b) $\lambda_i = (\eta_i - a)_+$ *for all* $i \in \tau$.

Proof. Let $\varphi(a) = \sum_{i\in\tau}(\eta_i - a)_+$. Then φ is non-increasing on $\mathbb{R}$. Let $\alpha = \sup_{i\in\tau} \eta_i$. Then $\varphi(0) \geq 1$, $\varphi(\alpha) = 0$ and φ is strictly decreasing on $(-\infty, \alpha]$. The first assertion and (a) follow.

To prove (b), let $\theta_j = (\eta_j - a)_+$. Note that for all j, θ_j is equal either to $\eta_j - a$ or to 0, hence $(\eta_j - \theta_j)\theta_j = a\theta_j$. Summing over j, one gets $\langle \eta - \theta, \theta\rangle = a$. Therefore,

$$\begin{aligned}\langle \eta - \theta, e_i - \theta\rangle &= \langle \eta - \theta, e_i\rangle - \langle \eta - \theta, \theta\rangle \\ &= \eta_i - \theta_i - a \leq 0.\end{aligned}$$

It follows that $|\tau|$ is in the half-space $\{\xi | \ \langle \eta - \theta, \xi - \theta\rangle \leq 0\}$, hence $\lambda = \theta$. □

Lemma 2.8. *With the notations of Lemma 2.7, let $f \subset \tau$ and $\eta, \zeta \in H$. Suppose that $\eta_i \leq 0$ for all $i \in \tau - f$ and that there exists $k \in f$ such that $\eta_k > 1 + \sqrt{2}\|\eta - \zeta\|$. Then $\zeta \in \Omega_{\tau,f}$.*

Proof. Let $\theta = q_\tau(\zeta)$. By Lemma 2.7, there exists $a \geq 0$ such that $\theta_j = (\zeta_j - a)_+$ for all j. Let $i \in \tau - f$. Let us show that $\theta_i = 0$.

$$\begin{aligned}\zeta_i - a &= [\eta_i + (\zeta_i - \eta_i)] + [(\zeta_k - a) + (\eta_k - \zeta_k) - \eta_k] \\ &\leq |\zeta_i - \eta_i| + [\theta_k + |\eta_k - \zeta_k| - \eta_k] \\ &\leq 1 + |\zeta_i - \eta_i| + |\eta_k - \zeta_k| - \eta_k && \text{since } \theta_k \leq 1 \\ &\leq 1 + \sqrt{2}\|\eta - \zeta\| - \eta_k && \text{by the Schwarz inequality} \\ &< 0,\end{aligned}$$

hence $\theta_i = (\zeta_i - a)_+ = 0$. This shows that $\zeta \in X_{\tau,f}$. In fact, since we obtained a strict inequality, it is not hard to see that $\zeta' \in X_{\tau,f}$ for $\|\zeta' - \zeta\|$ small enough, therefore $\zeta \in \Omega_{\tau,f}$. □

Lemma 2.9. *Let $\sigma \in \mathcal{F}_\Gamma$ and $(\mu, t) \in Z$ such that $\chi_{\mu,t}(\sigma) \neq 0$. Then the ball in $H_{\sigma \cup \mathrm{supp}(\mu)}$, centered at $\eta_{\sigma,\mu,t}$ and of radius one, satisfies*

$$B(\eta_{\sigma,\mu,t}, 1) \subset \Omega_{\sigma \cup \mathrm{supp}(\mu), \sigma - \mathrm{supp}(\mu)}.$$

Proof. Since $\chi_{\mu,t}(\sigma) \neq 0$, $c_{\sigma,\mu,t} \geq 1/2$, so there exists $g \in \sigma$ such that $c'_{\sigma,\mu,t,g} \geq 3 > 1 + \sqrt{2}$. The result follows from Lemma 2.8 applied to $\tau = \sigma \cup \mathrm{supp}(\mu)$, $f = \sigma - \mathrm{supp}(\mu)$, $\eta = \eta_{\sigma,\mu,t}$ and $k = g$. □

2.5. Construction of $\theta \in KK_\Gamma(C_0(\underline{E}\Gamma), C_0(\underline{E}\Gamma) \otimes A)$

As in [13, Theorem 2.3], we define an operator

$$P_{\mu,t} = \beta_{\mu,t} + \chi_{\mu,t}(1 - \beta_{\mu,t}^2)^{1/2}(1 \hat{\otimes} F_\mu) \tag{2.10}$$

acting on A_1 (recall notations (2.8), (2.9), (2.1); $\chi_{\mu,t}$ is considered as a diagonal operator in $\mathcal{L}(\mathcal{H}_\Gamma) \subset M(A_1)$).

Let $P = (P_{\mu,t})_{(\mu,t) \in Z}$ acting on $C_0(Z) \otimes A$. We show that the triple $(C_0(Z) \otimes A, \mathrm{Id} \otimes 1, P)$ defines an element $\theta \in KK_{Z \rtimes \Gamma}(C_0(Z), C_0(Z) \otimes A)$.

Let us first show that P is self-adjoint. It suffices to prove that for all $(\mu, t) \in Z$, $(1 \hat{\otimes} F_\mu)$, $\chi_{\mu,t}$ and $\beta_{\mu,t}$ commute in the graded sense.

$[\chi_{\mu,t}, \beta_{\mu,t}] = 0$ is obvious.

Let us show $[\beta_{\mu,t}, 1 \hat{\otimes} F_\mu] = 0$: for all $a \in \mathcal{A}(\ell^2_R(\Gamma))$ and $\sigma \in \mathcal{F}_\Gamma$, the vector $F_{\mu,t} e_\sigma = Qc(\xi_\mu) e_\sigma$ has the form $\sum_\tau \lambda_\tau e_\tau$, where the sum is supported on $\{\tau \in \mathcal{F}_\Gamma | \ \sigma \Delta \tau \subset \mathrm{supp}\, \mu\}$. Therefore, using the fact that $\sigma \Delta \tau \subset \mathrm{supp}\, \mu \implies \eta_{\sigma,\mu,t} = \eta_{\tau,\mu,t}$, we get

$$\begin{aligned}
\beta_{\mu,t}(1 \hat{\otimes} F_\mu)(a \otimes e_\sigma) &= (-1)^{\deg a} \sum_\tau \lambda_\tau \beta_{\mu,t}(a \otimes e_\tau) \\
&= (-1)^{\deg a} \sum_\tau \lambda_\tau i_{\eta_{\tau,\mu,t}}(f_0) a \otimes e_\tau \\
&= -(-1)^{\deg i_{\eta_{\sigma,\mu,t}}(f_0) a} \sum_\tau \lambda_\tau i_{\eta_{\sigma,\mu,t}}(f_0) a \otimes e_\tau \\
&= -(1 \hat{\otimes} F_\mu)(i_{\eta_{\sigma,\mu,t}}(f_0) a \otimes e_\sigma) \\
&= -(1 \hat{\otimes} F_\mu) \beta_{\mu,t}(a \otimes e_\sigma).
\end{aligned}$$

Let us show $[\chi_{\mu,t}, 1 \hat{\otimes} F_\mu] = 0$: with the same notations as above, using the fact that $\sigma \Delta \tau \subset \mathrm{supp}\, \mu \implies \chi_{\mu,t}(\sigma) = \chi_{\mu,t}(\tau)$, we get

$$\begin{aligned}
\chi_{\mu,t}(1 \hat{\otimes} F_\mu)(a \otimes e_\sigma) &= (-1)^{\deg a} a \otimes \sum_\tau \lambda_\tau \chi_{\mu,t}(\tau) e_\tau \\
&= (-1)^{\deg a} a \otimes \sum_\tau \lambda_\tau \chi_{\mu,t}(\sigma) e_\tau \\
&= (1 \hat{\otimes} F_\mu) \chi_{\mu,t}(a \otimes e_\sigma).
\end{aligned}$$

This completes the proof that $P = P^*$. Now, let us show that $P_{\mu,t} \in M(A)$. Suppose that $\sigma \neq \tau$, $x \in H_{\sigma\cup\tau}$ and $\langle e_\sigma, P_{\mu,t} e_\tau\rangle(x) \neq 0$. Since

$$\langle e_\sigma, P_{\mu,t} e_\tau\rangle = \chi_{\mu,t}(\sigma)\langle e_\sigma, F_\mu e_\tau\rangle\, i_{\eta_{\sigma,\mu,t}}(1 - f_0^2)^{1/2},$$

$\langle e_\sigma, F_\mu e_\tau\rangle \neq 0$, so σ and τ differ by an element of $\mathrm{supp}(\mu)$. Suppose for example that $\tau = \sigma \cup \{a\}$, $a \in \mathrm{supp}(\mu)$. Since $\chi_{\mu,t}(\sigma) \neq 0$, we have from Lemma 2.9 $B(\eta_{\sigma,\mu,t}, 1) \subset Y_{\sigma\cup\mathrm{supp}(\mu),\sigma-\mathrm{supp}(\mu)} \subset Y_{\sigma,\tau}$, therefore $x \in \Omega_{\sigma,\tau}$. The fact that $P_{\mu,t} \in M(A)$ follows from Lemma 2.5.

Let us show that $1 - P^2_{\mu,t} \in A$. Define first

$$P'_{\mu,t} = \beta_{\mu,t} + (1 - \beta^2_{\mu,t})^{1/2}(1\hat{\otimes}F_\mu) \in M(A_1). \tag{2.11}$$

Consider the operator $(1-\beta^2_{\mu,t})^{1/2}(1-\chi_{\mu,t})$. Its matrix representation is diagonal with entries $(1-\chi_{\mu,t}(\sigma))i_{\eta_{\sigma,\mu,t}}(1-f_0^2)^{1/2}$. From Lemma 2.6, there are with finitely many nonzero coefficients, all in $\mathcal{A}(\ell^2_{\mathbb{R}}(\Gamma))$, hence $(1 - \beta^2_{\mu,t})^{1/2}(1 - \chi_{\mu,t}) \in A_1$. It follows that $P_{\mu,t} - P'_{\mu,t} \in A_1$.

Likewise, since $F_\mu e_\sigma = F_\mu e_\sigma$ for all $\sigma \in \mathcal{F}_\Gamma$ such that $\sigma \not\subset \mathrm{supp}\,\mu$, the operator $1 - P'^2_{\mu,t} = (1 - \beta^2_{\mu,t})(1\hat{\otimes}(1 - F^2_\mu))$ has finitely many nonzero coefficients (indexed by (σ,τ) such that σ, $\tau \subset \mathrm{supp}\,(\mu)$), and all these coefficients belong to $\mathcal{A}(\ell^2_R(\Gamma))$. Therefore, $1 - P'^2_{\mu,t} \in A_1$, and thus, $1 - P^2_{\mu,t} \in A_1$. Since $A = M(A) \cap A_1$ (Lemma 2.5), we obtain $1 - P^2_{\mu,t} \in A$.

Finally, $(\mu, t) \mapsto P_{\mu,t}$ is continuous and Γ-equivariant, hence P defines an element $\theta \in KK_{\underline{E}\Gamma\rtimes\Gamma}(C_0(\underline{E}\Gamma), C_0(\underline{E}\Gamma) \otimes A)$.

2.6. Construction of $D \in KK_\Gamma(A, \mathbb{C})$

Let $i\colon A \to A_1$ be the natural inclusion. In [19, §8.5] is constructed an extension

$$0 \to (E_t)_{t\in(0,1]} \to E \xrightarrow{h_0} \mathcal{A}(\ell^2_{\mathbb{R}}(\Gamma)) \to 0,$$

where E is a field of C^*-algebras over $[0,1]$, $E_0 \simeq \mathcal{A}(\ell^2_{\mathbb{R}}(\Gamma))$ and $E_t \simeq \mathcal{S} \otimes \mathcal{K}$ for all $t \in (0,1]$. Let $h_1\colon E \to E_1$ be the evaluation at 1. Since Γ acts properly on A, exactly as in [19, §8.6] (see also [19, Corollaire 5.2] it can be shown that

$$(h_0)_*\colon KK_\Gamma(A, E) \to KK_\Gamma(A, \mathcal{A}(\ell^2_{\mathbb{R}}(\Gamma)))$$

is an isomorphism. Let $[i] \in KK_\Gamma(A, A_1) = KK_\Gamma(A, \mathcal{A}(\ell^2_{\mathbb{R}}(\Gamma)))$ be the KK-element induced by i, and let $D_0 = ((h_0)_*)^{-1}[i] \in KK_\Gamma(A, E)$. Define D by

$$D = D_0 \otimes_E [h_1] \otimes_{\mathcal{S}} [\mathrm{ev}_0], \tag{2.12}$$

where $\mathrm{ev}_0\colon \mathcal{S} \to \mathbb{C}$ is the evaluation $f \mapsto f(0)$.

2.7. Proof that $\theta \otimes_A D = 1$

Note that $\theta \otimes_A [i]$ can be represented by the bimodule $(C_0(Z) \otimes A_1, \mathrm{Id} \otimes 1, P')$ where $P' = (P'_{\mu,t})_{(\mu,t)\in Z}$, since $P_{\mu,t} - P'_{\mu,t} \in A_1$. Now, we can move the center of the Bott element $\beta_{\mu,t}$ to μ: let

$$\begin{aligned}\beta' &= (i_\mu(f_0))_{(\mu,t)\in Z} \in \mathcal{L}(C_0(Z)\otimes \mathcal{A}(\ell^2_{\mathbb{R}}(\Gamma)))\\ F &= (F_\mu)_{(\mu,t)\in Z} \in \mathcal{L}(C_0(Z)\otimes \mathcal{K}(\mathcal{H}_\Gamma)).\\ P'' &= \beta'\hat{\otimes}1 + (1-\beta'^2\hat{\otimes}1)^{1/2}(1\hat{\otimes}F).\end{aligned}$$

Then P' and P'' are homotopic through $(P'_{\mu,t,s})_{(\mu,t)\in Z, s\in[0,1]}$, where

$$\begin{aligned}\eta_{\sigma,\mu,t,s} &= (1-(1-s)\sum_{g\in\sigma} c'_{\sigma,\mu,t,g})\mu + (1-s)\sum_{g\in\sigma} c'_{\sigma,\mu,t,g} g\\ \beta_{\mu,t,s} &= \oplus_{\sigma\in\mathcal{F}_\Gamma} i_{\eta_{\sigma,\mu,t,s}}(f_0)\otimes \mathrm{Id}_{L_\sigma}\\ P'_{\mu,t,s} &= \beta_{\mu,t,s} + (1-\beta^2_{\mu,t,s})^{1/2}(1\hat{\otimes}F_\mu)\end{aligned}$$

Now, $\theta\otimes_A[i] = [(C_0(Z)\otimes A_1, \mathrm{Id}\otimes 1, P'')]$ is the Kasparov product $\eta_{Z\rtimes\Gamma}\otimes_{C_0(Z)}\gamma_0$ of the elements

$$\begin{aligned}\eta_{Z\rtimes\Gamma} &= [(C_0(Z)\otimes\mathcal{A}(\ell^2_{\mathbb{R}}(\Gamma)), \mathrm{Id}\otimes 1, \beta')] \in KK_{Z\rtimes\Gamma}(C_0(Z), C_0(Z)\otimes\mathcal{A}(\ell^2_{\mathbb{R}}(\Gamma)),\\ \gamma_0 &= [(C_0(Z)\otimes\mathcal{K}(\mathcal{H}_\Gamma), \mathrm{Id}\otimes 1, F)] \in KK_{Z\rtimes\Gamma}(C_0(Z), C_0(Z)).\end{aligned}$$

$\eta_{Z\rtimes\Gamma}$ is easily seen to be the dual-Dirac element constructed in [19, Section 7] for the groupoid $Z\rtimes\Gamma$. As for γ_0, $1-F^2_\mu$ is the projection onto the 1-dimensional space spanned by ξ_μ and $(\mu,t)\mapsto\xi_\mu$ is Γ-invariant, so $\gamma_0=1$.

Recall that the Dirac element $D_{Z\rtimes\Gamma}$ constructed in [19, §8.6] for the groupoid $Z\rtimes\Gamma$ is equal to

$$D'_0\otimes_E[h_1]\otimes_{\mathcal{S}}[\mathrm{ev}_0], \tag{2.13}$$

where $D'_0\in KK_{Z\rtimes\Gamma}(C_0(Z)\otimes\mathcal{A}(\ell^2_{\mathbb{R}}(\Gamma)), C_0(Z)\otimes E)$ is the unique element such that $D'_0\otimes_E[h_0]=1$.

Denote by $p\colon Z\rtimes\Gamma\to\Gamma$ the groupoid homomorphism $p(z,\gamma)=\gamma$ and by $p^*\colon KK_\Gamma(B_1,B_2)\to KK_{Z\rtimes\Gamma}(p^*B_1,p^*B_2)$ the induced morphism. As

$$\begin{aligned}(h_0)_*(p^*(D_0)) &= p^*(D_0)\otimes_E[h_0] = p^*(D_0\otimes_E[h_0])\\ &= p^*[i] = [i]\otimes_{A_1} 1_{C_0(Z)\otimes A_1}\\ &= [i]\otimes_{A_1}(D'_0\otimes_E[h_0])\\ &= ([i]\otimes_{A_1} D'_0)\otimes_E[h_0] = (h_0)_*([i]\otimes_{A_1} D'_0),\end{aligned}$$

and as $(h_0)_*\colon KK_{Z\rtimes\Gamma}(C_0(Z)\otimes A, C_0(Z)\otimes E)\to KK_{Z\rtimes\Gamma}(C_0(Z)\otimes A, C_0(Z)\otimes\mathcal{A}(\ell^2_{\mathbb{R}}(\Gamma)))$ is an isomorphism, it follows that $p^*(D_0)=[i]\otimes_{A_1}D'_0$. Therefore,

$$\begin{aligned}\theta\otimes_A D &= \theta\otimes_A D_0\otimes_E[h_1]\otimes_{\mathcal{S}}[\mathrm{ev}_0] && \text{by (2.12)}\\ &= \theta\otimes_{C_0(Z)\otimes A} p^*(D_0)\otimes_E[h_1]\otimes_{\mathcal{S}}[\mathrm{ev}_0]\\ &= (\theta\otimes_A[i])\otimes_{A_1}(D'_0\otimes_E[h_1]\otimes_{\mathcal{S}}[\mathrm{ev}_0]) && \text{by the above}\\ &= \eta_{Z\rtimes\Gamma}\otimes_{A_1} D_{Z\rtimes\Gamma} && \text{by (2.13)}\\ &= 1 && \text{since } Z\rtimes\Gamma \text{ is a proper groupoid.}\end{aligned}$$

3. The main theorem

This section is devoted to the proof of the main theorem (Theorem 3.3). The proof is divided into two steps.

Step 1: show (Proposition 3.2(ii) $\Longrightarrow$ (i)) that if there exists a compact space X with an action of Γ such that the groupoid $X \rtimes \Gamma$ has a gamma element, and such that X is F-contractible for every finite subgroup F of Γ, then Γ also has a gamma element. This step requires Theorem 2.1.

Step 2: show that if Γ admits a uniform embedding into Hilbert space then the assumption in Step 1 holds. This step uses the fact that Γ admits a uniform embedding into Hilbert space if and only if there exists a compact space Y with an action of Γ such that the groupoid $Y \rtimes \Gamma$ has the Haagerup property (i.e., has a proper negative type function).

To prove step 1, we first need a lemma:

Lemma 3.1. *Let Γ be a discrete group and X a Γ-space which is K-contractible for every compact subgroup K of Γ. Then for every locally compact, σ-compact proper Γ-space Z, there exists a continuous, Γ-equivariant map $Z \to X$.*

Proof. Analogous to [2, Proposition 1.8]. Let $\pi\colon Z \to Z/\Gamma$ be the quotient map. From [4, Theorem 1.8], There exists a locally finite cover $(Y_n)_{n\in\mathbb{N}}$ of Z by Γ-relatively compact open subspaces which admit Γ-equivariant maps $g_n\colon Y_n \to K_n\backslash\Gamma$, with K_n a compact subgroup. Let $Z_n = Y_0 \cup \cdots \cup Y_n$. Suppose we have found $f_{n-1}\colon Z_{n-1} \to X$ continuous Γ-equivariant. Let (φ, ψ) be a partition of unity of Z_n/Γ subordinate to the cover $(\pi(Z_{n-1}), \pi(Y_n))$. Choose a (not necessarily continuous) lifting $h\colon Y_n \to \Gamma$ of g_n and let $H\colon [0,1]\times X \to X$ be a K_n-equivariant map such that $H(0,x) = x$ for all $x \in X$ and $H(1,\cdot) = x_0$ is constant (necessarily, x_0 is K_n-invariant). Define

$$f_n(z) = \begin{cases} H(\psi(\pi(z)), f_{n-1}(z\cdot h(z)^{-1}))\cdot h(z) & \text{if } z \in Y_n \cap Z_{n-1},\\ x_0 h(z) & \text{if } z \in Y_n - Z_{n-1}\\ f_{n-1}(z) & \text{if } z \in Z_{n-1} - Y_n. \end{cases}$$

If h' is another lifting, then $h'(z) = k(z)h(z)$ where $z \mapsto k(z)$ is a K_n-valued function. Since x_0 is K_n-invariant and H is K_n-equivariant, the map f_n is independent of the choice of the lifting h.

Let $z_0 \in Z_n$ and $g_0 \in \Gamma$. To check that $f_n(z_0 g_0) = f_n(z_0)g_0$, it suffices to consider the case $z_0 \in Y_n$. Since f_n is independent of the choice of h, we may assume that $h(z_0 g_0) = h(z_0)g_0$. The verification is then immediate. Therefore, f_n is Γ-equivariant.

To show that f_n is continuous, note that the interior of $\psi^{-1}(0)$, the interior of $\psi^{-1}(1)$ and $Y_n \cap Z_{n-1}$ cover Z_n. It suffices to show that f_n is continuous on each of these subsets.

Since f_{n-1} is continuous and f_n coincides with f_{n-1} on $\psi^{-1}(0)$, the restriction of f_n to $\psi^{-1}(0)$ is continuous.

On $\psi^{-1}(1)$, we have $f_n(z) = x_0 h(z)$. Since the map

$$\Gamma \to X, \qquad \gamma \mapsto x_0\gamma$$

factors through a continuous map $P\colon K_n\backslash\Gamma \to X$, it follows that $z \mapsto f_n(z) = P \circ g_n(z)$ is continuous on $\psi^{-1}(1)$.

To prove that f_n is continuous on $Y_n \cap Z_{n-1}$, suppose that $z_i \in Y_n \cap Z_{n-1}$ converges to $z \in Y_n \cap Z_{n-1}$. Since $g_n(z_i)$ converges to $g_n(z)$, there exist $k_i \in K_n$ such that $k_i h(z_i)$ converges to $h(z)$. Since H is K_n-equivariant, one has $f_n(z_i) = H(\psi(\pi(z_i)), f_{n-1}(z(k_i h(z_i))^{-1})) \cdot k_i h(z_i)$, therefore $f_n(z_i)$ converges to $f_n(z)$.

Let $f(z) = \lim_{n\to\infty} f_n(z)$. It is not hard to check, using the fact that the cover (Y_n) is locally finite, that f is a continuous, Γ-equivariant map from Z to X. □

Proposition 3.2. *Let Γ be a discrete group. Using notations of Theorem 2.1, the following assertions are equivalent:*

(i) *Γ has a gamma element;*
(ii) *There exists a compact Γ-space X which is F-contractible for every finite subgroup F of Γ, such that the groupoid $X \rtimes \Gamma$ has a gamma element;*
(iii) *there exists $\eta \in KK_\Gamma(\mathbb{C}, A)$ such that $p^*(\eta) = \theta$.*

Furthermore, if η satisfies (iii), then $\eta \otimes_A D = \gamma$.

Proof. The implication (i) $\Longrightarrow$ (ii) is trivial.

To prove (iii) $\Longrightarrow$ (i), observe that $p^*(\eta \otimes_A D) = (p^*\eta) \otimes_A D = \theta \otimes_A D = 1$, hence Γ has gamma element and, by definition, $\gamma = \eta \otimes_A D$.

It remains to prove (ii) $\Longrightarrow$ (iii). By assumption, there exist a proper $X \rtimes \Gamma$-algebra B and elements

$$\eta' \in KK_{X\rtimes\Gamma}(C(X), B) \qquad d' \in KK_{X\rtimes\Gamma}(B, C(X))$$

such that $p^*(\eta' \otimes_B d') = 1 \in KK_{(X\times\underline{E}\Gamma)\rtimes\Gamma}(C_0(X \times \underline{E}\Gamma), C_0(X \times \underline{E}\Gamma))$. Let $\eta_0 = [q^*] \otimes_{C(X)} \eta \in KK_\Gamma(\mathbb{C}, B)$ where $q\colon X \to \mathrm{pt}$, and d_0 be obtained from d' by applying the forgetful functor $KK_{X\rtimes\Gamma} \to KK_\Gamma$.

From Lemma 3.1, there exists a Γ-equivariant map $\psi\colon \underline{E}\Gamma \to X$. Let $\varphi\colon \underline{E}\Gamma \to X \times \underline{E}\Gamma$ be defined by $\varphi(z) = (\psi(z), z)$. Consider the element

$$\sigma_{\underline{E}\Gamma,A}([\varphi^*]) \in KK_\Gamma(C(X) \otimes A, A)$$

and let

$$\eta = \eta_0 \otimes_B \sigma_{\underline{E}\Gamma,B}(\theta) \otimes_B d_0 \otimes_{C(X)\otimes A} \sigma_{\underline{E}\Gamma,A}([\varphi^*]) \in KK_\Gamma(\mathbb{C}, A).$$

Let $\mathrm{pr}_i\colon \underline{E}\Gamma \times \underline{E}\Gamma \to \underline{E}\Gamma$ be the two projections ($i = 1, 2$). As pr_1 and pr_2 are Γ-homotopic,

$$\begin{aligned}
\eta_0 \otimes_B p^*(\sigma_{\underline{E}\Gamma,B}(\theta)) &= \eta_0 \otimes_B \sigma_{\underline{E}\Gamma\times\underline{E}\Gamma,C_0(\underline{E}\Gamma)\otimes B}(\mathrm{pr}_2^*(\theta)) \\
&= \eta_0 \otimes_B \sigma_{\underline{E}\Gamma\times\underline{E}\Gamma,C_0(\underline{E}\Gamma)\otimes B}(\mathrm{pr}_1^*(\theta)) \\
&= \eta_0 \otimes_B \sigma_B(\theta) \\
&= \eta_0 \otimes_{\mathbb{C}} \theta \;=\; \theta \otimes_{\mathbb{C}} \eta_0,
\end{aligned}$$

hence

$$\begin{aligned} p^*(\eta) &= \eta_0 \otimes p^*(\sigma_{\underline{E}\Gamma,B}(\theta)) \otimes_B d_0 \otimes_{C(X)\otimes A} \sigma_{\underline{E}\Gamma,A}([\varphi^*]) \\ &= \theta \otimes_{\mathbb{C}} (\eta_0 \otimes_B d_0) \otimes_{C(X)\otimes A} \sigma_{\underline{E}\Gamma,A}([\varphi^*]) && \text{from the above} \\ &= \theta \otimes_{\mathbb{C}} [q^*] \otimes_{C(X)\otimes A} \sigma_{\underline{E}\Gamma,A}([\varphi^*]) && \text{since } \eta_0 = [q^*] \otimes_{C(X)} \eta \\ &= \theta \otimes_A \sigma_{\underline{E}\Gamma,A}([q^*] \otimes_{C(X)} [\varphi^*]) \\ &= \theta \otimes_{C_0(\underline{E}\Gamma)} ([q^*] \otimes_{C(X)} [\varphi^*]) \\ &= \theta \otimes_{C_0(\underline{E}\Gamma)} (p^*[q^*] \otimes_{C(X)\otimes C_0(\underline{E}\Gamma)} [\varphi^*]) \\ &= \theta \otimes_{C_0(\underline{E}\Gamma)} ([q'^*] \otimes_{C(X)\otimes C_0(\underline{E}\Gamma)} [\varphi^*]), \end{aligned}$$

where $q' \colon \underline{E}\Gamma \times X \to \underline{E}\Gamma$ is the first projection. Since $\varphi \circ q' = \mathrm{Id}_{\underline{E}\Gamma}$, we get $p^*(\eta) = \theta$. □

Here is the main theorem:

Theorem 3.3. *Let Γ be a discrete group which admits a uniform embedding into Hilbert space. Then Γ has a γ element.*

Proof. By [17], there exists a compact space Y acted upon by Γ, such that $Y \rtimes \Gamma$ has a proper negative type function. By [17], the space $X = \mathrm{Prob}(Y)$ of probability measures on Y with the weak*-topology also has a proper negative type function. By [19], the groupoid $X \rtimes \Gamma$ has a gamma element (equal to 1). Moreover, X is F-contractible for every finite subgroup F of Γ, so Proposition 3.2(ii $\Longrightarrow$ i) applies. □

In the corollary below, we say that a group G satisfies BC if the Baum–Connes map with coefficients $\mu_{\mathrm{red}} \colon K_*^{\mathrm{top}}(G; B) \to K_*(B \rtimes_r G)$ is an isomorphism for every G-algebra B.

Corollary 3.4. *Let Γ be a discrete group which admits a uniform embedding into Hilbert space, and let N be a normal subgroup. Denote by $q \colon \Gamma \to \Gamma/N$ the quotient map. Assume that*

(1) *for every finite subgroup F of Γ/N, the group $q^{-1}(F)$ satisfies BC;*
(2) *Γ/N satisfies BC.*

Then G satisfies BC.

Proof. See [3, Theorem 7.1] □

As noted by Chabert and Echterhoff, condition (1) is always satisfied if N is amenable ([3, Corollary 7.3]).

We also recall that every locally compact group with Haagerup's property (e.g., every amenable group, $SO(n,1)$, $SU(n,1)$) satisfy BC [7]. Moreover, P. Julg has recently shown that $Sp(n,1)$ satisfies BC [10], and thus every discrete subgroup of $Sp(n,1)$ satisfies BC.

4. Final remarks

(1) V. Lafforgue proved that if G is a locally compact group such that

(i) G admits an element γ, and $\gamma = 1$ in $KK_G^{\mathrm{ban}}(\mathbb{C}, \mathbb{C})$;
(ii) there exists a closed $*$-subalgebra $\mathcal{A} \subset C_r^*(G)$, closed under holomorphic functional calculus, which is an "unconditional" completion of $C_c(G)$,

then the Baum–Connes map for G is an isomorphism ([15, Proposition 1.7.3]). Thus, it would be of interest to determine whether $\gamma = 1$ in $KK_G^{\mathrm{ban}}(\mathbb{C}, \mathbb{C})$ for every discrete group G which admits a uniform embedding into Hilbert space. However, a more explicit construction of γ might be necessary.

(2) Let G be a totally disconnected group, and K be a compact open subgroup of G. Let $X = G/K$. Then, up to coarse equivalence, X may be endowed with a left invariant, bounded geometry metric, so it makes sense to say that X admits a uniform embedding into Hilbert space (moreover, this property does not depend on the choice of the subgroup K). So, suppose that such an embedding exists, then the above constructions probably still work, replacing Γ by G or by G/K when appropriate.

References

[1] C. Anatharaman-Delaroche, *Amenability and Exactness for Dynamical Systems and their C^*-Algebras.* Preprint, Université d'Orléans, 2000.

[2] P. Baum, A. Connes and N. Higson, *Classifying Space for Proper Actions and K-Theory of Group C^*-Algebras.* Contemporary Mathematics **167** (1994), 241–291.

[3] J. Chabert and S. Echterhoff, *Twisted Equivariant KK-Theory and the Baum-Connes Conjecture for Group Extensions.* K-Theory **23** (2001), no. 2, 157–200.

[4] J. Chabert, S. Echterhoff and R. Meyer, *Deux remarques sur l'application de Baum-Connes.* C. R. Acad. Sci. Paris Sér. I Math. **332** (2001), no. 7, 607–610.

[5] N. Higson, *Bivariant K-Theory and the Novikov Conjecture.* Geom. Funct. Anal. **10** (2000), no. 3, 563–581.

[6] N. Higson and G. Kasparov, *Operator K-Theory for Groups which Act Properly and Isometrically on Hilbert Space.* Electron. Res. Announc. Amer. Math. Soc. **3** (1997), 131–142 (electronic).

[7] N. Higson and G. Kasparov, *E-Theory and KK-Theory for Groups which Act Properly and Isometrically on Hilbert Space.* Invent. Math. **144** (2001), no. 1, 23–74.

[8] N. Higson and J. Roe, *Amenable Group Actions and the Novikov Conjecture.* J. Reine Angew. Math. **519** (2000), 143–153.

[9] P. Julg, *Travaux de N. Higson et G. Kasparov sur la conjecture de Baum-Connes. Séminaire Bourbaki.* Vol. 1997/98, Astérisque **252** (1998), Exp. No. 841, 4, 151–183.

[10] P. Julg, *La conjecture de Baum-Connes à coefficients pour le groupe* $\mathrm{Sp}(n, 1)$. C. R. Math. Acad. Sci. Paris **334** (2002), no. 7, 533–538.

[11] P. Julg and A. Valette, *K-Theoretic Amenability for* $\mathrm{SL}_2(Q_p)$*, and the Action on the Associated Tree.* J. Funct. Anal. **58** (1984), no. 2, 194–215.

[12] G. Kasparov, *Equivariant KK-Theory and the Novikov Conjecture.* Inv. Math. **91** (1988), 147–201.

[13] G. Kasparov and G. Skandalis, *Groups Acting on Buildings, Operator K-Theory and Novikov's Conjecture.* K-Theory **4** (1991), 303–337.

[14] G. Kasparov and G. Skandalis, *Groups Acting Properly on "Bolic" Spaces and the Novikov Conjecture.* Preprint, 2000.

[15] V. Lafforgue, *K-théorie bivariante pour les algèbres de Banach et conjecture de Baum-Connes.* Invent. Math. **149** (2002), no. 1, 1–95.

[16] P.Y. Le Gall, *Théorie de Kasparov équivariante et groupoïdes. I.* K-Theory **16** (1999), no. 4, 361–390.

[17] G. Skandalis, J.-L. Tu and G. Yu, *Coarse Baum-Connes Conjecture and Groupoids.* Preprint, 2000.

[18] J.-L. Tu, *La conjecture de Novikov pour les feuilletages hyperboliques.* K-Theory **16**, No. 2 (1999), 129–184.

[19] J.-L. Tu, *La conjecture de Novikov pour les feuilletages moyennables.* K-Theory **17**, No. 3 (1999), 215–264.

[20] J.-L. Tu, *The Baum-Connes Conjecture and Discrete Group Actions on Trees.* K-Theory **17** (1999), 303–318.

[21] G. Yu, *The Coarse Baum-Connes Conjecture for Spaces which Admit a Uniform Embedding into Hilbert Space.* Invent. Math. **139** (2000), no. 1, 201–240.

Jean-Louis Tu
Université Pierre et Marie Curie
Algèbres d'Opérateurs et Représentations
Institut de Mathématiques (UMR 7586)
4, place Jussieu
F-75252 Paris Cedex 05
France
e-mail: `tu@math.jussieu.fr`

Operator Theory:
Advances and Applications, Vol. 153, 287–306

Uniform Exponential Stability and Uniform Observability of Time-Varying Linear Stochastic Systems in Hilbert Spaces

Viorica Mariela Ungureanu

Abstract. The main object of this paper is to discuss the problem of the uniform exponential stability and uniform observability of time-varying linear stochastic equations in Hilbert spaces. We give a representation of the covariance operator associated to the mild solutions of these equations which allow us to obtain a characterization of the uniform exponential stability of uniformly observable systems in terms of Lyapunov equations.

Mathematics Subject Classification (2000). 60H15, 35B40, 93B07.

Keywords. Stochastic differential equation, Uniform exponential stability, Uniform observability, Lyapunov equation.

1. Preliminaries

Let H, V be separable real Hilbert spaces and let $L(H,V)$ be the Banach space of all bounded linear operators from H into V (if $H = V$ then $L(H,V) \overset{not}{=} L(H)$). We write $\langle \cdot, \cdot \rangle$ for the inner product and $\|\cdot\|$ for norms of elements and operators. We denote by $a \otimes b, a, b \in H$ the bounded linear operator of $L(H)$ defined by $a \otimes b(h) = \langle h, b \rangle a$ for all $h \in H$. The operator $A \in L(H)$ is said to be nonnegative and we write $A \geq 0$, if A is self-adjoint and $\langle Ax, x \rangle \geq 0$ for all $x \in H$. We denote by $L^+(H)$ the subset of $L(H)$ of nonnegative operators.

For $A \in L^+(H)$, $B \in L(H)$ we denote by $A^{1/2}$ the square root of A and by $|B|$ the operator $(B^*B)^{1/2}$.

If $A \in L(H)$ we put $\|A\|_1 = Tr(|A|) \leq \infty$ and we denote by $C_1(H)$ the set $\{A \in L(H) \mid \|A\|_1 < \infty\}$ (the *operators' trace class*). If $A \in C_1(H)$ we say that A is *nuclear* and it is not difficult to see that A is compact.

The definition of the nuclear operator introduced above is equivalent with that given in [2] and [5]. It is known (see [2]) that $C_1(H)$ is a Banach space endowed with the norm $\|\cdot\|_1$ and for all $A \in L(H)$ and $B \in C_1(H)$ we have $AB, BA \in C_1(H)$.

If $\|A\|_2 = (TrA^*A)^{1/2}$ we can introduce the *Hilbert Schmidt class* of operators, namely $C_2(H) = \{A \in L(H) \mid \|A\|_2 < \infty\}$ (see [12]). $C_2(H)$ is a Hilbert space with the inner product $\langle A, B\rangle_2 = TrB^*A$ (see [12]).

We denote by $\mathcal{H}_2$ the subspace of $C_2(H)$ of all self-adjoint operators. Since $\mathcal{H}_2$ is closed in $C_2(H)$ with respect to $\|.\|_2$ we deduce that it is a Hilbert space, too. It is known (see [5]) that for all $A \in C_1(H)$ we have

$$\|A\| \le \|A\|_2 \le \|A\|_1 . \tag{1.1}$$

For each interval $J \subset \mathbf{R}_+(\mathbf{R}_+ = [0,\infty))$ we denote by $C_s(J, L(H))$ the space of all mappings $G(t) : J \to L(H)$ that are strongly continuous and by $C_b(J, L(H))$ the subspace of $C_s(J, L(H))$, which consists of all mappings $G(t)$ such that $\sup_{t\in J} \|G(t)\| < \infty$. If E is a Banach space we also denote by $C(J, E)$ the space of all mappings $G(t) : J \to E$ that are continuous.

We need the following assumption:

P_1: a) $A(t)$, $t \in [0, \infty)$ is a closed linear operator on H with constant domain D dense in H.

b) there exist $M > 0$, $\eta \in (\frac{1}{2}\pi, \pi)$ and $\delta \in (-\infty, 0)$ such that $S_{\delta,\eta} = \{\lambda \in C;\ |\arg(\lambda - \delta)| < \eta\} \subset \rho(A(t))$, for all $t \ge 0$ and $\|R(\lambda, A(t))\| \le \frac{M}{|\lambda-\delta|}$ for all $\lambda \in S_{\delta,\eta}$.

c) there exist numbers $\alpha \in (0, 1)$ and $\widetilde{N} > 0$ such that $\left\|A(t)A^{-1}(s) - I\right\| \le \widetilde{N}\,|t - s|^{\alpha}, t \ge s \ge 0$, where we denote by $\rho(A)$, $R(\lambda, A)$ the resolvent set of A and respectively the resolvent of A.

It is known that if P_1 holds then the family $\{A(t)\}_{t\in\mathbf{R}_+}$ generates the evolution operator $U(t, s), t \ge s \ge 0$ (see [11]).

For any $n \in \mathbf{N}$ we have $n \in \rho(A(t))$. The operators $A_n(t) = n^2R(n, A(t))-nI$ are called the Yosida approximations of $A(t)$. If we denote by $U_n(t, s)$ the evolution operator generates by $A_n(t)$, which exists since $A_n(t)$ is bounded, then it is known (see [11]) that for each $x \in H$, one has $\lim_{n\to\infty} U_n(t, s)x = U(t, s)x$ uniformly on any bounded subset of $\{(t, s); t \ge s \ge 0\}$.

Let $(\Omega, F, \mathcal{F}_t, t\in[0,\infty), P)$ be a stochastic basis and $L_s^2(H) = L^2(\Omega, \mathcal{F}_s, P, H)$. We consider the stochastic equation

$$dy(t) = A(t)y(t)dt + \sum_{i=1}^{m} G_i(t)y(t)dw_i(t), \quad y(s) = \xi \in L_s^2(H), \tag{1.2}$$

where $A(t)$ satisfies the hypothesis P_1, $G_i \in C_s(\mathbf{R}_+, L(H))$, $i = 1, \ldots, m$ and w_i's are independent real Wiener processes relative to $\mathcal{F}_t$.

It is known (see [10]) that (1.2) has a unique mild solution in $C([s, T]; L^2(\Omega; H))$ that is adapted to $\mathcal{F}_t$; namely the solution of

$$y(t) = U(t, s)\xi + \sum_{i=1}^{m} \int_s^t U(t, r)G_i(r)y(r)dw_i(r). \tag{1.3}$$

We associate to (1.2) the approximating system:

$$dy_n(t) = A_n(t)y_n(t)dt + \sum_{i=1}^{m} G_i(t)y_n(t)dw_i(t), \quad y_n(s) = \xi \in L_s^2(H), \tag{1.4}$$

where $A_n(t), n \in \mathbf{N}$ are the Yosida approximations of $A(t)$ and $G_i(t), w_i(t)$ are defined as above. By convenience, we denote by $y(t,s;\xi)$ (respectively $y_n(t,s;\xi)$) the solution of (1.2) (respectively (1.4)) with initial condition $y(s) = \xi$ (respectively $y_n(s) = \xi$), $\xi \in L_s^2(H)$.

Lemma 1.1. ([11]) *Let P_1 hold. Then there exists a unique mild (respectively classical) solution to* (1.2) *(respectively* (1.4)*) and $y_n \to y$ in mean square uniformly on any bounded subset of* $[s,\infty]$.

Now we introduce the following Lyapunov equation:

$$\frac{dQ(s)}{ds} + A^*(s)Q(s) + Q(s)A(s) + \sum_{i=1}^{m} G_i^*(s)Q(s)G_i(s) + D(s) = 0, \quad s \geq 0 \tag{1.5}$$

where $A(s)$ satisfies $P_1, G_i \in C_s([0,\infty), L(H))$ and $D \in C_b([0,\infty), L(H))$.

According with [11], we say that Q is a mild solution on an interval $J \subset \mathbf{R}_+$ of (1.5), if $Q \in C_s(J, L^+(H))$ and if it satisfies

$$\begin{aligned} Q(s)x = U^*(t,s)Q(t)U(t,s)x + \int_s^t U^*(r,s)[\sum_{i=1}^{m} G_i^*(r)Q(r)G_i(r) \\ + D(r)]U(r,s)xdr \end{aligned} \tag{1.6}$$

for all $s \leq t$, $s,t \in J$ and $x \in H$. If $A_n(t), n \in \mathbf{N}$ are the Yosida approximations of $A(t)$ then we introduce the approximating equation:

$$\begin{aligned} \frac{dQ_n(s)}{ds} + A_n^*(s)Q_n(s) + Q_n(s)A_n(s) + \sum_{i=1}^{m} G_i^*(s)Q_n(s)G_i(s) + D(s) = 0, \\ s \geq 0. \end{aligned} \tag{1.7}$$

Lemma 1.2. ([11]) *Assume P_1. Let $0 < T < \infty$ and let $R \in L^+(H)$. Then there exists a unique mild (respectively classical) solution Q (respectively Q_n) of* (1.6) *(respectively* (1.7)*) on $[0,T]$ such that $Q(T) = R$ (respectively $Q_n(T) = R$). They are given by*

$$\begin{aligned} Q(s)x &= U^*(T,s)RU(T,s)x \\ &+ \int_s^T U^*(r,s)\left[\sum_{i=1}^{m} G_i^*(r)Q(r)G_i(r) + D(r)\right] U(r,s)xdr \\ Q_n(s)x &= U_n^*(T,s)RU_n(T,s)x \\ &+ \int_s^T U_n^*(r,s)\left[\sum_{i=1}^{m} G_i^*(r)Q_n(r)G_i(r) + D(r)\right] U_n(r,s)xdr \end{aligned}$$

and for each $x \in H$, $Q_n(s)x \to Q(s)x$ uniformly on any bounded subset of $[0,T]$. Moreover if we denote these solutions by $Q(T,s;R)$ and respectively $Q_n(T,s;R)$ then they are monotone in the sense that $Q(T,s;R_1) \leq Q(T,s;R_2)$ if $R_1 \leq R_2$.

In the subsequent considerations we assume that P_1 holds and

$$G_i \in C_s(\mathbf{R}_+, L(H)), \quad i = 1, \dots, m.$$

We will obtain a representation (see Theorem 3.3) of the covariance operator associated to the mild solution of stochastic equation (1.2) by using the mild solution of Lyapunov equation (2.6). We use this representation to obtain deterministic characterizations of the uniform exponential stability and uniform observability properties (Theorems 4.3, 4.4). The above results are infinite-dimensional versions of those obtained by T. Morozan (for the finite dimensional case) in [8].

We note that G. Da Prato and I. Ichikawa gave in [11] (Corollary 3) another different criterion for the uniform exponential stability in terms of Lyapunov equations.

The main result of this paper is Theorem 3.3 which establishes that the existence of a unique bounded and uniformly positive solution of Lyapunov equation (1.5) is a necessary and sufficient condition for the uniform exponential stability property of uniformly observable systems.

2. Differential equations on $\mathcal{H}_2$

For all $n \in \mathbf{N}$ and $t \geq 0$ we consider the mapping $L_n(t) : \mathcal{H}_2 \to \mathcal{H}_2$,

$$L_n(t)(P) = A_n(t)P + PA_n^*(t) + \sum_{i=1}^{m} G_i(t)PG_i^*(t), \quad P \in \mathcal{H}_2. \tag{2.1}$$

It is easy to verify that $L_n(t) \in L(\mathcal{H}_2)$ and the adjoint operator $L_n^*(t)$ is the linear and bounded operator on $\mathcal{H}_2$ given by

$$L_n^*(t)(R) = RA_n(t) + A_n^*(t)R + \sum_{i=1}^{m} G_i^*(t)RG_i(t) \tag{2.2}$$

for all $t \geq 0$, $P \in \mathcal{H}_2$.

Lemma 2.1. ([3]) *If P_1 holds and $G_i \in C_s([0,\infty), L(H))$, $i = 1, \dots, m$ then*

(a) *$A_n \in C([0,\infty), L(H))$ for all $n \in \mathbf{N}$ and*

(b) *$L_n \in C_s([0,\infty), L(\mathcal{H}_2))$ for all $n \in \mathbf{N}$.*

Proof. (a) Let $t, s \geq 0$. We have

$$\begin{aligned}
\|A_n(t) - A_n(s)\| &= n^2 \|R(n, A(t)) - R(n, A(s))\| \\
&= n^2 \|R(n, A(t))(nI - A(t))(R(n, A(t)) - R(n, A(s)))\| \\
&\leq n^2 \|R(n, A(t))\| \; \|I - [(nI - A(s)) + A(s) - A(t)]R(n, A(s))\| \\
&\leq n^2 \|R(n, A(t))\| \; \|I - I + [A(s) - A(t)]R(n, A(s))\| \\
&\leq n \|R(n, A(t))\| \; \left\|[A(s) - A(t)]A(s)^{-1}\right\| \; \|nA(s)R(n, A(s))\| \, .
\end{aligned}$$

Now we use P_1 ((b) and (c)) and we deduce that there exist $\delta < 0$, $\alpha \in (0,1)$, $M > 0$ and $\widetilde{N} > 0$ such that

$$\|nA(s)R(n, A(s))\| = \left\|n^2 R(n, A(s)) - nI\right\| \leq n^2 \frac{M}{n - \delta} + n$$

for any $s \geq 0$ and $\|A_n(t) - A_n(s)\| \leq n\frac{M}{n-\delta}(n^2 \frac{M}{n-\delta} + n)\widetilde{N}\, |t - s|^{\alpha}$. The proof of (a) is finished.

(b) We deduce from (a) that if $\mathcal{A}_n(t) : \mathcal{H}_2 \to \mathcal{H}_2$, $\mathcal{A}_n(t)(P) = A_n(t)P + PA_n^*(t), t \geq 0, n \in \mathbf{N}$ then $\mathcal{A}_n \in C([0,\infty), L(\mathcal{H}_2))$. We only have to prove that $\mathcal{G}_i \in C_s([0,\infty), L(\mathcal{H}_2))$, where $\mathcal{G}_i(t) : \mathcal{H}_2 \to \mathcal{H}_2$, $\mathcal{G}_i(t)(P) = G_i(t)PG_i^*(t)$, $i = 1, \ldots, m$. From Lemma 1 of [3] and since $G_i \in C_s([0,\infty), L(H))$, $i = 1, \ldots, m$ it follows $G_iP \in C([0,\infty), \mathcal{H}_2)$ and $PG_i^* \in C([0,\infty), \mathcal{H}_2)$ for all $P \in \mathcal{H}_2$ and $i = 1, \ldots, m$. For $s \geq 0, P \in \mathcal{H}_2$ fixed and for every $i \in \{1, \ldots, m\}$ we have

$$\begin{aligned}
&\|\mathcal{G}_i(t)(P) - \mathcal{G}_i(s)(P)\|_2 = \|G_i(t)PG_i^*(t) - G_i(s)PG_i^*(s)\|_2 \\
&\quad \leq \|G_i(t)PG_i^*(t) - G_i(t)PG_i^*(s)\|_2 + \|G_i(t)PG_i^*(s) - G_i(s)PG_i^*(s)\|_2 \, .
\end{aligned}$$

If $\widetilde{G}_{i,s} = \sup\limits_{t \in [0, s+1]} \|G_i(t)\|$ then, for all $t \in [0, s+1]$ we have

$$\begin{aligned}
&\|\mathcal{G}_i(t)(P) - \mathcal{G}_i(s)(P)\|_2 \\
&\quad \leq \widetilde{G}_{i,s} \|PG_i^*(t) - PG_i^*(s)\|_2 + \|G_i(t)PG_i^*(s) - G_i(s)PG_i^*(s)\|_2 \, .
\end{aligned}$$

As $t \to s$, we obtain $\lim\limits_{t \to s} \|\mathcal{G}_i(t)(P) - \mathcal{G}_i(s)(P)\|_2 = 0$.

If $s = 0$ we only have the limit from the right. □

If E is a Banach space and $\mathcal{L} \in C_s([0,\infty), L(E))$, we consider the initial value problem

$$\frac{\partial v(t)}{\partial t} = \mathcal{L}(t)v(t), \quad v(s) = x \in E, t \geq s \geq 0. \tag{2.3}$$

Let $T \geq s$. An E-valued function $v : [s, T] \to E$ is a classical solution of (2.3) if v is continuous on $[s, T]$, continuously differentiable on $[s, T]$ and satisfies (2.3). The following results have a standard proof (see [9]).

Lemma 2.2. *For every $x \in E$ the initial value problem* (2.3) *has a unique classical solution v.*

We define the "solution operator" of the initial value problem (2.3) by $V(t, s)x = v(t), x \in E$ for $0 \leq s \leq t \leq T$, where v is the solution of (2.3).

Proposition 2.3. *For every $0 \leq s \leq t \leq T$, $V(t,s)$ is a bounded linear operator and*

1) $\|V(t,s)\| \leq e^{\lambda(t-s)}$, *where* $\lambda = \sup_{t\in[0,T]} \|\mathcal{L}(t)\|$.
2) $V(s,s) = I$ *and* $V(t,s) = V(t,r)V(r,s)$ *for all* $0 \leq s \leq r \leq t \leq T$.
3) $V(t,s) \underset{t-s\to 0}{\to} I$ *in the uniform operator topology for all* $0 \leq s \leq t \leq T$.
4) $(t,s) \to V(t,s)$ *is continuous in the uniform operator topology on* $\{(t,s) \mid 0 \leq s \leq t \leq T\}$.
5) $\frac{\partial V(t,s)x}{\partial t} = \mathcal{L}(t)V(t,s)x$ *for all* $x \in E$ *and* $0 \leq s \leq t \leq T$.
6) $\frac{\partial V(t,s)x}{\partial s} = -V(t,s)\mathcal{L}(s)x$ *for all* $x \in E$ *and* $0 \leq s \leq t \leq T$. *We denote by* I *the identity operator on* E. *The operator* $V(t,s)$ *is called the evolution operator generated by the family* $\mathcal{L}$.

Let us consider the equation

$$\frac{dP_n(t)}{dt} = L_n(t)P_n(t), \quad P_n(s) = S \in \mathcal{H}_2,\ t \geq s \geq 0 \tag{2.4}$$

on $\mathcal{H}_2$, where L_n is given by (2.1).

From Lemma 2.1, Lemma 2.2 and the above proposition it follows that the unique classical solution of (2.4) is $P_n(t) = \mathcal{U}_n(t,s)(S)$, where $\mathcal{U}_n(t,s) \in L(\mathcal{H}_2)$ is the evolution operator generated by L_n and $\frac{\partial \mathcal{U}_n(t,s)S}{\partial s} = -\mathcal{U}_n(t,s)L_n(s)S$ for all $t \geq s \geq 0$, $S \in \mathcal{H}_2$. Now it is clear that

$$\frac{\partial}{\partial \sigma} \langle \mathcal{U}_n^*(t,\sigma)R, S\rangle_2 = \langle -L_n^*(\sigma)\mathcal{U}_n^*(t,\sigma)R, S\rangle_2, \quad S, R \in \mathcal{H}_2$$

for all $t \geq \sigma \geq 0$. Let us consider $S = x \otimes x$, $x \in H$. It is easy to see that $\langle Fx, x\rangle = TrF(x \otimes x)$ for all $F \in L(H)$ and $x \in H$. If $F \in \mathcal{H}_2$ then $\langle F, S\rangle_2 = \langle Fx, x\rangle$. Integrating from s to t, we have

$$\langle \mathcal{U}_n^*(t,s)Rx, x\rangle - \langle Rx, x\rangle = \int_s^t \langle L_n^*(\sigma)\mathcal{U}_n^*(t,\sigma)Rx, x\rangle\, d\sigma, \quad R \in \mathcal{H}_2. \tag{2.5}$$

Let us consider the Lyapunov equation

$$\frac{dQ(s)}{ds} + A^*(s)Q(s) + Q(s)A(s) + \sum_{i=1}^m G_i^*(s)Q(s)G_i(s) = 0 \tag{2.6}$$

on $L(H)$. The approximating equation of (2.6) is

$$\frac{dQ_n(s)}{ds} + A_n^*(s)Q_n(s) + Q_n(s)A_n(s) + \sum_{i=1}^m G_i^*(s)Q_n(s)G_i(s) = 0 \tag{2.7}$$

where $A_n(s)$ is the Yosida approximation of $A(s)$. Let $Q_n(t,s;R)$ be the unique classical solution of (2.7) such as $Q_n(t) = R$, $R \geq 0$. Hence we have

$$\langle Q_n(t,s;R)x, x\rangle - \langle Rx, x\rangle = \int_s^t \langle L_n^*(\sigma)Q_n(t,\sigma;R)x, x\rangle\, d\sigma, \quad R \geq 0. \tag{2.8}$$

If $R \in \mathcal{H}_2, R \geq 0$ it follows from (2.5) and (2.8)

$$\langle [\mathcal{U}_n^*(t,s)R - Q_n(t,s;R)]\,x,x\rangle = \int_s^t \langle L_n^*(\sigma)\,[\mathcal{U}_n^*(t,\sigma)R - Q_n(t,\sigma;R)]\,x,x\rangle\, d\sigma.$$

By the Uniform Boundedness Principle there exists $l_T > 0$ such that $\|L_n^*(t)P\| \leq l_T\,\|P\|$ for all $t \in [0,T]$, $P \in L(H)$ and we obtain

$$\|\mathcal{U}_n^*(t,s)R - Q_n(t,s;R)\| \leq \int_s^t l_T\,\|\mathcal{U}_n^*(t,\sigma)R - Q_n(t,\sigma;R)\|\, d\sigma.$$

Now we use Gronwall's inequality and we get

$$\mathcal{U}_n^*(t,s)R = Q_n(t,s;R) \quad \text{for all } R \in \mathcal{H}_2, R \geq 0, t \geq s. \tag{2.9}$$

From Proposition 2.3 and (1.1) we deduce that for all $R \in \mathcal{H}_2$ the map

$$(t,s) \to Q_n(t,s;R) \text{ is } \|\cdot\|\text{-continuous on } \{(s,t) \mid 0 \leq s \leq t\} \text{ and} \tag{2.10}$$

$$Q_n(t,s;\alpha R + \beta S) = \alpha Q_n(t,s;R) + \beta Q_n(t,s;S) \tag{2.11}$$

for all $\alpha, \beta \in \mathbf{R}_+$ and $R, S \in \mathcal{H}_2$, $R, S \geq 0$..

3. The covariance operator of the mild solutions of linear stochastic differential equations and the Lyapunov equations

Let $\xi \in L^2(\Omega, H)$. We denote by $E(\xi \otimes \xi)$ the bounded and linear operator which act on H given by $E(\xi \otimes \xi)(x) = E(\langle x, \xi\rangle\, \xi)$. The operator $E(\xi \otimes \xi)$ is called the covariance operator of ξ.

Lemma 3.1. ([6]) *Let V be another real, separable Hilbert space and $A \in L(H,V)$. If $\xi \in L^2(\Omega, H)$ then $E\,\|A(\xi)\|^2 = \|AE(\xi \otimes \xi)A^*\|_1 < \infty$. Particularly $E\,\|\xi\|^2 = \|E(\xi \otimes \xi)\|_1$.*

Proposition 3.2. *If $y_n(t,s;\xi)$, $\xi \in L_s^2(H)$ is the classical solution of (1.4) then $E[y_n(t,s;\xi) \otimes y_n(t,s;\xi)]$ is the unique classical solution of the following initial value problem*

$$\frac{dP_n(t)}{dt} = A_n(t)P_n(t) + P_n(t)A_n^*(t) + \sum_{i=1}^m G_i(t)P_n(t)G_i^*(t) \tag{3.1}$$

$$P_n(s) = E(\xi \otimes \xi), \xi \in L_s^2(H).$$

Proof. Let $u \in H$ and $T \geq 0$, fixed. We consider the function $F_u \overset{not}{=} F : \mathbf{R}_+ \times H \to \mathbf{R}$, $F(t,x) = \langle (x \otimes x)\,u, u\rangle$. By using Ito's formula for F and $y_n(t,s;\xi)$ we obtain for all $0 \leq s \leq t \leq T$

$$\begin{aligned}
&\langle [y_n(t,s;\xi) \otimes y_n(t,s;\xi)]\, u, u\rangle - \langle (\xi \otimes \xi)\, u, u\rangle \\
&\quad = \int_s^t \langle [A_n(r) y_n(r,s;\xi) \otimes y_n(r,s;\xi)]\, u, u\rangle + \langle [y_n(r,s;\xi) \otimes A_n(r) y_n(r,s;\xi)]\, u, u\rangle \\
&\qquad + \sum_{i=1}^m \langle [G_i(r) y_n(r,s;\xi) \otimes G_i(r) y_n(r,s;\xi)]\, u, u\rangle\, dr \\
&\qquad + \int_s^t \langle [y_n(r,s;\xi) \otimes G_i(r) y_n(r,s;\xi)]\, u, u\rangle \\
&\qquad + \langle [G_i(r) y_n(r,s;\xi) \otimes y_n(r,s;\xi)]\, u, u\rangle\, dw_i(r).
\end{aligned}$$

Taking expectations, we have

$$\begin{aligned}
&\langle E[y_n(t,s;\xi) \otimes y_n(t,s;\xi)] u, u\rangle - \langle E[\xi \otimes \xi] u, u\rangle \\
&= \int_s^t \langle E[y_n(r,s;\xi) \otimes y_n(r,s;\xi)] u, A_n^*(r) u\rangle + \langle E[y_n(r,s;\xi) \otimes y_n(r,s;\xi)] A_n^*(r) u, u\rangle \\
&\qquad + \sum_{i=1}^m \langle E[y_n(r,s;\xi) \otimes y_n(r,s;\xi)] G_i^*(r) u, G_i^*(r) u\rangle\, dr.
\end{aligned}$$

If $P_n(t) = E[y_n(t,s;\xi) \otimes y_n(t,s;\xi)]$ then

$$\begin{aligned}
&\langle P_n(t) u, u\rangle - \langle E[\xi \otimes \xi] u, u\rangle \\
&= \int_s^t \langle A_n(r) P_n(r) u, u\rangle + \langle P_n(r) A_n^*(r) u, u\rangle + \sum_{i=1}^m \langle G_i(r) P_n(r) G_i^*(r) u, u\rangle\, dr.
\end{aligned} \tag{3.2}$$

According with Lemmas 2.2, 2.1 and the statements of the last section, the equation (3.1) has a unique classical solution $\mathcal{U}_n(t,s) E\left(\xi \otimes \xi\right)$ in $\mathcal{H}_2$ and we have $\mathcal{U}_n(t,s) E\left(\xi \otimes \xi\right) = E\left(\xi \otimes \xi\right) + \int_s^t L_n(r) \mathcal{U}_n(r,s) E\left(\xi \otimes \xi\right) dr$. We note that $\mathcal{U}_n(t,s)$ is the evolution operator generated by L_n. Then $\langle \mathcal{U}_n(t,s) E\left(\xi \otimes \xi\right), u \otimes u\rangle_2 = \langle E\left(\xi \otimes \xi\right), u \otimes u\rangle_2 + \int_s^t \langle L_n(r) \mathcal{U}_n(r,s) E\left(\xi \otimes \xi\right), u \otimes u\rangle_2\, dr$ or equivalently

$$\langle \mathcal{U}_n(t,s) E\left(\xi \otimes \xi\right) u, u\rangle = \langle E\left(\xi \otimes \xi\right) u, u\rangle + \int_s^t \langle L_n(r) \mathcal{U}_n(r,s) E\left(\xi \otimes \xi\right) u, u\rangle\, dr.$$

From (3.2) and the last equality we obtain

$$\langle [\mathcal{U}_n(t,s) E\left(\xi \otimes \xi\right) - P_n(t)]\, u, u\rangle = \int_s^t \langle L_n(r) \left[\mathcal{U}_n(r,s) E\left(\xi \otimes \xi\right) - P_n(r)\right] u, u\rangle\, dr.$$

Since there exists $l_T > 0$ such that $\|L_n(t)\| \le l_T$ for all $t \in [0,T]$ and $\mathcal{U}_n(t,s)E(\xi \otimes \xi), P_n(t) \in \mathcal{E}$ we can use the Gronwall's inequality to deduce that

$$E[y_n(t,s;\xi) \otimes y_n(t,s;\xi)] = \mathcal{U}_n(t,s)E(\xi \otimes \xi) \tag{3.3}$$

for all $t \in [s,T]$. Since T is arbitrary we obtain the conclusion. □

The following theorem gives a representation of the covariance operator associated to the mild solution of (1.2), by using the mild solution of the Lyapunov equation (2.6).

Theorem 3.3. *Let V be another real separable Hilbert space and $B \in L(H,V)$. If $y(t,s;\xi), \xi \in L_s^2(H)$ is the mild solution of* (1.2) *and $Q(t,s,R)$ is the unique mild solution of* (2.6) *with the initial value $Q(t) = R \ge 0$ then:*

(a) $\langle E[y(t,s;\xi) \otimes y(t,s;\xi)]u,u\rangle = TrQ(t,s;u \otimes u)E(\xi \otimes \xi)$ *for all* $u \in H$;

(b) $E\|B\, y(t,s;\xi)\|^2 = TrQ(t,s;B^*B)E(\xi \otimes \xi)$.

Proof. (a) Let $u \in H, \xi \in L_s^2(H)$ and $y_n(t,s;\xi)$ be the classical solution of (1.4). By (3.3) we obtain successively

$$\begin{aligned}\langle E[y_n(t,s;\xi) \otimes y_n(t,s;\xi)]u,u\rangle &= \langle u \otimes u, \mathcal{U}_n(t,s)E(\xi \otimes \xi)\rangle_2 \\ &= \langle \mathcal{U}_n^*(t,s)(u \otimes u), E(\xi \otimes \xi)\rangle_2 \\ &= Tr\mathcal{U}_n^*(t,s)(u \otimes u)E(\xi \otimes \xi).\end{aligned}$$

From (2.9) we get

$$\langle E[y_n(t,s;\xi) \otimes y_n(t,s;\xi)]u,u\rangle = TrQ_n(t,s;u \otimes u)E(\xi \otimes \xi) \tag{3.4}$$

where $Q_n(t,s;u \otimes u)$ is the solution of (2.7) with $Q_n(t) = u \otimes u$. As $n \to \infty$ we get the conclusion. Indeed, since $Q_n(t,s;u \otimes u) \underset{n\to\infty}{\to} Q(t,s;u \otimes u)$ in the strong operator topology (Lemma 1.2) then it is not difficult to deduce from Lemma 1 of [3] that $Q_n(t,s;u \otimes u)E(\xi \otimes \xi) \underset{n\to\infty}{\to} Q(t,s;u \otimes u)E(\xi \otimes \xi)$ in $C_1(H)$. It is known that the map $Tr : C_1(H) \to \mathbf{C}$ is continuous. So we obtain

$$TrQ_n(t,s;u \otimes u)E(\xi \otimes \xi) \underset{n\to\infty}{\to} TrQ(t,s;u \otimes u)E(\xi \otimes \xi).$$

On the other hand, for all $u \in H$ we have

$$\begin{aligned}&|\langle\{E[y_n(t,s;\xi) \otimes y_n(t,s;\xi)] - E[y(t,s;\xi) \otimes y(t,s;\xi)]\}u,u\rangle| \\ &= \left|E(\langle y_n(t,s;\xi),u\rangle^2 - \langle y(t,s;\xi),u\rangle^2)\right| \\ &\le E\|y_n(t,s;\xi) - y(t,s;\xi)\|^2\|u\|^2 + 2E(\|y_n(t,s;\xi) - y(t,s;\xi)\|\,\|y(t,s;\xi)\|)\|u\|^2 \\ &\le \{E\|y_n(t,s;\xi) - y(t,s;\xi)\|^2 \\ &\qquad + 2(E\|y_n(t,s;\xi) - y(t,s;\xi)\|^2 E\|y(t,s;\xi)\|^2)^{1/2}\}\|u\|^2.\end{aligned}$$

From Lemma 1.1 and the last inequality we get

$$\langle E[y_n(t,s;\xi) \otimes y_n(t,s;\xi)]u,u\rangle \underset{n\to\infty}{\to} \langle E[y(t,s;\xi) \otimes y(t,s;\xi)]u,u\rangle$$

and the proof is finished.

(b) Let $\xi \in L_s^2(H)$ and $n \in \mathbf{N}$. It is sufficient to prove that

$$E\|By_n(t,s;\xi)\|^2 = TrQ_n(t,s;B^*B)E(\xi\otimes\xi).$$

By Lemma 3.1 we have

$$E\|By_n(t,s;\xi)\|^2 = \|BE[y_n(t,s;\xi)\otimes y_n(t,s;\xi)]B^*\|_1. \tag{3.5}$$

If $\{e_i\}_{i\in\mathbf{N}^*}$ is an orthonormal basis in V then we deduce from (a)

$$\begin{aligned}\|BE[\, y_n(t,s;\xi)\otimes y_n(t,s;\xi)]B^*\|_1 &= \sum_{i=1}^{\infty}\langle E[y_n(t,s;\xi)\otimes y_n(t,s;\xi)]B^*e_i, B^*e_i\rangle \\ &= \sum_{i=1}^{\infty} TrQ_n(t,s;B^*e_i\otimes B^*e_i)E(\xi\otimes\xi).\end{aligned}$$

Since $B^*e_i\otimes B^*e_i \in \mathcal{H}_2$ and $B^*e_i\otimes B^*e_i \geq 0$ for all $i\in\mathbf{N}^*$ we have by (2.11)

$$\begin{aligned}&\|BE[\, y_n(t,s;\xi)\otimes y_n(t,s;\xi)]B^*\|_1 \\ &\quad = \lim_{p\to\infty} TrQ_n(t,s;\sum_{i=1}^{p} B^*e_i\otimes e_iB)E(\xi\otimes\xi).\end{aligned} \tag{3.6}$$

The sequence $B_p = \sum_{i=1}^{p} B^*e_i\otimes e_iB$ is increasing and bounded above,

$$\langle B_p x, x\rangle = \sum_{i=1}^{p}\langle Bx, e_i\rangle^2 \leq \|Bx\|^2 = \langle B^*Bx, x\rangle.$$

Then $\{B_p\}_{p\in\mathbf{N}}$ converges in the strong operator topology to the operator $B^*B \in L^+(H)$. By Lemma 1.2 we deduce that the sequence $\{Q_n(t,s;B_p)\}_{p\in\mathbf{N}^*}$ is increasing and $Q_n(t,s;B_p) \leq Q_n(t,s;B^*B)$ for all $p\in\mathbf{N}^*$ and consequently it converges in the strong operator topology to the operator $Q_n(t,s)\in L^+(H)$.

If $U_n(t,s)$ is the evolution operator relative to $A_n(t)$ we have

$$\begin{aligned}&Q_n(t,s;B_p)x \\ &\quad = U_n^*(t,s)B_pU_n(t,s)x + \sum_{i=1}^{m}\int_s^t U_n^*(r,s)G_i^*(r)Q_n(t,r;B_p)G_i(r)U_n(r,s)x\,dr\end{aligned}$$

for all $x\in H$. We may write

$$\begin{aligned}\langle Q_n(t,s;B_p)x,x\rangle &= \langle B_pU_n(t,s)x, U_n(t,s)x\rangle \\ &\quad + \sum_{i=1}^{m}\int_s^t \langle Q_n(t,r;B_p)G_i(r)U_n(r,s)x, G_i(r)U_n(r,s)x\rangle\,dr.\end{aligned} \tag{3.7}$$

Since $B_p\in\mathcal{H}_2$ and $B_p\geq 0$ we deduce from (2.10) and the hypothesis that $r\to\langle Q_n(t,r;B_p)G_i(r)U_n(r,s)x, G_i(r)U_n(r,s)x\rangle$ is continuous.

On the other hand

$$\lim_{p\to\infty} \langle Q_n(t,r;B_p)G_i(r)U_n(r,s)x, G_i(r)U_n(r,s)x\rangle$$
$$= \langle Q_n(t,r)G_i(r)U_n(r,s)x, G_i(r)U_n(r,s)x\rangle$$

for all $r \in [s,t]$.

Thus it follows that $r \to \langle Q_n(t,r)G_i(r)U_n(r,s)x, G_i(r)U_n(r,s)x\rangle$ is a Borel measurable and nonnegative function defined on $[s,t]$ and bounded above by a continuous function, namely $r \to \langle Q_n(t,r;B^*B)G_i(r)U_n(r,s)x, G_i(r)U_n(r,s)x\rangle$.

From the Monotone Convergence Theorem we can pass to limit $p \to \infty$ in (3.7) and we have

$$\langle Q_n(t,s)x, x\rangle = \langle B^*BU_n(t,s)x, U_n(t,s)x\rangle$$
$$+\sum_{i=1}^{m}\int_s^t \langle Q_n(t,r)G_i(r)U_n(r,s)x, G_i(r)U_n(r,s)x\rangle\, dr,$$

where the above integral is in the Lebesgue sense. Since $Q_n(t,r;B^*B)$ also satisfies (3.7) we have

$$\langle [Q_n(t,s;B^*B) - Q_n(t,s)]x, x\rangle$$
$$= \sum_{i=1}^{m}\int_s^t \langle [Q_n(t,r;B^*B) - Q_n(t,r)]G_i(r)U_n(r,s)x, G_i(r)U_n(r,s)x\rangle\, dr. \tag{3.8}$$

The map $x \to \langle [Q_n(t,r;B^*B) - Q_n(t,r)]x, x\rangle$, $x \in H$ is continuous and $r \to \langle [Q_n(t,r;B^*B) - Q_n(t,r)]x, x\rangle$, $r \in [s,t]$ is a Borel measurable function.

Since $B_1 = \{x \in H, \|x\| = 1\}$ is separable [1] then there exists a net $\{y_n\}_{n\in\mathbf{N}} \subset B_1$ which is dense in B_1 and

$$\|Q_n(t,r;B^*B) - Q_n(t,r)\| = \sup_{y_n\in B_1} \langle [Q_n(t,r;B^*B) - Q_n(t,r)]y_n, y_n\rangle .$$

Thus $r \to \|Q_n(t,r;B^*B) - Q_n(t,r)\|$, $r \in [s,t]$ is a Borel measurable function.

Since $0 \le Q_n(t,r;B^*B) - Q_n(t,r) \le Q_n(t,r;B^*B)$ it is clear that $r \to \|Q_n(t,r;B^*B) - Q_n(t,r)\|\,\|U_n(r,s)\|^2$ is Lebesgue integrable. By (3.8) we have

$$\|Q_n(t,s;B^*B) - Q_n(t,s)\| \le \sum_{i=1}^{m}\widetilde{G}_i\int_s^t \|Q_n(t,r;B^*B) - Q_n(t,r)\|\,\|U_n(r,s)\|^2\, dr.$$

We use the Gronwall's inequality and we get $\|Q_n(t,s;B^*B) - Q_n(t,s)\| = 0$.

Thus $Q_n(t,s;B_p)x \underset{p\to\infty}{\to} Q_n(t,s;B^*B)x$ for all $x \in H$ and from Lemma 1 in [3] we can deduce that $Q_n(t,s;B_p)E(\xi\otimes\xi) \underset{p\to\infty}{\to} Q_n(t,s;B^*B)E(\xi\otimes\xi)$ in $\|.\|_1$. By (3.5), (3.6) and since Tr is continuous on $C_1(H)$ we obtain $E\|By_n(t,s;\xi)\|^2 = TrQ_n(t,s;B^*B)E(\xi\otimes\xi)$. As $n\to\infty$ we obtain the conclusion. □

4. The uniform exponential stability and uniform observability

We assume $C \in C_s([0,\infty), L(H,V))$.

Definition 4.1. [12] We say that (1.2) is uniformly exponential stable if there exist the constants $M \geq 1$, $\omega > 0$ such that

$$E\left\|y(t,s;x)\right\|^2 \leq Me^{-\omega(t-s)}\left\|x\right\|^2$$

for all $t \geq s \geq 0$ and $x \in H$.

We consider the equation (1.2) and the observation relation

$$z(t) = C(t)y(t,s,x) \tag{4.1}$$

The system (1.2), (4.1) will be denoted $\{A, C; G_i\}$.

Since $y(\cdot,s;x) \in C([s,T];L^2(\Omega,H))$ for all $x \in H$ it follows that $C(\cdot)y(\cdot,s;x) \in C([s,T];L^2(\Omega,V))$. We note that

$$t \to E\left\|C(t)y(t,s;x)\right\|^2 \text{ is continuous on } [s,T]. \tag{4.2}$$

Definition 4.2. ([7]) The system $\{A, C; G_i\}$ is uniformly observable if there exist $\tau > 0$ and $\gamma > 0$ such that

$$E\int_s^{s+\tau}\left\|C(t)y(t,s;x)\right\|^2 dt \geq \gamma\left\|x\right\|^2$$

for all s $\in \mathbf{R}_+$ and $x \in H$.

If $Q(t,s,R)$ is the unique mild solution of (2.6) such that $Q(t) = R, R \geq 0$ we have the following results.

Theorem 4.3. *The system* (1.2) *is uniformly exponentially stable if and only if there exist the constants* $M \geq 1$, $\omega > 0$ *such that*

$$Q(t,s;I) \leq Me^{-\omega(t-s)}I \quad \text{for all } t \geq s \geq 0,$$

where I *is the identity operator on* H.

Proof. From Theorem 3.3 (b) it follows $E\left\|y(t,s;x)\right\|^2 = \langle Q(t,s;I)x,x\rangle$ for all $x \in H$, where I is the identity operator. By Definition 4.1 we obtain the conclusion. □

Theorem 4.4. *The system* $\{A, C; G_i\}$ *is uniformly observable iff there exist* $\tau > 0$ *and* $\gamma > 0$ *such that* $\int_s^{s+\tau} Q(t,s;C^*(t)C(t))dt \geq \gamma I$ *for all* $s \in \mathbf{R}_+$, *where* I *is the identity operator on* H.

Proof. By Theorem 3.3 b) we get $E\left\|C(t)y(t,s;x)\right\|^2 = \langle Q(t,s;C^*(t)C(t))x,x\rangle$ for all $x \in H$. Because $t \to E\left\|C(t)y(t,s;x)\right\|^2$ is continuous we deduce

$$\int_s^{s+\tau} E\left\|C(t)y(t,s;x)\right\|^2 dt < \infty.$$

From Definition 4.2 and Fubini's theorem it follows the conclusion. □

5. The uniform observability and the Lyapunov equations

Let us assume

$$G_i \in C_b([0,\infty), L(H)),\ i = 1,\dots,m,$$
$$C \in C_b([0,\infty), L(H,V))$$

and

$$CC^* \in C_b([0,\infty), L(H)).$$

We put

$$\sup_{t\in\mathbf{R}_+} \|C(t)\| = \widetilde{C} < \infty \quad \text{and} \quad \sup_{t\in\mathbf{R}_+} \|G_i(t)\| = \widetilde{G}_i < \infty, i = 1,\dots,m.$$

The main result of this section is Theorem 5.2 which gives a characterization of the uniform exponential stability of linear stochastic systems, in terms of Lyapunov equations under uniform observability condition .

Lemma 5.1. *Let $T > 0$ and $0 \le s \le r \le T$. If $Q_n(r,s;C^*(r)C(r))$ is the classical solution of (2.7) such that $Q_n(r) = C^*(r)C(r)$ then*

(a) *the map $s \to Q_n(r,s;C^*(r)C(r))$ is continuous in the uniform operator topology on $[0,r]$ uniformly with respect to r for all $r \le T$.*
(b) *the map $r \to \langle Q_n(r,s;C^*(r)C(r))x, y\rangle$ is continuous on $[s,T]$ for all $x,y \in H$.*
(c) *the map $s \to \left\langle \frac{\partial Q_n(r,s;C^*(r)C(r))}{\partial s}x, y\right\rangle$ is continuous on $[0,r]$, uniformly with respect to r, for all $r \le T$ and $x,y \in H$.*
(d) *the map $r \to \left\langle \frac{\partial Q_n(r,s;C^*(r)C(r))}{\partial s}x, y\right\rangle$ is continuous on $[s,T]$ for all $x,y \in H$.*

Proof. (a) If $Q_n(r,s) = Q_n(r,s;C^*(r)C(r))$ we deduce by Lemma 1.2

$$Q_n(r,s)x = U_n^*(r,s)C^*(r)C(r)U_n(r,s)x + \sum_{i=1}^{m}\int_s^r U_n^*(p,s)G_i^*(p)Q_n(r,p)G_i(p)U_n(p,s)x\,dp.$$

It is not difficult to see that there exists a positive constant $\widetilde{M}_{T,1}$ such that $\|U_n(r,s)\| \le \widetilde{M}_{T,1}$ for all $0 \le s \le r \le T$ and $n \in \mathbf{N}$. Then we get

$$\|Q_n(r,s)x\| \le \widetilde{M}_{T,1}^2 \|x\| \left[\widetilde{C}^2 + \sum_{i=1}^{m}\widetilde{G}_i^2 \int_s^r \|Q_n(r,p)\|\,dp\right]$$

and by Gronwall's inequality it follows that there exists the positive constant q_T depending on $\widetilde{M}_{T,1}, \widetilde{C}, \widetilde{G}_i, i = 1,\dots,m$ and T such that $\|Q_n(r,s)\| \le q_T$ for all

$0 \le s \le r \le T$ and $n \in \mathbf{N}$. If $0 \le s_0 \le s \le r$ we have

$$\begin{aligned}
&Q_n(r,s)x - Q_n(r,s_0)x \\
&\quad = U_n^*(r,s)C^*(r)C(r)U_n(r,s)x - U_n^*(r,s_0)C^*(r)C(r)U_n(r,s_0)x \\
&\qquad + \sum_{i=1}^{m}\int_s^r U_n^*(p,s)G_i^*(p)Q_n(r,p)G_i(p)U_n(p,s)x - \\
&\qquad - U_n^*(p,s_0)G_i^*(p)Q_n(r,p)G_i(p)U_n(p,s_0)x dp \\
&\qquad + \sum_{i=1}^{m}\int_{s_0}^{s} U_n^*(p,s_0)G_i^*(p)Q_n(r,p)G_i(p)U_n(p,s_0)x dp
\end{aligned}$$

and

$$\begin{aligned}
&\|Q_n(r,s)x - Q_n(r,s_0)x\| \\
&\quad \le \widetilde{C}^2 \|U_n(r,s) - U_n(r,s_0)\|^2 \|x\| + 2\widetilde{C}^2\widetilde{M}_{T,1} \|U_n(r,s) - U_n(r,s_0)\| \|x\| \\
&\qquad + \sum_{i=1}^{m} \widetilde{G}_i^2 q_T \int_s^r \|U_n(p,s) - U_n(p,s_0)\|^2 \|x\| \\
&\qquad + 2\widetilde{M}_{T,1} \|U_n(r,s) - U_n(r,s_0)\| \|x\| dp \\
&\qquad + \sum_{i=1}^{m} \widetilde{G}_i^2 q_T \widetilde{M}_{T,1}^2 (s - s_0) \|x\| .
\end{aligned}$$

Since the map $(r,s) \to U_n(r,s)$ is continuous in the uniform operator topology (by Theorems 5.5.1 and 5.5.2 in [9]) on the compact set $\{(r,s), 0 \le s \le r \le T\}$ it follows that the map $s \to U_n(r,s)$ is continuous in the uniform operator topology on $[0,r]$, uniformly with respect to r, for all $r \le T$.

Thus, for any $\varepsilon \in (0,1)$ there exists $\delta_\varepsilon \in (0,\varepsilon)$ which not depend on r, such that $\|U_n(r,s) - U_n(r,s_0)\| < \varepsilon$ for all $0 < s - s_0 < \delta_\varepsilon$, $(r,s), (r,s_0) \in \{(r,s), 0 \le s \le r \le T\}$. Then for all $0 < s - s_0 < \delta_\varepsilon$ we have

$$\|Q_n(r,s) - Q_n(r,s_0)\| \le \varepsilon\left[\left(1 + 2\widetilde{M}_{T,1}\right)\left(\widetilde{C}^2 + T\sum_{i=1}^{m}\widetilde{G}_i^2 q_T\right) + \sum_{i=1}^{m}\widetilde{G}_i^2 q_T \widetilde{M}_{T,1}^2\right]$$

and $\lim\limits_{s \searrow s_0} \|Q_n(r,s) - Q_n(r,s_0)\| = 0$, uniformly with respect to r for all $r \le T$.

Similarly we can prove that $\lim\limits_{s \nearrow s_0} \|Q_n(r,s) - Q_n(r,s_0)\| = 0$, uniformly with respect to r for all $r \le T$, and the conclusion follows.

(b) follows by (4.2) and Theorem 3.3.

(c) With the notation of (a) we have

$$\frac{\partial Q_n(r,s)}{\partial s}x = -A_n^*(s)Q_n(r,s)x - Q_n(r,s)A_n(s)x - \sum_{i=1}^{m} G_i^*(s)Q_n(r,s)G_i(s)x. \quad (5.1)$$

The map $s \to -(A_n^*(s)Q_n(r,s) + Q_n(r,s)A_n(s))$ is continuous in the uniform operator topology, uniformly with respect to r, by (a) and by Lemma 2.1. So it is sufficient to prove that the maps $s \to -\langle G_i^*(s)Q_n(r,s)G_i(s)x, x\rangle, i = 1, \ldots, m$ are continuous in the uniform operator topology, uniformly with respect to r. If $0 \le s_0 \le r \le T$ we have

$$\begin{aligned}
&|\langle G_i^*(s)Q_n(r,s)G_i(s)x, x\rangle - \langle G_i^*(s_0)Q_n(r,s_0)G_i(s_0)x, x\rangle| \\
&\qquad \le q_T \|[G_i(s) - G_i(s_0)]\, x\|^2 + 2q_T\widetilde{G}_i \|[G_i(s) - G_i(s_0)]\, x\| \\
&\qquad\quad + \widetilde{G}_i^2 \|Q_n(r,s) - Q_n(r,s_0)\| \, \|x\|^2 .
\end{aligned}$$

By (a) we obtain the conclusion. Since the operators $G_i^*(s)Q_n(r,s)G_i(s)$ $i \in \{1, \ldots, m\}$ are self-adjoint we obtain (c). The statement d) follows by (5.1) and (b). □

P_2: The evolution operator $U(t,s)$ relative to $A(t)$ has an exponentially growth, that is, there exist the positive constants $M_1 \ge 1$ and ω_1 such that $\|U(t,s)\| \le M_1 e^{\omega_1(t-s)}$ for all $t \ge s \ge 0$.

Theorem 5.2. *We assume that P_2 holds and we consider the equations* (1.5), (1.7) *in the particular case when $D(s) = C^*(s)C(s), s \ge 0$. If $\{A, C; G_i\}$ is uniformly observable then the equation* (1.2) *is uniformly exponentially stable if and only if the equation* (1.5) *has a unique mild solution $\mathcal{Q}$ with the property that there exist the positive constants $\widetilde{m}$, $\widetilde{M}$ such that*

$$\widetilde{m} \|x\|^2 \le \langle \mathcal{Q}(s)x, x\rangle \le \widetilde{M} \|x\|^2 \tag{5.2}$$

for all $s \ge 0$ and $x \in H$.

Proof. $\Rightarrow$ Let $\mathcal{Q}$ be the linear, nonnegative operator, which is given by

$$\langle \mathcal{Q}(s)x, x\rangle = E \int_s^\infty \|C(r)y(r,s;x)\|^2 \, dr.$$

If M, ω are the constants introduced by Definition 4.1 then we have

$$0 \le \langle \mathcal{Q}(s)x, x\rangle \le \widetilde{C}^2 \int_s^\infty E \|y(r,s;x)\|^2 \, dr \le \widetilde{C}^2 \int_s^\infty M e^{-\omega(r-s)} \|x\|^2 \, dr < \widetilde{C}^2 \frac{M}{\omega} \|x\|^2 .$$

It is clear that $\mathcal{Q}(s) \in L^+(H)$ is well defined. Since $\{A, C; G_i\}$ is uniformly observable we deduce that there exist $\gamma > 0$ such that

$$\langle \mathcal{Q}(s)x, x\rangle \ge \gamma \|x\|^2 \quad \text{for all } s \ge 0 \text{ and } x \in H.$$

We take $\widetilde{m} = \gamma$ and $\widetilde{M} = \widetilde{C}^2 \frac{M}{\omega}$ and we obtain (5.2).

Let be $T > 0$. If $y_n(t, s; \xi)$ is the classical solution of (1.4) we consider the linear nonnegative operator $\mathcal{Q}_T^n(s)$ given by

$$\langle \mathcal{Q}_T^n(s)x, x\rangle = E \int_s^T \|C(r)y_n(r, s; x)\|^2 \, dr.$$

From Theorem 3.3 we deduce

$$\langle \mathcal{Q}_T^n(s)x, x\rangle = \int_s^T \langle Q_n(r, s; C^*(r)C(r))x, x\rangle \, dr,$$

where Q_n is the solution of the approximating Lyapunov equation (2.7). Since the hypotheses of Lemma 5.1 hold, it is not difficult to see that the function $\sigma \to \langle \mathcal{Q}_T^n(\sigma)x, x\rangle$ is differentiable on $[0, T]$ and

$$\frac{\partial}{\partial \sigma} \langle \mathcal{Q}_T^n(\sigma)x, x\rangle = -\langle L_n^*(\sigma)\mathcal{Q}_T^n(\sigma)x, x\rangle - \|C(\sigma)x\|^2,$$

where $L_n^*(s)$ is the operator introduced by (2.2). Integrating from s to T with respect to σ we get

$$\langle \mathcal{Q}_T^n(s)x, x\rangle = \int_s^T \langle [L_n^*(\sigma)\mathcal{Q}_T^n(\sigma) + C^*(\sigma)C(\sigma)]x, x\rangle \, d\sigma$$

The unique classical solution $\widehat{\mathcal{Q}}_T^n$ of (1.7) such that $\widehat{\mathcal{Q}}_T^n(T) = 0$ also satisfies the above integral equation. We have

$$\left\langle \left[\mathcal{Q}_T^n(s) - \widehat{\mathcal{Q}}_T^n(s)\right] x, x\right\rangle = \int_s^T \left\langle L_n^*(\sigma)\left[\mathcal{Q}_T^n(\sigma) - \widehat{\mathcal{Q}}_T^n(\sigma)\right] x, x\right\rangle d\sigma$$

Since $\mathcal{Q}_T^n(s)$ and $\widehat{\mathcal{Q}}_T^n(s)$, $s \in [0, T]$ are self-adjoint operators we use Gronwall's inequality and we obtain

$$\mathcal{Q}_T^n(s) = \widehat{\mathcal{Q}}_T^n(s) \tag{5.3}$$

for all $s \in [0, T]$. By Lemma 5.1 (b) and since there exists $q_T > 0$ such that $\|Q_n(r, s; C^*(r)C(r))\| \leq q_T$ for all $0 \leq s \leq r \leq T$ and $n \in \mathbf{N}$ we deduce that the map $r \to Q_n(r, s; C^*(r)C(r))x$ is Bochner integrable on $[s, T]$ and

$$\int_s^T Q_n(r, s; C^*(r)C(r))x dr < \infty.$$

Then

$$\mathcal{Q}_T^n(s)x = \int_s^T Q_n(r, s; C^*(r)C(r))x dr.$$

As $n \to \infty$ we have $\mathcal{Q}_T^n(s)x \underset{n\to\infty}{\longrightarrow} \mathcal{Q}_T(s)x$ for all $x \in H$, by Lemma 1.2. If Q is the mild solution of (2.6) then

$$\mathcal{Q}_T(s)x = \int_s^T Q(r,s;C^*(r)C(r))x dr$$

is the unique mild solution of (1.5) such that $\mathcal{Q}_T(T) = 0$.

Since the function $T \to \mathcal{Q}_T(s)$ is increasing and bounded above on $\mathbf{R}_+$, for every s fixed, it is clear that the sequence $\{\mathcal{Q}_T(s)\}_T$ converges (as $T \to \infty$) in the strongly operator topology to $\mathcal{Q}(s)$.

Let us consider the sequence $T_n \geq 0, n \in \mathbf{N}$, $T_n \to \infty$. If $s \leq t \leq T_n$ we have

$$\begin{aligned}\mathcal{Q}_{T_n}(s)x &= U^*(t,s)\mathcal{Q}_{T_n}(s)U(t,s)x \\ &+ \int_s^t U^*(r,s)[\sum_{i=1}^m G_i^*(r)\mathcal{Q}_{T_n}(r)G_i(r) + C^*(r)C(r)]U(r,s)x dr.\end{aligned}$$

As $T_n \to \infty$ it follows that $\mathcal{Q}(s)$ satisfies the integral equation

$$\begin{aligned}\mathcal{Q}(s)x &= U^*(t,s)\mathcal{Q}(s)U(t,s)x \\ &+ \int_s^t U^*(r,s)[\sum_{i=1}^m G_i^*(r)\mathcal{Q}(r)G_i(r) + C^*(r)C(r)]U(r,s)x dr.\end{aligned}$$

By Gronwall's inequality argument we see that $\mathcal{Q}(s)$ is a mild solution of (1.5).

For uniqueness, assume $\mathcal{Q}$, $\mathcal{Q}^1$ are solutions of (1.5) which satisfy (5.2). We consider the function $F(t,x) = \langle[\mathcal{Q}_n(t) - \mathcal{Q}_n^1(t)]x, x\rangle$, where $\mathcal{Q}_n(t)$ and $\mathcal{Q}_n^1(t)$ are solutions of (1.5) such that $\mathcal{Q}_n(T) = \mathcal{Q}(T)$ and respectively $\mathcal{Q}_n^1(T) = \mathcal{Q}^1(T)$. By using Ito's formula for $F(t,x)$ and the stochastic process $y_n(t,s;x)$, we obtain

$$\langle[\mathcal{Q}_n(s) - \mathcal{Q}_n^1(s)]x, x\rangle = E\langle[\mathcal{Q}(T) - \mathcal{Q}^1(T)]y_n(T,s;x), y_n(T,s;x)\rangle$$

for all $T \geq s$ and $x \in H$. As $n \to \infty$ we get

$$\langle[\mathcal{Q}(s) - \mathcal{Q}^1(s)]x, x\rangle = E\langle[\mathcal{Q}(T) - \mathcal{Q}^1(T)]y(T,s;x), y(T,s;x)\rangle.$$

Since (1.2) is uniformly exponentially stable and $\mathcal{Q}, \mathcal{Q}^1$ are bounded on $\mathbf{R}_+$, we pass to limit $(T \to \infty)$ in the last inequality and we deduce $\mathcal{Q}(s) - \mathcal{Q}^1(s) = 0$ for all $s \in \mathbf{R}_+$. We get the conclusion.

$\Leftarrow$ Let $\mathcal{Q}$ be the unique mild solution of equation (1.5) which satisfies (5.2). Let $T > 0$ and let $\mathcal{Q}_n(t) = \mathcal{Q}_n(T,t;\mathcal{Q}(T))$ be the classical solution of (1.7) such that $\mathcal{Q}_n(T) = \mathcal{Q}(T)$.

By using Ito's formula for function $F_n : \mathbf{R}_+ \times H \to R, F_n(r,x) = \langle \mathcal{Q}_n(r)x, x\rangle$ and the stochastic process $y_n(r,s;x)$, $x \in H$ and by taking expectations, we have

$$E\langle \mathcal{Q}_n(t)y_n(t,s;x), y_n(t,s;x)\rangle$$
$$= \langle \mathcal{Q}_n(s)x, x\rangle - E\int_s^t \langle C^*(r)C(r)y_n(r,s;x), y_n(r,s;x)\rangle\, dr$$

for any $x \in H, t \geq s \geq 0$. As $n \to \infty$, we obtain (by Lemma 1.2 and Lemma 1.1)

$$\langle \mathcal{Q}(s)x, x\rangle = E\langle \mathcal{Q}(t)y(t,s;x), y(t,s;x)\rangle + E\int_s^t \|C(r)y(r,s;x)\|^2\, dr$$

for any $x \in H$. Since T is arbitrary, the last equation holds for all $t \geq s \geq 0$.

Let τ and γ be the constants introduced by Definition 4.2. If $t = s+\tau,\ s \geq 0$ then we have

$$\langle \mathcal{Q}(s)x, x\rangle = E\langle \mathcal{Q}(s+\tau)y(s+\tau,s;x), y(s+\tau,s;x)\rangle + E\int_s^{s+\tau} \|C(r)y(r,s;x)\|^2\, dr.$$

By Theorem 3.3 and since $\{A, C; G_i\}$ is uniformly observable we get

$$\langle \mathcal{Q}(s)x, x\rangle \geq \langle Q(s+\tau,s;\mathcal{Q}(s+\tau))x, x\rangle + \gamma\|x\|^2, \quad x \in H.$$

Now we use (5.2) and we have

$$\left(1 - \frac{\gamma}{\widetilde{M}}\right)\langle \mathcal{Q}(s)x, x\rangle \geq \langle Q(s+\tau,s;\mathcal{Q}(s+\tau))x, x\rangle \quad \text{for all } x \in H.$$

If we replace x with $y(s,p;x), s \geq p \geq 0, x \in H$, we obtain

$$\left(1-\frac{\gamma}{\widetilde{M}}\right)E\langle \mathcal{Q}(s)y(s,p;x), y(s,p;x)\rangle \geq E\langle Q(s+\tau,s;\mathcal{Q}(s+\tau))y(s,p;x), y(s,p;x)\rangle$$

and $\left(1 - \frac{\gamma}{\widetilde{M}}\right)\langle Q(s,p;\mathcal{Q}(s))x, x\rangle \geq \langle Q(s,p;Q(s+\tau,s;\mathcal{Q}(s+\tau)))x, x\rangle$.

By Lemma 1.2 we get for all $s+\tau \geq p \geq 0$,

$$\left(1 - \frac{\gamma}{\widetilde{M}}\right)\langle Q(s,p;\mathcal{Q}(s))x, x\rangle \geq \langle Q(s+\tau,p;\mathcal{Q}(s+\tau))x, x\rangle$$

Let $t \in \mathbf{R}_+$. Then there exists $n \in \mathbf{N}$ such that $t-p = n\tau + r, 0 \leq r < \tau$ and we have (by induction)

$$\langle Q(t,p;\mathcal{Q}(t))x, x\rangle \leq \left(1 - \frac{\gamma}{\widetilde{M}}\right)^n \langle Q(r+p,p;\mathcal{Q}(r+p))x, x\rangle.$$

Since $Q(t,p;R)$ is monotone (Lemma 1.2) and (5.2) holds we obtain

$$\widetilde{m}\langle Q(t,p;I)x, x\rangle \leq \left(1 - \frac{\gamma}{\widetilde{M}}\right)^n \widetilde{M}\langle Q(r+p,p;I)x, x\rangle$$

for all $x \in H$ or equivalently

$$\widetilde{m}E\|y(t,p;x)\|^2 \left(1 - \frac{\gamma}{\widetilde{M}}\right)^n \widetilde{M}E\|y(r+p,p;x)\|^2.$$

From (1.3) and by using P_2 we have successively

$$E\left\|y(r+p,p;x)\right\|^2$$

$$\leq (m+1)\left[\left\|U(r+p,p)\right\|^2\left\|x\right\|^2+\sum_{i=1}^{m}\int_{p}^{r+p}\left\|U(r+p,s)\right\|^2 G_i^2 E\left\|y(s,p;x)\right\|^2 ds\right]$$

and

$$E\left\|y(r+p,p;x)\right\|^2 \leq (m+1)M_1^2 e^{2\omega_1\tau}\left[\left\|x\right\|^2+\sum_{i=1}^{m}\widetilde{G}_i^2\int_{p}^{r+p}E\left\|y(s,p;x)\right\|^2 ds\right].$$

for all $0 \leq r < \tau$.

Now we use the Gronwall's inequality and if

$$K=(m+1)M_1^2 e^{2\omega_1\tau}\exp\left[(m+1)\sum_{i=1}^{m}\widetilde{G}_i^2 M_1^2 e^{2\omega_1\tau}\tau\right]$$

we obtain $E\left\|y(r+p,p;x)\right\|^2 \leq K\left\|x\right\|^2$. We get

$$E\left\|y(t,p;x)\right\|^2 \leq \left[\left(1-\frac{\gamma}{\widetilde{M}}\right)^{1/\tau}\right]^{t-p} K\left(\frac{\widetilde{M}}{\widetilde{m}}\right)\left(\frac{1-\gamma}{\widetilde{M}}\right)^{-r/\tau}\left\|x\right\|^2$$

for all $x \in H$. The proof is complete. □

References

[1] W. Arveson, *An Invitation to C^*-Algebra.* Springer Verlag, New York, 1976.

[2] I. Gelfand, H. Vilenkin, *Funcţii generalizate. Aplicaţii ale analizei armonice.* Editura Ştiinţifică şi Enciclopedică, (romanian translation) Bucureşti, 1985.

[3] A. Germani, L. Jetto, M. Piccioni, *Galerkin Approximations for Optimal Linear Filtering of Infinite-Dimensional Linear Systems.* SIAM J. Control and Optim. **26** (1988), 1287–2305.

[4] I. Gohberg, S. Goldberg, *Basic Operator Theory.* Birkhäuser, 1981

[5] W. Grecksch, C. Tudor, *Stochastic Evolution Equations.* A Hilbert Space Approach Math. Res. vol. 75, Acad. Verlag, 1995.

[6] C.S. Kubrusly, *Mean Square Stability for Discrete Bounded Linear Systems in Hilbert Space*, SIAM J. Control and Optimization **23** (1985), 19–29.

[7] T. Morozan, *Stochastic Uniform Observability and Riccati Equations of Stochastic Control*, Rev. Roumaine Math. Pures Appl. **38** (1993), 771–481.

[8] T. Morozan, *On the Riccati Equation of Stochastic Control*, International Series on Numerical Mathematics, vol. 107, Birkhäuser Verlag Basel, 1992, pp. 231–236.

[9] A. Pazy, *Semigroups of Linear Operators and Applications to Partial Differential Equations.* Applied Mathematical Sciences, vol. 44, Springer-Verlag, Berlin–New York, 1983.

[10] G. Da Prato, A. Ichikawa, *Quadratic Control for Linear Time-Varying Systems.* SIAM. J. Control and Optimization **28** (1990), 359–381.

[11] G. Da Prato, A. Ichikawa, *Lyapunov Equations for Time-Varying Linear Systems.* Systems and Control Letters **9** (1987), 165–172.

[12] G. Da Prato, J. Zabczyc, *Stochastic Equations in Infinite Dimensions.* University Press Cambridge, 1992.

Viorica Mariela Ungureanu
Faculty of Engineering
"Constantin Brâncuşi" University,
Bul Republicii, nr.1
Târgu-Jiu
jud. Gorj 1400
România

Operator Theory:
Advances and Applications, Vol. 153, 307–320

$B(H)$-Commutators: A Historical Survey

Gary Weiss

Abstract. This is a historical survey that includes a progress report on the 1971 seminal paper of Pearcy and Topping and 32 years of subsequent investigations by a number of researchers culminating in a completely general characterization, for arbitrary ideal pairs, of their commutator ideal in terms of arithmetic means.

This characterization has applications to the study of generalized traces (linear functionals vanishing on the commutator ideal $[I, B(H)]$) and to the study of the $B(H)$-ideal lattice and certain special sublattices. The structure of commutator ideals is essential for investigating traces which in turn is relevant for the calculation of the cyclic homology and the algebraic K-theory of operator ideals.

Mathematics Subject Classification (2000). Primary: 47B47, 47B10, 47L20; Secondary: 47-02, 47L30.

Keywords. Commutators, commutator ideals, ideals, trace, trace class, Hilbert-Schmidt class, arithmetic means.

Commutators, linear operators of the form $AB - BA$, first appeared in physics, for instance in a mathematical formulation of Heisenberg's Uncertainty Principle [17]. A simple concrete example is the product rule in calculus applied to xf expressed in terms of operators: $I = \frac{d}{dx}M_x - M_x\frac{d}{dx}$ where the operators act on the class of differentiable functions. The situation changes in $B(H)$, that is, when the operators act boundedly on a Hilbert space. Wintner [32] and Wielandt [31] in 1947 and 1949, respectively, gave two elegant distinct proofs that the identity is not a commutator of two bounded linear operators on a Hilbert space. Both apply also to arbitrary complex normed algebras with unit, except Wintner's proof requires that the norm be complete. For the period preceding 1967, *A Hilbert space problem book* [15] (Chapter 24), provides a brief history of $B(H)$-commutators including some proofs. This survey starts with an elementary description of the subject similar to the viewpoint held by the author in the 1970's and continues with a report on the main contributions including references and some open problems spanning 1971–2003 from which our deeper understanding of the subject evolved.

$AB - BA$ (also denoted by $[A, B]$) represents in a sense the "degree" to which A and B do not commute, either via norm (the operator norm or some other norm, i.e., quantitative measures) or via the commutator's containment in a two-sided ideal (qualitative measures). (All ideals herein are two-sided $B(H)$-ideals where H is a separable, infinite-dimensional, complex Hilbert space.) For instance, one can study conditions for containment of a commutator in the class $F(H)$ of finite rank operators, the smallest nonzero ideal in the lattice of all $B(H)$-ideals, or containment in the class $K(H)$ of compact operators, the largest proper ideal. The latter is equivalent to the operators commuting when they are projected canonically into the Calkin algebra, $B(H)/K(H)$, commonly called "commuting modulo the compact operators." The Calkin algebra being a unital normed algebra, Wintner and Wielandt reveals that the identity of any such algebra is never a commutator. This translates in $B(H)$ immediately to the fact that the "thin" operators, $\lambda I + K$ where $\lambda \neq 0$ and $K \in K(H)$, are noncommutators.

Brown and Pearcy [5] in 1965 determined the structure of all commutators in $B(H)$ by proving that the thin operators are the only noncommutators. This then determines as well the commutators in the Calkin algebra – all except the nonzero scalars. (See also [15].)

Theorem 0.1 (Brown and Pearcy, 1965). *An operator is a commutator if and only if it is not thin.*

An approximation consequence of Theorem 0.1 is that thin operators are norm-close to commutators since thin operators are easily seen to be norm-close to non-thin operators. But trying to get norm-close to the identity with a sequence of commutators can only be achieved if the norms of the operators in the commutators are not all uniformly bounded. (Cf. [15].) In fact, $A_nB_n - B_nA_n \to I$ in the $B(H)$-norm implies $\max(\|A_nB_n\|, \|B_nA_n\|) \to \infty$. This provides a hint of what was to come. In a sense, it tells us that to obtain the identity as a commutator it is necessary in one's matrix design to "spread out" in order to exploit a lot of cancellation. An example of this phenomenon is described in the following paragraph. In later developments on solving commutator equations, various kinds of spreading out matrix designs were discovered that provided quantitative control (i.e., of various norms) and qualitative control (i.e., forcing membership in ideals-the smaller the ideal the better). (Cf. [27], [1], [29], [2], [3], [19] and [8].)

In finite-dimensional Hilbert space, the trace distinguishes between commutators and noncommutators. There the trace of a product of two operators is independent of their order (unlike three or more) and so, there, commutators have trace zero. Moreover, having a nonzero trace is the only obstruction to being a commutator. Likewise for infinite-dimensions for the analogous class, $F(H)$. That is, *on infinite dimensional Hilbert space, a finite rank operator is a commutator if and only if it has trace zero.* That the infinite dimensional case is different is evident from Theorem 0.1, but more simply from the fact that the self-commutator of the unilateral shift, $U^*U - UU^* = \operatorname{diag}(1, 0, \dots)$, the diagonal matrix with diagonal sequence $(1, 0, \dots)$, is a rank one projection operator with trace one. This commutator provided a beginning framework from which to approach many matrix design problems in the subject and is therefore one of the underlying themes of this survey. For instance, since the unilateral shift is not compact, in order to express the rank one projection diag$(1, 0, \dots)$ as a commutator or sum of commutators of compact operators, a key step in [25], another approach needed to be found. An illuminating and more transparent construction leading to its representation as such a sum follows from the identity illustrating the matricial spreading out phenomenon mentioned above:

$$\operatorname{diag}(1, 0, \dots) = 1 \oplus \Sigma^{\oplus}(-x_n D) + 0 \oplus \Sigma^{\oplus} x_n D$$

where $D := \operatorname{diag}(1, -1/2, -1/2)$ and $< x_n >$ is the sequence $\frac{1}{2^n}$-repeated 2^n-times for $n \geq 0$ (cf. [27], Chapter 1.3, pp. 34–40.) Noting that D is simply a self-commutator of a 3×3 weighted shift with weights $1, \frac{1}{\sqrt{2}}$ and that $\langle x_n \rangle \asymp \langle \frac{1}{n} \rangle$ (the harmonic sequence), each summand can be represented as a commutator of compact operators. Moreover, the summands are (I, J)-commutators when the diagonal operator with eigenvalues the harmonic sequence is contained in the ideal IJ (for notations see the third paragraph below and the explanation following Theorem 2.3). That containment of diag$(1, 1/2, 1/3, \dots)$ is necessary followed subsequently as one application of Theorem 2.3 (see the section below on Traces and Arithmetic Means, fourth paragraph).

When the trace of a trace class commutator must be zero is a deep question (cf. [13], Section III.8 for background). If AB and BA for instance are in the trace class (the trace ideal C_1), then their commutator has trace zero (cf. [21], Lemma 2.1). I know of no weaker general condition on AB and BA that would insure that $Tr\ (AB - BA) = 0$ whenever the commutator is trace class. At one extreme, it is easy to produce operators, even compact operators, with zero product but non-trace class commutator, e.g., $[\begin{pmatrix} 0 & A \\ 0 & 0 \end{pmatrix}, \begin{pmatrix} B & 0 \\ 0 & 0 \end{pmatrix}]$. But when the operators are both in the Hilbert-Schmidt class (the Hilbert-Schmidt ideal C_2) or one is a trace class operator and the other a $B(H)$-operator, then their products, in either order, are trace class operators so their commutator has trace zero. Hence the questions: *Is every trace class operator with trace zero representable as a commutator of Hilbert-Schmidt operators or as a commutator of a trace class operator with a* $B(H)$*-operator?* I.e., is the trace the only obstruction for a trace class operator to be either type of commutator?

In 1971 Pearcy and Topping [25] posed four seminal problems described below that have dominated the subject of $B(H)$-commutators ever since and that led to breakthroughs in the study of generalized traces with applications to algebraic K-theory for operator ideals, e.g., the computation of certain K-groups. (For further details see [8], [2] and [3].) Indeed, the study of generalized traces depends fundamentally on understanding particularly the structure of the commutator ideals $[I, B(H)]$. This is because ideals are the natural domains of generalized traces (algebraic linear functionals that are unitarily invariant) and for algebraic linear functionals, unitary invariance is equivalent to their vanishing on the commutator ideal $[I, B(H)]$. This equivalence follows from the commutator identity $[TU^*, U] = T - UTU^*$ for T arbitrary and U unitary and from the fact that every operator is the linear combination of four unitary operators.

Commutator ideals also play a role in the Fong, Miers, Sourour characterization of the Lie ideals of $B(H)$ [12] where they prove: if $\mathcal{L}$ is a Lie ideal of $B(H)$ (i.e., $\mathcal{L}$ is a linear subspace of $B(H)$ containing $[\mathcal{L}, B(H)]$), then $[I, B(H)] \subseteq \mathcal{L} \subseteq I + \mathbb{C}1$ for some $B(H)$-ideal I.

For each pair of not necessarily proper $B(H)$-ideals I, J, denote by $[I, J]$ their commutator ideal, that is, the algebraic linear span of the class $[I, J]_1$ of single (I, J)-commutators (the commutators $AB - BA$ with $A \in I$ and $B \in J$). Let $[I, J]_n$ denote the class of all n-sums of (I, J)-commutators.

In 1954 [14] Halmos proved $[B(H), B(H)]_2 = B(H)$ and thereafter investigators on the structure of commutator ideals, once a particular commutator ideal representation was proven, would try to reduce the number of commutators needed in the sum. To date, the strongest result known on this is: $[I, J] = [I, J]_4$ (cf. [8], Corollary 6.5). Under various additional conditions the number can be further reduced. For instance, $[I, B(H)] = [I, B(H)]_3$ (cf. [8], Theorem 6.1). For more results of this type see [8], Chapter 6.

The Pearcy-Topping problems and progress to date on them are as follows.

Problem 1. (Pearcy-Topping [25], 1971) Is $K(H) = [K(H), K(H)]_1$?
In particular, is $\mathrm{diag}(1, 0, \dots)$ and hence any rank one projection operator a commutator of compact operators?

Pearcy and Topping considered this a key test question presumably because representing a rank one projection as sums of appropriate types of commutators was at the core of their proofs in [25]: that $K(H) = [K(H), K(H)]$ and that $C_p = [C_{2p}, C_{2p}]$ when $p > 1$. Until then, matrix forms designed to solve commutator equations were bounded but not compact operators, even when the target operator was compact. Indeed, representing this simplest nonzero compact operator without trace zero, $\mathrm{diag}(1, 0, \dots)$, provided the first inroad into controlling the matrix forms' membership in prescribed ideals.

Problem 2. Is $C_p = [C_{2p}, C_{2p}]_1$ when $p > 1$?

When $I \subset C_1$ let I^o denote the class of all operators in I with trace zero.

Problem 3. Is $C_1^o = [C_2, C_2]_1$? I.e., is every trace class trace zero operator a commutator of Hilbert-Schmidt operators?

Problem 3′. Is $C_1^o = [C_2, C_2]$? I.e., is every trace class trace zero operator a finite sum of commutators of Hilbert-Schmidt operators?

Problem 1 remains open and the structure of $[C_2, C_2]_1$ and $[C_1, B(H)]_1$ remain unknown. Indeed, little progress has been made on characterizing any of the single commutator classes $[I, J]_1$ or even on understanding their structure with five notable exceptions: the fundamental work of Anderson in 1977 [1], related contrasting work of L.G. Brown in 1994 [4], work related to both in [8] (Chapter 7), and the negative solution to Problem 3 in 1980 [28] (respectively, Theorems 1.1–2.1 below). The fifth is a recent development (Spring 2004) on a positive solution to one of my early problems on single commutators not listed here. It was announced to me by Ken Davidson in a private communication without manuscript that, along with Marcoux and Radjavi and using methods from [1], they constructed strictly positive commutators of compact operators with eigenvalues of arbitrarily slow decrease. In my view, this problem up to now has presented an obstruction to further progress on Problem 1.

Commutator ideals, in contrast, have seen significant progress. Of Problems 3–3′ Pearcy and Topping wrote "The techniques involved in giving an affirmative answer to this question would likely enable us to solve some stubborn problems in the theory of commutators in finite von Neumann algebras (see [24]). Problem 3 is so intractable that we cannot even answer the weaker Problem 3′".

1. Single commutators

Theorem 1.1 (Anderson, [1], 1977). *Rank one projections and more generally operators whose kernels contain infinite-dimensional reducing subspaces are commutators of compact operators.*

Consequently, every compact operator is a $(K(H), B(H))$-commutator, that is, $K(H) = [K(H), B(H)]_1$. Moreover, $C_p \subset [C_{2p}, B(H)]_1$ when $p > 1$.

Theorem 1.2 (L.G. Brown, [4], 1994). *If $A \in C_p$, $B \in C_q$, $p^{-1} + q^{-1} \geq \frac{1}{2}$ and the commutator $[A, B]$ has finite rank, then $Tr\ [A, B] = 0$.*

Theorem 1.3 (ii) below is a paired down version of a theorem in [8] (the main theorem in Chapter 7). Chapter 7 is an outgrowth of a prior question of Wodzicki on when $[I, B(H)]_1$ can contain a finite rank operator with nonzero trace. In particular, is containment in I of the diagonal operator $\operatorname{diag}\langle 1/\sqrt{n}\rangle$ necessary and sufficient. Sufficiency follows from [1] (cf. [8] (Chapter 7)). Chapter 7 provides conditions on IJ for the necessity for general single commutator classes $[I, J]_1$. *The necessity in general remains an open question.* (The necessity of the weaker containment conditions: $\operatorname{diag}\langle 1/n\rangle \in I$, respectively IJ, follows by blending methods in [4] and [8].)

Problem 4. If $[I, J]_1$ contains a finite rank operator with nonzero trace, must IJ contain the diagonal operator $\operatorname{diag}\langle 1/\sqrt{n}\rangle$? If not, what about the cases $[I, B(H)]_1$?

Theorem 1.3 (Dykema, Figiel, Weiss and Wodzicki, [8], 2001, Theorems 7.1–7.3).

(i) *If the diagonal operator* $\operatorname{diag}\langle 1/\sqrt{n}\rangle \in I$, *then* $[I, B(H)]_1$ *contains a finite rank operator with nonzero trace.*

(ii) *If* $[I, B(H)]_1$ *contains a finite rank operator with nonzero trace, then* $\operatorname{diag}\langle 1/\sqrt{n}\rangle$ *is contained in the arithmetic mean closure of I (see below) and, in particular, in I itself when it is arithmetically mean closed.*

(iii) *For every compact operator T, $T \in [(T \otimes \operatorname{diag}\langle 1/\sqrt{n}\rangle), B(H)]_1$, the first ideal being the principal ideal generated by $T \otimes \operatorname{diag}\langle 1/\sqrt{n}\rangle$.*

The arithmetic mean (am) closure of an ideal I is the smallest enveloping ideal that is solid under domination by the induced arithmetic mean sequences of the s-numbers of its operators (i.e., $T \in I$ whenever $s(T)_a \leq s(A)_a$ for some $A \in I$). This condition is equivalent to I being solid under domination in the sense of the Hardy-Littlewood-Polya-Schur majorization ($\prec$) of operator s-number sequences, that is, $T \in I$ whenever $s(T) \prec s(A)$ for some $A \in I$. ($s(T) \prec s(A)$ means $\sum_1^n s(T)_j \leq \sum_1^n s(A)_j \ \forall n$). This ordering for finite sequences with equality at n provided the characterization of when two nonincreasing sequences x, y of length n satisfy $\sum_1^n \phi(x_j) \leq \sum_1^n \phi(x_j)$ for every convex function ϕ (cf. [16], Section 3.17 Theorem 108, and for the infinite case, [23] 1.A, 1.A.2, 1.D p. 16, 3.A.8, 3.C.1, and esp. 3.C.1.b).

An arithmetic mean closed ideal is one that is equal to its closure. (See the section below on arithmetic mean ideals for more details. Cf. [8] Theorem 7.3 and Section 2.8.) Theorem 1.3 strengthens some of the results in [1] blending some of Anderson's methods there with methods in [8].

Problems 3–3′ were settled in 1980 in the negative via Theorem 2.1 below [28]. A negative answer to Problem 3′ provides automatically the same for Problem 3. And since $C_1^o \supset [C_2, C_2]_1$ is automatic as discussed earlier, the problems reduce to whether or not $C_1^o \subset [C_2, C_2]_1$ (resp., $C_1^o \subset [C_2, C_2]$), that is, whether or not all trace class trace zero operators are commutators of Hilbert-Schmidt operators (respectively, sums of such commutators). An even more important question as it turned out is: if not, which ones are? Some of the important single commutator open problems are summarized below in Problem 5.

In Theorem 2.1, the sufficiency of the logarithmic condition for the special class of trace class trace zero diagonal operators given there was originally proved in 1973 [27] with a shortened proof in 1986 [29] with the help of E. Azoff. The necessity was originally proved in 1976 [28] providing a negative solution to Problems 3–3′.

2. Commutator ideals

Theorem 2.1 (Weiss [28], 1980). *Setting $d := \sum_1^\infty d_n$ for an arbitrary sequence $d_n \downarrow 0$, the following are equivalent.*

(i) $\operatorname{diag}(-d, d_1, d_2, \dots) \in [C_2, C_2]$
(ii) $\operatorname{diag}(-d, d_1, d_2, \dots) \in [C_1, B(H)]$
(iii) $\sum_1^\infty d_n \log n < \infty$.

In particular, if $\langle d_n \rangle = \langle \frac{1}{n \log^2 n} \rangle$, then $\operatorname{diag}(-d, d_1, d_2, \dots) \in C_1^o \setminus [C_2, C_2]$.

Theorem 2.1 (ii), although not included in [28], can be obtained from the same methods with minor modifications.

Diagonal operators of the form $\operatorname{diag}(-d, d_1, d_2, \dots)$ became the test cases inside C_1^o in [27]–[29] for early approaches to commutator problems where having trace zero was a necessary condition, and they continue to play an essential role in the C_1 study, for instance in the proof of the main summability theorem in [8], Theorem 5.11 (iii) (Theorem 2.4 below). Indeed, when $I \subset C_1$, i.e., inside C_1 they are the building blocks of I^o in that every I^o-operator has its real and imaginary parts easily decomposable as the sum of two compact selfadjoint operators each unitarily equivalent to diagonals of this form. Another important point about these diagonal forms is their resemblance to diag$(1,0,0,\dots)$ so that matricially they provide reasonable analogs to the rank one projection and its commutator problems mentioned earlier (for instance, Problem 1 and Problem 4 for the case diag$(1,0,0,\dots)$).

The study of this class began in 1973 (cf. [27] and [28]) by considering low dimension finite-dimensional cases (especially important were dimensions 2, 3 and 4), and this survey concludes with a discussion of some of the remaining open problems there that might lead to better insights into the structure of single commutator classes.

One negative weight and the rest positive in practice were the hardest operators to deal with in this investigation. In contrast, operators of the form diag$\langle \pm d_n \rangle$ are easy to represent as single commutators using tensors of diagonals and 2×2 scalar matrices with the best possible control on their norms or on their membership in prescribed ideals. Indeed, for every product representation $d_n = a_n b_n$:

$$\operatorname{diag}\langle \pm d_n \rangle = \left[\operatorname{diag}\langle a_n \rangle \otimes \begin{pmatrix} 0 & 1 \\ 0 & 0 \end{pmatrix}, \operatorname{diag}\langle b_n \rangle \otimes \begin{pmatrix} 0 & 0 \\ 1 & 0 \end{pmatrix}\right].$$

Indeed, from this it follows easily that IJ is the ideal generated by $[I, J]$. Since 1971, much of the work on commutators depended significantly on increasingly complicated methods of organizing cancellation in operator equations. Another elementary example of this was discussed above in the second paragraph following Theorem 0.1.

Regarding the monotone ordering on the sequence $\langle d_n \rangle$, since the summability condition in Theorem 2.1 along with other weighted summability conditions with nondecreasing weights appearing in [27] are minimal when the sequence $\langle d_n \rangle$ is arranged in nonincreasing order and since all the commutator classes under discussion are unitarily invariant, there is no loss of generality in assuming that the sequence is nonincreasing.

Anderson [3] in 1986 proved $C_p^o = [C_p, B(H)]$ when $p < 1$ and applied this to the computation of the K-groups, $K_1(C_p, B(H))$ (cf. [2]).

Kalton [19] in 1989 linked arithmetic means to $[C_1, B(H)]$ and to $[C_2, C_2]$ achieving the remarkable characterizations of the spaces:

Theorem 2.2 (Kalton, [19], 1989). *An operator $T \in C_1$ is in $[C_1, B(H)]$ (or in $[C_2, C_2]$) if and only if $\lambda(T)_a$ is absolutely summable.*

Here $\lambda(T) := \langle \lambda_n(T) \rangle$ denotes the sequence of eigenvalues counting algebraic multiplicity and arranged in order of nonincreasing moduli and $\lambda(T)_a := \langle \frac{\lambda_1(T)+\cdots+\lambda_n(T)}{n} \rangle$ denotes its averaging sequence.

Surprisingly and only indirectly via their same characterization did Kalton obtain: $[C_1, B(H)] = [C_2, C_2]$. No direct proof of this is known.

Theorem 2.2 provided some of the inspiration and methodology for the general characterization of commutator ideals in the following theorem (which in some respects is the central result of [8]). For more of Kalton's work on this subject, some joint with Dykema and some related to the following theorem; see [20] and [9].

Theorem 2.3 (Dykema, Figiel, Weiss and Wodzicki, [8], 2001). *If I, J are two arbitrary $B(H)$-ideals, at least one of which is proper, and $T = T^* \in IJ$, then*

$$T \in [I, J] \textit{ if and only if } \operatorname{diag} \lambda(T)_a \in IJ.$$

Consequently, $[I, J] = [IJ, B(H)]$.

For normal compact operators, $\lambda(T)$ is simply the sequence of eigenvalues counting ordinary multiplicity and arranged in order of decreasing moduli, as described above. (A stronger version of Theorem 2.3 requires only a more general condition than monotonization: that $|\lambda(T)| \le \nu$ for some $\nu \in \Sigma(I)$.) Moreover, IJ traditionally defined in ring theory as the ideal generated by all (I, J)-products, in $B(H)$ it is also precisely the class of all single (I, J)-products (cf. [8], Lemma 6.3). Alternatively, IJ is the class of compact operators T dominated by (I, J)-products in the sense of s-numbers, i.e., all compact operators T for which $s(T) \le s(A)s(B)$ for some $A \in I, B \in J$, and this reveals that the ideal product is a commutative operation. These facts follow in part from Calkin's inclusion preserving lattice isomorphism, $I \to \Sigma(I)$, from the class of $B(H)$-ideals onto the class of characteristic subsets of c_o^* (i.e., solid subsets of the class of nonincreasing nonnegative sequences tending to zero that are invariant under ampliation, $D_2(s(T)) := (s(T)_1, s(T)_1, s(T)_2, s(T)_2, \dots)$, and closed under addition) [6]. (See also [8], Chapter 2.)

Part of the value of Theorem 2.3 lies in the fact that, because IJ is a commutative ideal operation, the theorem reduces many noncommutative problems in this study to commutative problems involving c_o^*-sequences, often achieving greater accessibility. This is evidenced in [8] and [18].

Like Kalton's result, that $[I, J] = [IJ, B(H)]$ follows likewise in [8] only indirectly from their same characterization and likewise no direct proof is known. (The inclusion $[I, J]_2 \supset [IJ, B(H)]_1$ is straightforward.) Hence the primary general single commutator questions (cf. [8], Chapter 7):

Problem 5. (i) Characterize $[I, J]_1$.

(ii) Absent this, characterize $[I, J]_1$ for any pair $I, J \ne F(H)$ distinct from those covered in [8], Corollaries 7.2 and 7.6.

(iii) Is $[I, J]_1 = [IJ, B(H)]_1$?

(iv) In particular, is $[C_1, B(H)]_1 = [C_2, C_2]_1$?

In C_1, Theorem 2.3 also reveals the beginnings of a summability theory for trace class commutators as evidenced in the next theorem (see also [18]).

Theorem 2.4 (Dykema, Figiel, Weiss and Wodzicki [8], 2001, Theorem 5.11-(iii)). *If I, J, L are $B(H)$-ideals where $IJ \subset K(H)$, then*

(i) $\operatorname{diag}(-d, d_1, d_2, \ldots) \in [I, J]$ *if and only if* $\operatorname{diag}\langle d_n \rangle_{a_\infty} \in IJ$
(ii) *If $L \subset C_1$, then $L^o \subset [I, J]$ if and only if $L_{a_\infty} \subset IJ$.*
(iii) *If $I \subset C_1$, then $I^o = [I, B(H)]$ if and only if $I_{a_\infty} = I$*

(equivalently $I_{a_\infty} \subset I$ since the reverse inclusion is automatic).

Here the arithmetic mean at infinity is defined for summable sequences λ by: $\lambda_{a_\infty} := \frac{\lambda_{n+1} + \lambda_{n+2} + \cdots}{n}$. See the third paragraph below for the definition of the arithmetic mean at infinity I_{a_∞}.

3. Traces and arithmetic mean ideals

Consequences of Theorem 2.3 can be found in [8] and in [18]. It provides a new point of view on traces and arithmetic mean ideals and a brief introduction including a few of its consequences are presented below. Applications of Theorem 2.3 to cyclic homology and the algebraic K-theory of operator ideals are described in [8], Introduction.

Theorem 2.3 characterizes when traces and nonsingular traces exist. Traces (also called generalized traces), as mentioned earlier, are unitarily invariant linear functionals on an ideal I and unitary invariance is equivalent to the linear functional vanishing on the commutator ideal $[I, B(H)]$. Alternatively, traces are simply natural liftings to I of linear functionals on the quotient $\frac{I}{[I,B(H)]}$, so they exist precisely when $I \neq [I, B(H)]$, or equivalently, by Theorem 2.3, when $I \neq I_a$ (i.e., when the characteristic sets $\Sigma(I_a) \not\subset \Sigma(I)$).

I_a and ${}_aI$, the basic arithmetic mean ideals called respectively the arithmetic mean ideal and the pre-arithmetic mean ideal of I, are defined as:

$$I_a := \{T \in K(H) \mid s(T) \leq s(A)_a \text{ for some } A \in I\} \text{ and}$$
$${}_aI := \{T \in K(H) \mid s(T)_a \leq s(A) \text{ for some } A \in I\}.$$

Likewise I_{a_∞} and ${}_{a_\infty}I$, the basic arithmetic mean ideals at infinity called respectively the arithmetic mean ideal at infinity and the pre-arithmetic mean ideal at infinity of I, are defined as:

$$I_{a_\infty} := \{T \in K(H) \mid s(T) \leq s(A)_{a_\infty} \text{ for some } A \in I\} \text{ and}$$
$${}_{a_\infty}I := \{T \in K(H) \mid s(T)_{a_\infty} \leq s(A) \text{ for some } A \in I\}.$$

Nonsingular traces exist precisely when $\operatorname{diag}(1, 1/2, 1/3, \ldots) \notin I$. Nonsingular traces are those generalized traces nonvanishing on $F(H)$, or equivalently, whose restrictions to $F(H)$ up to scalar multiplication are simply the classical trace. The necessity of $\operatorname{diag}(1, 1/2, 1/3, \dots) \notin I$ follows by applying Theorem 2.3 to obtain the key idea that

$$\operatorname{diag}(1, 0, 0, \dots) \in [I, B(H)] \text{ if and only if } \operatorname{diag}\langle 1/n\rangle \in I.$$

Examples of nonsingular traces are: the classical trace on $F(H)$ or C_1; and an example of a singular trace is the Dixmier trace on the Köthe dual of the Macaev ideal (cf. [7] and [22]). In the language of arithmetic mean ideals (see below), the Macaev ideal is the Lorentz ideal $\mathcal{L}(\log\, n)$ and its Köthe dual is the pre-arithmetic mean ideal of a principal ideal, ${}_a(\operatorname{diag}\langle\frac{\log\, n}{n}\rangle)$, also known as the Marcinkiewicz ideal $\mathcal{M}(\langle\frac{n}{\log\, n}\rangle)$, and which is also the arithmetic mean closure of a principal ideal, ${}_a((\operatorname{diag}\langle 1/n\rangle)_a)$. (Cf. [8], Sections 4.7 and 4.10. The arithmetic mean (am) closure was discussed earlier in the paragraph following Theorem 1.3.) Indeed many classical ideals arising from classical spaces in the literature (e.g., Lorentz, Marcinkiewicz and Orlicz sequence spaces) fit this new context identified in [8] and the additional notions of "softness" and "soft enveloping ideals" in [18] arose from Theorem 2.3 applied to the work of Dixmier [7] and Varga [26].

One consequence of Theorem 2.3 for nonsingular traces is that the classical trace extends beyond the trace class to the strictly larger ideal of operators T whose s-numbers $s(T) = o(1/n)$ (see also [10]). This is the soft interior of the principal ideal generated by $\operatorname{diag}\langle 1/n\rangle$.

By Theorem 2.3, $({}_aI)^+ = [I, B(H)]^+$ and the ideal ${}_aI \subset [I, B(H)]$ consequently. As mentioned earlier, the ideal generated by $[I, B(H)]$ is I. So ${}_aI \subset [I, B(H)] \subset I$ where ${}_aI, I$ form the optimal upper and lower ideal envelopes for $[I, B(H)]$ and the inclusions become equalities if and only if $[I, B(H)]$ is itself an ideal.

In c_0^*, the new double inequality linking tensors with arithmetic means (see [8], Proposition 3.14): $(x \otimes \omega)^* \leq x_a \leq 2\ (x \otimes \omega)^*$ provides an alternate characterization of the arithmetic mean ideal of I: $I_a = I \otimes (\operatorname{diag}\ \omega)$. Here $*$ means monotonization and $\omega := \langle 1/n\rangle$.

Am-stability, $I = I_a$, is equivalent to ${}_aI = I$, and from the previous comment, am-stability is equivalent to the condition $I = I \otimes (\operatorname{diag}\ \omega)$ (i.e., $I \supset I \otimes (\operatorname{diag}\ \omega)$ since the reverse inclusion is automatic). This is related to the tensor product closure property (TPCP), $I = I \otimes I$, investigated in [27]. (See [8], Sections 2.8 and 4.3 for more details on the tensor operation for ideals.)

The following 5-chain with some arithmetic mean relations links the basic arithmetic mean ideals: ${}_aI \subset ({}_aI)_a \subset I \subset {}_a(I_a) \subset I_a$. Denoting $A(I) := I_a$ and $A^{-1}(I) := {}_aI$ for convenience of notation, $\forall\, n \in \mathbb{Z}$ am-ideals satisfy the relations $A^nA^{-n}A^n = A^n$ which imply the 5-chain and that A^nA^{-n} and $A^nA^{-2n}A^n$ are idempotents. The interior $I^o := ({}_aI)_a$ and the closure $I^- := {}_a(I_a)$ reveal a topological like structure (except of course the complement of an ideal is not an ideal) where the am-closed ideals are those equal to their closures (equivalently as mentioned earlier, those solid under Hardy-Littlewood-Polya-Schur majorization), and the am-open ideals are those equal to their interiors. More generally, am-interiors/am-closures form the optimal am-open/am-closed inner and outer envelopes, respectively. For am-stability, all the inclusions become equality. An open question here is *whether or not there exists a clopen ideal that is not am-stable.*

Three notable features of this structure are that every ideal contains a largest am-closed ideal and is contained in a smallest am-open ideal; that they can be expressed in terms of certain inner and outer convex envelopes; and that the am-closure operation distributes over finite sums of ideals. (Cf. [18] for details on this topological like structure.)

There is also an emerging summability theory with analogous phenomena related to the slightly subtler 5-chain for arithmetic mean ideals at infinity: ${}_{a_\infty}I \subset ({}_{a_\infty}I)_{a_\infty} \subset I \subset {}_{a_\infty}(I_{a_\infty}) \subset I_{a_\infty}$.

4. Single commutator problems on $\operatorname{diag}(-d, d_1, d_2, \dots)$

The development leading to Theorem 2.1 left some interesting open questions. This is a brief description of the related work and problems from [27], Section 1.8, pp. 122–136, [28], Section I, pp. 576–580, and [29], p. 885.

An early construct used by the author in this study was that, using the standard notation $[A, B] := AB - BA$ and a parameter $0 \le t \le 1$:

$$\begin{pmatrix} -\sum_1^\infty d_n & 0 & 0 & \cdots \\ 0 & d_1 & 0 & \cdots \\ 0 & 0 & d_2 & \ddots \\ \vdots & \vdots & \ddots & \ddots \end{pmatrix}$$

$$= \left[\begin{pmatrix} 0 & 0 & 0 & \cdots \\ (\sum_1^\infty d_n)^{1-t} & 0 & 0 & \cdots \\ 0 & (\sum_2^\infty d_n)^{1-t} & 0 & \cdots \\ \vdots & \ddots & \ddots & \ddots \end{pmatrix}, \begin{pmatrix} 0 & (\sum_1^\infty d_n)^t & 0 & \cdots \\ 0 & 0 & (\sum_2^\infty d_n)^t & \cdots \\ 0 & 0 & 0 & \ddots \\ \vdots & \vdots & \vdots & \ddots \end{pmatrix} \right].$$

Here the IJ-norm in both the $[C_1, B(H)]_1$ and $[C_2, C_2]_1$ contexts is

$$\sum_1^\infty \sum_k^\infty d_n = \sum_1^\infty n d_n.$$

That is, a sufficient condition that

$$\operatorname{diag}(-d, d_1, d_2, \dots) \in [C_2, C_2]_1 \text{ or } [C_1, B(H)]_1 \text{ is that } \sum_1^\infty n d_n < \infty.$$

Weighted shifts such as these play a central role throughout the theory in building solution operators for commutator equations. For instance, they were essential for the sufficiency part of Theorem 2.1 (iii). The strategy back then was to look for suitable matrix forms with smaller trace norms in the form of weighted sums where the weights increase slower than n, say like $o(n)$. The point of

Theorem 2.1 (iii) was that the sufficiency of the weights n were strengthened to the sufficiency of the weights log n thereby weakening the condition on $\langle d_n\rangle$ and hence expanding the class of achievable diag$(-d, d_1, d_2, \dots)$. Attempts to reduce further the summability condition qualitatively failed, but quantitatively, the following construct succeeded.

First note that

$$\operatorname{diag}(-d, d_1, d_2, \dots) \cong \operatorname{diag}(-d, d_1, d_3, \dots) \oplus \operatorname{diag}(d_2, d_4, \dots) = [A, B]$$

where

$$A = \begin{pmatrix} U_t & -\left(\sum_{n=1}^{\infty} d_{2n}\right)^t P \\ 0 & -V_t^* \end{pmatrix} \quad \text{and} \quad B = \begin{pmatrix} U_{1-t}^* & 0 \\ \left(\sum_{n=1}^{\infty} d_{2n}\right)^{1-t} P & V_{1-t} \end{pmatrix}$$

with $P := \operatorname{diag}(1, 0, \dots)$, the parameter $0 \le t \le 1$, the weighted shift

$$U_t := U\Big(\Big(\sum_1^{\infty} d_{2n-1}\Big)^t, \Big(\sum_2^{\infty} d_{2n-1}\Big)^t, \dots\Big)$$

and the weighted shift

$$V_t := U\Big(\Big(\sum_2^{\infty} d_{2n}\Big)^t, \Big(\sum_3^{\infty} d_{2n}\Big)^t, \dots\Big).$$

When $t = 1$, $\|A\|_1 = \sum_1^{\infty} [\frac{n+1}{2}]\, d_n$ where $[x]$ denotes the greatest integer function, and $\|B\| = 1$. When $t = \frac{1}{2}$, $\|A\|_2^2 = \|B\|_2^2 = \sum_1^{\infty} [\frac{n+1}{2}]\, d_n$. And for any $0 \le t \le 1$, $\|AB\|_1 = \|BA\|_1 = \sum_1^{\infty} [\frac{n+1}{2}]\, d_n$. These are all quantitative improvements over $\sum_1^{\infty} n d_n$ for the respective norms, but not qualitative improvements. Hence the following test questions for investigating the structure of $[C_1, B(H)]_1$ and $[C_2, C_2]_1$:

Problem 6. Is $\sum_1^{\infty} [\frac{n+1}{2}]\, d_n$ minimal for each of these three contexts $(\max(\|AB\|_1, \|BA\|_1)$, $\|A\|_2\|B\|_2$ and $\|A\|_1\|B\|)$?

For which of these three contexts is the condition $\sum_1^{\infty} n d_n < \infty$ necessary?

Finally, the problems on diag$(-d,\ d_1,\ d_2,\ \dots)$ all have finite-dimensional analogs by simply taking the sequence $\langle d_n\rangle$ finite. Indeed Theorem 2.1 owes its discovery as described in [28] to the test cases $d_1 = d_2 = \cdots = d_N = \frac{1}{N}$. The formula in Problem 6 for the cases diag$(-1, \frac{1}{N}, \frac{1}{N}, \dots, \frac{1}{N})$ is $\frac{N+2}{4}$ for N even and $\frac{(N+1)^2}{4N}$ for N odd. Since $\|\operatorname{diag}(-1, d_1, d_2, \dots)\|_1 = 2$ (by normalizing, $d = 1$), we see that diag$(-1, d_1, d_2, \dots) = AB - BA$ implies

$$2 = \|AB - BA\|_1 \le \|AB\|_1 + \|BA\|_1 \text{ hence by Hölder's inequality,}$$

$\|A\|_2\|B\|_2$, $\|A\|_1\|B\|$, $\max(\|AB\|_1, \|BA\|_1) \ge 1$.

For $N = 1, 2$, the inequality

$$2 = \|AB - BA\|_1 \le \|AB\|_1 + \|BA\|_1 \le 2\ \|A\|_2\|B\|_2$$

in the "minimal" case is actually equality. But for $N = 3$, the $\|A\|_2\|B\|_2$-minimum turned out strictly larger than 1, namely 4/3, and for $N = 4$, the minimum is 3/2. So for $1 \le N \le 4$, the formula in Problem 6 is sharp for minimizing $\|A\|_2\|B\|_2$ and perhaps something interesting is occurring at $N = 5$ like new solution operators providing smaller Hilbert-Schmidt norms or a new stronger proof of sharpness. The proof for $1 \leq N \leq 4$ provided the keys in [28] to proving Theorem 2.1.

Problem 7. Minimize $\|A\|_2\|B\|_2$ for $N \geq 5$ for the target diagonal operators $\operatorname{diag}(-1, \frac{1}{N}, \frac{1}{N}, \ldots, \frac{1}{N})$ and $\operatorname{diag}(-1, \frac{1}{N}, \frac{1}{N}, \ldots, \frac{1}{N}) \oplus 0$ where 0 is finite and infinite-dimensional?

References

[1] J. Anderson, *Commutators of Compact Operators.* J. Reine Angew. Math. **291** (1977), 128–132.

[2] J. Anderson, L.N. Vaserstein, *Commutators in Ideals of Trace Class Operators.* Indiana University Mathematics Journal **35** (2) (1986), 345–372.

[3] J. Anderson, *Commutators in Ideals of Trace Class Operators.* II, Indiana University Mathematics Journal **35** (2) (1986), 373–378.

[4] L.G. Brown, *Traces of Commutators of Schatten-von Neumann Class Operators.* J. Reine Angew. Math. **451** (1994), 171–174.

[5] A. Brown, C. Pearcy, *Structure of Commutators of Operators,* Ann. of Math. (2) **82** (1965), 112–127.

[6] J.M. Calkin, *Two-Sided Ideals and Congruences in the Ring of Bounded Operators in Hilbert Space,* Ann. of Math. (2) **42** (1941), 839–873.

[7] J. Dixmier, *Existence de Traces non Normales.* C. R. Acad. Sci. Paris Sér. A-B **262** (1966), A1107–A1108.

[8] K. Dykema, T. Figiel, G. Weiss, M. Wodzicki, *The Commutator Structure of Operator Ideals,* Adv. Math. **185** (2004), no. 1, 1–79.

[9] K.J. Dykema, N.J. Kalton, *Spectral Characterization of Sums of Commutators.* II, J. Reine Angew. Math. **504** (1998), 127–137.

[10] K. Dykema, G. Weiss, M. Wodzicki, *Unitarily Invariant Trace Extensions Beyond the Trace Class,* Complex analysis and related topics (Cuernavaca, 1996), Oper. Theory Adv. Appl. **114** (2000), 59–65.

[11] K. Fan, *Maximum Properties and Inequalities for the Eigenvalues of Completely Continuous Operators.* Proc. Natl. Acad. Sci. USA **37** (1951), 760–766.

[12] C.K. Fong, C.R. Miers, A.R. Sourour, *Lie and Jordan Ideals of Operators on Hilbert Space.* Proc. Amer. Math. Soc. **84** (1982), no. 4, 516–520.

[13] I.C. Gohberg, M.G. Kreĭn, *Introduction to the Theory of Linear Nonselfadjoint Operators.* American Mathematical Society (1969).

[14] P.R. Halmos, *Commutators of Operators.* II, Amer. J. Math. **76** (1954), 191–198.

[15] P.R. Halmos, *A Hilbert Space Problem Book,* 2nd Edition, Graduate Texts in Mathematics (19), Springer-Verlag (1982). From 1st Edition, D. Van Nostrand Co., Inc., Princeton, N.J.-Toronto, Ont.-London (1967).

[16] G.H. Hardy, J.E. Littlewood, G. Pólya, *Inequalities.* Cambridge University Press (1934, 1952, 1988).

[17] W. Heisenberg, *Über den anschaulichen Inhalt der quantentheoretischen Kinematik und Mechanik,* Zeitschrift für Physik **43** (1927), 172–198.

[18] V. Kaftal, G. Weiss, *Traces, Ideals, and Arithmetic Means.* Proc. Natl. Acad. Sci. USA **99** (11) (2002), 7356–7360.

[19] N.J. Kalton, *Trace-Class Operators and Commutators.* J. Funct. Anal. **86** (1989), 41–74.

[20] N.J. Kalton, *Spectral Characterization of Sums of Commutators.* I, J. Reine Angew. Math. **504** (1998), 115–125.

[21] C. Laurie, E. Nordgren, H. Radjavi, P. Rosenthal, *On Triangularization of Algebras of Operators,* Journal für die Reine und Angewandte Mathematik **327** (1981), 143–155.

[22] V.I. Macaev, *A class of Completely Continuous Operators.* Dokl. Akad. Nauk SSSR **139** (1961), 548–551.

[23] A.W. Marshall, I. Olkin, *Inequalities: Theory of Majorization and its Applications,* Academic Press Inc. [Harcourt Brace Jovanovich Publishers], Mathematics in Science and Engineering **143** (1979).

[24] C. Pearcy, D. Topping, *Commutators and Certain* II_1*-factors.* J. Functional Analysis **3** (1969), 69–78.

[25] C. Pearcy, D. Topping, *On Commutators in Ideals of Compact Operators.* Michigan Math. J. **18** (1971), 247–252.

[26] J. Varga, *Traces on Irregular Ideals.* Proc. Amer. Math. Soc. **107** 3 (1989), 715–723.

[27] G. Weiss, *Commutators and Operators Ideals.* dissertation (1975), University of Michigan Microfilm.

[28] G. Weiss, *Commutators of Hilbert-Schmidt Operators. II.* Integral Equations Operator Theory **3** (4) (1980), 574–600.

[29] G. Weiss, *Commutators of Hilbert-Schmidt Operators. I.* Integral Equations Operator Theory **9** (1986), 877–892.

[30] G. Weiss, *Classification of Certain Commutator Ideals and the Tensor Product Closure Property.* Integral Equations Operator Theory **12** (1989), 99–128.

[31] H. Wielandt, *Über die Unbeschränktheit der Operatoren der Quantenmechanik.* Math. Ann. **121** (1949) 21.

[32] A. Wintner, *The Unboundedness of Quantum-Mechanical Matrices.* Phys. Rev. **71** (1947), 738–739.

Gary Weiss
University of Cincinnati
Department of Mathematics
Cincinnati, OH, 45221-0025
USA
e-mail: gary.weiss@math.uc.edu

Operator Theory:
Advances and Applications, Vol. 153, 321–344

Semiclassical Weyl Formula for Elliptic Operators with Non-Smooth Coefficients

Lech Zielinski

Abstract. We consider the Weyl formula for the asymptotic number of eigenvalues of self-adjoint elliptic differential operators with coefficients which have Hölder continuous first-order derivatives. Our aim is to prove that the Weyl formula holds with a remainder usually considered in the case of operators with smooth coefficients.

Mathematics Subject Classification (2000). 35P20.

Keywords. Spectral asymptotics, Semiclassical approximation, Elliptic operator, Weyl formula.

1. Introduction

This paper presents a generalization of our earlier results described in [14], [16], [17] and motivated by the well-known result of L. Hörmander ([5]) on spectral asymptotics of a self-adjoint elliptic operator A with smooth coefficients on a smooth (boundaryless) manifold M. If M is a compact manifold of dimension d and $2m$ is the order of A, then the associated counting function $\mathcal{N}(A,\lambda)$, defined as the number of eigenvalues (counted with their multiplicities) smaller than λ, satisfies the Weyl formula

$$\mathcal{N}(A,\lambda) = c\lambda^{\frac{d}{2m}} + O(\lambda^{\frac{d-\mu}{2m}}), \tag{1.1}$$

where $c = c_{A,M}$ is a constant and $\mu = 1$.

It is natural to ask if (1.1) still holds for the operators with non-smooth coefficients. In [16] we proved that for coefficients which are Hölder continuous of exponent $r \in]0;\ 1]$, the formula (1.1) holds if $\mu < r$. In [17] we proved (1.1) with $\mu = 1$ assuming that the first-order derivatives of coefficients are Lipschitz continuous and in this paper we show a way of replacing the hypothesis of Lipschitz continuity by Hölder continuity. Moreover we use the semiclassical framework, which allows us to obtain the classical Weyl formula (1.1) as a corollary.

Let $\mathcal{B}^r$ denote the set of bounded, Hölder continuous functions on $\mathbb{R}^d$, i.e. $a \in \mathcal{B}^r$ means that $a \in L^\infty(\mathbb{R}^d)$ and there is $C > 0$ such that

$$|a(x) - a(y)| \leq C|x-y|^r \quad (x, y \in \mathbb{R}^d).$$

Further on $m \in \mathbb{N} \setminus \{0\}$ and we fix $r \in]0;\ 1]$. For $\nu, \bar{\nu} \in \mathbb{N}^d$, $|\nu|, |\bar{\nu}| \leq m$, let $a_{\nu,\bar{\nu}}$ be real-valued coefficients satisfying $a_{\nu,\bar{\nu}} = a_{\bar{\nu},\nu}$ and

$$\partial_{x_j} a_{\nu,\bar{\nu}} \in \mathcal{B}^r \quad (j = 1, \ldots, d). \tag{1.2}$$

We assume moreover the existence of a constant $c > 0$ such that

$$\sum_{|\nu|=|\bar{\nu}|=m} a_{\nu,\bar{\nu}}(x)\xi^{\nu+\bar{\nu}} \geq c|\xi|^{2m} \quad (x, \xi \in \mathbb{R}^d) \tag{1.3}$$

and for $h > 0$ we introduce the quadratic form $\mathcal{A}_h$ defined for $\varphi, \psi \in C_0^m(\mathbb{R}^d)$ by the formula

$$\mathcal{A}_h[\varphi, \psi] = \sum_{|\nu|,|\bar{\nu}|\leq m} (a_{\nu,\bar{\nu}}(hD)^\nu \varphi, (hD)^{\bar{\nu}} \psi), \tag{1.4}$$

where $(\cdot,\cdot)$ is the scalar product of $L^2(\mathbb{R}^d)$ and $(hD)^\nu = (-ih)^{|\nu|}\partial^\nu/\partial x^\nu$.

The ellipticity hypothesis (1.3) ensures the fact that $\mathcal{A}_h$ is bounded from below and its closure defines a self-adjoint operator A_h. We introduce

$$a(x, \xi) = \sum_{|\nu|,|\bar{\nu}|\leq m} a_{\nu,\bar{\nu}}(x)\xi^{\nu+\bar{\nu}} \tag{1.5}$$

and for $E \in \mathbb{R}$ we denote

$$\Gamma_E = a^{-1}(]-\infty;\ E[) = \{(x, \xi) \in \mathbb{R}^{2d} : a(x, \xi) < E\}. \tag{1.6}$$

We will present a proof of

Theorem 1.1. *Let $E_0, E \in \mathbb{R}$ be such that $E_0 < E$ and Γ_E is bounded. Then the operators A_h described above have discrete spectrum in $]-\infty;\ E_0]$ for $h \in]0;\ h_0]$, if $h_0 > 0$ is fixed small enough. We assume moreover*

$$a(x, \xi) = E_0 \Rightarrow \nabla_\xi a(x, \xi) \neq 0. \tag{1.7}$$

Then, for $h \in]0;\ h_0]$, the counting function $\mathcal{N}(A_h, E_0)$ satisfies the asymptotic formula

$$\mathcal{N}(A_h, E_0) = |\Gamma_{E_0}|(2\pi h)^{-d} + O(h^{1-d}), \tag{1.8}$$

where $|\Gamma_{E_0}|$ denotes the Lebesgue measure of Γ_{E_0}.

In the case of smooth coefficients the assertion of Theorem 1.1 is well known from numerous papers (cf., e.g., [1], [4]) and books (cf. [2], [7], [13]). The case of non-smooth coefficients can be reduced to a special "smooth" problem. This problem can be studied by means of a microlocal analysis and it is possible to adopt the theory developed in the book of V. Ivrii ([7]) (cf. [8]), but the aim of this paper is to give a self-contained presentation based on a modification of the approach we used in [17] according to the plan exposed below.

In Section 2 we describe the regularization of non-smooth coefficients giving the operators P_h with smooth coefficients. In Section 3 we introduce pseudodifferential operators to describe smooth functions of P_h and in Section 4 we consider refinements needed to study functions of P_h/h which appear in the standard Tauberian approach used to obtain the asymptotic formula with the remainder estimate $O(h^{1-d})$. In comparison with [17] the principal difficulty comes from the fact that in the case of the Hölder continuous Hamiltonian flow, the dependence on initial conditions cannot be controlled as in the standard situation of Lipschitz continuity considered in [17]. To overcome this difficulty we can write every coefficient as a sum of a regular term and a remainder similarly as in [17]. However instead of a study of the corresponding operator decomposition considered in [17], we introduce it in Section 5 as a tool to obtain suitable propagation estimates in a very simple way. Then an easy reasoning described in Section 6 gives the property of the finite propagation speed and the asymptotic trace formula follows similarly as in [7].

Remark (A). Let $\tilde{\mathcal{A}}_h = \mathcal{A}_h + h\mathcal{A}_{h,1}$ where $\mathcal{A}_h$ is like in Theorem 1.1 and

$$\exists\, C_0 > 0 \forall \varphi \in C_0^m(\mathbb{R}^d), \quad |\mathcal{A}_{h,1}[\varphi,\varphi]| \leq C_0 \mathcal{A}_h[\varphi,\varphi].$$

Then $\tilde{\mathcal{A}}_h[\varphi,\varphi]^{1/2}$ and $\|(I-h^2\Delta)^{m/2}\varphi\|$ are equivalent norms if $h < 1/C_0$ and we can define $\tilde{A}_h$, the associated self-adjoint operator in $L^2(\mathbb{R}^d)$. The reasoning described in Section 2 ensures that the assertion of Theorem 1.1 still holds with $\tilde{A}_h$ instead of A_h.

Remark (B). Let M be a compact (boundaryless) manifold with a density dx of class C^m and let $\mathcal{A}_{M,h}$ be a quadratic form on $C^m(M) \times C^m(M)$ satisfying

$$\operatorname{supp} \tilde{\varphi} \cap \operatorname{supp} \tilde{\psi} = \emptyset \Rightarrow \mathcal{A}_{M,h}[\tilde{\varphi}, \tilde{\psi}] = 0.$$

Assume that in local coordinates on $\mathcal{U} \subset \mathbb{R}^d$ the form $\mathcal{A}_{M,h}$ is acting on $\varphi, \psi \in C^m(\mathcal{U})$ according to the formula (1.4) with all the hypotheses of Theorem 1.1 satisfied. Then following a standard reasoning one can affirm that the counting function of the associated self-adjoint operator $A_{M,h}$ satisfies

$$\mathcal{N}(A_{M,h}, E_0) = ch^{-d} + O(h^{1-d}). \tag{1.9}$$

Remark (C). Let M and $A_{M,h}$ be as in (B) and the operator is homogeneous, i.e.,

$$|\nu| + |\bar{\nu}| < 2m \Rightarrow a_{\nu,\bar{\nu}} = 0. \tag{1.10}$$

Then the ξ-homogeneity in local coordinates ensures

$$a(x,\xi) = 1 \Rightarrow |\nabla_\xi a(x,\xi)| \geq \xi \cdot \nabla_\xi a(x,\xi) = 2ma(x,\xi) = 2m > 0$$

and Theorem 1.1 with $E_0 = 1$, $h = \lambda^{-1/(2m)}$ gives the classical formula

$$\mathcal{N}(A_{M,1}, \lambda) = \mathcal{N}(A_{M,\lambda^{-1/(2m)}}, 1) = c\lambda^{\frac{d}{2m}} + O(\lambda^{\frac{d-1}{2m}}). \tag{1.11}$$

Moreover, it is easy to see that the assertion given in Remark (A) ensures that (1.11) still holds if the condition (1.10) is replaced by the condition

$$|\nu| + |\bar{\nu}| < 2m \Rightarrow a_{\nu,\bar{\nu}} \in L^\infty. \tag{1.12}$$

Remark (D). The regularity hypothesis (1.2) is really essential only for x such that that $(x,\xi) \in \Gamma_E$ with a certain $E > E_0$ and the behavior of coefficients for other values of x can be more general: the main requirement is the possibility of reducing the problem by adding an auxiliary cut-off supported in Γ_E as in Lemma 3.1. In particular we have assumed $a_{\nu,\bar\nu} \in L^\infty(\mathbb{R}^d)$ for sake of simplicity, but it is possible to consider unbounded coefficients in a framework of tempered variation models on $T^*\mathbb{R}^d$ (cf., eg., [4]).

Replacing (1.2) by the condition $a_{\nu,\bar\nu} \in \mathcal{B}^r$ (for a certain $r \in]0;\ 1]$), we can prove that the asymptotic formula (1.8) holds with the remainder $O(h^{\mu-d})$ (instead of $O(h^{1-d})$), where μ is an arbitrary number strictly smaller that r.

Another paper will present possibilities of weakening the condition (1.7) in the cases of weaker remainder estimates and stronger regularity hypotheses.

Finally we note that the classical Weyl formula for operators with non-smooth coefficients has been also considered for the boundary value problems (cf. [8–11], [15]).

2. Regularization of coefficients

Let $\gamma_1 \in C_0^\infty(\mathbb{R}^d)$ be such that $\gamma_1(-x) = \gamma_1(x)$, $\int \gamma_1(x)\,\mathrm{d}x = 1$ and for $\varepsilon > 0$ set $\gamma_\varepsilon(x) = \varepsilon^{-d}\gamma_1(x/\varepsilon)$. We fix $\delta \in]0;\ 1[$ and introduce

$$a_{\nu,\bar\nu,h}(x) = (a_{\nu,\bar\nu} * \gamma_{h^\delta})(x) = \int a_{\nu,\bar\nu}(y)\gamma_1(h^{-\delta}(x-y))\,h^{-\delta d}\,\mathrm{d}y. \tag{2.1}$$

It is easy to see (cf. Section 7(A)) that (1.2) ensures

$$|a_{\nu,\bar\nu}(x) - a_{\nu,\bar\nu,h}(x)| \le Ch^{\delta(1+r)}, \tag{2.2}$$

$$|\partial_x^\alpha a_{\nu,\bar\nu,h}(x)| \le C_\alpha h^{-\delta(|\alpha|-1-r)_+}, \tag{2.3}$$

where $s_+ = (s+|s|)/2$ is the positive part of $s \in \mathbb{R}$. We define

$$p_h(x,\xi) = \sum_{|\nu|,|\bar\nu|\le m} a_{\nu,\bar\nu,h}(x)\xi^{\nu+\bar\nu} \tag{2.4}$$

and assume further on $(1+r)\delta \ge 1$, hence (2.2) ensures

$$|\partial_\xi^\alpha(a-p_h)(x,\xi)| \le C_\alpha h\langle\xi\rangle^{2m-|\alpha|}, \tag{2.5}$$

where $\langle\xi\rangle = (1+|\xi|^2)^{1/2}$. Moreover the operator

$$P_h = \sum_{|\nu|,|\bar\nu|\le m} (hD)^\nu a_{\nu,\bar\nu,h}(x)(hD)^{\bar\nu} \tag{2.6}$$

is an approximation of A_h satisfying

$$\pm(A_h - P_h) \le Ch\langle hD\rangle^{2m} = Ch(I - h^2\Delta)^m$$

and $P_h^- \le A_h \le P_h^+$ holds (in the sense of quadratic forms) with

$$P_h^\pm = P_h \pm Ch\langle hD\rangle^{2m}. \tag{2.7}$$

Our principal aim is to show

Theorem 2.1. *The assertion of Theorem* 1.1 *holds if* A_h *is replaced by* $P_h^{\pm}$.

First of all we remark that the discreteness of the spectrum of $P_h^{\pm}$ is a standard result and analogical reasoning can be applied to A_h (cf. Section 7(B) for details). Then the asymptotic formula (1.8) follows from the estimates of the counting functions for $P_h^{\pm}$ due to the min-max principle (cf. [12]), which ensures

$$P_h^- \leq A_h \leq P_h^+ \Rightarrow \mathcal{N}(P_h^+, E) \leq \mathcal{N}(A_h, E) \leq \mathcal{N}(P_h^-, E). \tag{2.8}$$

In the remaining part of this section we describe a version of Theorem 2.1 formulated suitably for the proof that will follow.

Let $\mathbf{I}_Z : \mathbb{R} \to \{0, 1\}$ be the characteristic function of $Z \subset \mathbb{R}$ and set

$$\mathcal{N}_h(p_h, E) = \int_{p_h(x,\xi) \leq E} \frac{dx d\xi}{(2\pi h)^d} = \int \frac{dx d\xi}{(2\pi h)^d} \mathbf{I}_{]-\infty;\, E]} \circ p_h. \tag{2.9}$$

Further on $c_0 > 0$ is such that $\Gamma_{E_0+4c_0}$ is bounded. Due to (1.7) we can choose $c_0 > 0$ small enough to ensure

$$E_0 - 3c_0 \leq a(x, \xi) \leq E_0 + 3c_0 \Rightarrow |\nabla_\xi a(x, \xi)| \geq 2c_0 \tag{2.10}$$

and (2.10) allows us to estimate

$$\mathcal{N}_h(p_h, E) = |\Gamma_E|(2\pi h)^{-d} + O(h^{1-d}) \tag{2.11}$$

uniformly with respect to $E \in [E_0 - 2c_0;\ E_0 + 2c_0]$. Indeed, it is easy to see that (2.10) implies the existence of $C > 0$ such that

$$-C|E - E'| \leq |\Gamma_{E'}| - |\Gamma_E| \leq C|E - E'| \tag{2.12}$$

holds for $E, E' \in [E_0 - 3c_0;\ E_0 + 3c_0]$. Then (2.5) ensures the implication $p_h(x, \xi) \leq E \Rightarrow a(x, \xi) \leq E + C_0 h$ for a certain $C_0 > 0$, hence

$$\mathcal{N}_h(p_h, E) \leq |\Gamma_{E+C_0 h}|(2\pi h)^{-d} \leq |\Gamma_E|(2\pi h)^{-d} + C_1 h^{1-d} \tag{2.13}$$

and we complete the proof of (2.11) by a similar reasoning ensuring

$$\mathcal{N}_h(p_h, E) \geq |\Gamma_{E-C_0 h}|(2\pi h)^{-d} \geq |\Gamma_E|(2\pi h)^{-d} - C_1 h^{1-d}. \tag{2.14}$$

Further on we also write $\mathbf{I}_Z(P_h^{\pm})$ to denote the spectral projector of $P_h^{\pm}$ on a Borel set $Z \subset \mathbb{R}$ and due to (2.8) and (2.11), to obtain the asymptotic formula (1.8) it suffices to show

$$\operatorname{tr} \mathbf{I}_{]-\infty;\, E_0]}(P_h^{\pm}) = \mathcal{N}(P_h^{\pm}, E_0) = \mathcal{N}_h(p_h, E_0) + O(h^{1-d}). \tag{2.15}$$

To show (2.15) we introduce

$$\begin{cases} g \in C_0^\infty(]E_0 - 2c_0;\ E_0 + 2c_0[), g = 1 \text{ on } [E_0 - c_0;\ E_0 + c_0], \\ \tilde{g} \in C_0^\infty(] - \infty;\ E_0 - c_0[) \end{cases} \tag{2.16}$$

and assume $\tilde{g} + g^2 = 1$ on $[-E_0';\ E_0 + c_0]$ where $E_0' > 0$ is chosen sufficiently large to ensure $-E_0' \leq \inf p_h^{\pm}$ and $-E_0' \leq \inf \sigma(P_h^{\pm})$ for $h \in]0;\ h_0]$.

Then for $E \in [E_0 - c_0;\ E_0 + c_0]$ we can write the decompositions

$$\mathcal{N}_h(p_h, E) = \int \frac{\mathrm{d}x\mathrm{d}\xi}{(2\pi h)^d} (\tilde{g} + g^2 \mathbf{1}_Z)(p_h(x,\xi)),$$

$$\mathbf{1}_{]-\infty;\ E]}(P_h^{\pm}) = (\tilde{g} + g^2 \mathbf{1}_Z)(P_h^{\pm}),$$

where $Z = [E_0 - 2c_0;\ E]$. Now it is clear that (2.15) is a consequence of

Theorem 2.2. *One has*

$$\operatorname{tr} \tilde{g}(P_h^{\pm}) = \int \frac{\mathrm{d}x\mathrm{d}\xi}{(2\pi h)^d} \tilde{g} \circ p_h + O(h^{1-d}), \tag{2.17}$$

$$\operatorname{tr} (g^2 \mathbf{1}_Z)(P_h^{\pm}) = \int \frac{\mathrm{d}x\mathrm{d}\xi}{(2\pi h)^d} (g^2 \mathbf{1}_Z) \circ p_h + O(h^{1-d}). \tag{2.18}$$

Further on $\bar{c} > 0$ is small enough and $f_0 \in C_0^\infty(]\bar{c}/2;\ \bar{c}/2[)$. Then we define $f_1 = f_0 * f_0 \in C_0^\infty(]\bar{c};\ \bar{c}[)$ and

$$\tilde{f}_h(\lambda) = \mathcal{F}_h^{-1} f_1(\lambda) = \int_{-\infty}^{\infty} \mathrm{e}^{i\lambda t/h} f_1(t) \frac{\mathrm{d}t}{2\pi h}. \tag{2.19}$$

We note that $\tilde{f}_h(\lambda) = \tilde{f}_1(\lambda/h)/h$, $\tilde{f}_1$ is rapidly decreasing and $\tilde{f}_1 > 0$. Moreover we can choose f_0 such that $\int_{-\infty}^{\infty} \tilde{f}_h = f_1(0) = 1$. Then we use

$$\tilde{f}_h^Z(\lambda) = (\mathbf{1}_Z * \tilde{f}_h)(\lambda) = \int_Z \mathrm{d}\lambda'\, \tilde{f}_h(\lambda - \lambda') \tag{2.20}$$

as an approximation of $\mathbf{1}_Z$. In Section 7(C) we recall a proof of

Proposition 2.3. *The estimate* (2.18) *follows if*

$$\operatorname{tr} (g^2 \tilde{f}_h^Z)(P_h^{\pm}) = \int \frac{\mathrm{d}x\mathrm{d}\xi}{(2\pi h)^d} (g^2 \tilde{f}_h^Z) \circ p_h + O(h^{1-d}) \tag{2.21}$$

holds uniformly with respect to $Z = [E';\ E]$.

3. Pseudodifferential approximation

In this section we show the estimate (2.17). We use pseudodifferential operators in a reasoning which will be developed in Section 4 to treat (2.18).

We denote by $\|\cdot\|$ the norm of bounded operators $B(L^2(\mathbb{R}^d))$ and $\|B\|_{\mathrm{tr}} = \operatorname{tr}(B^*B)^{1/2}$ is the trace class norm. If $b_h \in C_0^\infty(\mathbb{R}^{2d})$ then we denote by $b_h(x, hD)$ the pseudodifferential operator acting on $\varphi \in \mathcal{S}(\mathbb{R}^d)$ according to the formula

$$(b_h(x, hD)\varphi)(x) = \int \frac{\mathrm{d}y\mathrm{d}\xi}{(2\pi h)^d} \mathrm{e}^{i(x-y)\xi/h} b_h(x,\xi)\varphi(y). \tag{3.1}$$

If $b_h, l_h \in C_0^\infty(\mathbb{R}^{2d})$, then $b_h(x, hD) l_h(x, hD)^*$ has the integral kernel

$$\mathcal{K}_h(x, y) = \int \frac{\mathrm{d}\xi}{(2\pi h)^d} \mathrm{e}^{i(x-y)\xi/h} b_h(x,\xi)\overline{l_h(y,\xi)},$$

hence

$$\operatorname{tr} b_h(x,hD)l_h(x,hD)^* = \int \mathrm{d}x\, \mathcal{K}_h(x,x) = \int \frac{\mathrm{d}x\mathrm{d}\xi}{(2\pi h)^d}\, b_h \bar{l}_h. \tag{3.2}$$

In this section we consider $\tilde{g} \in C_0^\infty(]-\infty;\ E_1[)$ with E_1 such that $\Gamma_{E_1+2c_1}$ is bounded for a certain $c_1 > 0$. Then using (2.5) we can find $h_0 > 0$ small enough to ensure

$$0 < h < h_0 \Rightarrow \operatorname{supp}(\tilde{g} \circ p_h) \subset \Gamma_{E_1+c_1}. \tag{3.3}$$

Next we fix an auxiliary h-independent cut-off function $l \in C_0^\infty(\mathbb{R}^{2d})$ such that $l = 1$ on a neighborhood of $\overline{\Gamma_{E_1+c_1}}$ and $l \geq 0$. We denote $L_h = l(x,hD)$ and recall standard estimates of pseudodifferential operators:

$$\|L_h\| = O(1), \|L_h\|_{\mathrm{tr}} = O(h^{-d}). \tag{3.4}$$

Moreover (3.2) allows us to write

$$\int \frac{\mathrm{d}y\mathrm{d}\xi}{(2\pi h)^d}\, \tilde{g} \circ p_h = \int \frac{\mathrm{d}y\mathrm{d}\xi}{(2\pi h)^d}\, l(\tilde{g} \circ p_h) = \operatorname{tr}(\tilde{g} \circ p_h)(x,hD)L_h^* \tag{3.5}$$

and since $|\operatorname{tr} B| \leq \|B\|_{\mathrm{tr}}$, it is clear that (2.17) follows from

$$\|\tilde{g}(P_h^\pm) - L_h^*(\tilde{g} \circ p_h)(x,hD)\|_{\mathrm{tr}} = O(h^{1-d}). \tag{3.6}$$

Our proof of (3.6) is based on an auxiliary lemma (cf. Section 7(D)):

Lemma 3.1. *If $\tilde{g}$ and L_h are as above, then for every $N \in \mathbb{N}$ one has*

$$\|\tilde{g}(P_h^\pm)(I - L_h)\|_{\mathrm{tr}} = O(h^N). \tag{3.7}$$

Since $\|\tilde{g}(P_h^\pm)L_h\|_{\mathrm{tr}} \leq \|\tilde{g}(P_h^\pm)\|\, \|L_h\|_{\mathrm{tr}} = O(h^{-d})$ due to (3.4), Lemma 3.1 ensures $\|\tilde{g}(P_h^\pm)\|_{\mathrm{tr}} = O(h^{-d})$. Another immediate corollary of (3.7) is

$$\|\tilde{g}(P_h^\pm) - L_h^*\, \tilde{g}(P_h^\pm)L_h\|_{\mathrm{tr}} = O(h^N)$$

and we can conclude that instead of (3.6) it suffices to show

$$\|L_h^*\tilde{g}(P_h^\pm)L_h - L_h^*(\tilde{g} \circ p_h)(x,hD)\|_{\mathrm{tr}} = O(h^{1-d}). \tag{3.6'}$$

Further on $g_1 \in \mathcal{S}(\mathbb{R})$ denotes the Fourier transform of $\tilde{g}$, i.e., we have $\tilde{g}(\lambda) = \int_{-\infty}^{\infty} \frac{\mathrm{d}t}{2\pi}\, g_1(t)\, \mathrm{e}^{it\lambda}$ and

$$(\tilde{g} \circ p_h)(x,\xi) = \tilde{g}(p_h(x,\xi))l(x,\xi) = \int_{-\infty}^{\infty} \frac{\mathrm{d}t}{2\pi}\, g_1(t)\, (\mathrm{e}^{itp_h} l)(x,\xi), \tag{3.8}$$

$$\tilde{g}(P_h^\pm) = \int_{-\infty}^{\infty} \frac{\mathrm{d}t}{2\pi}\, g_1(t)\, \mathrm{e}^{itP_h^\pm}. \tag{3.9}$$

Therefore we can write

$$\tilde{g}(P_h^\pm)L_h - (\tilde{g} \circ p_h)(x,hD) = \int_{-\infty}^{\infty} \frac{\mathrm{d}t}{2\pi}\, g_1(t)\, R_{h,t}$$

with

$$R_{h,t} = \mathrm{e}^{itP_h^\pm} L_h - (\mathrm{e}^{itp_h} l)(x,hD) \tag{3.10}$$

and the left-hand side of (3.6′) can be estimated by

$$\int_{-\infty}^{\infty} \frac{\mathrm{d}t}{2\pi}\, |g_1(t)|\, \|L_h^* R_{h,t}\|_{\mathrm{tr}}. \tag{3.11}$$

However $\|L_h^* R_{h,t}\|_{\mathrm{tr}} \le \|L_h^*\|_{\mathrm{tr}}\, \|R_{h,t}\| \le Ch^{-d}\|R_{h,t}\|$ and g_1 is rapidly decaying, hence in order to estimate the quantity (3.11) by $O(h^{1-d})$ it suffices to show

Lemma 3.2. *There is $C > 0$ such that $\|R_{h,t}\| \le Ch\langle t\rangle^C$.*

Before starting the proof of Lemma 3.2 it is useful to introduce some notations. We write the differential operators $P_h^{\pm}$ in the standard form

$$P_h^{\pm} = \sum_{|\alpha|\le 2m} a^{\pm}_{\nu,\bar{\nu},h}(x)(hD)^{\alpha} = p_h^{\pm}(x, hD), \tag{3.12}$$

$$p_h^{\pm}(x,\xi) = \sum_{|\alpha|\le 2m} a^{\pm}_{\nu,\bar{\nu},h}(x)\xi^{\alpha} \tag{3.13}$$

and it is easy to check that (2.5) still holds with $p_h^{\pm}$ instead of p_h.

Let $\tilde{m} \in \mathbb{R}$ and $\delta_1 \ge \delta_0 \ge 0$ be such that $\delta_0 + \delta_1 < 1$. If $b = \{b_h\}_{0<h\le h_0}$ is a family of smooth functions satisfying the estimates

$$|\partial_\xi^\alpha \partial_x^\beta b_h(x,\xi)| \le C_{n,\alpha,\beta} h^{-\tilde{m}-|\alpha|\delta_0-|\beta|\delta_1}\langle\xi\rangle^{-n} \tag{3.14}$$

(for every $\alpha, \beta \in \mathbb{N}^d$ and $n \in \mathbb{N}$), then we write $b \in S^{\tilde{m}}_{\delta_0,\delta_1}$.

If $l \in C_0^\infty(\mathbb{R}^{2d})$ is h-independent, then the families lp_h, $l\tilde{p}_h^{\pm}$, $l\mathrm{e}^{itp_h}$ belong to $S^0_{0,\delta}$ and well-known L^2-estimates of pseudodifferential operators (cf. [2],[6]) give

$$b \in S^{\tilde{m}}_{0,\delta} \Rightarrow \exists C > 0, \|(\mathrm{e}^{itp_h} l b_h)(x, hD)\| \le Ch^{-\tilde{m}}\langle t\rangle^C. \tag{3.15}$$

If $s \in \mathbb{R}$ then we write $B_h = L_h + O(h^s)$ if and only if there is a constant $C > 0$ such that $\|B_h - L_h\| \le Ch^s$ holds for all $h \in]0;\ h_0]$. Moreover further on we adopt the convention to drop the index h whenever there is no confusion. Instead of $P_h^{\pm}$, L_h, $R_{h,t}$, p_h, $p_h^{\pm}$, we are going to write simply P, L, R_t, p, $p^{\pm}$.

Proof of Lemma 3.2 We are going to obtain suitable estimates of

$$\tilde{R}_t = (i\partial_t + P)(\mathrm{e}^{itp} l)(x, hD) \tag{3.16}$$

and the assertion of Lemma 3.2 will follow from

$$R_t = -[\mathrm{e}^{i(t-\tau)P}(\mathrm{e}^{i\tau p} l)(x, hD)]_{\tau=0}^{\tau=t} = i\int_0^t \mathrm{d}\tau\, \mathrm{e}^{i(t-\tau)P}\tilde{R}_\tau. \tag{3.17}$$

The standard formula of the composition with a differential operator gives

$$p^{\pm}(x, hD)(\mathrm{e}^{itp} l)(x, hD) = \tilde{q}_t(x, hD), \tag{3.18}$$

with

$$\tilde{q}_t = \sum_{|\alpha|\le 2m} \frac{(-ih)^{|\alpha|}}{\alpha!} \partial_x^\alpha(\mathrm{e}^{itp} l)\partial_\xi^\alpha p^\pm = \mathrm{e}^{itp} \sum_{k=0}^{2m} t^k q_k, \tag{3.19}$$

$$q_k = \sum_{k\le|\alpha|\le 2m} \sum_{\substack{\alpha_0+\dots+\alpha_k=\alpha \\ \alpha_j\neq 0 \text{ if } j\neq 0}} c_{\alpha_0,\dots,\alpha_k}\, \partial_x^{\alpha_0} l\, \partial_x^{\alpha_1} p \dots \partial_x^{\alpha_k} p\, \partial_\xi^\alpha p^\pm h^{|\alpha|} \tag{3.20}$$

and $q_k \in S_{0,\delta}^{-k}$ follows from $-|\alpha| + \sum_{j=1}^k \delta(|\alpha_j|-1) \le -k-(1-\delta)(|\alpha|-k)$.

However (3.15) ensures

$$q_k \in S_{0,\delta}^{-k} \Rightarrow (\mathrm{e}^{itp} q_k)(x, hD) = O(h^k \langle t\rangle^{C_k}),$$

hence

$$P(\mathrm{e}^{itp} l)(x, hD) = (\mathrm{e}^{itp} q_0)(x, hD) + O(h\langle t\rangle^C).$$

To complete the proof we note that $l(p^\pm - p) \in S_{0,\delta}^{-1}$ and

$$q_0 - lp^\pm = \sum_{1\le|\alpha|\le 2m} \frac{(-ih)^{|\alpha|}}{\alpha!} \partial_x^\alpha l\, \partial_\xi^\alpha p^\pm \in S_{0,\delta}^{-1} \tag{3.21}$$

ensure $q_0 - lp \in S_{0,\delta}^{-1}$, hence

$$i\partial_t(\mathrm{e}^{itp} l)(x, hD) + (\mathrm{e}^{itp} q_0)(x, hD) = (\mathrm{e}^{itp}(q_0 - lp))(x, hD) = O(h\langle t\rangle^C).$$

4. Refinement

Due to Proposition 2.3, to complete the proof of Theorem 2.2 it suffices to show (2.21). We will use a reasoning of Section 3 in a refined form. We keep notations introduced before and in particular we drop the index h.

To begin we remark that replacing $\tilde{g}$ by $g \in C_0^\infty(]E_0 - c_0;\ E_0 + c_0[)$ in (3.6), we can introduce $l_0 = g \circ p$, $L_0 = l_0(x, hD)$ and write

$$\operatorname{tr}(g^2 \tilde{f}_h^Z)(P) = \operatorname{tr} L_0^* \tilde{f}_h^Z(P) L_0 + O(h^{1-d}). \tag{4.1}$$

As before $Z = [E';\ E]$ and we use the Fourier transform

$$\begin{aligned} f_h^Z(t) &:= \mathcal{F}_h \tilde{f}_h^Z(t) = \int_{-\infty}^{\infty} \mathrm{d}\lambda\, \mathrm{e}^{-it\lambda/h} \tilde{f}_h^Z(\lambda) \\ &= f_1(t) \int_{E'}^{E} \mathrm{d}\lambda \mathrm{e}^{-it\lambda/h} = f_1(t) \frac{\mathrm{e}^{-it(E+E')/(2h)}}{t/(2h)} \sin\frac{(E'-E)t}{2h} \end{aligned}$$

to write

$$\operatorname{tr} L_0^* \tilde{f}_h^Z(P) L_0 = \int_{-\infty}^{\infty} \frac{\mathrm{d}t}{2\pi h} f_h^Z(t) \operatorname{tr} L_0^* \mathrm{e}^{itP/h} L_0, \tag{4.2}$$

$$\int \frac{\mathrm{d}x\mathrm{d}\xi}{(2\pi h)^d} l_0^2(\tilde{f}_h^Z \circ p) = \int \frac{\mathrm{d}x\mathrm{d}\xi}{(2\pi h)^d} \int_{-\infty}^{\infty} \frac{\mathrm{d}t}{2\pi h} f_h^Z(t)\, l_0^2 \mathrm{e}^{itp/h}. \tag{4.3}$$

However (3.2) allows us to express (4.3) as

$$\int_{-\infty}^{\infty} \frac{\mathrm{d}t}{2\pi h} f_h^Z(t) \operatorname{tr} L_0^*(l_0 \mathrm{e}^{itp/h})(x, hD) \tag{4.3'}$$

and introducing

$$R_{t/h}^0 = \mathrm{e}^{itP/h} L_0 - (\mathrm{e}^{itp/h} l_0)(x, hD) \tag{4.4}$$

we find that (2.21) is equivalent to the estimate

$$\int_{-\infty}^{\infty} \frac{\mathrm{d}t}{2\pi h} f_h^Z(t) \operatorname{tr} L_0^* R_{t/h}^0 = O(h^{1-d}). \tag{4.5}$$

Next we observe that (2.5) and (2.10) ensure

$$E_0 - 2c_0 \leq p(x,\xi) \leq E_0 + 2c_0 \Rightarrow |\nabla_\xi p(x,\xi)| \geq c_0 \tag{4.6}$$

for $h \in]0;\ h_0]$ if $h_0 > 0$ is fixed small enough. The condition (4.6) ensures

Lemma 4.1. *Assume that $b \in S_{0,\delta}^0$. Then for every $n \in \mathbb{N}$ one can find $b_n \in S_{0,\delta}^0$ such that*

$$t^n \int \mathrm{d}\xi b\, \mathrm{e}^{itp/h} = h^n \int \mathrm{d}\xi b_n \, \mathrm{e}^{itp/h}. \tag{4.7(n)}$$

Proof. Using (4.6) in a standard reasoning we obtain

$$\forall b \in S_{0,\delta}^0 \exists \tilde{b}_j \in S_{0,\delta}^0, \quad b = \sum\nolimits_{j=1}^d \tilde{b}_j \partial_{\xi_j} p. \tag{4.8}$$

Therefore performing the integration by parts

$$t \int \mathrm{d}\xi \sum\nolimits_{j=1}^d b_j \partial_{\xi_j} p \, \mathrm{e}^{itp/h} = ih \int \mathrm{d}\xi \sum\nolimits_{j=1}^d \partial_{\xi_j} \tilde{b}_j \, \mathrm{e}^{itp/h}$$

we obtain the assertion of Lemma 4.1 for $n = 1$ with $b_1 = \sum_{j=1}^d i\partial_{\xi_j} \tilde{b}_j$.

Obviously the general case follows by induction with respect to n. □

We will use Lemma 4.1 to replace $R_{t/h}^0$ in (4.5) by

$$R_{t/h} = \mathrm{e}^{itP/h} L_0 - (\mathrm{e}^{itp/h} l_t)(x, hD), \tag{4.4'}$$

where $l_t = l_0 + tl_0' + \frac{1}{2}t^2 l_0''$ with certain $l_0' \in S_{0,\delta}^0$ and $l_0'' \in S_{0,\delta}^1$.

In fact $\operatorname{tr} L_0^*(R_{t/h} - R_{t/h}^0)$ can be expressed as

$$\int \frac{\mathrm{d}x\mathrm{d}\xi}{(2\pi h)^d} l_0(l_t - l_0)\mathrm{e}^{itp/h} = h^{1-d} \int \mathrm{d}x\mathrm{d}\xi\, \tilde{l}\, \mathrm{e}^{itp/h},$$

with a ceratin $\tilde{l} \in S_{0,\delta}^0$ (due to Lemma 4.1 with $n = 1$ and 2). Thus

$$\int_{-\infty}^{\infty} \frac{\mathrm{d}t}{2\pi h} f_h^Z(t) \operatorname{tr} L_0^*(R_{t/h} - R_{t/h}^0) = h^{1-d} \int \mathrm{d}x\mathrm{d}\xi\, \tilde{l}(\tilde{f}_h^Z \circ p)$$

is clearly $O(h^{1-d})$, hence (4.5) is equivalent to

$$\int_{-\infty}^{\infty} \frac{\mathrm{d}t}{2\pi h} f_h^Z(t) \operatorname{tr} L_0^* R_{t/h} = O(h^{1-d}). \tag{4.5'}$$

Our next remark is that $\operatorname{supp} f_h^Z = \operatorname{supp} f_1 \subset [-\bar{c};\ \bar{c}]$ and

$$\int_{-\bar{c}}^{\bar{c}} \frac{\mathrm{d}t}{h} |f_h^Z(t)| \leq C \int_{-C/h}^{C/h} \mathrm{d}s\, \frac{|\sin s|}{|s|} = O(\log(1/h)),$$

hence to obtain (4.5′) it suffices to show that

$$\sup_{|t|\leq\bar{c}} |\operatorname{tr} L_0^* R_{t/h}| \leq C h^{1-d+\varepsilon} \tag{4.9}$$

holds if $\bar{c}, \varepsilon > 0$ are small enough. The main result of this section is

Proposition 4.2. *Let $\delta_1 \in]0;\ 1[$ and $\varepsilon > 0$. If $1-\delta_1$ and ε are small enough, then one can find $l_0' \in S^0_{0,\delta}$ and $l_0'' \in S^1_{0,\delta}$ such that $R_{t/h}$ given by (4.4′) with $l_t = l_0 + t l_0' + \frac{1}{2} t^2 l_0''$ satisfies*

$$\sup_{|t|\leq h^{\delta_1}} |\operatorname{tr} L_0^* R_{t/h}| \leq C h^{1-d+\varepsilon}. \tag{4.10}$$

Before starting the proof of Proposition 4.2 we introduce some conventions serving to simplify notations further on. For given numbers $c, c', s, s', \tilde{s} \in \mathbb{R}$ and operators $B_{h,t}, \tilde{B}_{h,t} \in B(L^2(\mathbb{R}^d))$ we write

$$ch^s \leq t \leq c'h^{s'} \Rightarrow B_{h,t} = \tilde{B}_{h,t} + O(h^{\tilde{s}}) \tag{4.11}$$

if and only if there is $C > 0$ such that $\sup_{t\in[ch^s;\ c'h^{s'}]} \|B_{h,t} - \tilde{B}_{h,t}\| \leq C h^{\tilde{s}}$.

If $b_t = \{b_{t,h}\}_{0<h\leq h_0}$ is a family of smooth functions satisfying the estimates

$$\sup_{t\in[ch^s;\ c'h^{s'}]} |\partial_\xi^\alpha \partial_x^\beta b_{t,h}(x,\xi)| \leq C_{n,\alpha,\beta} h^{-\tilde{m}-|\alpha|\delta_0-|\beta|\delta_1} \langle\xi\rangle^{-n} \tag{4.12}$$

(for all $\alpha, \beta \in \mathbb{N}^d$, $n \in \mathbb{N}$), then we write $ch^s \leq t \leq c'h^{s'} \Rightarrow b_t \in S^{\tilde{m}}_{\delta_0,\delta_1}$.

We write $ch^s \leq t \leq c'h^{s'} \Rightarrow B_t \in \mathrm{Op}(S^{\tilde{m}}_{\delta_0,\delta_1})$ if and only if there exists $b_t = \{b_{t,h}\}_{0<h\leq h_0}$ such that $B_t = \{b_{t,h}(x,hD)\}_{0<h\leq h_0}$ and $ch^s \leq t \leq c'h^{s'} \Rightarrow b_t \in S^{\tilde{m}}_{\delta_0,\delta_1}$.

Proof of Proposition 4.2. To begin we introduce

$$\tilde{R}_{t/h} = (ih\partial_t + P)(\mathrm{e}^{itp/h} l_t)(x, hD) \tag{4.13}$$

and check that (4.10) follows from

$$\sup_{|t|\leq h^{\delta_1}} \|\tilde{R}_{t/h}\| \leq C h^{2+\varepsilon-\delta_1}. \tag{4.14}$$

Indeed, writing

$$R_{t/h} = -[\mathrm{e}^{i(t-\tau)P/h}(\mathrm{e}^{i\tau p/h} l_t)(x,hD)]_{\tau=0}^{\tau=t} = ih^{-1}\int_0^t \mathrm{d}\tau\, \mathrm{e}^{i(t-\tau)P/h} \tilde{R}_{\tau/h}$$

and using (4.14) we obtain

$$|t| \leq h^{\delta_1} \Rightarrow \|R_{t/h}\| \leq h^{\delta_1 - 1} \sup_{|\tau|\leq h^{\delta_1}} \|\tilde{R}_{\tau/h}\| \leq C h^{1+\varepsilon},$$

which implies $\|L_0^* R_{t/h}\|_{\mathrm{tr}} \leq C h^{-d} \|R_{t/h}\| \leq C h^{1+\varepsilon-d}$ when $|t| \leq h^{\delta_1}$.

Using l_0 instead of l in (3.18) we still have (3.19) with some $q_k \in S^{-k}_{0,\delta}$ and

$$|t| \leq h^{\delta_1} \Rightarrow (t/h)^k q_k \in S^{-k\delta_1}_{0,\delta}. \tag{4.15}$$

Without writing we assume all the time $|t| \leq h^{\delta_1}$. Replacing t by t/h in (3.18–3.19) and (3.15) we observe that $\langle t/h\rangle^C \leq h^{-\varepsilon_1}$ holds with $\varepsilon_1 = C(1-\delta_1)$ and we obtain

$$P(\mathrm{e}^{itp/h}l_0)(x,hD) = \Big(\mathrm{e}^{itp/h}(q_0 + \frac{t}{h}q_1)\Big)(x,hD) + O(h^{2\delta_1-\varepsilon_1}). \tag{4.16}$$

Similarly using $l_0' \in S^0_{0,\delta}$ instead of l we obtain (3.18)–(3.19) with some $q_k' \in S^{-k}_{0,\delta}$ instead of q_k and instead of (4.16), a similar formula without $(t/h)q_1' \in S^{-\delta_1}_{0,\delta}$ gives

$$P(\mathrm{e}^{itp/h}l_0')(x,hD) = (\mathrm{e}^{itp/h}q_0')(x,hD) + O(h^{\delta_1-\varepsilon_1}). \tag{4.17}$$

If we use $l_0'' \in S^1_{0,\delta}$ instead of $l_0' \in S^0_{0,\delta}$ in a formula analogical to (4.17) we obtain an additional factor of order h^{-1}, i.e.

$$P(\mathrm{e}^{itp/h}l_0'')(x,hD) = (\mathrm{e}^{itp/h}q_0'')(x,hD) + O(h^{\delta_1-1-\varepsilon_1}) \tag{4.18}$$

holds with a certain $q_0'' \in S^1_{0,\delta}$. Therefore multiplying (4.17) by $t = O(h^{\delta_1})$ and (4.18) by $t^2 = O(h^{2\delta_1})$, we obtain

$$P(\mathrm{e}^{itp/h}l_t)(x,hD) = (\mathrm{e}^{itp/h}(q_0 + t(q_1/h + q_0') + t^2 q_0''/2))(x,hD) + O(h^{s'}), \tag{4.19}$$

with $s' = \min\{2\delta_1, 3\delta_1 - 1\} - \varepsilon_1$. Taking $l_0' = i(q_0 - l_0 p)/h$, $l_0'' = -ih^{-2}q_1$, we find that $\tilde{R}_{t/h}$ can be expressed as

$$(\mathrm{e}^{itp/h}(q_0 - l_0 p + ihl_0' + t(q_0' - l_0'p + q_1/h + ihl_0''p) + \frac{t^2}{2}(q_0'' - l_0''p))(x,hD) + O(h^{s'})$$

$$= (\mathrm{e}^{itp/h}(t(q_0' - l_0'p) + \frac{t^2}{2}(q_0'' - l_0''p))(x,hD) + O(h^{s'}) \tag{4.20}$$

and since using $l_0' \in S^0_{0,\delta}$, $l_0'' \in S^1_{0,\delta}$ instead of $l \in S^0_{0,0}$ in (3.21) leads to

$$q_0' - l_0'p \in S^{\delta-1}_{0,\delta}, \quad q_0'' - l_0''p \in S^{\delta}_{0,\delta}, \tag{4.21}$$

(4.20) gives $\tilde{R}_{t/h} = O(h^s)$ with $s = \min\{s', \delta_1 + 1 - \delta - \varepsilon_1, 2\delta_1 - \delta - \varepsilon_1\}$. To complete the proof we observe that choosing $1 - \delta_1$ and ε small enough we can ensure $s \geq 2 + \varepsilon - \delta_1$. □

Due to Proposition 4.1 it remains to consider the estimate (4.9) if $h^{\delta_1} \leq |t| \leq \bar{c}$. However for every $N \in \mathbb{N}$ we can find a constant C_N such that

$$\sup_{h^{\delta_1} \leq |t| \leq 1} \Big| \mathrm{tr}\, L_0^*(\mathrm{e}^{itp/h}l_0)(x,hD)\Big| \leq C_N h^N. \tag{4.22}$$

Indeed, Lemma 4.1 ensures the existence of $\tilde{l}_n \in S^0_{0,\delta}$ such that

$$\mathrm{tr}\, L_0^*(\mathrm{e}^{itp/h}l_0)(x,hD) = \frac{h^n}{t^n}\mathrm{tr}\,(\mathrm{e}^{itp/h}\tilde{l}_n)(x,hD)$$

and the last quantity is obviously $O(h^{-d}(h/t)^n)$, i.e., (4.22) follows.

Now it is clear that to complete the proof of (2.21) it suffices to prove

Proposition 4.3. *If $\bar{c} > 0$ is small enough, then*

$$\sup_{h^{\delta_1} \le |t| \le \bar{c}} \left| \operatorname{tr} L_0^* \, \mathrm{e}^{itP/h} \, L_0 \right| \le C_N h^N. \tag{4.23}$$

The proof of Proposition 4.3 will be given in Sections 5 and 6.

5. Approximation of the Hamiltonian flow

Further on we abbreviate $U_t = \mathrm{e}^{itP/h}$ and we begin the proof of Proposition 4.3 writing (4.23) in the following symbolic form

$$h^{\delta_1} \le |t| \le \bar{c} \Rightarrow \operatorname{tr} L_0^* \, U_t \, L_0 = O(h^N). \tag{5.1}$$

Further on $C_{\mathrm{ef}} > 0$ is a fixed large constant and $\theta_{\mathrm{ef}} \in C_0^\infty(\mathbb{R}^d)$ is a fixed h-independent cut-off function satisfying $\theta_{\mathrm{ef}}(\xi) = 1$ for $|\xi| \le C_{\mathrm{ef}} + 1$.

We fix $0 < \delta_0 < 1/2$ small enough to ensure $\delta_0 + \delta \le \delta_0 + \delta_1 < 1$ (where $\delta < \delta_1 < 1$ are as in Section 4) and define

$$\tilde{p}(x, \xi) = \int p(y, \xi) \theta_{\mathrm{ef}}(\xi) \, \gamma_{h^{\delta_0}}(x - y) \mathrm{d}y, \tag{5.2}$$

where γ_ε is as in Section 2. Further on $\sigma > 0$ is fixed small enough to ensure $\sigma \le \min\{\delta_0 r, \, \delta - \delta_0, 1 - 2\delta_0\}$. Then it is easy to check (cf. Section 7(E)) that

$$|\alpha| \le 1 \Rightarrow \partial_x^\alpha \tilde{p} \in S^0_{0,\delta_0}, l \, \partial_x^\alpha (p - \tilde{p}) \in S^{-\sigma}_{0,\delta} \tag{5.3}$$

if $l \in S^0_{0,0}$ satisfies $l(x, \xi) = 0$ for $|\xi| \ge C_{\mathrm{ef}}$.

Let $\vartheta_t = \exp(t H_{\tilde{p}})$ denote the Hamiltonian flow of $\tilde{p}$, i.e.,

$$\begin{cases} (\mathrm{d}\vartheta_t / \mathrm{d}t)(x, \xi) = H_{\tilde{p}}(\vartheta_t(x, \xi)), \\ \vartheta_t(x, \xi)|_{t=0} = (x, \xi), \end{cases} \tag{5.4}$$

where $H_{\tilde{p}} = (\nabla_\xi \tilde{p}, -\nabla_x \tilde{p})$ and introduce

$$\bar{C} = \sup_{x, \xi \in \mathbb{R}^d} |\nabla_{x,\xi} \tilde{p}(x, \xi)| + 1. \tag{5.5}$$

Further on we assume $0 < \bar{c} < 1/(2\bar{C})$, hence

$$|t| \le \bar{c} \Rightarrow |\vartheta_t(x, \xi) - (x, \xi)| \le \bar{C}|t| \le 1/2, \tag{5.6}$$

i.e., ϑ_t is invertible. Moreover

$$|t| \le 2h^{\delta_0} \Rightarrow \nabla_{x,\xi} \vartheta_t \in S^0_{\delta_0,\delta_0}. \tag{5.7}$$

Indeed, if $\delta_0 = 0$ then (5.7) becomes a well-known property of solutions of ordinary differential equations with all derivatives bounded and the general case follows easily by the change of variables $(t', x', \xi') = h^{\delta_0}(t, x, \xi)$. Similar reduction to the case $\delta_0 = 0$ gives the following property

$$\forall l \in S^{\tilde{m}}_{\delta_0,\delta_0}, |t| \le 2h^{\delta_0} \Rightarrow l \circ \vartheta_t \in S^{\tilde{m}}_{\delta_0,\delta_0}. \tag{5.8}$$

We denote by b_h^w the Weyl quantization of b_h, i.e., the pseudodifferential operator acting on $\varphi \in \mathcal{S}(\mathbb{R}^d)$ according to the formula

$$(b_h^w \varphi)(x) = \int \frac{\mathrm{d}y\mathrm{d}\xi}{(2\pi h)^d}\, \mathrm{e}^{i(x-y)\xi/h} b_h((x+y)/2, \xi)\varphi(y). \tag{5.9}$$

Further on we usually drop the index h and we write $B \cong \tilde{B}$ if $B = \tilde{B} + O(h^{\tilde{s}})$ holds for every $\tilde{s} \in \mathbb{R}$. In particular if $b, \tilde{b} \in S^{\tilde{m}}_{\delta_0,\delta_1}$ are such that $\operatorname{supp} b \cap \operatorname{supp} \tilde{b} = \emptyset$ then $b^w \tilde{b}^w \cong 0$ (clearly $b(x, hD)\tilde{b}(x, hD) \cong 0$ and $b(x, hD)\tilde{b}^w \cong 0$ hold as well). Moreover we write

$$ch^s \le t \le c'h^{s'} \Rightarrow B_t \cong \tilde{B}_t \tag{5.10}$$

if (4.11) holds for every $\tilde{s} \in \mathbb{R}$. We introduce formally

$$\mathbb{D}_{P/h} B_t = \frac{\mathrm{d}}{\mathrm{d}t} B_t + i[B_t,\, P/h] \tag{5.11}$$

and observe that $\frac{\mathrm{d}}{\mathrm{d}t}(U_{-t} B_t U_t) = U_{-t}(\mathbb{D}_{P/h} B_t) U_t$.

Lemma 5.1. *Let $l \in S^0_{\delta_0,\delta_0}$ and σ as above (5.3). If $l_t = l \circ \vartheta_t$, then*

$$|t| \le 2h^{\delta_0} \Rightarrow t\, \mathbb{D}_{P/h} l_t^w \in \operatorname{Op}(S^{-\sigma}_{\delta_0,\delta}). \tag{5.12}$$

Proof. In the first step we check that

$$|t| \le 2h^{\delta_0} \Rightarrow \frac{\mathrm{d}}{\mathrm{d}t} l_t^w + i[l_t^w,\, \tilde{p}^w/h] \in \operatorname{Op}(S^{\delta_0-(1-2\delta_0)}_{\delta_0,\delta_0}). \tag{5.13}$$

Without writing we assume all the time $|t| \le 2h^{\delta_0}$. By the definition of the Hamiltonian flow we have $\frac{\mathrm{d}}{\mathrm{d}t} l_t + \{l_t,\, \tilde{p}\} = 0$, where the Poisson bracket $\{l_t,\, \tilde{p}\}$ is the leading term of the asymptotic formula giving the symbol of the commutator $i[l_t^w,\, \tilde{p}^w/h]$. The remaining terms appear with multi-indices $\alpha, \tilde{\alpha} \in \mathbb{N}^{2d}$ of order 2 at least and to obtain (5.13) we note that

$$|\alpha| = |\tilde{\alpha}| = n \ge 2 \Rightarrow h^{|\alpha|-1} \partial^\alpha l_t\, \partial^\alpha \tilde{p} \in S^{\delta_0-(1-2\delta_0)}_{\delta_0,\delta_0} \tag{5.14}$$

due to $1 - n + \delta_0 n + \delta_0(n-1) = \delta_0 - (1-2\delta_0)(n-1)$.

In the next step we observe that $[l_t^w,\, (p-\tilde{p})^w/h] \in \operatorname{Op}(S^{\delta_0-\sigma}_{\delta_0,\delta})$. Indeed,

$$|\alpha| = |\tilde{\alpha}| = n \ge 1 \Rightarrow h^{|\alpha|-1} \partial^\alpha l_t\, \partial^\alpha (p - \tilde{p}) \in S^{\delta_0-\sigma}_{\delta_0,\delta}, \tag{5.15}$$

holds due to $1 - n + \delta_0 n + \delta(n-1) - \sigma = \delta_0 - \sigma - (1 - \delta_0 - \delta)(n-1)$ (cf. (5.3)). Thus we obtain $\mathbb{D}_{P/h} l_t^w \in S^{\delta_0-\sigma}_{\delta_0,\delta}$, which completes the proof due to the condition $|t| \le 2h^{\delta_0}$. □

Proposition 5.2. *Let $b, \tilde{b} \in S^0_{\delta_0,\delta_0}$ be such that*

$$\operatorname{dist}(\operatorname{supp} \tilde{b},\, \operatorname{supp}(1-b)) \ge ch^{\delta_0} \tag{5.16}$$

holds with a fixed $c > 0$. Then

$$|t| \le 2h^{\delta_0} \Rightarrow U_t \tilde{b}^w \cong (b \circ \vartheta_t)^w\, U_t \tilde{b}^w. \tag{5.17}$$

Proof. Using the induction with respect to $n \in \mathbb{N}$ we show that

$$|t| \le 2h^{\delta_0} \Rightarrow U_t \tilde{b}^w = (b \circ \vartheta_t)^w U_t \tilde{b}^w + O(h^{n\sigma}) \tag{5.18(n)}$$

holds for all $b, \tilde{b} \in S^0_{\delta_0,\delta_0}$ satisfying (5.16) with some $c > 0$.

Without writing we assume all the time $|t| \le 2h^{\delta_0}$. Obviously (5.18(n)) holds for $n = 0$. The hypothesis (5.16) allows us to find $l, \tilde{l} \in S^0_{\delta_0,\delta_0}$ such that $\operatorname{supp} l \cap \operatorname{supp}(1 - \tilde{l}) = \emptyset$ and

$$\operatorname{dist}(\operatorname{supp} \tilde{b}, \operatorname{supp}(1 - l)) \ge ch^{\delta_0}/4. \tag{5.19}$$

$$\operatorname{dist}(\operatorname{supp} \tilde{l}, \operatorname{supp}(1 - b)) \ge ch^{\delta_0}/4. \tag{5.20}$$

Thus we have $\tilde{b}^w \cong l^w \tilde{b}^w$ and using (5.20) with $l_t = l \circ \vartheta_t$, $b_t = b \circ \vartheta_t$ instead of l, b, we can write $(I - b_t^w) l_t^w \cong 0$ and

$$\begin{aligned} (I - b_t^w) U_t \tilde{b}^w &\cong (I - b_t^w)(U_t l^w - l_t^w U_t)\tilde{b}^w \\ &\cong -i \int_0^1 \mathrm{d}s\, (I - b_t^w) U_{t(1-s)}\, t\,(\mathbb{D}_{P/h} l_{ts}^w) U_{ts}\, \tilde{b}^w. \end{aligned} \tag{5.21}$$

By the induction hypothesis we can use (5.18(n)) with ts, l_{ts} instead of t, b_t, i.e.,

$$U_{ts}\, \tilde{b}^w = l_{ts}^w\, U_{ts}\, \tilde{b}^w + O(h^{n\sigma}) \tag{5.22}$$

and since $t\,(\mathbb{D}_{P/h} l_{ts}^w) = O(h^\sigma)$, we obtain

$$\begin{aligned} &(I - b_t^w)\, U_{t(1-s)}\, t\,(\mathbb{D}_{P/h} l_{ts}^w) U_{ts}\, \tilde{b}^w = \\ &(I - b_t^w)\, U_{t(1-s)}\, t\,(\mathbb{D}_{P/h} l_{ts}^w) l_{ts}^w\, U_{ts}\, \tilde{b}^w + O(h^{(n+1)\sigma}). \end{aligned} \tag{5.23}$$

Since $\operatorname{supp} l_{ts} \cap \operatorname{supp}(1 - \tilde{l}_{ts}) = \emptyset \Rightarrow t\,(\mathbb{D}_{P/h} l_{ts}^w)\, l_{ts}^w \cong \tilde{l}_{ts}^w\, t\,(\mathbb{D}_{P/h} l_{ts}^w)\, l_{ts}^w$, the expression (5.23) can be written as

$$(I - b_t^w)\, U_{t(1-s)}\, \tilde{l}_{ts}^w\, t\,(\mathbb{D}_{P/h} l_{ts}^w)\, l_{ts}^w\, U_{ts}\, \tilde{b}^w + O(h^{(n+1)\sigma}). \tag{5.24}$$

However (5.16) holds with b_{ts}, $\tilde{l}_{ts}$, $c/4$ instead of b, $\tilde{b}$, c and replacing t, $\tilde{b}$ by $t - ts$, $\tilde{l}_{ts}$ in (5.18(n)) we obtain

$$(I - b_t^w) U_{t(1-s)}\, \tilde{l}_{ts}^w = O(h^{n\sigma}). \tag{5.25}$$

Using (5.25) and $t\,(\mathbb{D}_{P/h} l_{ts}^w) = O(h^\sigma)$ we find that the expressions (5.24) and (5.23) are $O(h^{(n+1)\sigma})$, completing the proof of (5.18($n+1$)) due to (5.21). □

6. End of the proof

We begin by the property of a finite propagation speed formulated in

Proposition 6.1. *Let $l, \tilde{l} \in C_0^\infty(\mathbb{R}^{2d})$ be h-independent. If $\tilde{l} = 1$ on a neighborhood of $\operatorname{supp} l$, then one can find $\bar{c} > 0$ such that*

$$|t| \le \bar{c} \Rightarrow U_t\, l^w \cong \tilde{l}^w\, U_t\, l^w. \tag{6.1}$$

Notation:

a) Further on $Z = \{Z_h\}_{0<h<h_0}$ is a family of subsets $Z_h \subset \mathbb{R}^{2d}$ and without writing we always assume that $(x,\xi) \in Z_h \Rightarrow |\xi| \le C_{\mathrm{ef}}$.

b) If $0 < s \le 2$ then we define $Z(s) = \{Z_h(s)\}_{0<h<h_0}$ by

$$Z_h(s) = \{v \in \mathbb{R}^{2d} : \operatorname{dist}(v, Z_h) < s\}. \tag{6.2}$$

In particular using (5.6) we can write

$$\operatorname{supp} l \subset Z \Rightarrow \operatorname{supp}(l \circ \vartheta_t) \subset Z(\bar{C}|t|). \tag{6.3}$$

c) For $\varepsilon > 0$ let $f_Z^{\delta_0,\varepsilon} = \{f_{Z_h}^{\delta_0,\varepsilon}\}_{0<h<h_0} \in S^0_{\delta_0,\delta_0}$ be such that

$$\operatorname{supp} f_{Z_h}^{\delta_0,\varepsilon} \subset Z_h(\varepsilon h^{\delta_0}) \text{and} f_{Z_h}^{\delta_0,\varepsilon} = 1 \text{on} Z_h(0{,}9\,\varepsilon h^{\delta_0}). \tag{6.4}$$

d) We denote by $F_{Z_h}^{\delta_0,\varepsilon}$ the corresponding operators $\left(f_{Z_h}^{\delta_0,\varepsilon}\right)^w$.

Lemma 6.2. *Let $0 < \varepsilon < 1$, $N, n \in \mathbb{N}$. If $2^n \bar{C} h^{\delta_0} \le 1$, then*

$$\sup_{|\tau| \le 2h^{\delta_0}} \|(I - F_{Z_{\tau,n}}^{\delta_0,\varepsilon})\, U_{2^n\tau}\, F_Z^{\delta_0,\varepsilon}\| \le C_{N,\varepsilon}\, 4^n h^N \tag{6.5(n)}$$

holds with $Z_{\tau,n} = Z(2^n(\varepsilon h^{\delta_0} + \bar{C}|\tau|))$.

Proof. In the first step we check the assertion of lemma for $n = 0$.

We use Proposition 5.2 with $\tilde{b} = f_Z^{\delta_0,\varepsilon}$ and $b = f_Z^{\delta_0,\frac{6}{5}\varepsilon}$. Since $f_{Z_{\tau,0}}^{\delta_0,\varepsilon} = 1$ on a neighborhood of $Z(\frac{6}{5}\varepsilon h^{\delta_0} + \bar{C}|\tau|) \supset \operatorname{supp} b_\tau$, we have $(I - F_{Z_{\tau,0}}^{\delta_0,\varepsilon}) b_\tau^w \cong 0$ and it is clear that (5.17) implies (6.5(0)).

To complete the proof we proceed by induction with respect to $n \in \mathbb{N}$.

We assume that the assertion of the lemma holds for a given $n \in \mathbb{N}$. Hence

$$U_{2^{n+1}\tau}\, F_Z^{\delta_0,\varepsilon} = U_{2^n\tau}\, F_{Z_{\tau,n}}^{\delta_0,\varepsilon}\, U_{2^n\tau}\, F_Z^{\delta_0,\varepsilon} + O(4^n h^N). \tag{6.5}$$

Next we observe that applying (6.5(n)) to $Z_{\tau,n}$ instead of Z we can write $Z_{\tau,n+1} = Z_{\tau,n}(2^n(\varepsilon h^{\delta_0} + \bar{C}|\tau|))$ and replace (6.6) by

$$F_{Z_{\tau,n+1}}^{\delta_0,\varepsilon}\, U_{2^n\tau}\, F_{Z_{\tau,n}}^{\delta_0,\varepsilon}\, U_{2^n\tau}\, F_Z^{\delta_0,\varepsilon} + O(2 \cdot 4^n h^N). \tag{6.6}$$

To complete the proof of (6.5($n+1$)) we use once more (6.5(n)) to write (6.7) as

$$F_{Z_{\tau,n+1}}^{\delta_0,\varepsilon}\, U_{2^{n+1}\tau}\, F_Z^{\delta_0,\varepsilon} + O(3 \cdot 4^n h^N).$$

Proof of Proposition 6.1. Let $n \in \mathbb{N}$ be such that $2^{-n} \le h^{\delta_0} \le 2^{1-n}$ and $\tau = 2^{-n}t$. Then $|t| \le \bar{c} \le 1 \Rightarrow |\tau| \le h^{\delta_0}$ and (6.5(n)) gives

$$U_t\, F_Z^{\delta_0,\varepsilon} \cong F_{Z(2\varepsilon + \bar{C}|t|)}^{\delta_0,\varepsilon}\, U_t\, F_Z^{\delta_0,\varepsilon}. \tag{6.7}$$

If $Z = \operatorname{supp} l$ and $\operatorname{dist}(Z, \operatorname{supp}(1 - \tilde{l})) \ge c > 0$, then $l^w \cong F_Z^{\delta_0,\varepsilon} l^w$ and $(I - \tilde{l}^w) F_{Z(c/2)}^{\delta_0,\varepsilon} \cong 0$, hence (6.8) implies (6.1) if $2\varepsilon + C|t| \le c/3$. □

Next we consider families of intervals $Z = [s;\ \tilde{s}] = \{[s_h;\ \tilde{s}_h]\}_{0<h<h_0}$ and for $\varepsilon > 0$ let $\tilde{f}^{\delta_1,\varepsilon}_{Z,j} \in S^0_{0,\delta_1}$ be of the form $\tilde{f}^{\delta_1,\varepsilon}_{Z,j}(x,\xi) = f^{\delta_1,\varepsilon}_Z(x_j)\theta_{\mathrm{ef}}(\xi)$ with

$$\mathbf{1}_{[s_h-0{,}9\varepsilon h^{\delta_1};\ \tilde{s}_h+0{,}9\varepsilon h^{\delta_1}]} \le f^{\delta_1,\varepsilon}_{Z_h} \le \mathbf{1}_{[s_h-\varepsilon h^{\delta_1};\ \tilde{s}_h+\varepsilon h^{\delta_1}]}. \tag{6.8}$$

We denote by $f^{\delta_1,\varepsilon}_Z(x_j)$ the corresponding multiplication operators and we include the case $s_h = \tilde{s}_h$ corresponding to $Z_h = \{s_h\}$.

Proposition 6.3. *Assume that $c > 0$, $l \in C_0^\infty(\mathbb{R}^{2d})$ are h-independent and*

$$\bar{C}_{-j} + 4c \le \partial_{\xi_j} p(x,\xi) \le \bar{C}_j - 4c \tag{6.9}$$

holds for $(x,\xi) \in \operatorname{supp} l$. Assume also $0 \notin [\bar{C}_{-j};\ \bar{C}_j]$. Then

$$h^{\delta_1} \le |t| \le \bar{c} \Rightarrow U_t f^{\delta_1,\varepsilon}_Z(x_j) l^w \cong f^{\delta_1,\varepsilon}_{Z_t}(x_j) U_t f^{\delta_1,\varepsilon}_Z(x_j) l^w \tag{6.10}$$

holds with $Z_t = Z + t[\bar{C}_{-j};\ \bar{C}_j]$ if $\varepsilon, \bar{c} > 0$ are small enough.

Proof that (6.11) holds with $h^{\delta_0} \le |t| \le \bar{c}$

Adopting the idea of the proof of Lemma 6.2 we introduce the notations $\tilde{Z} = \operatorname{supp} l$, $Z^{(j)} = \tilde{Z} \cap \{(x,\xi) :\ s \le x_j \le \tilde{s}\}$,

$$Z^{(j)}_{\tau,n} = \tilde{Z}(2^n(\varepsilon h^{\delta_0} + |\tau|\bar{C})) \cap \{(x,\xi) :\ x_j \in [s;\ \tilde{s}] + 2^n\tau[\bar{C}_{-j} + c;\ \bar{C}_j - c]\}$$

and using the induction with respect to $n \in \mathbb{N}$ we are going to show

$$\sup_{h^{\delta_0} \le |\tau| \le \min\{2h^{\delta_0},\, 2^{-n}\bar{c}\}} \|(I - F^{\delta_0,\varepsilon}_{Z^{(j)}_{\tau,n}}) U_{2^n\tau} F^{\delta_0,\varepsilon}_{Z^{(j)}}\| \le C_{N,\varepsilon}\, 4^n h^N. \tag{6.5'(n)}$$

If $h^{\delta_0} \le |t| \le \bar{c}$ and $n \in \mathbb{N}$ is chosen such that $2^n \le h^{-\delta_0}|t| < 2^{n+1}$, then taking $\tau = 2^{-n}t$ we obtain $h^{\delta_0} \le |\tau| \le \min\{2h^{\delta_0},\ 2^{-n}\bar{c}\}$ and (6.5′(n)) implies (6.11). Indeed, to obtain (6.11) from (6.5′(n)) we observe that $\delta_0 \le \delta_1$ gives $f^{\delta_1,\varepsilon}_Z(x_j)l^w \cong F^{\delta_0,\varepsilon}_{Z^{(j)}} f^{\delta_1,\varepsilon}_Z(x_j) l^w$ and $F^{\delta_0,\varepsilon}_{Z^{(j)}_{\tau,n}} \cong f^{\delta_1,\varepsilon}_{Z_{2^n\tau}}(x_j) F^{\delta_0,\varepsilon}_{Z^{(j)}_{\tau,n}}$ follows from

$$(x,\xi) \in \operatorname{supp} f^{\delta_0,\varepsilon}_{Z_{\tau,n}} \Rightarrow x_j \in [s - \varepsilon h^{\delta_0};\ \tilde{s} + \varepsilon h^{\delta_0}] + t[\bar{C}_{-j} + c;\ \bar{C}_j - c] \subset Z_t,$$

where $t = 2^n\tau$, $\varepsilon \le c$ and the last inclusion holds due to $\varepsilon h^{\delta_0} \le c|t|$ and $0 \notin [\bar{C}_{-j};\ \bar{C}_j]$. Further on $h \in]0;\ h_0]$ with $h_0 > 0$ small enough to ensure

$$(x,\xi) \in \tilde{Z} \Rightarrow \bar{C}_{-j} + 3c \le \partial_{\xi_j}\tilde{p}(x,\xi) \le \bar{C}_j - 3c, \tag{6.10'}$$

which is possible due to (5.3). Since $2^n h^{\delta_0} \le 2^n|\tau| \le \bar{c} \Rightarrow Z_{\tau,n} \subset \tilde{Z}(\bar{c}(\bar{C} + \varepsilon))$, choosing $\bar{c} > 0$ small enough we ensure

$$(x,\xi) \in Z_{\tau,n} \Rightarrow \bar{C}_{-j} + 2c \le \partial_{\xi_j}\tilde{p}(x,\xi) \le \bar{C}_j - 2c. \tag{6.10''}$$

Let $(x(t),\xi(t)) = \vartheta_t(x(0),\xi(0))$. If $(x(0),\xi(0)) \in \tilde{Z}(\varepsilon h^{\delta_0})$ and $|t| \le \bar{c}$, then

$$x_j(t) - x_j(0) = t\int_0^1 \mathrm{d}\rho\, \partial_{\xi_j}\tilde{p}(x(t\rho),\xi(t\rho)) \in t[\bar{C}_{-j} + 2c;\ \bar{C}_j - 2c],$$

hence the conditions $|t| \ge h^{\delta_0}$, $2\varepsilon \le c$ and $0 \notin [\bar{C}_{-j};\ \bar{C}_j]$ ensure

$$s - 2\varepsilon h^{\delta_0} \le x_j(0) \le \tilde{s} + 2\varepsilon h^{\delta_0} \Rightarrow x_j(t) \in [s;\ \tilde{s}] + t[\bar{C}_{-j} + c;\ \bar{C}_j - c]. \tag{6.11}$$

Thus taking $\tilde b = f^{\delta_0,\varepsilon}_{Z^{(j)}}$, $b = f^{\delta_0,2\varepsilon}_{Z^{(j)}}$ we have $\operatorname{supp} b_t = \vartheta_{-t}(\operatorname{supp} b) \subset Z^{(j)}_{t,0}$ and (6.5′(0)) follows from Proposition 5.2. Then reasoning as in the proof of Lemma 6.2 we deduce (6.5′($n+1$)) from (6.5′(n)), because the replacement of $Z^{(j)}$ by $Z^{(j)}_{\tau,n}$ corresponds to the replacement of $Z^{(j)}_{\tau,n}$ by $Z^{(j)}_{\tau,n+1}$ in (6.5′(n)).

Proof that (6.11) holds with $h^{\delta_1} \le |t| \le h^{\delta_0}$. It uses

Lemma 6.4. *Let us denote $\bar v'_j = \partial_{\xi_j} p(\bar v)$ for $\bar v \in \operatorname{supp} l$. Then*

$$|t| \le 2h^{\delta_1} \Rightarrow (I - f^{\delta_1,2\varepsilon}_{Z+t\bar v'_j}(x_j))\, U_t\, f^{\delta_1,\varepsilon}_Z(x_j)\, F^{\delta_0,\varepsilon}_{\{\bar v\}} \cong 0. \tag{6.12}$$

Proof. Clearly $f^{\delta_1,\varepsilon}_Z(x_j) F^{\delta_0,\varepsilon}_{\{\bar v\}} \cong F^{\delta_0,2\varepsilon}_{\{\bar v\}} f^{\delta_1,\varepsilon}_Z(x_j) F^{\delta_0,\varepsilon}_{\{\bar v\}}$ and we check that

$$|t| \le 2h^{\delta_1} \Rightarrow (I - F^{\delta_0,4\varepsilon}_{\{\bar v\}}) U_t\, F^{\delta_0,2\varepsilon}_{\{\bar v\}} \cong 0 \tag{6.13}$$

using (6.5(n)) with $n = 0$ and $Z = \{\bar v\}$. Indeed, since $\delta_0 < \delta_1$, we find $|t| \le 2h^{\delta_1} \le h^{\delta_0} \Rightarrow Z_{t,0} = B(\bar v, 2\varepsilon h^{\delta_0} + \bar C|t|) \subset B(\bar v, 3\varepsilon h^{\delta_0})$ for $h < h_0$ if h_0 is small enough. Reasoning as in the proof of Proposition 5.2 we take

$$\tilde b(x,\xi) = f^{\delta_1,\varepsilon}_Z(x_j)\theta_{\mathrm{ef}}(\xi), \qquad b_t(x,\xi) = f^{\delta_1,2\varepsilon}_Z(x_j - t\bar v'_j)\theta_{\mathrm{ef}}(\xi)$$

and $b = b_t|_{t=0}$, we define ϑ_t as the change of variable $x_j \to x_j - t\bar v'_j$ instead of $\exp(tH_{\tilde p})$ considered in Proposition 5.2, the classes of symbols $S^{\tilde m}_{\delta_0,\delta_1}$ replacing the classes $S^{\tilde m}_{\delta_0,\delta_0}$ of Proposition 5.2.

The proof of Proposition 5.2 was based on the estimate of the Heisenberg derivative of Lemma 5.1. However due to the property (6.13), the presence of $F^{\delta_0,\varepsilon}_{\{\bar v\}}$ in (6.12) allows us to use the auxiliary cut-off $F^{\delta_0,4\varepsilon}_{\{\bar v\}}$, i.e., to complete the proof of Lemma 6.3 it suffices to show that

$$|t| \le 2h^{\delta_1} \Rightarrow (t\,\mathbb{D}_{P/h} l^w_t) F^{\delta_0,4\varepsilon}_{\{\bar v\}} \in \mathrm{Op}(S^{-\sigma}_{\delta_0,\delta_1}), \tag{6.14}$$

where $\sigma > 0$ and $l_t \in S^0_{\delta_0,\delta_1}$ has the form $l_t(x,\xi) = f(x_j - t\bar v'_j)\theta_{\mathrm{ef}}(\xi)$.

Since $f'(x_j - t\bar v'_j)\theta_{\mathrm{ef}}(\xi)$ are symbols belonging to $S^{\delta_1}_{0,\delta_1}$, writing only the leading term of the commutator we have

$$\big(\mathbb{D}_{P/h} l^w_t - f'(x_j - t\bar v'_j)(\partial_{\xi_j} p^w - \bar v'_j)\big)\, F^{\delta_0,4\varepsilon}_{\{\bar v\}} \in \mathrm{Op}(S^{\delta_1-\sigma}_{\delta_0,\delta_1}).$$

Then $(\partial_{\xi_j} p^w - \bar v'_j) F^{\delta_0,4\varepsilon}_{\{\bar v\}} \in \mathrm{Op}(S^{-\delta_0}_{\delta_0,\delta_1}) \Rightarrow (\mathbb{D}_{P/h} l^w_t) F^{\delta_0,4\varepsilon}_{\{\bar v\}} \in \mathrm{Op}(S^{\delta_1-\sigma}_{\delta_0,\delta_1})$ and $|t| \le 2h^{\delta_1}$ ensures (6.14), completing the proof of lemma. □

To prove (6.11) with $h^{\delta_1} \le |t| \le h^{\delta_0}$ we introduce the notations $\tilde Z = \operatorname{supp} l$, $\tilde Z_n = \tilde Z(2\bar C h^{\delta_0}(\log_2(1/h) - n))$ for $n < \log_2(1/h) - 1$ and $Z_{\tau,n} = Z + 2^n\tau[\bar C_{-j};\ \bar C_j]$. Then it suffices to show the estimates

$$\sup_{h^{\delta_1} \le |\tau| \le \min\{2h^{\delta_1},\, 2^{-n}h^{\delta_0}\}} \|(I - f^{\delta_1,\varepsilon}_{Z_{\tau,n}}(x_j))\, U_{2^n\tau}\, f^{\delta_1,\varepsilon}_Z(x_j) F^{\delta_0,1}_{\tilde Z_n}\| \le C_{N,\varepsilon}\, 4^n h^N. \tag{6.15(n)}$$

Indeed, if $n \in \mathbb{N}$ satisfies $2^n \le h^{-\delta_1}|t| \le 2^{n+1}$ and $h^{\delta_1} \le |t| \le h^{\delta_0}$, then $n \le (\delta_1 - \delta_0)\log_2(1/h)$ and (6.11) follows from (6.15(n)) with $\tau = 2^{-n}t$.

In the first step we check that Lemma 6.4 implies (6.15(n)) for $n = 0$.

Let $\{l_k\}_{k=0,\dots,K_h}$ be a uniformly bounded family of symbols $S^0_{\delta_0,\delta_0}$ (i.e., l_k satisfy the estimates (3.14) with constants $C_{n,\alpha,\beta}$ independent of k) satisfying

$$F^{\delta_0,1}_{\tilde{Z}_0} = \sum_{0\le k\le K_h} l_k \quad \text{with } \operatorname{supp} l_k \subset B(\bar{v}(k), \varepsilon h^{\delta_0}/2) \tag{6.16}$$

where $\bar{v}(k) \in \tilde{Z}_0$ and $K_h \le Ch^{-3d\delta_0}$. Since $l_k^w \cong F^{\delta_0,\varepsilon}_{\{\bar{v}(k)\}} l_k^w$, Lemma 6.4 ensures

$$\sup_{|\tau|\le 2h^{\delta_1}} \|(I - f^{\delta_1,2\varepsilon}_{Z+\tau\bar{v}'_{j,k}}(x_j))U_\tau\, f^{\delta_1,\varepsilon}_Z(x_j) l_k^w\| \le C_{\bar{N}} h^{\bar{N}}, \tag{6.17}$$

where $\bar{v}'_{j,k} = \partial_{\xi_j} p(\bar{v}(k)) \in [\bar{C}_{-j} + c;\ \bar{C}_j - c]$ and $\bar{N} \in \mathbb{N}$ is arbitrary.

However the conditions $|\tau| \ge h^{\delta_1}$, $c \ge 2\varepsilon$ and $0 \notin [\bar{C}_{-j};\ \bar{C}_j]$ ensure

$$[s + \tau\bar{v}'_{j,k} - 2\varepsilon h^{\delta_1};\ \tilde{s} + \tau\bar{v}'_{j,k} + 2\varepsilon h^{\delta_1}] \subset [s;\ \tilde{s}] + \tau[\bar{v}'_{j,k} - c;\ \bar{v}'_{j,k} + c] \subset Z_{\tau,0},$$

which means that $f^{\delta_1,\varepsilon}_{Z_{\tau,0}}(x_j) = 1$ on $\operatorname{supp} f^{\delta_1,2\varepsilon}_{Z+\tau\bar{v}'_{j,k}}$. Thus $I - f^{\delta_1,2\varepsilon}_{Z+\tau\bar{v}'_{j,k}}(x_j)$ in (6.17) can be replaced by $I - f^{\delta_1,2\varepsilon}_{Z_{\tau,0}}(x_j)$ if $|\tau| \ge h^{\delta_1}$ and summing up with respect to k we obtain (6.15(0)) with the error term $K_h h^{\bar{N}} = O(h^{\bar{N}-2d\delta_0})$.

Finally we show (6.15(n)) by induction with respect to $n \in \mathbb{N}$. Since $\tilde{Z}_n \supset \tilde{Z}_{n+1}(2\bar{C}h^{\delta_0})$, using (6.5(0)) with $Z = \tilde{Z}_{n+1}$ we find $|t| \le h^{\delta_0} \Rightarrow U_t F^{\delta_0,1}_{\tilde{Z}_{n+1}} \cong F^{\delta_0,1}_{\tilde{Z}_n} U_t F^{\delta_0,1}_{\tilde{Z}_{n+1}}$. We use the last property for $t = 0$ and $t = 2^n\tau$ to insert the additional cut-off $F^{\delta_0,1}_{\tilde{Z}_n}$, allowing us to make the induction step reasoning similarly as in the proof of Lemma 6.2. □

End of the proof. As we noticed at the end of Section 4, it remains to prove Proposition 4.3, i.e., the estimate (5.1). To begin we observe that (4.6) allows us to find $l_j \in C_0^\infty(\mathbb{R}^{2d})$, $j = \pm 1, \dots, \pm d$, such that

$$\pm\partial_{\xi_{\pm j}} p(x,\xi) \ge c_0 \quad \text{for } (x,\xi) \in \operatorname{supp} l_{\pm j} \tag{6.18}$$

and $L_0 L_0^* \cong \sum_{1\le j\le d}(l_j^w + l_{-j}^w)$, i.e. for every $\bar{N} \in \mathbb{N}$ we have

$$\operatorname{tr} L_0^* U_t L_0 = \operatorname{tr} U_t L_0 L_0^* = \sum_{1\le j\le d} \operatorname{tr} U_t(l_j^w + l_{-j}^w) + O(h^{\bar{N}}).$$

Let $\{l_{j,k}\}_{k=0,\dots,K_h}$ be a uniformly bounded family of symbols S^0_{0,δ_1} with $K_h \le Ch^{-d\delta_1}$, satisfying

$$\sum_{0\le k\le K_h} l_{j,k} = l_j, \quad \text{and} \quad |x_j - \bar{v}_{j,k}| \ge \varepsilon h^{\delta_1}/2 \Rightarrow l_{j,k}(x,\xi) = 0$$

for certain $\bar{v}_{j,k} \in \mathbb{R}$. Thus $(I - f^{\delta_1,\varepsilon}_{\{\bar{v}_{j,k}\}}(x_j))l^w_{j,k} \cong 0$ and

$$\begin{aligned} \operatorname{tr} U_t l^w_{j,k} &= \operatorname{tr} U_t f^{\delta_1,\varepsilon}_{\{\bar{v}_{j,k}\}}(x_j) l^w_{j,k} f^{\delta_1,\varepsilon}_{\{\bar{v}_{j,k}\}}(x_j) + O(h^{\bar{N}}) \\ &= \operatorname{tr} f^{\delta_1,\varepsilon}_{\{\bar{v}_{j,k}\}}(x_j) U_t f^{\delta_1,\varepsilon}_{\{\bar{v}_{j,k}\}}(x_j) l^w_{j,k} + O(h^{\bar{N}}). \end{aligned} \tag{6.19}$$

Since the conditions (6.18) imply (6.10) with $0 \notin [\bar{C}_{-j};\ \bar{C}_j]$, Proposition 6.3 allows us to replace $f^{\delta_1,\varepsilon}_{\{\bar{v}_{j,k}\}}(x_j)$ by $f^{\delta_1,\varepsilon}_{\{\bar{v}_{j,k}\}} f^{\delta_1,\varepsilon}_{\bar{v}_{j,k}+t[\bar{C}_{-j};\ \bar{C}_j]}(x_j) = 0$ if $|t| \geq h^{\delta_1}$ and $2\varepsilon \leq c_0$, i.e., the whole expression (6.19) is $O(h^{\bar{N}})$ if $h^{\delta_1} \leq |t| \leq \bar{c}$.

To complete the proof it remains to sum up with respect to k, which gives (5.1) with the error term $K_h h^{\bar{N}} = O(h^{\bar{N}-d\delta_1})$. □

7. Appendix

(A) Proof of (2.2) and (2.3) Dropping the indices $\nu, \bar{\nu}$ we write

$$\partial_x^\alpha \partial_{x_j} a_h(x) = \int \partial_{x_j} a(y) \gamma^{(\alpha)}_{h^\delta}(x-y) h^{-\delta|\alpha|}\, dy. \tag{7.1}$$

Further on we assume $|\alpha| \geq 1$, hence $\int \gamma^{(\alpha)}_{h^\delta}(x-y) dy = 0$ and (7.1) still holds if $\partial_{x_j} a(y)$ is replaced by $\partial_{x_j} a(y) - \partial_{x_j} a(x)$. Therefore

$$\begin{aligned} |\partial_x^\alpha \partial_{x_j} a_h(x)| &\leq \int |\partial_{x_j} a(y) - \partial_{x_j} a(x)| |\gamma^{(\alpha)}_{h^\delta}(x-y)| h^{-\delta|\alpha|}\, dy \\ &\leq C \int |y-x|^r |\gamma^{(\alpha)}_{h^\delta}(x-y)| h^{-\delta|\alpha|}\, dy \;=\; C h^{(r-|\alpha|)\delta} \int |y|^r |\gamma^{(\alpha)}_1(y)|\, dy \end{aligned}$$

completes the proof of (2.3). Next we are going to estimate

$$a_h(x) - a(x) = \int (a(y) - a(x)) \gamma_{h^\delta}(x-y)\, dy, \tag{7.2}$$

using the fact that the expression

$$a(y) - a(x) - (y-x)\cdot\nabla a(x) = \int_0^1 ds\, s(y-x)\cdot(\nabla a(x+s(y-x)) - \nabla a(x)) \tag{7.3}$$

is $O(|y-x|^{1+r})$. Moreover $\gamma_1(-x) = \gamma_1(x) \Rightarrow \int (y-x)\gamma_{h^\delta}(x-y)\, dy = 0$ and (7.2) still holds if $a(y) - a(x)$ is replaced by (7.3). Therefore

$$|a_h(x) - a(x)| \leq C \int |y-x|^{1+r} |\gamma_{h^\delta}(x-y)|\, dy \;=\; C_r h^{(1+r)\delta}.$$

(B) Proof of the discreteness of the spectrum

Since $|x| \geq \bar{C} \Rightarrow a(x,\xi) \geq E_0 + 4c_0$, we can find $\theta \in C_0^\infty(\mathbb{R}^d)$ such that

$$\theta \geq 0 \quad \text{and} \quad a(x,\xi) + \theta(x) \geq E_0 + 3c_0.$$

Following [6] it is possible to introduce the symbol p'_h such that its Weyl quantization (defined in Section 5) is ${p'_h}^w = P_h^-$. Then (2.5) still holds with p'_h instead of p_h, hence

$$ {p'_h}^w(x,\xi) + \theta(x) \geq a(x,\xi) + \theta(x) - Ch \geq E_0 + c_0 \tag{7.4}$$

if $h \in]0;\ h_0]$ and h_0 is fixed small enough. Let Θ denote the operator of multiplication by θ. Due to (7.4) the Garding inequality (cf. [6]) ensures

$$A_h + \Theta \geq P_h^- + \Theta - Ch \geq E_0 + 2c_0 - C'h \geq E_0 + c_0$$

if $h \in]0;\ h_0]$ with h_0 small enough. Since Θ is a relatively compact perturbation of A_h, the essential spectrum does not change (cf. [12]) and $A_h + \Theta \geq E_0 + c_0$ ensures

$$\sigma_{\text{ess}}(A_h) = \sigma_{\text{ess}}(A_h + \Theta) \subset \sigma(A_h + \Theta) \subset [E_0 + c_0;\ \infty[.$$

(C) We check that (2.21) implies (2.18). It suffices to show

$$\int \frac{dxd\xi}{(2\pi h)^d} (g^2(\tilde{f}_h^Z - \mathbf{1}_Z)) \circ p_h = O(h^{1-d}), \tag{7.5}$$

$$\operatorname{tr}(g^2(\tilde{f}_h^Z - \mathbf{1}_Z)(P_h^\pm) = O(h^{1-d}), \tag{7.6}$$

where instead of $Z = [E';\ E]$ we will consider $Z =]-\infty;\ E]$ with $E \in [E_0 - 2c_0;\ E_0 + 2c_0]$. We begin by writing the following easy estimate

$$|(\tilde{f}_h^Z - \mathbf{1}_Z)(\lambda)| \leq C_N \langle (\lambda - E)/h \rangle^{-N}, \tag{7.7}$$

which holds for every $N \in \mathbb{N}$. Using (7.7) with $N = 2$ we can estimate the left-hand side of (7.5) by

$$\int \frac{dxd\xi}{(2\pi h)^d} g^2(p_h(x,\xi))\, C_2 \langle (p_h(x,\xi) - E)/h \rangle^{-2} \leq h^{-d} \sum_{k\in\mathbb{Z}} C'_2 \langle k \rangle^{-2} |\Gamma_{E,k}^h|,$$

$\Gamma_{E,k}^h = \{(x,\xi) : p_h(x,\xi) \in [E + kh;\ E + (k+1)h] \cap [E_0 - 2c_0;\ E_0 + 2c_0]\}$.

To complete the proof of (7.5) we observe that $|\Gamma_{E,k}^h| \leq Ch$ holds due to (2.12)–(2.14).

In the next step we use the hypothesis that (2.21) holds in the particular case $Z = [E;\ E + h]$ allowing us to write

$$\operatorname{tr}(g^2 \tilde{f}_h^{[E;\ E+h]})(P_h^\pm) = \int \frac{dxd\xi}{(2\pi h)^d} (g^2 \tilde{f}_h^{[E;\ E+h]}) \circ p_h + O(h^{1-d}). \tag{7.8}$$

Due to (7.5) the integral in (7.8) is $O(h^{1-d})$, hence the left-hand side of (7.8) is $O(h^{1-d})$. Since $\tilde{f}_h^Z > 0$, we have $\mathbf{1}_{[E;\ E+h]} \leq C\tilde{f}_h^{[E;\ E+h]}$ with a certain constant $C > 0$ and

$$\operatorname{tr}(g^2 \mathbf{1}_{[E;\ E+h]})(P_h^\pm) \leq C\operatorname{tr}(g^2 \tilde{f}_h^{[E;\ E+h]})(P_h^\pm) = O(h^{1-d}). \tag{7.9}$$

Due to (7.7) the left-hand side of (7.6) can be estimated by

$$\operatorname{tr} g^2(P_h^\pm)\, C_2\langle (P_h^\pm - E)/h\rangle^{-2} \le \sum_{k\in\mathbb{Z}} \langle k\rangle^{-2} C_2' \operatorname{tr} g^2 \mathbf{1}_{[E+kh;\, E+(k+1)h]}(P_h^\pm),$$

which is $O(h^{1-d})$ due to (7.9).

(D) Proof of Lemma 3.1. Let $l_1 \in C_0^\infty(\mathbb{R}^{2d})$ be h-independent and such that $\operatorname{supp} l_1 \cap \operatorname{supp}(1-l) = \emptyset$. We introduce $P_1 = \tilde{l}_1^w (P - E_1)\tilde{l}_1^w + E_1$, where $\tilde{l}_1^w$ is the Weyl quantization (described in Section 5) of $\tilde{l}_1 = 1 - l_1$. Denoting $\tilde{L} = I - L = (1-l)(x, hD)$ we obtain

$$\tilde{L}(P_1 - P) = \tilde{L}(l_1^w(P-E_1)l_1^w - Pl_1^w - l_1^w P) = O(h^{\bar{N}})$$

for every $\bar{N} \in \mathbb{N}$. We denote $\mathrm{ad}^0_{P,P_1}(\tilde{L}) = \tilde{L}$ and define

$$\mathrm{ad}^{k+1}_{P,P_1}(\tilde{L}) = P\mathrm{ad}^k_{P,P_1}(\tilde{L}) - \mathrm{ad}^k_{P,P_1}(\tilde{L})P_1$$

by induction with respect to $k \in \mathbb{N}$. Then Taylor's formula gives

$$\tilde{L}e^{itP} = e^{itP_1} \sum_{0\le k\le N-1} \frac{(-it)^k}{k!} \mathrm{ad}^k_{P,P_1}(\tilde{L}) + R_N(t), \tag{7.10}$$

$$R_N(t) = \frac{(-it)^N}{(N-1)!} \int_0^1 ds\, (1-s)^{N-1} e^{i(1-s)tP} \mathrm{ad}^N_{P,P_1}(\tilde{L}) e^{itsP_1}.$$

However $\mathrm{ad}^1_{P,P_1}(\tilde{L}) = [P, L] - \tilde{L}(P_1 - P) = [P, L] + O(h^{\bar{N}})$ and by induction

$$\mathrm{ad}^k_{P,P_1}(\tilde{L}) = \mathrm{ad}^k_{P,P}(\tilde{L}) + O(h^{\bar{N}}) = O(h^{k(1-\delta)}) + O(h^{\bar{N}}).$$

If $\hat{f}$ is the Fourier transform of $f \in C_0^\infty(\mathbb{R})$, then $(-it)^k \hat{f} = (f^{(k)})\hat{}$ and using $f(T) = \int_{-\infty}^{\infty} \frac{dt}{2\pi} \hat{f}(t) e^{itT}$ for $T = P$ or $T = P_1$, we note that (7.10) and $R_N(t) = O(h^{N(1-\delta)})$ give

$$\tilde{L}f(P) = \sum_{0\le k\le N-1} \frac{f^{(k)}(P_1)}{k!} \mathrm{ad}^k_{P,P_1}(\tilde{L}) + O(h^{N(1-\delta)}). \tag{7.11}$$

We assume $l_1 = 1$ on $p^{-1}(]-\infty, E_1])$. Let $\theta_n \in C_0^\infty(\mathbb{R}^d)$ be such that $0 \le \theta_n \le 1$ and $\theta_n(\xi) = 1$ if $|\xi| \le n$. If n is large enough and $\tilde{\theta}_n = 1 - \theta_n$ then $P'' = (P\theta_n(hD) + \theta_n(hD)P)/2 \ge -Ch$ and we can decompose $P - E_1 = p'^w + P''$ with $p' \in S^0_{0,\delta}$ satisfying $p'\tilde{l}_1^2 \ge -Ch$ due to $(p' - (p - E_1))\tilde{l}_1^2 \in S^{-1}_{0,\delta}$ and $(p - E_1))\tilde{l}_1^2 \ge 0$. Therefore

$$P_1 - E_1 = \tilde{l}_1^w (P - E_1)\tilde{l}_1^w \ge \tilde{l}_1^w p'^w \tilde{l}_1^w - Ch \ge -C'h^\delta \tag{7.12}$$

holds for a certain $C' > 0$ due to the Garding inequality (cf. [6]).

We complete the proof using (7.11) with $f = \tilde{g}$, hence $\operatorname{supp} f \subset]-\infty, E_1[$) and $f^{(k)}(P_1) = 0$ for $h < h_0$ if h_0 is small enough due to (7.12).

(E) We check the estimates (5.3). Let $|\alpha| \leq 1$, $\beta, \beta' \in \mathbb{N}^d$. Then

$$\partial_\xi^{\beta'} \partial_x^\beta \partial_x^\alpha \tilde{p}(\cdot, \xi) = \partial_\xi^{\beta'} (\partial^\beta \gamma_{h^{\delta_0}} * \partial_x^\alpha p(\cdot, \xi) \theta_{\mathrm{ef}}(\xi))$$

can be estimated by $C_{\alpha,\beta'} \int |\partial^\beta \gamma_{h^{\delta_0}}| = O(h^{-|\beta|\delta_0})$.

To show the second estimate (5.3) we note that

$$\partial_\xi^{\beta'} \partial_{x_j} p(x, \xi) - \partial_\xi^{\beta'} \partial_{x_j} p(x', \xi) = \int (\partial_\xi^{\beta'} \partial_{x_j} a(x - y, \xi) - \partial_\xi^{\beta'} \partial_{x_j} a(x' - y, \xi)) \gamma_{h^\delta}(y) \, dy$$

can be estimated by $C\langle\xi\rangle^{2m-|\beta'|}|x - x'|^r$ and for $|\xi| \leq C_{\mathrm{ef}}$ we have

$$|\partial_\xi^{\beta'} \partial_{x_j} (\tilde{p} - p)(x, \xi)| \leq C_{\beta'} \int |y - x|^r |\gamma_{h^{\delta_0}}(x - y)| \, dy = O(h^{r\delta_0})$$

It is clear that similar estimates hold without ∂_{x_j}. Since we have already proved $\partial_\xi^{\beta'} \partial_x^\beta \partial_{x_j} \tilde{p} = O(h^{-\delta_0|\beta|})$ and $l\partial_\xi^{\beta'} \partial_x^\beta \partial_{x_j} p = O(h^{-\delta(|\beta|-r)})$ if $|\beta| \geq 1$, the remaining estimates of $\partial_\xi^{\beta'} \partial_x^\beta \partial_{x_j} (\tilde{p} - p)$ follow.

References

[1] J. Chazarain, *Spectre d'un hamiltonien quantique et mécanique classique.* Comm. Partial Diff. Eq. (1980), 595–644.

[2] M. Dimassi, J. Sjöstrand, *Spectral Asymptotics in the Semiclassical Limit.* London Mathematical Society Lecture Note Series 268, Cambridge University Press, 1999.

[3] B. Helffer, D. Robert, *Comportement semi-classique du spectre des hamiltoniens quantiques elliptiques.* Ann. Inst. Fourier, Grenoble **31 (3)** (1981), 169–223.

[4] B. Helffer, D. Robert, *Calcul fonctionnel par la transformation de Mellin et opérateurs admissibles.* J. Funct. Anal. **53 (3)** (1983), 246–268.

[5] L. Hörmander, *The Spectral Function of an Elliptic Operator.* Acta Math. **121** (1968), 173–218.

[6] L. Hörmander, *The Analysis of Linear Partial Differential Operators.* Springer-Verlag, vol. 1, 2, 3, 4, Berlin–Heidelberg–New York–Tokyo, 1983, 1985.

[7] V. Ivrii, *Microlocal Analysis and Precise Spectral Asymptotics.* Springer-Verlag, Berlin, 1998.

[8] V. Ivrii, *Sharp Spectral Asymptotics for Operators with Irregular Coefficients.* Internat. Math. Res. Notices (2000), 1155–1166.

[9] G. Métivier, *Valeurs propres des problèmes aux limites irréguliers.* Bull. Soc. Math. France Mem. **51–52** (1977), 125–219.

[10] Y. Miyazaki, *A Sharp Asymptotic Remainder Estimate for the Eigenvalues of Operators Associated with Strongly Elliptic Sesquilinear Forms.* Japan J. Math. **15** (1989), 65–97.

[11] Y. Miyazaki, *The Eigenvalue Distribution of elliptic Operators with Hölder Continuous Coefficients.* Osaka J. Math. **28** (1991), 935–973; Part 2, Osaka J. Math. **30** (1993), 267–302.

[12] M. Reed, and B. Simon, *Methods of Modern Mathematical Physics.* vol. I-IV, Academic Press, New York–San Francisco–London, 1972, 1975, 1979.

[13] D. Robert, *Autour de l'approximation semi-classique.* Birkhäuser, Boston 1987.

[14] L. Zielinski, *Asymptotic Behavior of Eigenvalues of Differential Operators with Irregular Coefficients on a Compact Manifold.* C. R. Acad. Sci. Paris Sér. I Math. **310** (1990), 563–568.

[15] L. Zielinski, *Asymptotic Distribution of Eigenvalues for Elliptic Boundary Value Problems.* Asymptot. Anal. **16** (1998), 181–201.

[16] L. Zielinski, *Asymptotic Distribution of Eigenvalues of Some Elliptic Operators with Intermediate Remainder Estimates.* Asymptot. Anal. **17** (1998), 93–120.

[17] L. Zielinski, *Sharp Spectral Asymptotics and Weyl Formula for Elliptic Operators with Non-Smooth Coefficients.* Math. Phys. Anal. Geom. **2** (1999), 291–321; Part 2: Colloq. Math. **92** (2002), 1–18.

Lech Zielinski
LMPA, Centre Mi-Voix
Université du Littoral
B.P. 699
62228 Calais Cedex
France
e-mail: `Lech.Zielinski@lmpa.univ-littoral.fr`

GPSR Compliance

The European Union's (EU) General Product Safety Regulation (GPSR) is a set of rules that requires consumer products to be safe and our obligations to ensure this.

If you have any concerns about our products, you can contact us on ProductSafety@springernature.com

In case Publisher is established outside the EU, the EU authorized representative is:

Springer Nature Customer Service Center GmbH
Europaplatz 3
69115 Heidelberg, Germany

Batch number: 10732470

Printed by Printforce, the Netherlands